LACUSTRINE FACIES ANALYSIS

LACUSTRINE FACIES ANALYSIS

Lacustrine Facies Analysis

EDITED BY
P. ANADÓN, Ll. CABRERA
AND K. KELTS

SPECIAL PUBLICATION NUMBER 13 OF THE
INTERNATIONAL ASSOCIATION OF SEDIMENTOLOGISTS
PUBLISHED BY BLACKWELL SCIENTIFIC PUBLICATIONS
OXFORD LONDON EDINBURGH BOSTON
MELBOURNE PARIS BERLIN VIENNA

DISTRIBUTORS

Marston Book Services Ltd
PO Box 87
Oxford OX2 0DT
(*Orders*: Tel: 0865 791155
        Fax: 0865 791927
    Telex: 837515)

USA
Blackwell Scientific Publications, Inc.
3 Cambridge Center
Cambridge, MA 02142
(*Orders*: Tel: 800 759-6102)

Canada
Oxford University Press
70 Wynford Drive
Don Mills
Ontario M3C 1J9
(*Orders*: Tel: 416 441-2941)

Australia
Blackwell Scientific Publications
(Australia) Pty Ltd
54 University Street
Carlton, Victoria 3053
(*Orders*: Tel: 03 347-0300)

British Library
Cataloguing in Publication Data

Lacustrine facies analysis.
  1. Lakes
  I. Anadon, P. II. Cabrera, Ll. III. Kelts,
  K. IV. Series 551.482

  ISBN 0-632-03149-2

Library of Congress
Cataloging in Publication Data

Lacustrine facies analysis/edited by
  P. Anadon, Ll. Cabrera
  and K. Kelts.
        p.   cm. —
  (Special publication no. 13 of the
  International Association of Sedimentologists)
    ISBN 0-632-03149-2
    1. Lake sediments.   2. Sedimentology.
  3. Facies (Geology)
  I. Anadon, P. II. Cabrera, Ll.
  III. Kelts, K. IV. Title:
  Lacustrine facies analysis.
  V. Series: Special publication . . .
  of the International Association of
  Sedimentologists; no. 13.
  QE472.A34 1991
  551.3'04 — dc20

Contents

Preface

Lake basins are fascinating systems that define palaeoenvironments on a regional scale. The interactions between biospheric, geospheric, hydrospheric and atmospheric elements produce unique fingerprints. Large and small lakes have response rates greater than the oceans and two further degrees of freedom with varied chemistries and salinities. Long-lived lakes are common in tectonic basins and thus record overprints of tectonic and climatic change. How can these be separated? What are the criteria for recognition of subfacies? Lake basins also present difficult problems of correlation, without global index fossils and with a plethora of endemic species.

It has been 13 years since the publication of International Association of Sedimentology (IAS) Special Publication Number 2 *Modern and Ancient Lake Sediments* (Matter and Tucker, eds, 1978). Much has happened since. Lacustrine sedimentology has developed from a marginal curiosity into a frontier area of geology. Many lake-basin sequences contain hydrocarbon source rocks; others contain pristine archives of environmental dynamics. The field has become increasingly diversified. In 1984 the International Geological Correlation Program (IGCP) established IGCP project Number 219 on *Comparative Lacustrine Sedimentology in Space and Time*. This volume presents selected papers derived from two meetings co-sponsored by the IAS and IGCP 219. One was a field symposium in Rubielos, Spain (3–6 October 1988) on *Lacustrine facies models in rift systems and related natural resources*. The other was in the context of an IAS Symposium in Beijing, China (30 July–4 August, 1988) on *Sedimentology related to mineral deposits*. The reader is also referred to related presentations on lacustrine basins given at the large Beijing meeting which appear in two 1991 issues of the *Journal of China Earth Sciences*.

The first group of papers in this volume focus on natural resources in lacustrine basins, and underline the economic potential of lacustrine sequences. Tiercelin summarizes the diversity of lacustrine environments found in the great East African rift system, and reviews their known links to mineral deposits. Ordoñez, Calvo, García del Cura, Alonso-Zarza and Hoyos synthesize data on the occurrence of sodium sulphate and unusual clay minerals in lacustrine evaporite sequences from the Tertiary Madrid Basin, an intra-plate, compressional depression. Platt and Wright propose a general framework of diverse lacustrine facies in the context of source rocks and reservoirs and hydrocarbon exploration. This theme is further developed by Lin, Yang and Li with a perspective on a Tertiary strike-slip, deep-lake basin in southeast China. Kasiński shows the importance and environment of coal deposition in Cenozoic rift basins of central and eastern Europe. His emphasis is on links with the timing of tectonic events. Coal deposition is also a theme for Gierlowski-Kordesch, Gómez Fernández and Meléndez but in the context of the interplay of carbonate and alluvial sediments from Cretaceous lake basins in central Spain. They caution on the impact of diagenesis for the interpretation of lacustrine carbonates.

Modern processes in lakes of the East African rift are at the focus of a second theme with three papers. Scott, Ng'ang'a, Johnson and Rosendahl interpret new high-resolution seismic profiles and cores from deep Lake Malawi, showing the interplay of active faulting, hiatus, and sedimentation patterns. Seismic studies of the East African lakes are providing new insights into the mechanisms of rifting. Baltzer demonstrates the impact of muddy turbidity currents on the basinal facies of Lake Tanganyika. In contrast, Renaut and Owen dissect nearshore sediments from shallow, brackish alkaline Lake Bogoria, Kenya with a model for the dynamics of beach development.

A third set of four papers examines sequence stratigraphy and rhythmicity in lacustrine deposits. These show the problems of separating signatures of orbital forcing from tectonic movements. Rogers and Astin propose a reinterpretation of cyclic lacustrine sequences from the extensive Middle Devonian, Orcadian deposits of northern Britain with playa features. These are considered in terms of orbital forcing of palaeoclimate. Martel and Gibling stress the tectonic component in Early Carboniferous lacustrine rhythms from Nova Scotia, Canada. Gómez Fernández and Meléndez discuss a mixture of high-frequency climate variability and lower frequency tectonic pulsing illustrated by carbonate rhythms in Lower Cretaceous lacustrine sequences of northeast Spain. Anadón, Cabrera, Julià and Marzo describe well-developed, multiple lacustrine rhythms comprising deeper-water laminated carbonates, oil shales and shallow-water clastics from the middle Miocene Rubielos Basin, northeast Spain. They also emphasize tectonics overprinted on climatic fingerprints for the interpretation of sequence characteristics.

A fourth set of papers link lacustrine sedimentology and other palaeoenvironmental approaches. Szulc, Roger, Mouline and Lenguin include new isotopic data on the origin of sediments in the late Oligocene Narbonne Basin, southwest France. Robbins, Zhou and Zhou apply organic facies analyses and petrography toward palaeoenvironmental interpretations of calcareous lacustrine deposits from one of the Tertiary strike-slip basins of eastern China.

The papers of this volume thus represent a mixture of syntheses and case studies. They expand our knowledge of the regional distribution of diverse lacustrine deposits and further document the tendency for occurrences linked to preferred geological periods and zones of active tectonics. Many of the lacustrine sequences display the typical threshold behaviour of lacustrine systems. The central theme illustrates the interplay of climate changes modulated within megasequences controlled by regional tectonics. We need better, integrated tools which can provide sufficient time-resolution to identify unequivocally orbital-scale forcing. As our knowledge of sequence stratigraphy and palaeoenvironments in ancient lacustrine basins expands, the results can be applied to improve modelling of deposition in continental basins. Better exploration models for the diverse lacustrine resources will result. The papers of this volume also illustrate how the interest in lacustrine systems has expanded to a larger, more international, sedimentology community.

We are very appreciative of the efforts of reviewers. Each paper was examined by at least two. In particular we wish to acknowledge the following: P.A. Allen, B. Alonso, J.P. Calvo, C. Dabrio, M. Esteban, R. Flores, E. Gierlowski-Kordesch, T. Kreuser, M.R. Leeder, N. Platt, C. Pumot, E.I. Robbins, M.R. Talbot, J.-J. Tiercelin, N.H. Trewin, M.E. Tucker, and others remaining anonymous. We benefited greatly from the technical assistance of E. Clavero, C. Docherty, R. Ramis, and Béatrice Schwertfegger.

Pere Anadón, Lluís Cabrera, Kerry Kelts
January 1991

Natural Resources in Lacustrine Basins

Spec. Publs Int. Ass. Sediment. (1991) **13**, 3–37

Natural resources in the lacustrine facies of the Cenozoic rift basins of East Africa

J.-J. TIERCELIN

Centre National de le Recherche Scientifique, Unité de Recherche Associée 1278 et Groupement de Recherche 'Genèse et Evolution des Domaines Océaniques', Université de Bretagne occidentale, 29287 Brest Cedex, France

ABSTRACT

Ancient lacustrine sediments deposited in continental rift systems contain large quantities of economic resources. The world's largest active continental rift system, the East African rift system, is characterized by a combination of troughs of various tectonic styles occupied by small modern to large ancient 'tectonic' lakes, up to late Cenozoic or older. Large lakes are generally fresh water, fed by permanent streams; others are saline, fed by groundwater or hot springs, with ephemeral surface water inflow. At least eight primary lacustrine deposits and associated minerals are considered in terms of natural resources. Several are mined. These are organic-rich oozes, diatomites, zeolites and sodium silicates, metallic proto-ores, clays, phosphates, and coals. Other facies, such as stromatolites and lake-deposited pyroclastics, have primary economic significance or later importance, such as hydrocarbon reservoirs. Occurrences of such natural resources within the East African rift system are described in terms of processes of formation, industrial uses, and mining operations.

INTRODUCTION

Hydrocarbons, metallic minerals, and non-metallic resources commonly occur in large quantities in the sediments of modern and ancient continental rifts. Robbins (1983) noted that such resources accumulate in rift systems as a result of interactions between rift forming and sedimentation processes. These controls include tectonic, magmatic, climatic, hydrological, chemical, biological, and sedimentological factors. The world's largest active continental rift, the East African rift system, illustrates well examples of rift deposition and natural resources.

THE EAST AFRICAN RIFT SYSTEM ARCHITECTURE

The architecture of the East African rift system comprises a series of troughs and faults. Spanning the Equator, the system is 4000 km long. From north to south, there are two extensional branches, the eastern and western, three major NW−SE transcurrent fault zones, Aswa, Tanganyika−Rukwa−Malawi (TRM), and Zambezi−Mwanza, several hundred kilometres long; and a 'southern complex' of NE−SW troughs related to the western branch, extending from the Luangwa Trough of Zimbabwe to the Okavango Basin of South Africa (Chorowicz & Mukonki, 1980; Kazmin, 1980; Chorowicz *et al.*, 1983; Rosendahl, 1987; Tiercelin *et al.*, 1988a; Daly *et al.*, 1989; Ebinger, 1989b; Mondeguer *et al.*, 1989) (Fig. 1). Depressions or rift basins along the eastern and western branches form by links between smaller basins (mean size 50 × 30 km). These are generally asymmetric half-grabens, organized in line or 'en échelon'. These half-grabens are linked together by various structures, which generally follow the fabric of the pre-rift rocks (McConnell, 1967,1972; Daly *et al.*, 1989; Versfelt & Rosendahl, 1989). Transfer faults (Gibbs, 1984) occur between half-grabens, whereas the major Aswa, TRM, and Zambezi−Mwanza fault zones are areas of strike-slip (transform) motion between rift segments. These zones are considered as large sutures in the pre-rift rocks that have been reactivated during successive African orogenies and rifting episodes (Tiercelin *et al.*, 1988a; Daly *et al.*, 1989).

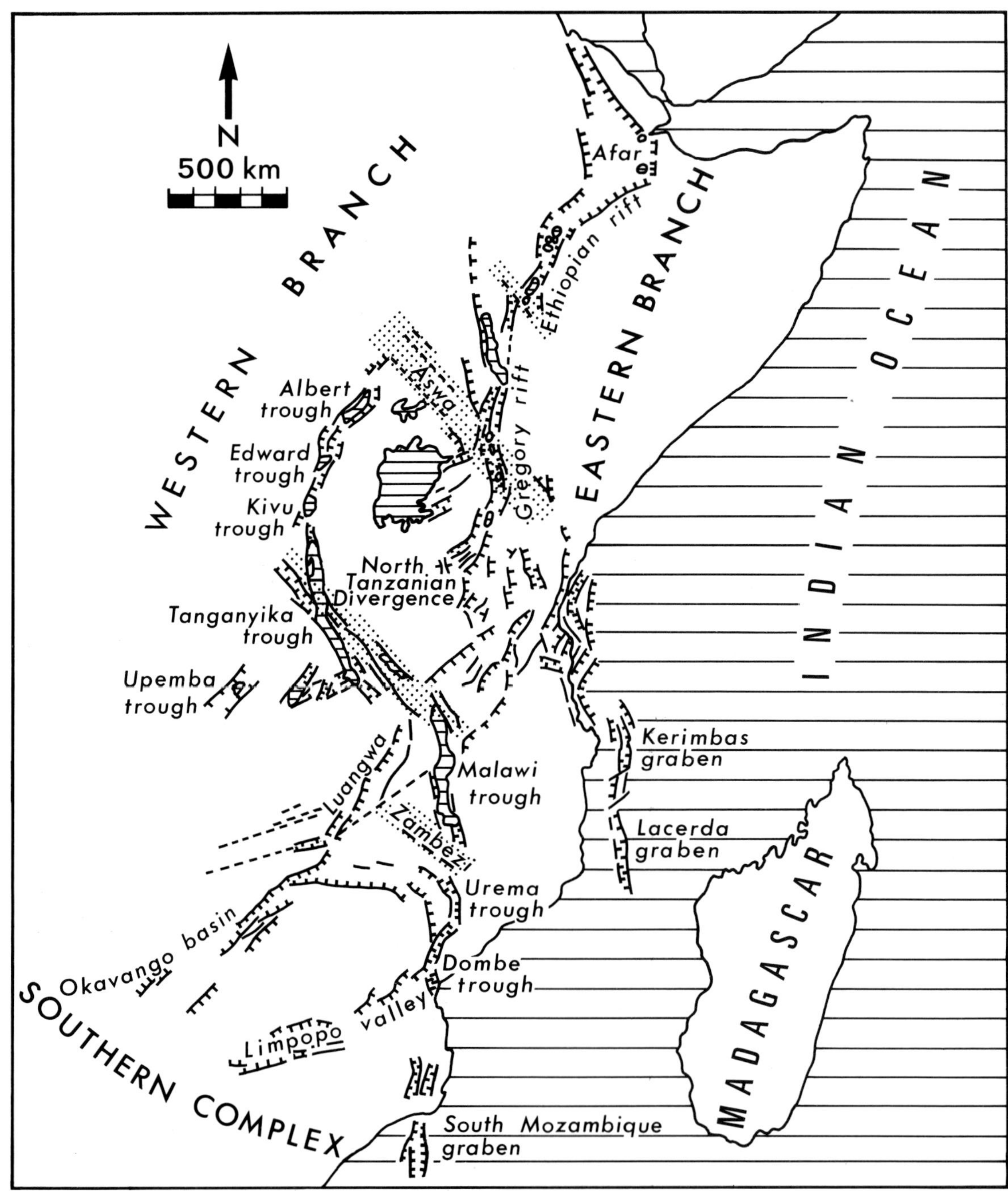

Fig. 1. Sketch showing the fundamental architecture of the East African rift system, eastern branch, western branch and seaward extensions, and southern complex (after Mougenot *et al.*, 1986; Mondeguer *et al.*, 1989).

The eastern branch

The eastern branch is divided into three rift segments from north to south: the Afar Depression, the Ethiopian rift, and the Gregory rift. The Gregory rift intersects the NW−SE Aswa transform fault zone near the Equator, and extends to the south along the North Tanzanian Divergence. On the continental margin of northern Mozambique, the N−S trending Kerimbas and Lacerda graben form a seaward extension of the North Tanzanian Divergence (Mougenot *et al.*, 1986,1989) (Fig. 1). One major characteristic of the eastern branch is the voluminous syntectonic alkaline volcanism, which filled depressions nearly as fast as they subsided (Fig. 2).

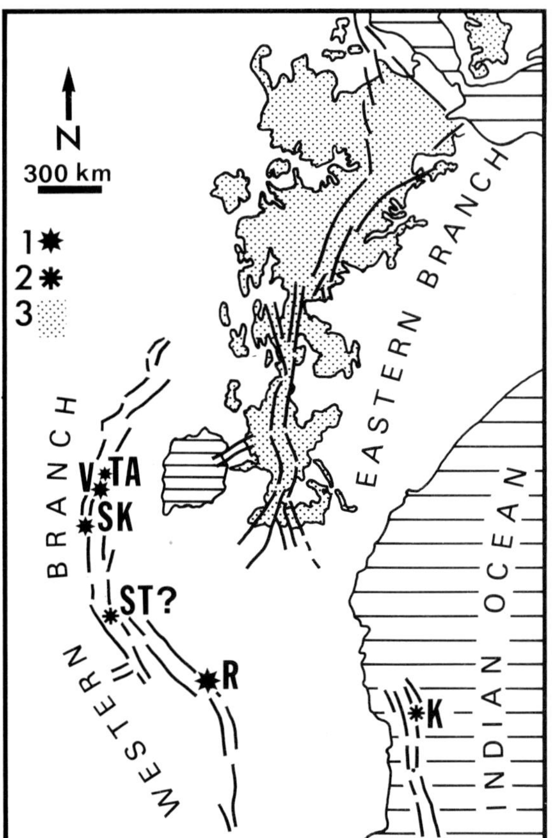

Fig. 2. Generalized map showing distribution of volcanic rocks of the East African rift system. (1) Volcanic fields of the western branch: TA = Toro-Ankole; V = Virunga; SK = South Kivu; R = Rungwe. (2) Volcanic intrusions: ST? = possible intrusions in southern Tanganyika (after Burgess, 1985); K = intrusions of the Kerimbas graben (after Mougenot *et al.*, 1989). (3) Volcanic rocks (Miocene to present) of the eastern branch.

Estimated volumes of the eastern branch flood volcanics range up to $470\,000$ km^3 (Baker *et al.*, 1972). The major volcanic episodes in the Ethiopian and Somalian plateaus were Eocene−Oligocene Trap Series fissure basalts. In the northern Gregory rift, volcanism was early Miocene. In the Ethiopian and Gregory rifts, volcanic episodes range from late Miocene to early Pliocene flood phonolites to Pliocene to early Pleistocene fissure and central eruptions (Baker *et al.*, 1971). Middle Pleistocene and later volcanism formed an axial chain of central volcanoes along the floor of the rift from northern Tanzania to Afar. These divide the eastern branch into separate valleys.

The western branch

The western branch is divisible into two main segments oriented NNE−SSW to N−S. From north to south, these are the Lake Albert and Lake Edward Troughs, the Kivu and northern Tanganyika Troughs, and the northern Malawi Trough down to the Urema and Dombe Troughs (Fig. 1). Its northern and southern terminations correspond to intersections with the NW−SE Aswa and Zambezi−Mwanza fault zones. Two main segments are connected through the major structure of the TRM dextral transform fault zone. The extreme difference in the quantity of volcanics in the two branches is characteristic. There are only four main centres of Eocene to Recent volcanism: the Rungwe massif in the northern Malawi Trough, the south and north (Virunga) Kivu volcanic fields, and the Toro−Ankole volcanic field east of Lake Edward (Fig. 2). The western branch is considered as volcanically 'dry' compared with the 'wetter' eastern branch (Mohr, 1982).

The transverse fault zones

From north to south the transverse fault zones are (Fig. 1):
1 The Aswa lineament is an important fault zone 1200 km long by 60 km wide, underlined by mylonites of Pan-African age (Cahen, 1970). It links the western branch of the rift (Lake Albert Trough) with the eastern branch (central Gregory rift). Transcurrent movements varied during its period of activity (Vidal, 1985).
2 The 1000 km long by 200 km wide TRM fault zone intersects the western branch of the rift. This mobile belt has the imprint of the Ubendian Orogeny

(Dodson *et al.*, 1975), and has acted during the Cenozoic as a dextral strike-slip system (Tiercelin *et al.*, 1988a).

3 The Zambezi–Mwanza Fault ends the rift system to the south. It defines a regional transfer zone that links displacement to the Urema and Dombe grabens (Daly *et al.*, 1989).

The sedimentary basins of the East African rift system

The geological history of rift basins and lakes in the East African rift system is complex. It is remarkably different depending on location in the eastern or western branches. The large quantity of volcanic rocks results in the eastern branch being characterized by shallow and ephemeral sedimentary basins; the western branch has had little volcanism, leading to deep, long-lived basins and thick sedimentary fill (Tiercelin, 1981,1990).

The major sedimentary basins within the eastern branch today have small (maximum 30 × 20 km), shallow (maximum 10 m) tectonic lakes (Hutchinson, 1957). The exception is Lake Turkana at the northern end of the Gregory rift, which is 250 km long, 15–30 km wide, and has a maximum water depth of 125 m. Some of the lakes are fresh water or weakly saline, fed by permanent streams, such as Lakes Abhe and Langano in the Afar Depression and the Ethiopian rift, and Lakes Turkana, Baringo, and Naivasha in the Gregory rift. Others are saline, fed by groundwater or hot springs having temporary inflows from ephemeral streams. These are Lake Asal in the Afar Depression, Lakes Ziway, Abiyata, and Shala in the Ethiopian rift, Lakes Bogoria, Nakuru, Elmenteita, Magadi, and Little Magadi in the Gregory rift, and Lakes Natron, Eyasi, and Manyara in the North Tanzanian Divergence. All these present-day water-bodies are remnants of larger and fresher Pliocene or Pleistocene–Holocene precursors (Grove & Goudie, 1971; Grove *et al.*, 1975). The total thickness of sediment packages within each basin is generally unknown, as there have been few deep drilling or seismic surveys. In Lake Turkana, 24-fold reflection data indicated a maximum sediment thickness of about 4 km (Dunkelman *et al.*, 1988). A magnetotelluric study in the Baringo–Bogoria half-graben of the Gregory rift indicated a sediment thickness of 500–1000 m (Rooney & Hutton, 1977). Ancient Miocene or Pliocene lake sediments are present in the eastern branch as outcrops of sedimentary packages of vary-

ing thickness, situated in grabens beyond the limits of present-day tectonic and magmatic activity. Sediments of lacustrine origin, several tens to a hundred metres thick, occur in the Tugen (Kamasia) Hills sequence within the northern Gregory rift. These bury a large part of the tectonic fabric of this part of the rift from older than 14 Ma to younger than 4 Ma (Hill *et al.*, 1986). Mio-Pliocene lacustrine series, several tens or a hundred metres thick, occur in the southern and central Afar Depression, such as at Ch'orora, Bodo, and Hadar (Tiercelin *et al.*, 1979; Tiercelin, 1986; Williams *et al.*, 1986) (Fig. 3).

In contrast to these short-lived (0.5–3 Ma) lacustrine episodes, the major basins of the western branch contain long-lived, large and deep lakes, including 1470-m-deep Lake Tanganyika and 770-m-deep Lake Malawi (Fig. 3). Recent seismic investigations indicate a sedimentary sequence 6 km thick beneath Lake Tanganyika (Rosendahl *et al.*, 1986)

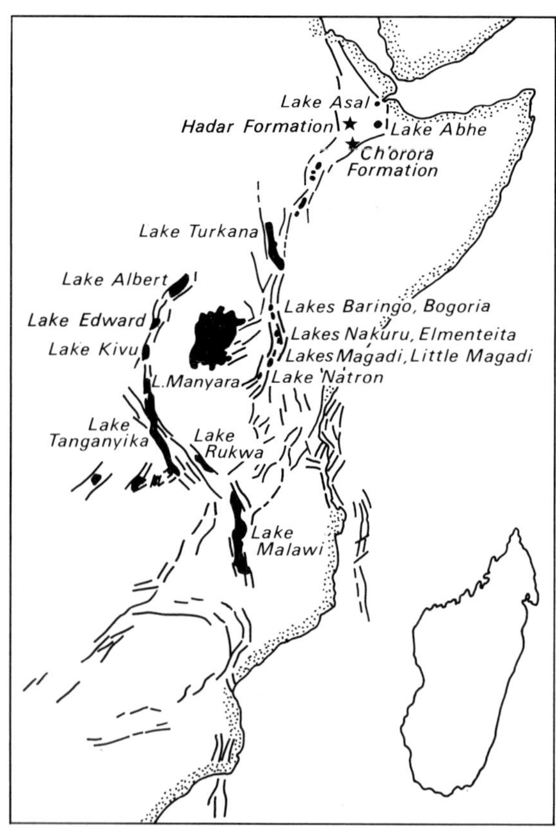

Fig. 3. Lacustrine basins and Cenozoic sedimentary formations of the East African rift system cited in the text.

and a maximum of 4.5 km beneath Lake Malawi (Livingstone Basin) (Johnson & Ng'ang'a, 1990). Lake Kivu is 485 m deep, with approximately 0.5 km of infill (Degens *et al.*, 1973). Extrapolating modern sedimentation rates and correcting for compaction implies that Lake Tanganyika would date to the Miocene epoch, Lake Malawi to late Miocene, and Lake Kivu to mid-Pleistocene. Further north, Lakes Edward and Albert are less well known. In contrast to the Tanganyika, Malawi and Kivu Basins, where very little sediment crops out along the basin margins, the Edward and Albert basins contain sedimentary packages several hundreds of metres thick, which crop out along faulted basin margins and are dated as lower Miocene to lower Pliocene (Pickford *et al.*, 1988,1989). Boreholes drilled in the northeastern Albert Basin penetrated 1200 m of sediment overlying crystalline basement; geophysical evidence suggests a total thickness of at least 2400 m of sediments may fill the depression (Harris *et al.*, 1956).

In the structural context of basins in the western branch, the Lake Rukwa Basin (Fig. 3) is the only one strictly controlled on its southwest and northeast borders by NW−SE trending faults belonging to the TRM transform fault zone, which follow a Precambrian basement and later the Karoo structural trend (Peirce & Lipkov, 1988; Tiercelin *et al.*, 1988a; Morley *et al.*, 1989). Present Lake Rukwa is large (120 × 16 km) and very shallow (3 m maximum). Extensive gravity and seismic reflection surveys and two wells drilled by the Amoco Production Company provide much new information (Morley *et al.*, 1989). The sedimentary package is divided into Karoo sedimentary rocks, late Miocene fluvial red beds, and late Miocene−Recent lacustrine sediments (Haberyan, 1987). The thickness of the Tertiary−Recent sediments is 6−7 km, and the base of the section is only 7−5 Ma old.

LACUSTRINE FACIES OF THE EAST AFRICAN RIFT VALLEY IN TERMS OF NATURAL RESOURCES

Among the sedimentary facies and mineralizations associated with rift lakes, at least eight may be considered as important in terms of natural resources. These are primary deposits, such as organic-rich (or hydrocarbon-rich) oozes, diatomites, evaporites, zeolites and sodium silicates, metallic proto-ores, clays, phosphates, and coals. Other sediments, such as stromatolites and lake-deposited pyroclastics, may have immediate industrial uses or hydrocarbon reservoir properties (Fig. 4A,B).

Hydrocarbon-rich oozes

Organic-rich lacustrine sediments are good source rocks for hydrocarbons. Oilfields occur in lacustrine rocks in Mesozoic continental rift basins along the Atlantic margins of Africa and South America, and other fields in China, Europe or North America (e.g. Talbot, 1988).

Oil seeps and gas emanations have long been known from the Lake Albert Trough. Oil shales of presumed Miocene age were drilled in the Lake Albert depression by the Anglo-Persian Oil Company in 1940, and asphaltic oil was recovered from some wells (Harris *et al.*, 1956). No significant oil has yet been discovered in the East African rift system. Chevron made discoveries in troughs from a Cretaceous rift system in Sudan (Schull, 1984). Explorations, including seismic surveys and onshore drilling, have been carried out by the Amoco Production Company in northern Tanganyika Trough, where sublacustrine asphalt seeps are known at Cape Kalamba on the Zaire side of the lake (Le Mut, 1983; Tiercelin *et al.*, 1989a), and in the Rukwa Trough (Morley *et al.*, 1989). Sediments with a rich hydrocarbon potential have been identified in several East African rift lakes (Talbot, 1988). Three provide models for lake basins: Lake Bogoria in the eastern branch, and Lakes Tanganyika and Kivu in the western branch.

Lake Bogoria (formerly Hannington): a shallow, hypersaline, meromictic, organic-rich lake (Figs 3, 5)

This small (17 × 3.5 km), shallow (10−12 m), meromictic, hypersaline, sodium carbonate lake is located in the Gregory rift (Kenya) immediately north of the Equator (Fig. 5). It sits in a marked asymmetrical, N−S trending half-graben bounded by abrupt fault escarpments (Bogoria and Emsos Faults) to the east, and to the west by a zone of N−S ribbon-like structures cutting Pleistocene trachyphonolitic lavas (Griffiths, 1977; Tiercelin *et al.*, 1980). This half-graben was formed during the major volcano-tectonic phase that affected the northern Gregory rift in the middle Pleistocene. Strong seismic activity is still felt in the vicinity of Lake Bogoria (Loupekine, 1971) and there are almost 200 hydrothermal springs (Tiercelin *et al.*, 1980,1987). Modern

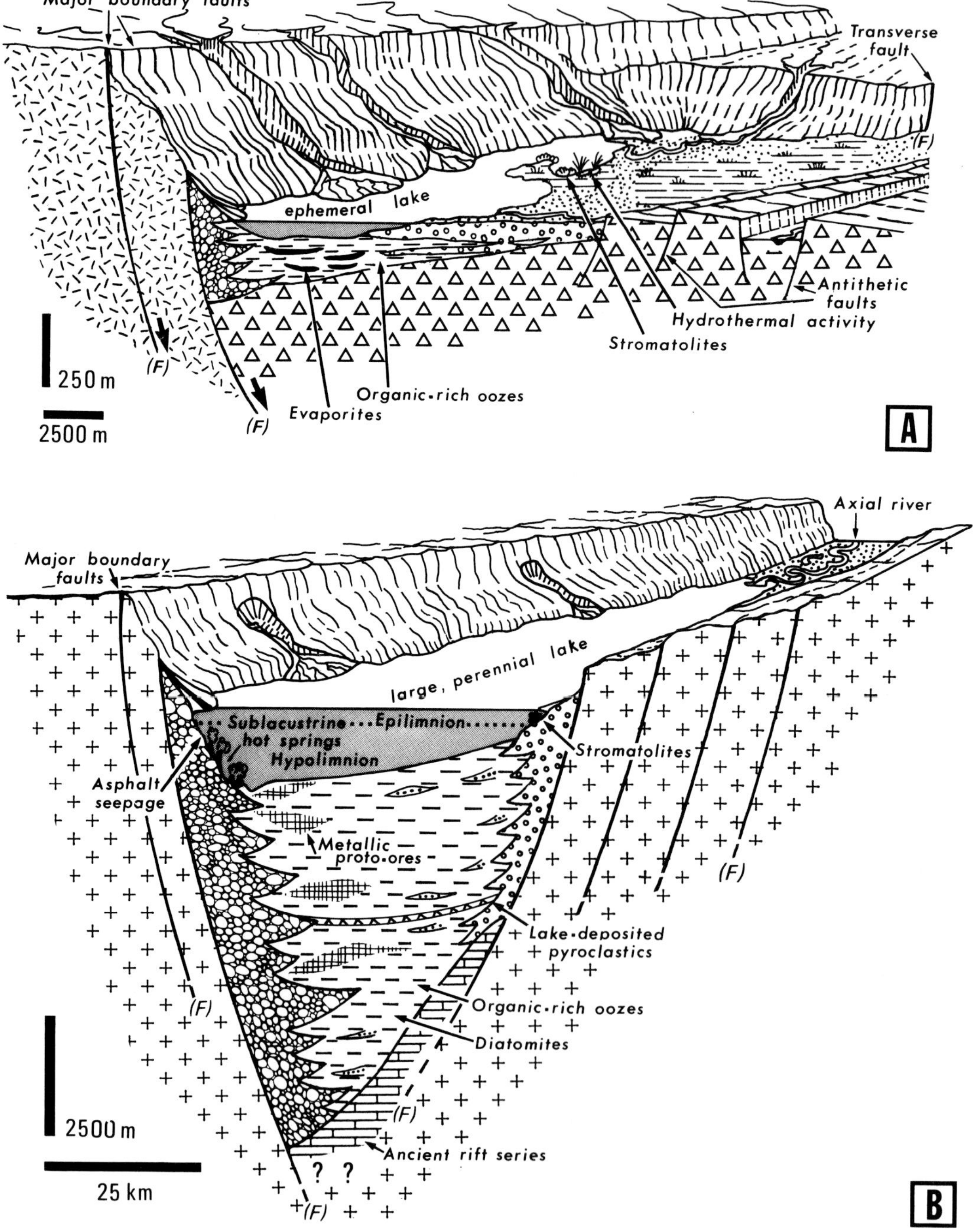

Fig. 4. Block diagrams showing distribution of potential facies along lakes in hypothetical segments of a continental rift. (A) Young, shallow half-graben occupied by a small ephemeral lake. The sedimentary fill, a few hundred metres thick, is mainly formed by cycles of organic-rich oozes and evaporites. (B) Ancient, deep half-graben occupied by a large, permanent lake, often stratified. The sedimentary fill is thick (several thousands of metres) and mainly formed by organic-rich oozes and diatomites, sometimes interbedded with metallic proto-ores or lake-deposited pyroclastics (modified from Frostick & Reid, 1987).

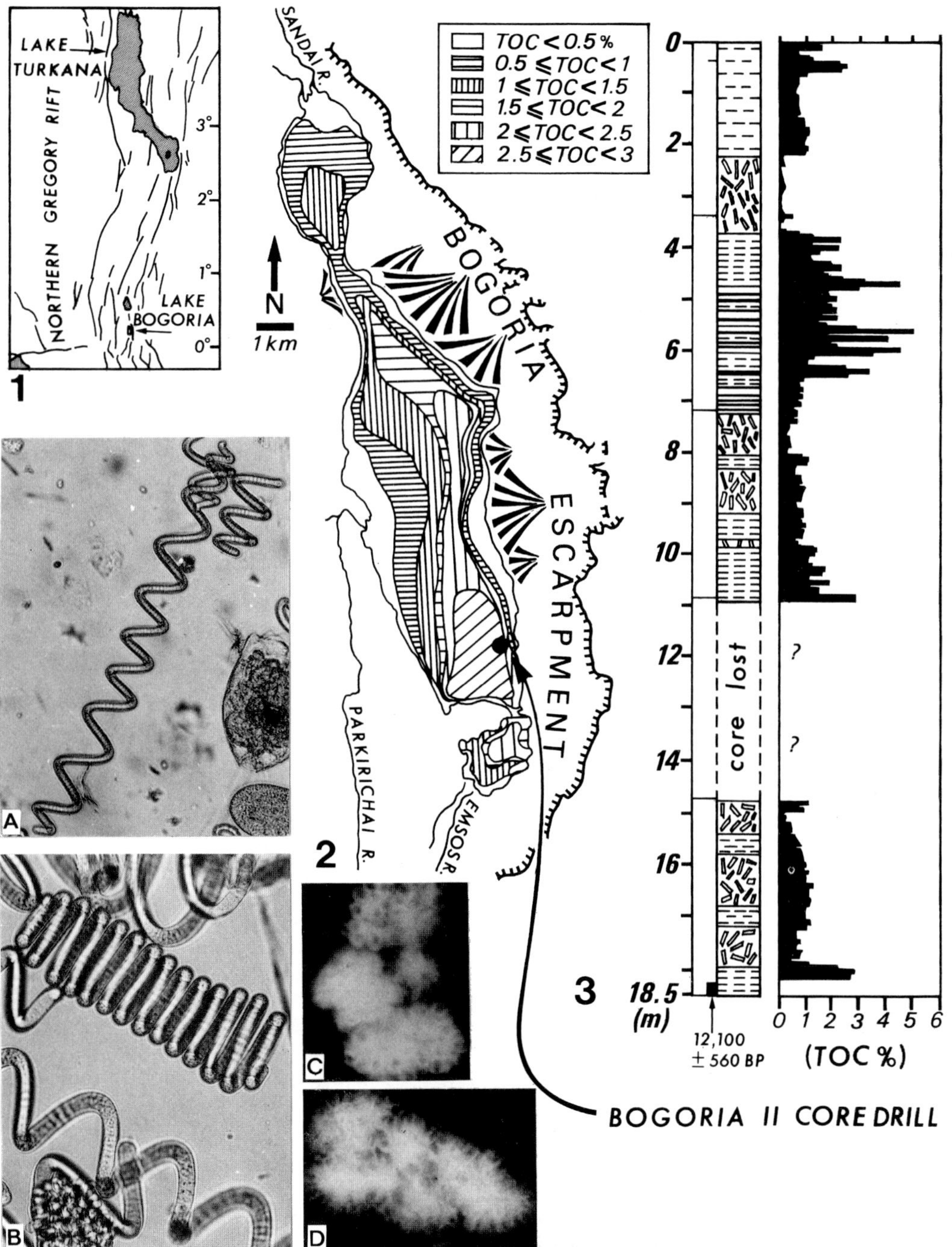

Fig. 5. Lake Bogoria, northern Gregory rift. (1) Location of Lake Bogoria within the northern Gregory rift. (2) Regional distribution of the total organic carbon (TOC) in its modern sediments (after Herbin, 1979): A, the cyanobacteria *Oscillatoria (= Spirulina = Arthrospira) platensis* (Gomont) Bourrelly, main contributor to modern organic matter in Lake Bogoria (length 100 μm); B, specimen with very unrolled spiral (length 80 μm). (3) Vertical distribution of the TOC in the organic-rich oozes — evaporite sequence in Bogoria II drill core: C and D, *Botryococcus* algae, contributor to organic matter in upper Pleistocene to Holocene sediments of Lake Bogoria (from Tiercelin *et al.*, 1987).

organic-rich oozes have been collected from the deeper parts (6–12 m) of Lake Bogoria. They are generally homogeneous, but some are laminated, black or dark-green, very fine-grained muds. Total organic carbon (TOC; wt. % of dry sediment) contents vary from 1.5 to 3%. Hydrogen index (HI) values vary between 150 and 400 (Herbin, 1979). The horizontal distribution of organic carbon reveals a very strong relationship to the topography of the lake floor, in that a significant increase of organic carbon content is seen towards the deeper parts of the lake (Fig. 5). Amorphous organic matter is dominant in most samples (Palacios *et al.*, 1987a). Contributing to this organic matter is mainly phytoplankton, which in Lake Bogoria consists predominantly of cyanobacteria (Type I kerogen) (Talbot, 1988), the prevailing species being *Oscillatoria platensis* (Gomont) Bourrelly (Fig. 5A,B). Cyanobacteria are now recognized as forming the kerogen of some oil shales, such as the Green River Shale, the Belt Series (USA), or the Fig Tree Series (South Africa) (Péniguel *et al.*, 1989). Their production has been studied by various authors (Milbrink, 1977; Vareschi, 1978; Melack, 1979). Organic production can reach the exceptional level of 100 g of *Oscillatoria* per square metre per day, or 3500 tons of fresh weight of cyanobacteria per day for the lake (Tiercelin *et al.*, 1987).

Organic-rich oozes have been encountered also in the form of metre-thick, homogeneous or finely laminated, black or black-green muds interbedded with evaporites dominated by sodium carbonate in the 18.5 m long Bogoria II drill core (Tiercelin *et al.*, 1982; Renaut *et al.*, 1986; Tiercelin *et al.*, 1987). The TOC values in these upper Pleistocene to Holocene sediments are moderate, 1–3 or 2–6% (Fig. 5). Amorphous organic matter is dominant. *Botryococcus* is abundant among identifiable phytoplankton remains (Fig. 5C,D). Locally, cyanobacteria and some higher plant debris are also seen (Palacios *et al.*, 1987b). This suite characterizes sediments that accumulated during the dilute phases of Lake Bogoria. In terms of an oil source rock, the Bogoria organic oozes show a good petroleum potential, ranging up to 10 kg of hydrocarbons per ton of sediment (Herbin, 1979), in basin filling sediments tens of metres thick. Accumulation and preservation of high TOC, high HI sediments are mainly controlled by environmental conditions, essentially nutrients, lake levels, and winds. Frequent turnover in shallow water-bodies favours higher bottom water oxygenation and more persistent oxidizing con-

ditions, thereby greatly reducing the preservation potential of the more labile organic components (Talbot, 1988). Lake Bogoria appears to have been characterized by a stable stratification regime during several late Pleistocene to Holocene intervals. Stratification was independent of low or high lake levels (Tiercelin *et al.*, 1987), thus enhancing the preservation potential. Maturation of petroleum-precursor organisms requires temperatures in the range of 66–132°C, at burial depths greater than 1200 m, and some minimal amount of time, usually cited as 10 000 years for petroleum generation (Tissot & Welte, 1978; Robbins, 1983). Some of these requirements are satisfied in the Bogoria half-graben, where surface sediment temperatures are greater than 60°C as a consequence of hydrothermal activity (Naylor, 1972). Up to 20 m of organic-rich sediments have accumulated in the Bogoria depression in the last 30 000 years. The estimated thickness of the Bogoria sedimentary fill is only 250 m (Tiercelin, 1981; Renaut, 1982), but greater temperatures at depth might compensate a shallow burial depth.

Lake Tanganyika: a deep fresh water, anoxic, and organic-rich lake (Figs 3, 6)

Lake Tanganyika is a stratified lake 1470 m deep (Coulter, 1968), the second largest tectonic lake in the world after Lake Baikal (Hutchinson, 1957). Demaison and Moore (1980) used it as the type basin for a deep anoxic lake model of lacustrine source rock accumulation. It lies between 3°30′ and 9°S. At 5°S, it intersects the NW–SE Tanganyika–Rukwa–Malawi dextral transform fault zone (Tiercelin *et al.*, 1988a), which delineates a 250 km long, N–S trending North Tanganyika Basin, and a 400 km long, NNW–SSE trending South Tanganyika Basin (Mondeguer *et al.*, 1989). These basins are subdivided into a mosaic of seven asymmetric, rectangular-shaped sub-basins separated by ridges of basement rocks (Rosendahl *et al.*, 1986; Tiercelin & Mondeguer, 1991) (Fig. 6A). Recent organic-rich oozes dredged in northern and southern Tanganyika Basin during the Géorift Project of the Elf-Aquitaine oil company comprise black or black-green, homogeneous or finely laminated muds. The TOC contents are from 6 to 9.7%, and the HI is as much as 600 (mean value 387) (Huc *et al.*, 1987,1990). The main source for this organic matter is diatoms. Primary productivity is high (800 mg C/m^2/day) (De Bont, 1972; Hecky & Kling, 1981). In contrast to Lake Bogoria, the lateral distribution of organic matter is

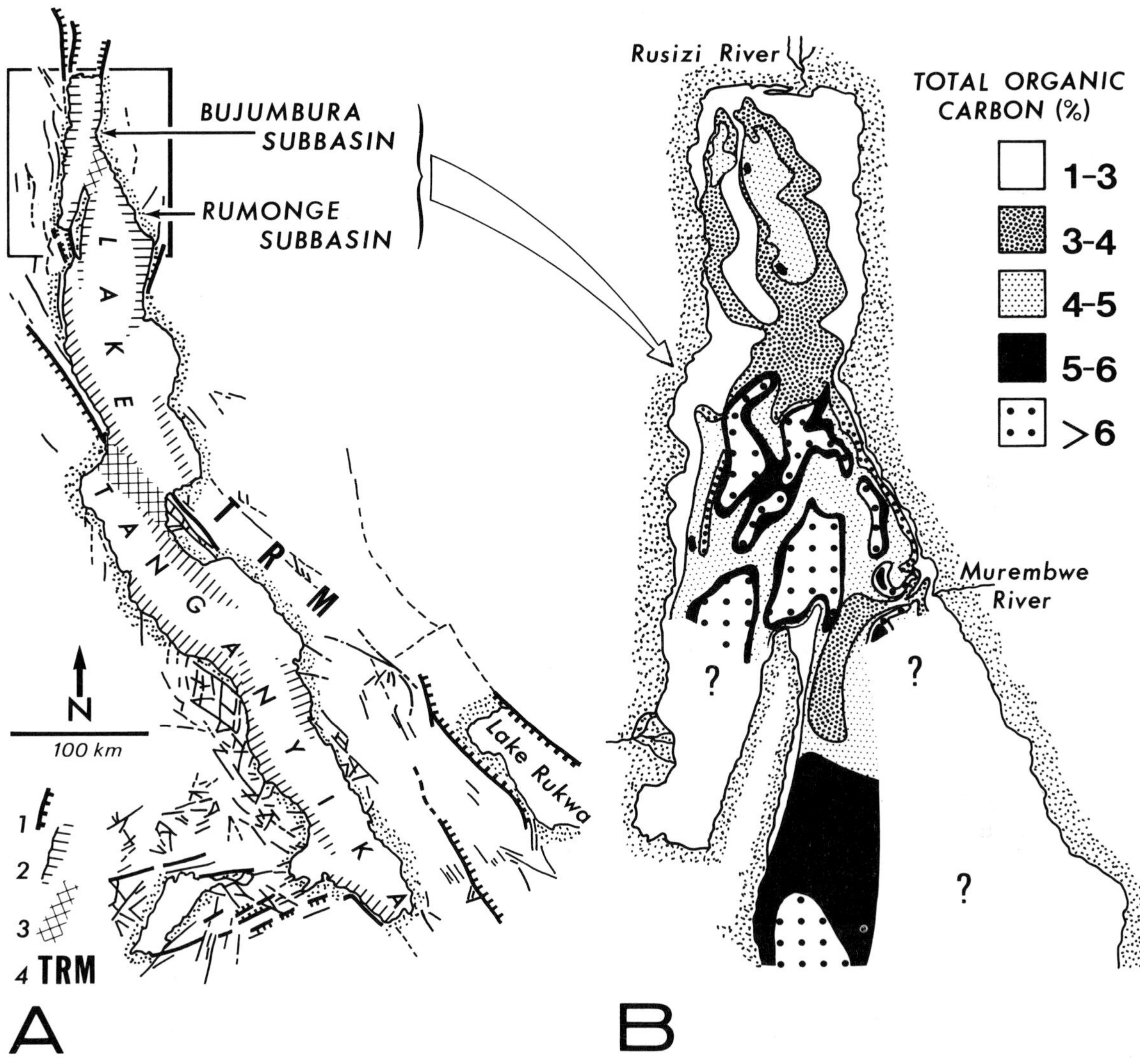

Fig. 6. Lake Tanganyika, western branch of the East African rift system. (A) Structural sketch map: 1, major normal faults; 2, sublacustrine fault escarpments; 3, transverse shoals; 4, TRM transcurrent fault zone. (B) Regional distribution of TOC in modern sediments from the northern end of the lake (Bujumbura and Rumonge sub-basins) (simplified from Huc *et al.*, 1986).

characterized by considerable heterogeneity. The TOC content increases significantly below water depths of about 100 m, which corresponds to the thermocline (Coulter, 1968). The highest TOC values occur on top of sublacustrine shoals (structural blocks), such as the sublacustrine extension of the Ubwari Peninsula in the northern basin (Fig. 6B). Organic matter accumulates because it is located below the thermocline and protected from disturb-

ance by waves and currents and clastic dilution. Cores from these raised areas have uniform pelagic sequences. Winnowing processes sweep low-density organic matter from the shallow waters (above the thermocline) towards the deeper parts of the basin. Gravity transport mechanisms are generally related to major pulses of river-fed suspensions. Turbidity currents fill deeper depocentres. Cores of green to black clays and diatom-rich laminated muds penetrate up to 40 000 years (Tiercelin *et al.*, 1989b). They are characterized by TOC contents as much as 10% and HI as high as 632 (Huc *et al.*, 1986). Downcore parameters vary in relation to shifting environmental conditions, such as lake levels and wind regimes. Meromixis probably persisted during middle–late Pleistocene, when levels were as low as the 300–600 m depth contour (Tiercelin *et al.*, 1989b).

Thus, in terms of source rock, the Tanganyika sediments appear extremely heterogeneous, with sediments having petroleum potential and yielding as much as 35 kg of hydrocarbons per ton of rock only a few metres from sediments having low petroleum potential and yielding as little as 3 kg of hydrocarbons per ton of rock (Huc *et al.*, 1987). In terms of preservation potential, Lake Tanganyika offers optimal conditions in the form of a thick, almost permanently anoxic hypolimnion. Maturation prerequisites are satisfied with sediment thickness as much as 6000 m, ranging in age to the Miocene (Rosendahl *et al.*, 1986). Sufficient temperature and heat flow values probably relate to sublacustrine hydrothermal activity (Le Douaran, 1986; Tiercelin *et al.*, 1989a). Asphalt occurs in the form of floating fragments along the eastern shore of the Ubwari Pcninsula as far as thc dclta of thc Rusizi River. A geochemical analysis of this asphalt yielded 56% resins + asphaltenes and 44% hydrocarbons (Le Mut, 1983). Local fisherman use it as caulking. Asphalt appears to be related to sublacustrine hydrothermal activity located at Cape Kalamba on the Ubwari Peninsula (Fig. 7). Similar occurrences of hydrothermally generated petroleum have been noted in the Guaymas Basin, Gulf of California (Simoneit, 1985). Gaseous hydrocarbons (methane and heavier hydrocarbons) also have been discovered in hydrothermal fluids at the Pemba and Cape Banza sites north of Cape Kalamba (Fig. 7). The Pemba site is located along the major, N–S

trending normal Uvira Fault, which forms the western boundary of the 250-m-deep Bujumbura sub-basin and is a zone of strong seismic activity (Wohlenberg, 1969). The Cape Banza site is located along the N–S trending fault escarpment of the Ubwari Peninsula, which bounds the 1200-m-deep Rumonge sub-basin (Figs 6, 7). At Pemba, abundant gas and hydrothermal fluids having a measured temperature of 65°C escape through sandy bottom-sediments or basement fractures (Fig. 7A). At Cape Banza, fluids having temperatures of 70–80°C flow through orificial chimneys constructed of aragonite in various sizes and morphologies (Fig. 7D,E,F) (Tiercelin *et al.*, 1989a). Methane and heavier hydrocarbons are present in the Pemba and Cape Banza hydrothermal fluids. Maximum values have been measured at Cape Banza, 842 μl/l CH_4, which are equivalent to those found at the latitude 21°N hydrothermal sites of the East Pacific Rise (Welhan & Lupton, 1987). Tiercelin *et al.* (1989a) suggest a thermocatalytic origin for the North Tanganyika hydrocarbons or a mixture of gases having both thermocatalytic and biogenic origins.

Methane added to the oxic zone, such as at the Pemba and Cape Banza sites is rapidly oxidized. At 360 m depth, methane concentrations reach 70–80 μmol/l (Rudd, 1980), and increase downward through the hypolimnion (Craig & Craig, 1981). Craig (1973) and Hecky (1978) suggest that Lake Tanganyika's hypolimnion may be a relict water mass from a cooler, drier period, and some of the methane would therefore derive from old carbon deposited during previous intervals of the lake's history. Craig *et al.* (1974) suggest emissions from geothermal sources at depth, because the gas helium is highly supersaturated below the metalimnion. High heat-flow values measured during the Géorift Project in the northern basin (Le Douaran, 1986) may be linked to these sources, which supply heat and thermocatalytic methane to the hypolimnion.

Fig. 7. The Pemba, Cape Banza, and Cape Kalamba sublacustrine hydrothermal sites of the northern Tanganyika Trough. (A) Gas (H_2S, CH_4) bubbling from sandy bottom-sediment and basement fractures at Pemba site, 10 m water depth. (B) Sulphide block at a depth of 10 m at Pemba site. (C) Section of block fragment showing an assemblage of corrugated flakes formed by striped massive and porous sulphides (pyrite and marcasite) (a); enclosed translucent patches formed by quartz, kaolinite and acicular crystals of barite (b); and encrusted detrital grains (c). (D), (E) Hydrothermal aragonite chimneys having hot fluid plumes, Cape Banza site, 6 m water depth. (F) 'Big Chimney': multiple orifice hydrothermal vent, Cape Banza site, 4 m water depth. Photograph (G) and sketch (H) of circulation pipes of hydrothermal fluids after removal of an aragonite chimney: (S) rock substratum. (OC) old chimneys. (Ac) aragonite coating; (P) circulation pipes. The walls of these pipes and entrapped detrital grains (I), mainly quartz, show thin coatings of pyrite (J): SEM photographs — scale bar is 100 μm for (I), 10 μm for (J).

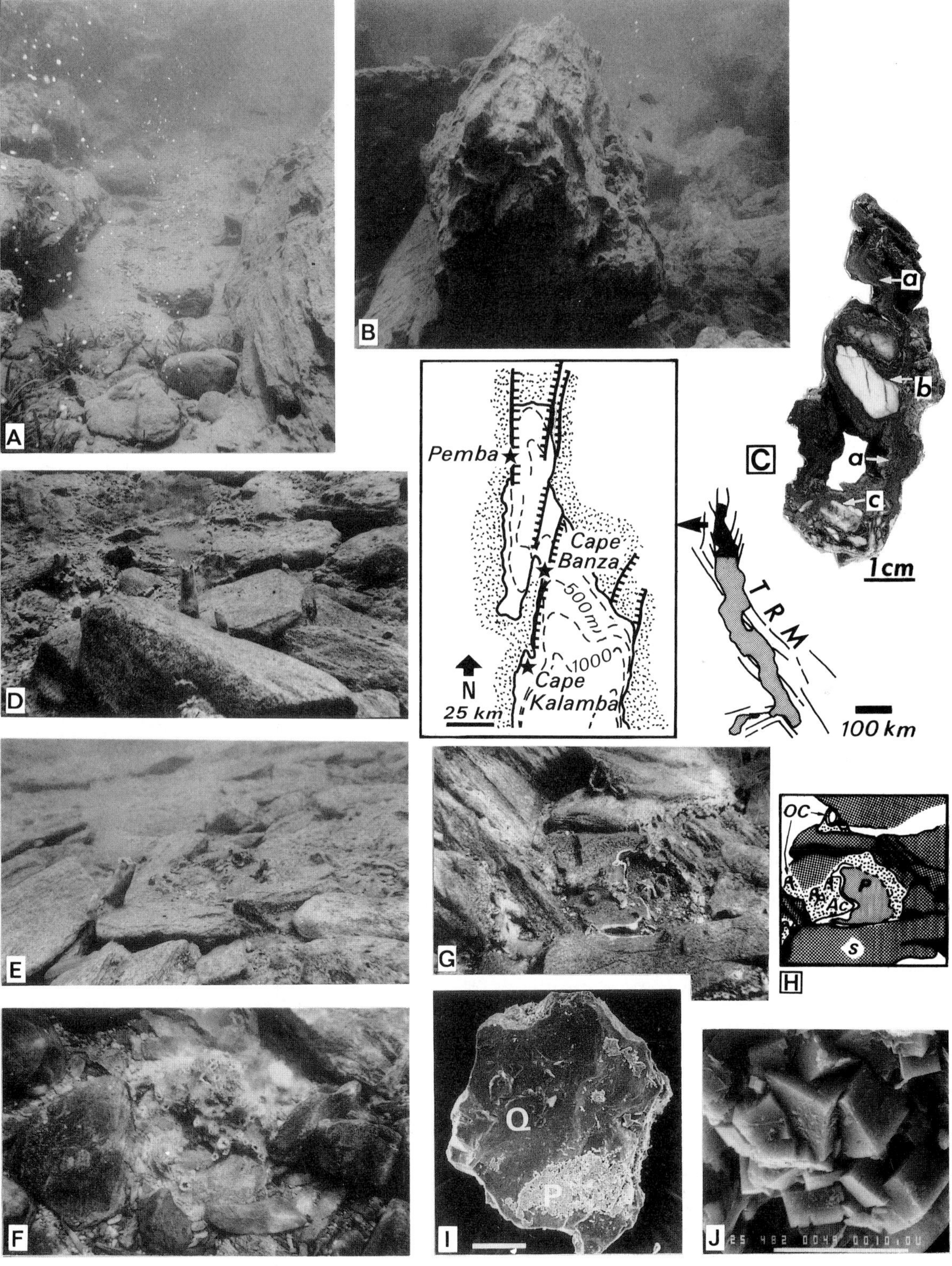
a
b
c
C
1cm
Pemba
Cape
Banza
500 m
1000
N
25 km
Cape
Kalamba
T R M
100 km
OC
AC
P
S
H
Q
P
I
J
25 482 0043 0010 10U

Lake Kivu: a deep meromictic, methane-rich lake
(Figs 3, 8)

Situated between 1°40′ and 2°30′S, Lake Kivu has a depth of 485 m and an area of 2060 km². It lies in a NE–SW trending graben (Fig. 8A) that formed in the late Miocene to early Pliocene, following two major stages of volcanic activity at 10–6 Ma and 8–4 Ma (Ebinger, 1989a). Lacustrine sedimentation began approximately 3 Ma ago, and has accumulated 0.5 km (Degens *et al.*, 1971,1973; Wong & Von Herzen, 1974). Hydrologically, Lake Kivu is meromictic with an unusual thermohaline structure showing increasing temperature and salinity with depth. Its hypolimnion is anoxic (Degens *et al.*, 1973). Large quantities of economic hydrocarbons (methane and higher gaseous hydrocarbons) and other gases (hydrogen sulphide, carbon dioxide) are stored in the hypolimnion (Damas, 1937; Tietze *et al.*, 1980). The total amount of methane is about 63×10^9 m³ at standard temperature and pressure (0°C, 1 atm). No comparable methane accumulation occurs in any other lake. These large quantities are the result of stable density layering over a long period.

Models for the origin of Kivu methane have for long been controversial (Deuser *et al.*, 1973; Jannasch, 1975). Recent work by Tietze *et al.* (1980) confirms that most of the Kivu methane is bacterially generated from the organic matter in the sediment (Fig. 8B). As much as 200 ppm of gaseous hydrocarbons with higher carbon numbers suggests small amounts of thermocatalytic methane in addition to bacterial methane. According to Wong and Von Herzen (1974), a sedimentary package 0.5 km thick and active volcanoes provide heat sources. Hydrothermal inputs, mainly characterized by metallic deposits, also have been identified in Lake Kivu for the late Pleistocene to Holocene period (Degens & Kulbicki, 1973). Such sources, similar to northern

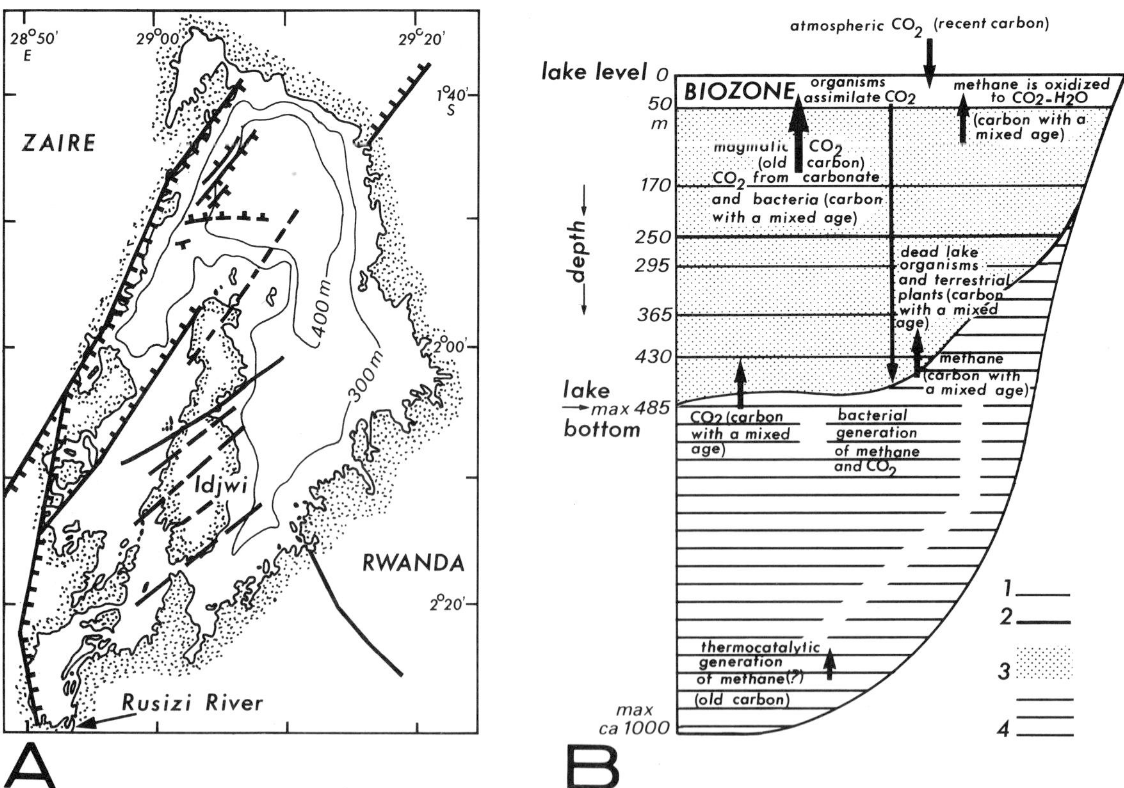

Fig. 8. Lake Kivu, western branch of the East African rift system. (A) Structural sketch map and bathymetry. (B) Model of methane genesis: 1, water gradient boundaries; 2, main gradient boundary; 3, water containing gas; 4, sediment (after Tietze *et al.*, 1980).

Lake Tanganyika (Tiercelin *et al.*, 1989a), may inject significant quantities of methane in the deep waters of Lake Kivu.

The practicability of extracting economic methane from Lake Kivu has been studied recently at the request of the Communauté Economique des Pays des Grands Lacs (CEPGL) by the French group Technip and associates (Sogreah, Sedes, and Bureau de Recherches Geólogiques et Minières — BRGM). The rate of replenishment of the methane has been estimated to be about 400 years (Jannasch, 1975). An environmental impact statement remains to be completed before exploitation of methane begins. The fear of oxygenation of the bottom leading to lake overturn in the manner of Lake Nyos in Cameroon (Kling *et al.*, 1987) must also be addressed.

Other organic-rich lake deposits in the East African rift system

Other lakes in the rift have deposits with high to very high TOC values. Lake Edward has an 11.6% mean TOC in premodern sediments. Lake Albert, which has intermittent stratification, and Lake Malawi, which is permanently stratified in the north, are accumulating sediments with TOCs of 1.4−4.0% and 3−6%, respectively (Talbot, 1988; Johnson & Ng'ang'a, 1990). In sharp contrast, Lakes Turkana and Baringo (northern Gregory rift) have less than 1% TOC preserved in the modern sediments (Yuretich, 1979; Tiercelin & le Fournier, 1980). High sedimentation rates may, in part, be responsible for the low TOC, which would then be dominated by autochthonous material, in the case of Lake Turkana (Kelts, 1988). The permanently well-mixed waters of shallow Lake Baringo are the principal limitation on organic matter preservation. Despite low TOC, Lake Turkana sediments are methane-rich (Johnson & Davis, 1989).

Meromictic conditions appear favourable but not essential for the accumulation of organic-rich sediments. Rift valley lacustrine sediments are characterized by Type II kerogens, which result from a mixture of alginite and exinite (Talbot, 1988). The hydrocarbon potential of the East African rift system lakes also relates to quality and extent of reservoir beds. Multifold seismic investigations of thick sedimentary fills in Lakes Albert, Kivu, Tanganyika, and Malawi should define the geometry of such bodies. Deep drilling will be required to ascertain the quality of potential reservoir beds. Recent exploration using extensive gravity and seismic reflection surveys and wells by the Amoco Production Company in the North Tanganyika and Lake Rukwa portions of the East African rift show the complexity of the regional tectonics and stratigraphy (Morley *et al.*, 1989), and suggest occurrences of hydrocarbon source rocks within pre-Cenozoic rift series, mainly of Karoo age.

Diatomites

Diatomite is derived from marine or lacustrine diatom frustules. Chemically it consists mainly of silica. Several hundred diatomite products are used commercially, including filter aids, thermal and sound insulation, absorbents, insecticide carriers and dilutents, fertilizer conditioners, abrasives, ceramics, and drilling mud thickener (Kirk-Othmer, 1985).

Pure diatomites of great thickness are rare in geological sequences. Controls include: a diatom-dominated phytoplanktonic ecosystem, a high and consistent rate of surface productivity, a regular source of nutrients, and a lack of detrital sediment in the area of diatom deposition.

East African rift water-bodies span an enormous range of ecological and sedimentological conditions, and their diatom flora is often abundant and highly diversified. Large, deep, freshwater rift lakes, having regular hydrodynamic regimes, such as Lakes Tanganyika and Malawi, are characterized by high biological productivity rates. In such environments, diatom productivity is largely controlled by competition with other algal species (Hecky & Kling, 1981). Wide, open water areas in these environments are generally protected from the influence of detrital inputs associated with permanent streams feeding the lake. Very fine-grained sedimentation, mainly dominated by diatom frustules (up to 10^9 frustules per gram of sediment), results. Small, shallow, often saline, rift lakes are affected by drastic changes in lake levels as a consequence of seasonal climatic variations (or major climatic changes throughout the Quaternary). Diatom communities vary strongly according to changes in water volume and salinity. High pH values, which characterize most of the saline lakes of the East African rift system, greatly reduce the preservation potential of opaline diatom frustules.

High diatom productivity is stimulated by high dissolved-silica content in lake waters, provided by weathering of volcanic ashfalls, or directly through hydrothermal springs, which are abundant in rift systems (Kilham *et al.*, 1986). An ashfall-produced *Nitzschia* bloom was suggested for an 11 000 yr BP layer in a southern Tanganyika Basin core (Haberyan & Hecky, 1987; Tiercelin *et al.*, 1988b; Mondeguer *et al.*, 1989).

Diatomite occurrences in the East African rift system

Pure diatomites of Pliocene to lower Pleistocene and Holocene age were formerly extensive in the Abhe Basin, northern Afar Depression (Republic of Djibouti) (Fig. 9). Present or past erosion has removed most of the diatomites, leaving very small residual hills along the palaeoshorelines (Fig. 9A). Pliocene to lower Pleistocene diatomites generally underlie basaltic lavas (Fig. 9B). Because of the small size of these outcrops, no economic exploitation can be foreseen for the Abhe diatomites. Other occurrences of diatomite exist as decimetres to metres thick beds interbedded with pumice, ash, and clays in the Miocene Ch'orora Formation (Tiercelin *et al.*, 1979) (Fig. 9C,D) and in the Plio-Pleistocene Bodo Formation (Williams *et al.*, 1986) of the southern Afar Depression (Ethiopia). The bed thickness and transport problems limit exploitation at present.

Diatomite occurrences of the Gregory rift

Among the numerous sedimentary basins of the East African rift system, only a few contain thick (several metres to tens of metres) pure diatomites. The only occurrences of pure diatomite of a workable thickness are at Kariandusi (reserves estimated at over 1 500 000 tons) and at Kockum and Brown's working on the Soysambu Estate (reserves estimated at nearly 4 000 000 tons), both middle Pleistocene, near Lakes Nakuru and Elmenteita (McCall, 1967) (Figs 3, 9). Kariandusi diatomites are now quarried by the E.A. Diatomite Syndicate. Production first began around 1940 to supply the local soap and sugar industries and the Indian market. The best diatomite outcrops as a layer over 30 m thick. Difficulty of quarrying is increased by thick over-

burden, and the presence of interbedded pumice and clay layers (Fig. 9E). Other diatomite beds of Pleistocene age exist in the Gregory rift as discontinuous horizons in the Lake Naivasha Basin (Thompson & Dodson, 1963), or as widespread beds in the Olorgesailie Formation (Magadi Basin), where diatomite was mined by the Magadi Soda Company for filtration in its sodium bicarbonate plant (Baker, 1958).

Evaporites

Continental evaporites are generally associated with desert regions having internal drainage basins. Economic evaporitic lacustrine minerals include gypsum, halite, trona, and associated minerals (fluorite and villiaumite, gaylussite, sodium silicates). Trona is the most important form of sodium carbonate. Its primary economic use is in the glass industry. In particular, sodium silicates and sodium phosphates are used in the production of chemicals for the pulp and paper industries, detergents and cleaners, and water treatment. Calcium fluoride or fluorite is the principal source of fluorine and its compounds, and is mainly used in ceramics, electric arc welders, certain cements, dentifrices, paint pigment, and as a catalyst in wood preservatives. Sodium fluoride or villiaumite is used for fluoridation of municipal water, wood preservative, insecticide, and glass manufacture. Gypsum is used for numerous purposes, including Portland cement retarder, as a source of sulphur and sulphuric acid, for paints and paper, and for metallurgy. Halite or sodium chloride is used by the chemical industry, for metallurgy, in mineral waters and soap manufacture, and for road deicing (Hawley, 1981).

Rift systems favour the formation of partitioned basins, as the result of interval faulting. Structures act as morphological barriers, which lead to basin isolation by controlling the course of potential tributary rivers. Volcanic activity produces large volcanic cones or lava flows, which alter drainage patterns and contribute to basin isolation. Such isolation enhances evaporation of rift lakes in areas where

Fig. 9. Main diatomite occurrences in the East African rift system. (1) Lake Abhe Basin, northern Afar, Republic of Djibouti: (A) residual hills of diatomite of Holocene age; (B) Pliocene to lower Pleistocene(?) diatomite overlain by basaltic lavas. (2) The Ch'orora Formation, southern Afar, Ethiopia: (C) faulted contact between basalts of the Lower Afar Series and Miocene diatomites, Arba River; (D) 10.5 Ma diatomites interbedded with pumice beds at the Ch'orora type locality. (3) The Kariandusi quarry, central Gregory rift, Nakuru area, Kenya: (E) the Pleistocene diatomites of Kariandusi, interbedded with pumice beds and faulted.

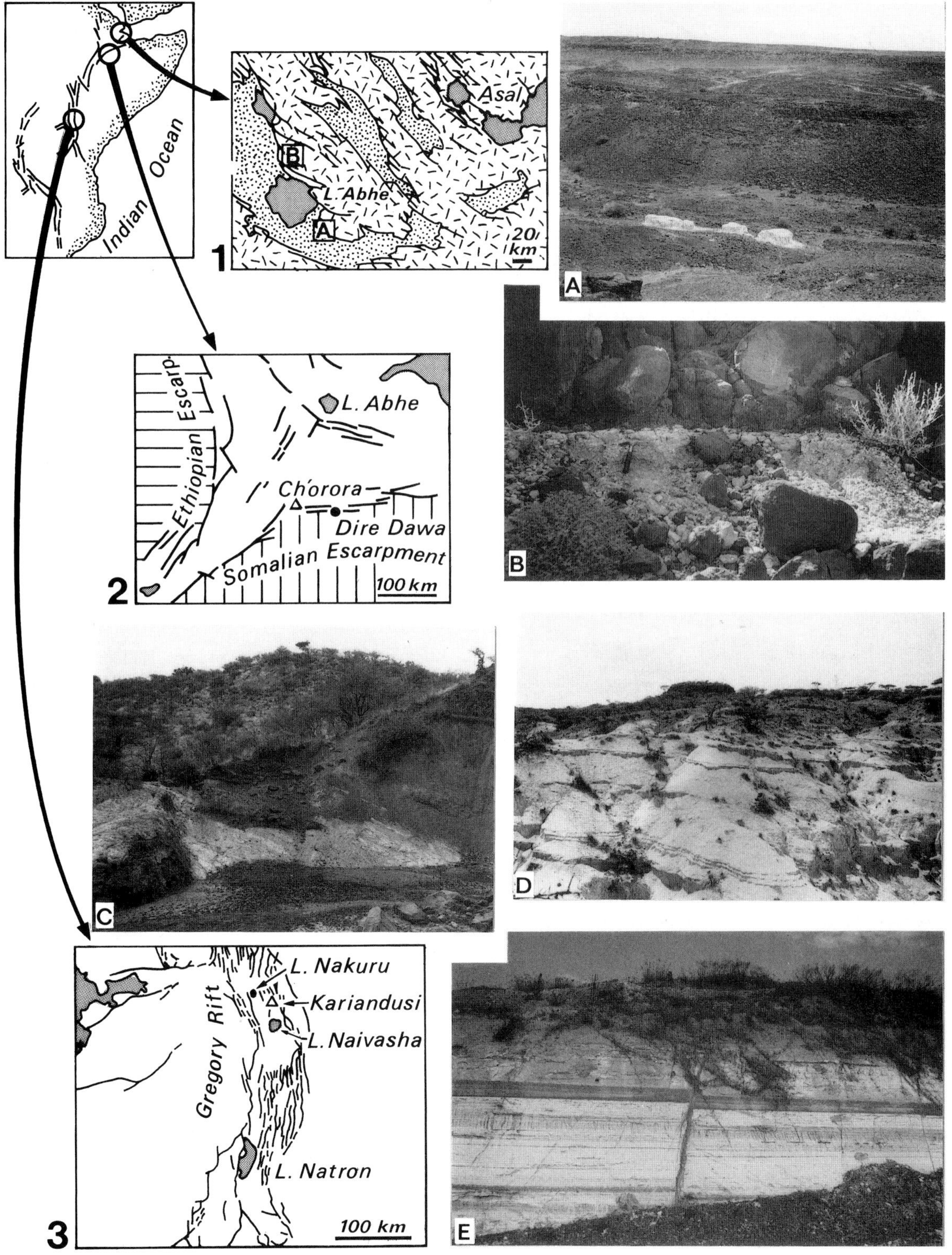

Indian Ocean
Asal
L. Abhe
B
A
20 km
1
Ethiopian Escarp.
L. Abhe
Ch'orora
Dire Dawa
Somalian Escarpment
100 km
2
Gregory Rift
L. Nakuru
Kariandusi
L. Naivasha
L. Natron
100 km
3
A
B
C
D
E

geographical position and rift structure contribute to increasing aridity. Hydrothermal circulation favoured by earthquake rupturing (Sibson, 1981, 1987) contributes to the input of solutes.

Only a few East African basins contain evaporites. All are in the eastern branch, mainly as a consequence of its more arid conditions and of its 'wet' volcanic character (Mohr, 1982; Yuretich, 1982; Tiercelin, 1990). These evaporitic basins include, from north to south, the Asal Basin (11°40′N) and the Abhe Basin (11°10′N) in the Afar Depression, the Bogoria (0°15′N), Nakuru (0°30′S), Elmenteita (0°25′S), Magadi, and Little Magadi (1°50′S) Basins in the Gregory rift, and the Natron (2°30′S) and Manyara (3°35′S) Basins in the North Tanzanian Divergence (Fig. 3). Among these basins, Asal can be distinguished by the chemistry of its brines, which are of chloride−calcium sulphate type, and by its associated deposits of halite and gypsum. All the others have sodium carbonate type brines and associated salts and silicates (Jones *et al.*, 1977). Only two sites, Nakuru and Magadi, have been worked for industrial purposes.

Lake Magadi: an evaporitic basin with sodium carbonate brines and salts (Figs 3, 10)

The Lake Magadi sodium carbonate deposits were first surveyed in 1904, and later in 1908 (Anonymous, 1923). The first concession was to a private prospector; in 1911 the Magadi Soda Co. Ltd was created (Baker, 1958). Today the Magadi Soda Company is part of the Imperial Chemical Industries PLC group of the UK.

Lake Magadi occupies a narrow (164 km^2) graben defined by a complex system of small N−S, NNW−SSE, and NNE−SSW faults in the axial part of the Gregory rift floor (Figs 10, 11A). Volcanism in this

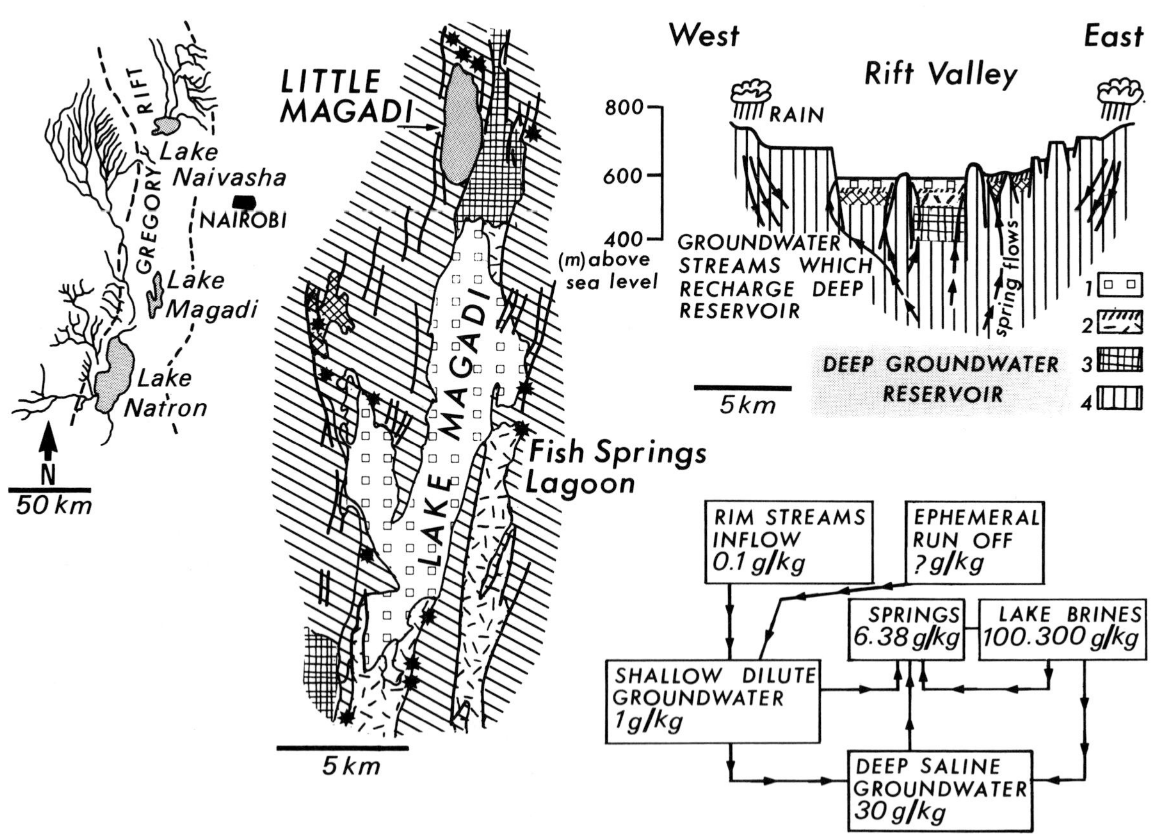

Fig. 10. Lake Magadi, southern Gregory rift, Kenya. Structural and sedimentological sketch map and schematic cross-section through the Magadi graben: 1, trona; 2, chert and evaporites; 3, volcanic tuffs and silts; 4, bedrock (trachytes); bold star, thermal springs. Schematic diagram showing stream inflow to closed brine system (courtesy of Magadi Soda Company Limited).

part of the rift floor ceased by 0.8 Ma, after which many small faults created the Magadi and Natron Basins (Baker, 1963,1986). Deposits from two Pleistocene precursors of Lake Magadi can be recognized, dated to >780 000 yr BP for Lake Oloronga and 10 000 yr BP for High Magadi Lake phases (Eugster, 1986). Both lakes had alkaline brines throughout their existence, without ever reaching the stage of extensive trona precipitation of the early Holocene to modern Lake Magadi. As much as 40 m of the Holocene evaporite series has been drilled in the Magadi Basin (Eugster, 1980). The chief mineral is trona, which occurs as a very porous, sparry framework of large blades; the interstices are filled by saturated brines (Fig. 11B). The lake covers about 80 km^2 (Fig. 10) and contains enough raw material to extract 100 million tons of soda ash. At the present time, during the dry season, the level of the interstitial brine falls several metres below the trona surface, whereas during the rainy season, flooding occurs, dissolving the surface trona crust. After a few months, this flood layer evaporates, precipitating and recrystallizing the surface layer at an estimated accumulation rate of 2−3 mm/year (Eugster, 1980) (Fig. 11C). Additional solutes are contributed to the interstitial brines by ephemeral storm runoff dissolving the evaporitic crusts (trona−thermonatrite−halite) that coat the Pleistocene lake deposits, and by perennial saline springs, which issue through the sediments and trachyte lava flows surrounding the lake (Eugster, 1970; Jones *et al.*, 1977) (Fig. 10).

Today, the Magadi Soda Company exploits raw trona with two dredges. Because the brine level is never far below the trona surface, the dredges remain more or less at ground-level and are designed to cut to 8 ft. below the trona surface (Fig. 11D,E,F). Crushed trona is mixed with brine, thus forming a slurry that is pumped through a pipeline to the factory. After dewatering and washing, the trona mush is passed to calciners and transformed into soda ash or anhydrous sodium carbonate (Na_2CO_3), which is bagged and loaded into railway trucks for the port at Mombasa. Magadi Soda is planning to increase its production to 550 000 tons/year.

Soda ash has always been the company's main product. It also manufactures sodium chloride by pumping a liquor rich in sodium chloride from a depth of 8−10 ft. in a series of wells, and concentrating it into 'salt-making ponds' (Fig. 11G). Magadi Soda produces as much as 40 000 tons of salt a year, enough to meet Kenya's demands for industrial and edible salt. Sodium fluoride (villiaumite) was also extracted from the trona slurry, particularly during World War II for use by air forces in Africa and India (Baker, 1958).

Among the three other soda lakes in the Gregory rift, Elmenteita, Nakuru, and Bogoria (Fig. 3), only Lake Nakuru has been worked for the extraction of soda. Nakuru Lake Syndicate was granted a licence in 1952 to remove and process soda goods upon the lake shore. Solar ponds were constructed for this purpose. Limited production of soda ash was carried out for the fertilizer industry. From 1955, a rapid rise of the lake in a series of wet years prevented the formation of a trona crust. By 1959, the buildings and installations had been removed and the lake water almost covered the site of the original establishment. Lake Elmenteita and Lake Bogoria, which is known to contain metre-thick trona beds interbedded with organic oozes (Tiercelin *et al.*, 1982), have never been worked. It seems doubtful if such operations could compete economically with Lake Magadi.

Other evaporite deposits, while unlikely to prove commercially exploitable, play an important part in the life of the Kenya tribes. In the Suguta valley (northern Gregory rift), evaporite deposits consist of variable-sized layers of fine acicular crystals contained in a clay about 0.15 m thick. Although the composition of this salt is unlike the trona from Lake Magadi, it is used for the same purposes as unprocessed trona, namely, as an addition to finely ground tobacco to produce snuff, which is extensively used by most tribes in Kenya, and particularly by Turkana tribesmen who travel for considerable distances to collect supplies of the salt (Dodson, 1963).

Gaylussite, a hydrated double carbonate of sodium and calcium, occurs in large quantities in a black or dark green host clay in the Amboseli Lake Beds of upper Pleistocene age in southern Kenya (Williams, 1972). Proved resources of 15 000 000 tons of gaylussite were found in an area of 15 square miles. An increase to 16 500 000 tons is possible if particles of −30 mesh size are recovered during working of the ore. Despite this amount of reserves, preliminary investigations were proved disappointing. Gaylussite deposits also have been recorded from drill core SIII from the northern basin of Lake Bogoria (Tiercelin *et al.*, 1982; Renaut *et al.*, 1986). There, crystals (2−10 mm) occur at several levels in grey to brown clays in beds a few decimetres to a metre thick; the beds have been dated between 14 000 and 8000 yr BP (Tiercelin *et al.*, 1987). Small gaylussite euhedra were also observed on the margins of the lake near

freshwater springs (Cerling, 1979), and as surficial efflorescent crusts on the Sandai delta-plain north of Lake Bogoria. The small proportion of gaylussite in the sediment and the remoteness of the outcrops mean that these occurrences have no commercial interest.

Lake Asal: an evaporitic basin having chloride and calcium-sulphate brines and salts (Figs 3, 12)

The Asal Basin (Republic of Djibouti) is within the NW−SE tectonic system of the Asal−Ghoubbat al Kharab rift, which extends to the northeast via the oceanic structure of the Gulf of Aden (Fig. 12A). This system is considered to be the type spreading segment in the world rift system and is characterized by an extension rate of 30−60 mm/yr (McKenzie *et al.*, 1970; Ruegg & Kasser, 1987). Lake Asal is shallow (maximum depth 30 m; mean depth 7.4 m), having an area of 55 km^2, and is located at an altitude of −155 m below sea-level (Langguth & Pouchan, 1975). Its waters reach salinities of 350 g/l and are rich in Cl, Na, K, SO_4, and Mg ions (Lopoukhine, 1973). Local rainfall is insignificant. Several springs, hot and cold, of continental and marine origin, located along the southeastern, northern, eastern, and southwestern shores compensate for this rainfall deficit (Valette, 1975); these springs supply an estimated 45×10^6 m^3/year of water (Lopoukhine, in Demange *et al.*, 1971). The age of the spring waters are of the order of tens of years, or, at most, one or two centuries (Fontes *et al.*, 1980). Modern sedimentation in Lake Asal is essentially evaporitic; gypsum precipitates in the southeastern part of the lake because of the mixing of lake brines with sea water rising up from fractures through the lake bottom (Fig. 12B,C). In the northeastern part of the lake, halite is precipitating and forming a wide (60 km^2) and thick (20−80 m) salt plain rising 0.30−0.80 m above the lake level. Bedded gypsum crops out around the lake as high as 55 m above the present lake level (Stieltjes, 1973) (Fig. 12D,F,G). The salt plain was dry until volcano-tectonic activity

in 1978, which gave rise to fractures and fault openings. The resulting increase of sea water inflow from the Ghoubbat al Kharab partially submerged the salt plain (Ruegg *et al.*, 1979) (Fig. 12E,H). Lake Asal has not been worked for the extraction of gypsum or halite except for local salt use by the Afar tribesmen. This indifference to such a substantial evaporite accumulation can be explained by the difficult access to the area, which should only be undertaken by 4-wheel drive vehicles, and also by the wide abundance of halite and gypsum resources throughout the world.

Another evaporite sequence is known to the north-northwest of the Asal Basin in the Danakil area of Tigre and Eritrea provinces of Ethiopia (northern Afar Depression). There, 960 m of bedded halite, including two potash-bearing horizons and a sylvite-rich member, a kainite-rich member ($MgSO_4 . KCL . 3H_2O$), and gypsum, anhydrite, and shale interbeds, were drilled in the area of Dallol (Holwerda & Hutchinson, 1968). This sylvite-bearing sequence, which has been known for at least 80 years, was further explored by the Ralph M. Parsons Company of Los Angeles in 1954. Pre-production and development working of the sylvite and kainite, used as a source of potassium compounds for fertilizers, were completed in 1965.

In contrast to the Asal evaporite sequence, the Danakil strata are of exclusively marine origin. They were laid down during a Pleistocene high stand in a NNW−SSE trending, asymmetric basin that formed during late Miocene to Pleistocene rifting and subsidence related to the opening of the southern Red Sea (Hutchinson & Engels, 1970). These data serve to explain that continental rift basins may also evolve into marine environments.

Zeolites and sodium silicates

Large-scale precipitation of authigenic minerals is produced by the interaction of alkaline carbonate

Fig. 11. (A) General view of the Magadi salt pan. (B) Block of trona crystals (blades up to 10 cm long) interbedded with organic oozes. (C) Trona crust at the surface of salt pan at the end of the dry season, showing 'tepee structures' — which are the consequence of water pumping mechanisms. (D), (E), (F) Distant view and details of the Magadi Soda Company dredges. From the dredge buckets (E) or mill (F), the raw trona passes down to breakers that crush it to two-inch fragments, thence to vibrating screen and crushing rolls for oversize material. Next the crushed trona is mixed with lake liquor and pumped through a pipeline to the factory. (G) General view of the 'salt-making ponds' used for sodium chloride production.

A
Red Sea
Gulf of Aden
LAKE
ASAL
B
5 km
1
2
3
4
LAKE
ASAL
-155
m
C
L. Asal
Asal
rift
Ghoubbat
al - Kharab
D
E
F
G
H

brines with the volcanogenic sediment that fills evaporitic rift basins. These minerals include zeolites, such as erionite or analcime and their precursors the Na−Al−Si gels, and sodium silicates, such as magadiite, kenyaite or kanemite, all described in the lacustrine sediments of the Bogoria and Magadi−Natron Basins (Gregory rift) (Eugster & Jones, 1968; Périnet *et al.*, 1982; Renaut *et al.*, 1986), or of the Manyara Basin (North Tanzanian Divergence) (Eugster, 1967; Surdam & Eugster, 1976). Zeolites are used industrially for water softening and as detergent builders, but they are not economically workable in the Bogoria, Magadi or Manyara Basins. Sodium silicate is industrially known as the simplest form of glass. The magadiite and kenyaite forms, which often transform into chert (Eugster, 1967,1969), have never been worked for industrial uses.

Metalliferous sediments and metallic ores

Metalliferous sedimentation has become linked to spreading on oceanic ridges. Such deposits are the metalliferous sediments and encrustations and massive sulphide bodies that share a common origin as precipitates from hydrothermal solutions in regions of volcanism and high heat flux at oceanic spreading centres (Rona, 1984). However, the occurrence of systems capable of concentrating hydrothermal mineral deposits is not limited to sea-floor spreading centres, but may occur wherever the components are present: volcanogenic heat source, parent fluid, and permeable rocks or sediments. As an extension of oceanic structures of the Red Sea and the Gulf of Aden, the East African rift system would appear to be an environment quite propitious to the formation of metalliferous sediments. Differences might exist because of changes in the spreading rates, the petrological composition of the underlying magma and surrounding rocks, and the geochemistry of parent fluids. Wide geothermal fields exist all along the rift and are characterized by various mineral deposits (Fig. 13). Several fields have been explored recently for geothermal energy, Asal in the Afar Depression, Galla Lakes in the Ethiopian rift, Bogoria−Loburu and Olkaria in the Gregory rift (Fig. 14A,B).

Among the wide variety of hydrothermal metallic minerals in East African rift lakes, pyrite and marcasite, limonite, siderite, nontronite (Fe silicate), and vivianite (Fe phosphate) are the most abundant and occur as single crystals concentrated in the sediment or as metallic ores, and are of great economic interest. Large amounts of other elements are also concentrated in lake sediments, including Ni, Al, Mn, Pb, Ti, Ba, Cr, Cu, Zn, V; all of these present an increasing interest because of modern alloy metallurgy. Pyrite (and marcasite) is used for manufacture of sulphur, sulphuric acid and sulphur dioxide, cheap jewelry, and recovery of other metals. Limonite and siderite are major ores of iron. Sphalerite, a natural zinc sulphide, is the most important ore of zinc, which is used for alloys, galvanizing iron and other metals, and electroplating. Lead is used for storage batteries, gasoline additive, radiation shielding, and cable covering. Barite is used for weighting mud in oil-drilling, paper coatings, paints, plastics, and X-ray photography.

Among the deepest lakes of the East African rift system, Lakes Kivu and Malawi have been described as containing metalliferous sediments and encrustations, the origin of which would be bound to hydrothermal activity developing along fractures within the lakes. Concerning Lake Tanganyika, mineralization of the massive sulphide type occurs in areas having sublacustrine hydrothermal activity. Although no economic extraction has been undertaken on these metallic ores and metalliferous sediments, wide attention has been devoted recently to Lakes Kivu and Malawi. The recent discovery of sublacustrine hydrothermal activity and associated massive sulphides in Lake Tanganyika will undoubtedly increase the interest of at least geologists and geochemists.

Fig. 12. Lake Asal, northern Afar Depression, Republic of Djibouti. (A) Location of the lake at the 'triple junction' of the Red Sea, Gulf of Aden, and East African rift. (B) Simplified geological map of the Asal Basin: 1; stratified basalts and rhyolites (4−1 Ma): 2, late Pleistocene to present basaltic flows of the Asal rift: 3, lacustrine carbonates, early and mid-Holocene; 4, halite, late Holocene to present (after Stieltjes, 1973). (C) Structural connection between Lake Asal and the Ghoubbat al-Kharab−Gulf of Aden through the intensively fissured and active Asal rift. (D) Aerial view of Lake Asal and the salt plain in the direction of the Asal−Ghoubbat al-Kharab rift structure. (E) '*Le Petit Rift*', junction of the Asal rift and Ghoubbat al-Kharab oceanic structures. (F) Halite crystals on the salt plain surface. (G) Aerial view of the salt plain at the northwest end of the lake. (H) Open cracks in the Asal rift floor formed during the November 1978 seismo-tectonic crisis allowing intense water circulation between the Ghoubbat al-Kharab and Lake Asal, which is −155 m below sea-level.

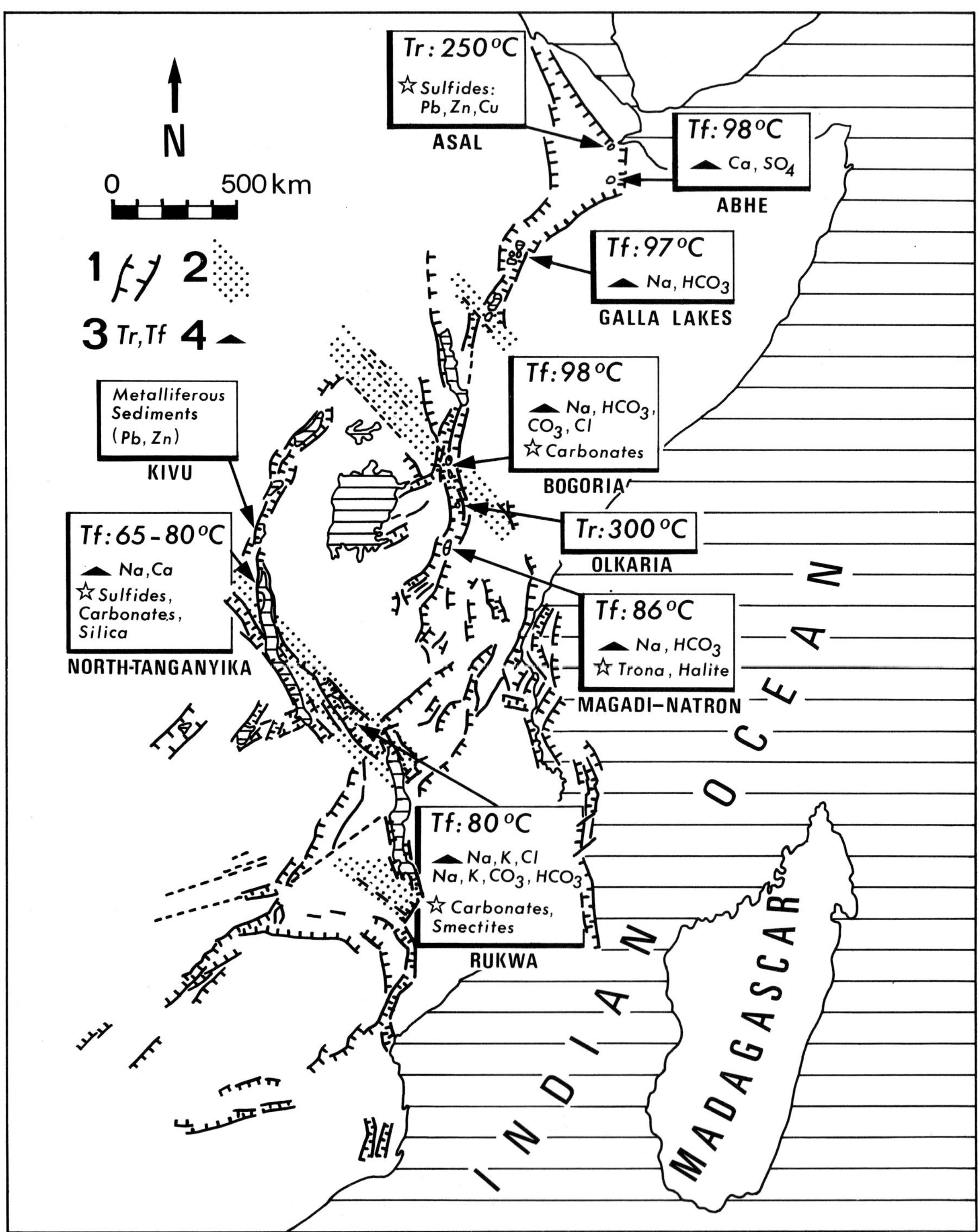

Fig. 13. Major known geothermal areas and associated mineralizations of the East African rift system: 1, major normal faults; 2, transcurrent fault zones; 3, temperature of reservoir (Tr) or fluid (Tf); 4, chemical composition of fluids.

Fig. 14. (A) The Loburu geothermal field, Lake Bogoria, northern Gregory rift. (B) The Olkaria geothermal field, Lake Naivasha Basin, central Gregory rift (September 1977). A 45 MW(e) (megawatts electricity) power plant has been constructed in the eastern part of the field and started power production in 1981 (Bodvarsson *et al.*, 1987).

Lakes Kivu and Malawi: metalliferous sediment basins and associated hydrothermalism
(Figs 3, 8, 15)

The particular hydrological characters of Lake Kivu — a meromictic regime with an unusual ther-

mohaline structure (Degens *et al.*, 1973), and an anoxic hypolimnion containing large quantities of methane, hydrogen sulphide, and carbon dioxide (Deuser *et al.*, 1973; Tietze *et al.*, 1980) — are partly related to the volcanic and geothermal environment of the lake. Active volcanoes are present to the north (Virunga volcanic field) (Fig. 3); the Nyamuragira and Nyiragongo volcanoes have been active since the latest Pleistocene (Tazieff, 1984; Aoki *et al.*, 1985). Hydrothermal springs are abundant around and within the lake (Degens *et al.*, 1973; Wong & Von Herzen, 1974), and periods of high hydrothermal activity are recorded in the sediments (Stoffers & Hecky, 1978).

Sediments cored in the lake show high concentrations of oxides, sulphides, silicates, and carbonates. The more abundant minerals are pyrite (as framboids), siderite, and manganosiderite. These sediments also contain significant amounts of elements such as Ni, Al, Mn, Ti, Cr, Cu, Zn, V, and, particularly, Fe and Pb, which reach a content of 5800 ppm (Hecky & Degens, 1973). Thus 250×10^3 tons of Pb, and 60×10^6 tons of Fe are present in the 1.2-m-thick sedimentary column deposited in Lake Kivu during the last 5000 years, using a sedimentation rate of 250 mm/1000 yr (Degens *et al.*, 1971; Degens & Kulbicki, 1973). Large amounts of such metals also are present in the lake waters, particularly zinc, which is present in the form of microcrystalline sphalerite in resin globules suspended in lake waters. The average Lake Kivu water thus has about 2 ppm zinc; that is, 1 million tons of zinc are contained in Lake Kivu in the form of sphalerite (Degens *et al.*, 1972).

The origin of these metals entrapped in the Kivu waters and sediments is related to hydrothermal activity localized around the lake and within the deep basinal zones, where hydrothermal 'jets' have been detected on seismic reflection profiles (Degens *et al.*, 1973). This abundant hydrothermal activity is related directly to the extreme volcanic character of the Kivu graben (Bellon & Pouclet, 1980). Metals originate in the rock strata through which the hydrothermal solutions pass before discharging into the lake. On the basis of stable isotope studies, the parent fluids are local rain-water and ground-water (Degens *et al.*, 1972). Sediment cores give clear evidence not only that such processes are proceeding today, but that within the last 12 000 years, three major hydrothermal events are recorded. This activity resulted in inducing an increase in lake salinity and the formation of a thermohaline meromictic regime, which is the most typical feature of the present Lake Kivu (Haberyan & Hecky, 1987).

Lake Malawi ($28\,000$ km^2, 770 m deep) is located between $9°40'$ and $14°30'$S in the western branch (Figs 3, 15A). It lies in a graben that extends over 900 km, and is morphologically divided from north to south into four depositional provinces, the Livingstone Basin, the Usisya–Mbamba Basin, the Bandawe–Metangula Basin, and the Mwanjage–MtaKaKata Basin, which are structurally half-grabens (Johnson & Ng'ang'a, 1990). Maximum sediment thickness observed in seismic records is 4.5 km, suggesting a late Miocene age for the Malawi Basin (Scott *et al.*, this volume; Johnson & Ng'ang'a, 1990). Hydrologically, the northern part of the lake is stably stratified below 250 m and the hypolimnion is anoxic (Eccles, 1974).

Iron-rich sediments were encountered in cores collected in the southern part of the lake, where thicknesses ranged from 0.13 m to at least 0.80 m. These sediments include nontronite-rich sediments, having nontronite as the prevailing mineral, limonite in minor amounts, and vivianite; limonite-rich sediments with minor nontronite and opal, composed of well-sorted oolites; vivianite-rich sediments with crusts and hemispheres covering the surface of quartz pebbles (Müller & Förstner, 1973).

The recent formation of iron proto-ores in the form of nontronite, limonite, and vivianite-rich sediments at the bottom of Lake Malawi is related to strong hydrothermal input within the lake. This indicates that iron is being leached from the sediments and rocks of the basin by SiO$_2$-rich geothermal springs. Subsequent precipitation of nontronite or limonite occurs at the sediment–water interface in the oxic parts of the lake (depth less than 250 m). Vivianite forms in the uppermost layers of sediments where Ca-phosphates from fish teeth, fish bones, and zooplankton fecal pellets (Haberyan, 1984) are dissolved and mix with dissolved Fe^{2+} from hydrothermal input (Fig. 15B). Corrected heat flow values of 171.8 and 170.1 mW/m^2 (Von Herzen & Vacquier, 1967; Ebinger *et al.*, 1987) suggest hydrothermal

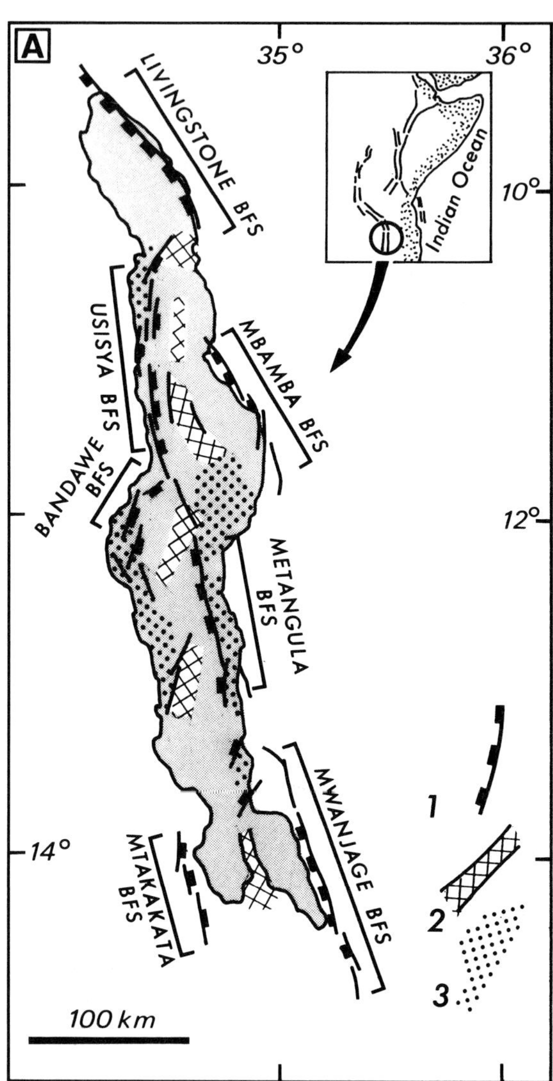

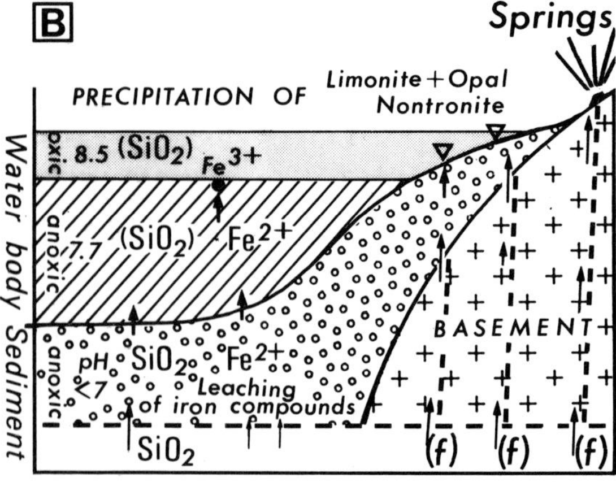

Fig. 15. Lake Malawi, western branch of the East African rift system. (A) Structural sketch map: 1, border fault system; 2, 'accommodation' zone; 3, structural platform (from Johnson & Ng'ang'a, 1990). (B) Simplified model of processes leading to the formation of iron-rich sediments in Lake Malawi; (f) faults and basement fractures (modified from Müller & Förstner, 1973).

circulation patterns are present within the graben; numerous hot springs also occur along the border faults (Kirkpatrick, 1969). Sublacustrine hydrothermal activity in Lake Malawi has not been investigated *in situ* at this time (Johnson & Ng'ang'a, 1990).

Lake Tanganyika: iron ores, massive sulphides, and associated hydrothermal activity (Figs 3, 7)

The occurrence of nontronite in a form known as 'pellet type' has been identified in Lake Tanganyika (Stoffers, in Müller & Förstner, 1973). Vivianite also has been found in the shallow water environment of the Mpulungu sub-basin (southern Lake Tanganyika) (Mondeguer *et al.*, 1989). Its origin must be similar to that in Lake Malawi, which also has geothermal springs around and within the lake (Tiercelin *et al.*, 1989a), or it may be related to normal low temperature, early diagenetic processes in organic-rich sediments with low-salinity pore-waters (M. Talbot, pers. comm., 1989).

Besides the occurrence of iron proto-ores in Lake Tanganyika, iron sulphide mineralization is also associated with important sublacustrine hydrothermal activity along the Zairian margin of northern Lake Tanganyika (Fig. 7B,C) (Vaslet *et al.*, 1987a,b; Tiercelin *et al.*, 1988c).

The Pemba site (Fig. 7C) is characterized on land and under water by hydrothermal activity. On land, the intensively fractured Precambrian basement has numerous sulphide occurrences (predominantly pyrite, with minor marcasite) and quartz veins, thus forming a stockwork 100 m long. In the area of greatest alteration, the basement is kaolinized to a depth of 5 m (Vaslet, 1987). Under water, metre-sized sulphide blocks overlie the bottom sediment at a depth of 10 m. Below 15 m, the slope increases rapidly to at least 100 m depth, along a cliff that may represent part of the Uvira fault plane, which bounds the basin. Sulphide mineralization is present all along this cliff to a minimum depth of 20 m. Active hydrothermal activity in the area is characterized by 65°C water and abundant H_2S and CH_4 emanations from sandy bottom-sediments and basement fractures (Fig. 7A). Sulphide blocks show a complex morphology of corrugated flakes of alternating layers of pyrite and marcasite, which in some cases include quartz, kaolinite, and acicular crystals of barite (Fig. 7B,C). Sulphide precipitation as colloidal pyrite and marcasite possibly results from thermal fluid circulation through bottom sediment, linked to microbial actions and films (Honnorez, 1969; Tiercelin

et al., 1989a; J. Casanova, pers. comm., 1989).

At Cape Banza, the hydrothermal field is essentially sublacustrine. These fluids flow at a temperature of about 70–80°C, through several tens of vents characterized by orificial aragonite chimneys of various sizes and morphologies (Fig. 7D,E,F). Sulphide precipitation also occurs at the Cape Banza site as thin pyrite coatings on the walls of fluid circulation pipes within the substrate (Fig. 7G,H,I,J).

Massive sulphide occurrences have been observed in the nearshore area of Lake Tanganyika to a depth of 20 m. In the areas of the geothermal sites, the lake depth is 250 and 1000 m for the Pemba and Cape Banza occurrences, respectively. High-temperature fluids and mineralization also can be expected to depths of 1000 m.

Lake Albert: the ironstone horizons of the Pliocene Nkondo Formation (Figs 3, 16A,B)

In the Kaiso region, on the southeast side of the Lake Albert Basin (western branch, Uganda) (Fig. 16A), the Lower Pliocene Nkondo Formation is characterized by the abundance of thin (up to 1 m) ironstone horizons, which extend south to the Lake Edward Basin. Principal types include oolitic (the most abundant), shaley, and scoriaceous ironstones (Fig. 16B). Iron deposition occurred from at least 8 Ma to about 1.5 Ma, their maximum development being between 7 and 3 Ma (Pickford *et al.*, 1988).

Such iron concentrations are unknown from other areas of the eastern and western branches of the East African rift system. They show conspicuous similarities with the middle Eocene to pre-Quaternary 'Sidérolithique' facies observed in West Africa (Nigeria, Niger, Mali, Chad) (Lang *et al.*, 1986). An origin by intense weathering and reworking of Upper Cretaceous to Miocene marine sediments to form detrital accumulations is generally considered for these formations. Nevertheless a continental, lacustrine origin is proposed for a 'Sidérolithique' facies series (ferruginous oolite-rich clays) of upper Eocene age in Mali (Radier, 1957). A lacustrine origin is demonstrated for the ironstone deposits of the Albert Basin, which contain freshwater fossils such as fish, molluscs, and crustaceans. The articulated nature of crustacean remains, the existence of fossilized fruits and abundant burrow casts in the ironstone bands suggest that depositional environments must have been very calm (Pickford *et al.*, 1990) (Fig. 16B). Iron-rich ground-water may have discharged in a fringing wetland of the lake, and iron

Fig. 16. Lake Albert. (A) View of the lake from the southeastern bank in the area of Nkondo. On the horizon, the northeast trending Zairian escarpment. (B) The oolitic iron ore facies of the Pliocene Nkondo Formation, here showing abundant burrows.

bacteria may have precipitated the iron to form bog iron-ore (E.I. Robbins, pers. comm., 1989). A similarity with the origin of the 'Sidérolithique' facies of West Africa may be considered. In such cases ironstone formation appears widely independent from the rift context of the Albert–Edward Basins. Volcanic and hydrothermal influences associated to rifting processes also may be considered in the genesis of such metalliferous sediments.

Oolitic iron ores of various ages have been worked in several parts of the world. Among these deposits, one of the most famous is the 'Minette oolithique' from the Upper Lias of Lorraine (France). Such industrially worked formations generally have a wide horizontal extent (several tens to a hundred kilometres) and are some tens of metres thick.

Despite the wide distribution of the Albert–Edward ironstones, the thinness of the beds, the remoteness from any large market, and their general inaccessibility means that there will be no important commercial interest in the ironstones.

Clays

Clays are typically a major component of lacustrine rift sequences. They are often impure and organic-rich, as a result of deposition in shallow water environments (shallow lakes, swamps, or fluvio-deltaic plains). At several places in East Africa clays are used to manufacture pottery or in brick-making by mixing with kaolinitic clays. The kaolinitic clays worked for pottery manufacture are not of lacustrine origin, but instead are formed by the action of steam-jets or geothermal fluids on acid lavas or metamorphic basement, mainly gneiss. Such occurrences are known at Eburru in the Lake Naivasha area, where deposits have been worked for a number of years. Reserves are estimated to 69 000 tons, and in 1956 the total extraction of kaolin amounted to 1499 tons (Thompson & Dodson, 1963). Kaolin has numerous industrial uses, such as filler and coatings for paper and in the manufacture of rubber, ceramics, cements, and fertilizers.

Other occurrences of kaolin are in areas of geothermal activity, especially at the hydrothermal site of Pemba, northern Lake Tanganyika (Fig. 7), where the gneissic basement is kaolinized to a depth of 5 m (Tiercelin *et al.*, 1989a). Kaolinite is worked by local people to whiten house walls. Associated with hydrothermal kaolinite is alunite or natroalunite $(NaK)Al_3(SO_4)_2(OH)_6$, which is used industrially for production of alun potassium compounds, decolorizing and deodorizing agents, and fertilizer.

Smectites, also known as bentonites or montmorillonites, have important uses for industry: oil-well drilling fluids, cement slurries for oil-well casings, and bonding agent in foundry sands. The occurrence of bentonite has been described at Amboseli Lake in southern Kenya (Williams, 1972). Analyses carried out on the Amboseli clays proved it to be sepiolite contaminated by carbonate minerals, which were impossible to separate because of the poor filtration characteristics of the clay. Pure sepiolite or meerschaum has been found associated with the Amboseli lacustrine marl and clayey limestone, and mechanically mined at the Sinya mine for several years (Williams, 1972). Most of the meerschaum recovered at Sinya was used in the production of smoking pipes at the Arusha factory (Tanzania). Meerschaum was also used in limited quantities in the cosmetics industry, and in the preparation of a grease-spot removal powder.

Large amounts of pure smectites are known in the Hadar Formation of Ethiopia. Located between 11° and 11°20′N in the Afar Depression (Fig. 3), the 150−300-m-thick Hadar Formation comprises fine to coarse fluvio-lacustrine sediments of Pliocene age (4.2−2.4 Ma) (Fig. 17). These lacustrine clays are mainly smectite with very small amounts of illite (up to 5%) and kaolinite (Tiercelin, 1986). The Hadar clays crop out over a wide (30 × 50 km) desert area. Despite the extent and good quality of the outcrops, their remoteness from any market and access difficulties, along with poor security, makes it very doubtful that they will ever be worked economically.

Phosphates

Biological concentrations of zooplankton fecal pellets and fossil debris, such as fish teeth or fish bones, are sources of phosphate accumulation in lacustrine sediments (Porter & Robbins, 1981; Haberyan, 1984). Phosphatic material is generally carbonate apatite, which is used industrially as a source of phosphorus and phosphoric acid and in the manufacture of fertilizers. When present in the sediment it can contribute to the formation of minerals, such as vivianite encountered in recent sediments of Lakes Malawi and Tanganyika (Müller & Förstner, 1973). Carbonate apatite also forms nodules, spherulites, and concretions, frequently found in shale facies,

Fig. 17. View of outcrops of smectitic clays of the Pliocene fluvio-lacustrine Hadar Formation, Afar Depression of Ethiopia.

such as in the fluvio-lacustrine Hadar Formation in the Afar Depression of Ethiopia (Tiercelin, 1986). Richer concentrations may occur in the form of phosphatic beds, such as the Plio-Pleistocene lacustrine deposits of Minjingu in the Lake Manyara Basin (Tanzania) (Casanova, 1986a; Schlüter, 1986a,b) (Fig. 3). The 15-m-thick Minjingu Formation is composed of clays, bedded cherts, and phosphatic beds, rich in fish and bird debris, and is extensively worked.

Coals

Accumulations of economic peat and coal are mined from active continental rifts, such as the Rhine Rift or the Dead Sea Rift, or ancient rifts, such as the Karoo (Permo-Triassic) rifts of southern Tanzania and Zaire. In the East African rift system, only small coal beds are known within the Pliocene lacustrine Hadar Formation in the Afar Depression (Tiercelin, 1986), or the late Miocene Chilga fluvio-lacustrine deposit in the northwestern Ethiopian highlands (Yemane *et al.*, 1987). The peat bodies are too small for mining.

Stromatolites

The East African rift system contains many lacustrine basins having biologically precipitated stromatolite carbonates (Fig. 18). Despite the absence of living examples, stromatolites have colonized these basins for the last 10 Ma. In the eastern branch, fossil stromatolites may be associated with the hydrothermal and/or lacustrine environments, such as those at Lake Abhe (Fontes & Pouchan, 1975) (Fig. 18A,B,C,D) or at Lake Bogoria (Renaut, 1982; Casanova, 1986a,b) (Fig. 18E,F). In the western branch, stromatolites have been documented only from Lake Tanganyika from water depths ranging between 8 and 45 m (Cohen & Thouin, 1987; Cohen & Awramik, 1988; Casanova & Thouin, 1990) (Fig. 18G).

These stromatolites provide evidence for former levels of freshwater palaeolakes. Although brief lacustrine episodes may have resulted in the development of surface crusts on hard substrates (boulders, pebbles, vegetal remains) (Fig. 18A,B,F), stromatolitic 'reefs' (2 × 0.6 m) also develop during more prolonged periods of lacustrine stability (Casanova, 1986a). 'Reefs' have been observed along the eastern margin of Lake Manyara, in the form of massive and distinct bioherms (Fig. 18H,I). Such bioherms are

1
2
3
4
INDIAN OCEAN
A
B
C
D
E
F
G
H
I

actually worked for building material by local artisans. Their outcrop quality, density, and ease of access may suggest a semi-industrial mining for calcium carbonate, which has many economical uses. Reserves in the Manyara Basin are estimated at over 1 000 000 tons (Casanova, 1986a). The Abhe, Suguta, Bogoria, or Magadi-Natron stromatolite occurrences, even though their geographical development is quite large, consist of only millimetre to centimetre-thick crusts on hard substrates, thus constituting small and hardly workable quantities of carbonate of low economic interest (Casanova & Tiercelin, 1982; Casanova, 1986a,b).

Stromatolitic bioherms, as well as marine reefs, are known to possess good hydrocarbon reservoir characteristics. The importance of marine reefs in petroleum accumulation is owed to the fact that they furnish good reservoirs closely adjacent to favourable source areas. Petroleum occurrences associated with fresh or brackish lacustrine sediments are known to exist in continental rifts of various ages, especially in the East African rift system where oil and gas seepages are related to the Miocene freshwater lacustrine sediments of Lake Albert (western Uganda) (Harris *et al.*, 1956), or in Lake Tanganyika (Tiercelin *et al.*, 1989a). Stromatolitic bioherms discovered in such favourable petroleum source areas (Cohen & Thouin, 1987) may be considered as potential reservoirs. Nevertheless, as in industrial mining, the limited size of these potential oil-bearing bioherms may be a limiting factor to future oil prospection.

Lake-deposited pyroclastics

Pyroclastic material is abundant in the eastern branch of the East African rift system, particularly in the Ethiopian and Kenya segments. The excavation and dressing of blocks of volcanic tuffs for building stone forms a considerable industry, particularly in Kenya where volcanics outcrop extensively. Tuffs occur both as terrestrially laid and lacustrine beds. There are ample reserves of such material in the area of Lakes Nakuru and Bogoria in central Kenya (Fig. 3). Such tuffs are sometimes mineralized by secondary processes, and may contain manganese ore (McCall, 1967). Pumice also occurs extensively as lacustrine beds, and is used mainly in the production of concrete blocks. It also can be used as an abrasive, or is included in plaster for acoustic purposes or mixed with Portland cement to produce an hydraulic cement. Such pyroclastic beds may also be considered as potential hydrocarbon reservoirs when associated with propitious petroleum source areas.

CONCLUSIONS

Active and ancient continental rift systems around the world are known for their economic deposits. A large majority of newly discovered oilfield regions are associated with fresh or brackish lacustrine sediments within continental rift systems. The major oilfields in Mesozoic continental rift basins along the South Atlantic margins of Africa and South America are among the best examples of such petroleum occurrences (Brice & Pardo, 1981). Metallic ores have been identified or mined in rifts of various ages, such the lead, zinc, and barite deposits of Cretaceous age in the Benue rift of West Africa, the phosphate and coal mines in the Newark rift of eastern North America, and many other examples around the world (Robbins, 1983).

As shown in this paper, the East African rift system offers a wide range of potential natural resources associated with a lacustrine environment in rift basins. In terms of economic substances, these deposits form a dramatic reference series that can be used as models for additional resources in structurally comparable environments. The diversity of the factors likely to influence the type of sedimentation and the formation of deposits with an economic

Fig. 18. Map of some stromatolite occurrences in the East African rift system. (1) The Lake Abhe Basin, northern Afar Depression (Republic of Djibouti): (A), (B) stromatolites encrusting hard substrate (pebbles, blocks), dated radiometrically at 100 yr BP; (C) hydrothermal chimneys built along fracture zones at the bottom of Lake Abhe during the major mid-Holocene high stand; (D) stromatolites encrusting the hydrothermal chimneys and dated at 100 yr BP (from Fontes & Pouchan, 1975). (2) The Lake Bogoria Basin, Gregory rift (Kenya): (E) carbonate constructions of stromatolitic types associated with geyser and hydrothermal pool, Loburu hydrothermal field; (F) encrusted surface (blocks and twigs) from the Holocene shoreline of Mwanasis Peninsula. (3) Lake Manyara, North Tanzanian Divergence (Tanzania): (H), (I) Pleistocene stromatolitic bioherms (from Casanova, 1986a); (4) northern Lake Tanganyika, western branch: (G) Holocene stromatolites, −25 m water depth (from Casanova & Thouin, 1990).

interest sometimes make prospecting hazardous and not economically viable. A detailed knowledge of the tectonic, volcanic, and climatic evolution of a rift system should permit the identification of industrial targets with a high accuracy and guarantee successful results for the prospecting projects.

In the particular case of the East African rift system, much progress has been made in the knowledge of rift evolution as a result of international and multidisciplinary projects using the most recent methods in geology and geophysics. Among the interacting processes that form or take place in continental rift systems, those encountered most frequently in the mechanisms of formation of natural resources are volcanism and associated hydrothermal activity. Thus, the eastern branch of the East African rift system, described as volcanologically 'wet' by Mohr (1982) with 470 000 km^3 of extrusive volcanics from Eocene to the present (Baker *et al.*, 1972) (Fig. 2), appears to be richer in terms of salts than the volcanologically 'dry' western branch. However, evaporite formation also is a consequence of basin partitioning, which is the result of volcano-tectonic events in arid climatic zones. In contrast, the 'dry' volcanic character and tropical wet climatic conditions in the western branch result in the development of large and deep basins filled with wide and almost permanent bodies of fresh water (Fig. 4B). Such basins are eminently favourable for the accumulation of hydrocarbon precursors, having abundant production of organic matter of the favourable types, as well as favourable conditions for preservation of source material, and readily available reservoirs of sufficient size, as shown by the occurrence of petroleum in the Albert and Tanganyika Basins. Hydrothermal activity is also present in these large basins, and it contributes to early hydrocarbon maturation. A study of the distribution of geothermal sites and volcanoes in the East African rift system suggests that volcanic intrusions are present in these basins (Mohr & Wood, 1976; Crane & O'Connell, 1983; Tiercelin *et al.*, 1989a). This abundant geothermal activity also results in the accumulation in the lake sediments and waters of metallic elements that form ores and proto-ores of economic interest. Such accumulations, as in Lake Kivu or, less evidently, in Lake Malawi, are strongly favoured by the meromictic regime of these deep lakes. Other economic substances, such as diatomites and clays, may form under the influence of various interacting processes characterizing the rift environment. Finally, biogenic and volcanogenic sediments having

reservoir potential are also formed.

Knowledge of the formation and geological evolution of such lacustrine rocks in terms of natural resources shows that the East African rift system has much potential. Its natural resources are the result of processes that evolve through time, are subject to migration, or affected by tectonic, volcanic, or hydrothermal activity, and even are affected by changes in climate.

ACKNOWLEDGEMENTS

I am very grateful to P. Anadón, R. Julia and Ll. Cabrera for inviting me to the opening conference of the international workshop−field seminar 'Lacustrine Facies Models in Rift Systems and Related Natural Resources', held in Barcelona−Rubielos de Mora, Spain, 3−6 October 1988. I would like also to thank E.I. Robbins and M. Talbot for their constructive review of the manuscript, M. Briand from Savubo for preparing photographic plates, and K. Kelts for reworking the last draft.

REFERENCES

ANONYMOUS (1923) The soda deposits of Lake Magadi in Kenya. *Bull. Imp. Inst. London* **21**, 431−444.

AOKI, K., YOSHIDA, T., YUSA, K. & NAKAMURA, Y. (1985) Petrology and geochemistry of the Nyamuragira volcano, Zaire. *J. Volcanol. Geotherm. Res.* **25**, 1−28.

BAKER, B.H. (1958) Geology of the Magadi area. *Geol. Surv. Kenya Rep.* **42**, 82 pp.

BAKER, B.H. (1963) Geology of the area south of Magadi. *Geol. Surv. Kenya Rep.* **61**, 27 pp.

BAKER, B.H. (1986) Tectonics and volcanism of the Southern Kenya Rift Valley and its influence on rift sedimentation. In: *Sedimentation in the African Rifts* (Eds Frostick, L.E., Renaut, R.W., Reid, I. & Tiercelin, J.-J.), Geol. Soc. London Spec. Publ. **25**, 45−57.

BAKER, B.H., WILLIAMS, L.A.J., MILLER, J.A. & FITCH, F.J. (1971) Sequence and geochronology of the Kenya Rift volcanics. *Tectonophysics* **11**, 191−215.

BAKER, B.H., MOHR, P.A. & WILLIAMS, L.A.J. 1972. Geology of the Eastern Rift System of Africa. *Geol. Soc. Am. Spec. Pap.* **136**, 67 pp.

BELLON, H. & POUCLET, A. (1980) Datations K−Ar de quelques laves du Rift-ouest de l'Afrique Centrale; implications sur l'évolution magmatique et structurale. *Geol. Rundsch.* **69**(1), 49−62.

BODVARSSON, G.S., PRUESS, K., STEFANSSON, V., BJORNSSON, S. & OJIAMBO, S.B. (1987) East Olkaria geothermal field, Kenya. 2. History match with production and pressure decline data. *J. Geophys. Res.* **92B**(1), 521−539.

BRICE, S.E. & PARDO, G. (1981) Oil exploration in the

nonmarine rift sequence of the South Atlantic: Cabinda Angola (abstr.) *Geophysics* **46**, GE-4.

BURGESS, C.F. (1985) *The structural and stratigraphic evolution of Lake Tanganyika: a case study of continental rifting*. Thesis, Duke University, Durham, NC, 46 pp.

CAHEN, L. (1970) Igneous activity and mineralization episodes in the evolution of the Kibaride and Katangide orogenic belts of Central Africa. In: *African Magmatism and Tectonics* (Eds Clifford, T.N. & Gass, I.G.), pp. 97–118. Oliver & Boyd, Edinburgh.

CASANOVA, J. (1986a) *Les stromatolites continentaux: Paléoécologie, Paléohydrologie, Paléoclimatologie. Application au Rift Gregory*. Thèse Etat Sciences, Université Aix-Marseille II, 256 pp.

CASANOVA, J. (1986b) East African Rift stromatolites. In: *Sedimentation in the African Rifts* (Eds Frostick, L.E., Renaut, R.W., Reid, I. & Tiercelin, J.-J.), Geol. Soc. London Spec. Publ. **25**, 201–210.

CASANOVA, J. & THOUIN, C. (1990) Biosédimentologie des carbonates microbiens du lac Tanganyika (Burundi). Implications hydrologiques. *Bull. Soc. Géol. Fr.* **6**(8), 647–656.

CASANOVA, J. & TIERCELIN, J.-J. (1982) Constructions stromatolitiques en milieu carbonaté sodique: les oncolites des plaines inondables du lac Magadi (Kenya). *C. R. Acad. Sci.* **295**, 1139–1144.

CERLING, T.E. (1979) Paleochemistry of Plio-Pleistocene Lake Turkana, Kenya. *Palaeogeogr. Palaeoclimatol. Palaeoecol.* **27**, 247–285.

CHOROWICZ, J. & MUKONKI, M. NA BANTU (1980) Linéaments anciens, zones transformantes récentes et géotectonique des fossés de l'Est africain, d'après la télédétection et la microtectonique. *Museé Royal de l'Afrique Centrale, Département Geologie–Minéralogie, rapport annuel 1979*, Tervuren, pp. 143–167.

CHOROWICZ, J., LE FOURNIER, J., LE MUT, C., RICHERT, J.P., SPY-ANDERSON, F.L. & TIERCELIN, J.-J. (1983) Observation par télédétection et au sol de mouvements décrochants NW–SE dextres dans le secteur transformant Tanganyika-Rukwa-Malawi du rift est-africain. *C. R. Acad. Sci.* **296**, 997–1002.

COHEN, A.S. & AWRAMIK, S. (1988) Recent stromatolites from Lake Tanganyika, Africa (abstr.). *International Association of Sedimentologists, Abstracts of International Symposium on Sedimentology Related to Mineral Deposits*, Beijing, China, p. 40.

COHEN, A. & THOUIN, C. (1987) Nearshore carbonate deposits in Lake Tanganyika. *Geology* **15**, 414–418.

COULTER, G.W. (1968) Hydrological processes and primary production in Lake Tanganyika. *Proceedings of 11th Conference on Great Lakes Research*, pp. 609–626. International Association of Great Lakes Research.

CRAIG, H. (1973) *Lake Tanganyika Geochemical and Hydrographic Survey*. United Nations Food and Agriculture Organisation, Report F.I.: DP/URT/71/012/14, 1–11.

CRAIG, H. & CRAIG, V. (1981) Geochemical studies of Lake Tanganyika (abstr.). In: *The Ecology and Utilization of African Inland Waters* (Eds Symoens, J.-J., Burgis, M. & Gaudet, J.-J.), UNEP Reports and Proceedings, Series 1, Nairobi, Kenya, December 1979.

CRAIG, H., DIXON, F., CRAIG, V., EDMOND, J. & COULTER, G. (1974) Lake Tanganyika geochemical and hydrographic study: 1973 expedition. *Scripps Inst. Oceanogr., Ser. 75* **5**, 1–83.

CRANE, K. & O'CONNELL, S. (1983) The distribution and implications of heat flow from the Gregory Rift in Kenya. In: *Processes of Continental Rifting* (Eds Morgan, P. & Baker, B.H.), Tectonophysics **94**, 253–275.

DALY, M.C., CHOROWICZ, J. & FAIRHEAD, J.D. (1989) Rift basin evolution in Africa: the influence of reactivated steep basement shear zones. In: *Inversion Tectonics* (Eds Cooper, M.A. & Williams, G.D.), Geol. Soc. London Spec. Publ. **44**, 309–334.

DAMAS, H. (1937) La stratification thermique et chimique des lacs Kivu, Edouard et Ndalager (Congo Belge). *Verh. Int. Ver. Theor. Angew. Limnol.* **8**, 51–68.

DE BONT, A.F. (1972) La productivité du lac Tanganyika. *Verh. Int. Ver. Theor. Angew. Limnol.* **18**, 656–660.

DEGENS, E.T. & KULBICKI, G. (1973) Hydrothermal origin of metals in some East African Rift Lakes. *Mineral. Deposita* **8**, 388–404.

DEGENS, E.T., DEUSER, W.G., VON HERZEN, R.P., WONG, H.K., WOODING, F.B., JANNASCH, H.W., KANWISHER, J.W. (1971) Lake Kivu Expedition: Geophysics, hydrography, sedimentology. *Tech. Rep. Woods Hole Oceanogr. Inst.* **71-52**, 1–20.

DEGENS, E.T., OKADA, H., HONJO, S. & HATAWAY, J.S. (1972) Microcrystalline sphalerite in resin globules suspended in lake Kivu, East Africa. *Mineral. Deposita* **7**, 1–12.

DEGENS, E.T., VON HERZEN, R.P., WONG, H.K. & JANNASCH, H.W. (1973) Lake Kivu: structure, chemistry and biology of an East African rift lake. *Geol. Rundsch.* **61**, 245–277.

DEMAISON, G.J. & MOORE, G.T. (1980) Anoxic environments and oil source bed genesis. *Organic Geochem.* **2**, 9–31.

DEMANGE, J., DI PAOLA, G.M., LAVIGNE, J., LOPOUKHINE, M. & STIELTJES, L. (1971) *Etude géothermique du Territoire Français des Afars et des Issas*. Nov. 70, Avr. 71, Rapp. 71, SGN 262 GTM Bureau de Recherches Géologiques et Minières, Orléans.

DEUSER, W.G., DEGENS, E.T., HARVEY, G.R. & RUBIN, M. (1973) Methane in lake Kivu: new data bearing on its origin. *Science* **181**, 51–54.

DODSON, R.G. (1963) Geology of the South Horr area. *Geol. Surv. Kenya Rep.* **60**, 53 pp.

DODSON, M.H., CAVANAGH, B.J., THATCHER, E.C. & AFTALION, M. (1975) Age limits for the Ubendian metamorphic episode in Northern Malawi. *Geol. Mag.* **112**, 403–410.

DUNKELMAN, T.J., KARSON, J.A. & ROSENDAHL, B.R. (1988) Structural style of the Turkana Rift, Kenya. *Geology* **16**, 258–261.

EBINGER, C.J. (1989a) Geometric and kinematic development of border faults and accommodation zones, Kivu–Rusizi rift, Africa. *Tectonics* **8**(1), 117–133.

EBINGER, C.J. (1989b) Tectonic development of the western branch of the East African rift system. *Geol. Soc. Am. Bull.* **101**, 885–903.

EBINGER, C.J., ROSENDAHL, B.R. & REYNOLDS, D.J. (1987) Tectonic model of the Malawi rift, Africa. *Tectonophysics* **141**, 215–235.

ECCLES, D.H. (1974) An outline of the physical limnology of Lake Malawi (Lake Nyasa). *Limnol. Oceanogr.* **19**(5), 730–742.

EUGSTER, H.P. (1967) Hydrous sodium silicates from Lake Magadi, Kenya; precursors of bedded chert. *Science*

157, 1177–1180.

EUGSTER, H.P. (1969) Inorganic bedded cherts from the Magadi area, Kenya. *Contrib. Mineral. Petrol.* **22**, 1–31.

EUGSTER, H.P. (1970) Chemistry and origin of the brines of Lake Magadi, Kenya. *Spec. Publ. Mineral. Soc. Am.* **3**, 215–235.

EUGSTER, H.P. (1980) Lake Magadi, Kenya and its precursors. In: *Hypersaline Brines and Evaporites* (Ed. Nissenbaum, A.), Dev. Sedimentol. **28**, 195–232.

EUGSTER, H.P. (1986) Lake Magadi, Kenya: a model for rift valley hydrochemistry and sedimentation? In: *Sedimentation in the African Rifts* (Eds Frostick, L.E., Renaut, R.W., Reid, I. & Tiercelin, J.-J.), Geol. Soc. London Spec. Publ. **25**, 177–189.

EUGSTER, H.P. & JONES, B.F. (1968) Gels composed of sodium aluminum silicate, Lake Magadi, Kenya. *Science* **161**, 160–164.

FONTES, J.C. & POUCHAN, P. (1975) Les cheminées du lac Abhé (T.F.A.I.): stations hydroclimatiques de l'Holocène. *C. R. Acad. Sci.* **280**, 383–386.

FONTES, J.C., POUCHAN, P., SALIEGE, J.F. & ZUPPI, G.M. (1980) Environmental isotope study of groundwater systems in the Republic of Djibouti. In: *Arid-zone Hydrology*, Proceedings of Advisory Group Meeting, Vienna, 2–9 September 1978, pp. 237–262. International Atomic Energy Agency, Vienna.

FROSTICK, L.E. & REID, I. (1987) A new look at rifts. *Geology Today* **3**, 122–126.

GIBBS, A.D. (1984) Structural evolution of extensional basin margins. *J. Geol. Soc. London* **141**, 609–620.

GRIFFITHS, P. (1977) *Geology of the Lake Hannington–Perkerra River area, Kenya Rift Valley.* Unpublished PhD thesis, University of London, 187 pp.

GROVE, A.T. & GOUDIE, A.S. (1971) Late Quaternary lake levels in the Rift Valley of southern Ethiopia and elsewhere in tropical Africa. *Nature* **234**, 403–405.

GROVE, A.T., STREET, F.A. & GOUDIE, A.S. (1975) Former lake levels and climatic change in the Rift Valley of southern Ethiopia. *Geogr. J.* **141**, 177–202.

HABERYAN, K.A. (1984) *Copepod fecal pellets and microfossil deposition in Lake Tanganyika.* Thesis, University of Georgia, Athens, 151 pp.

HABERYAN, K.A. (1987) Fossil diatoms and the paleolimnology of Lake Rukwa, Tanzania. *Freshwater Biol.* **17**, 429–436.

HABERYAN, K.A. & HECKY, R.E. (1987) The Late Pleistocene and Holocene stratigraphy and paleolimnology of lakes Kivu and Tanganyika. *Palaeogeogr. Palaeoclimatol. Palaeoecol.* **61**, 169–197.

HARRIS, N., PALLISTER, J.W. & BROWN, J.M. (1956) Oil in Uganda. *Geol. Surv. Uganda* **9**, 33 pp.

HAWLEY, G.G. (1981) *The Condensed Chemical Dictionary.* Van Nostrand Reinhold, New York, 1135 pp.

HECKY, R.E. (1978) The Kivu–Tanganyika basin: the last 14 000 years. *Polskie Archiv. Hydrobiol.* **25**, 159–165.

HECKY, R.E. & DEGENS, E.T. (1973) Late Pleistocene–Holocene chemical stratigraphy and paleolimnology of the rift valley lakes of central Africa. *Tech. Rep. Woods Hole Oceanogr. Inst.* **73–28**, 93 pp (unpublished).

HECKY, R.E. & KLING, H.J. (1981) The phytoplankton and protozooplankton of the euphotic zone of Lake Tanganyika: species composition, biomass, chlorophyll content, and spatio-temporal distribution. *Limnol. Oceanogr.* **26**, 548–564.

HERBIN, J.P. (1979) Sédimentation de rift: géochimie organique des sédiments récents du lac Bogoria. *Rapp. Inst. Fr. Pétrole* **26881**, 20 pp.

HILL, A., CURTIS, G. & DRAKE, R. (1986) Sedimentary stratigraphy of the Tugen Hills, Baringo, Kenya. In: *Sedimentation in the African Rifts* (Eds Frostick, L.E., Renaut, R.W., Reid, I. & Tiercelin, J.-J.), Geol. Soc. London Spec. Publ. **25**, 285–295.

HOLWERDA, J.G. & HUTCHINSON, R.W. (1968) Potash-bearing evaporites in the Danakil area, Ethiopia. *Econ. Geol.* **63**, 124–150.

HONNOREZ, J. (1969) La formation actuelle d'un gisement sous-marin de sulfures fumerolliens à Vulcano (mer tyrrhénienne). Partie I. Les minéraux sulfurés des tufs immergés à faible profondeur. *Mineral. Deposita* **4**, 114–131.

HUC, A.Y., VANDENBROUCKE, M., BESSEREAU, G. & FABRE, M. (1986) Etude de la matière organique sédimentaire dans la partie nord du lac Tanganyika (Projet GEORIFT). *Rapp. Inst. Fr. Pétrole* **34617**, 24 pp.

HUC, A.Y., VANDENBROUCKE, M., BESSEREAU, G. & LE FOURNIER, J. (1987) Distribution of organic facies in the recent sediments of the northern part of Lake Tanganyika (abstr.) *American Association of Petroleum Geologists Annual Convention*, Los Angeles, 7–10 June 1987.

HUC, A.Y., LE FOURNIER, J., VANDENBROUCKE, M., BESSEREAU, G., BERNON, M., DA SILVA, M. & FABRE, M. (1990) Northern Lake Tanganyika: a conceptual model of organic sedimentation in a rift lake. In: *Lacustrine Basin Exploration: Case Studies and Modern Analogs* (Ed. Katz, B.J.). Am. Assoc. Petrol. Geol. Memoir 50, Tulsa.

HUTCHINSON, G.E. (ed.) (1957) *A Treatise on Limnology*, vol. I, part I: *Geography and Physics of Lakes.* Wiley, New York, 540 pp.

HUTCHINSON, R.W. & ENGELS, G.G. (1970) Tectonic significance of regional geology and evaporite lithofacies in northeastern Ethiopia. *Philos. Trans. R. Soc. London* **267**, 313–329.

JANNASCH, H.W. (1975) Methane oxidation in Lake Kivu (Central Africa). *Limnol. Oceanogr.* **20**, 860–864.

JOHNSON, T.C. & DAVIS, T.W. (1989) High resolution seismic profiles from Lake Malawi, East Africa. *J. Afr. Earth Sci.* **8**, 2/3/4, 383–392.

JOHNSON, T.C. & NG'ANG'A, P. (1990) Reflections on a rift lake. In: *Lacustrine Basin Exploration: Case studies and Modern Analogs* (Ed. Katz, B.J.). Am. Assoc. Petrol. Geol. Memoir 50, Tulsa, 113–135.

JONES, B.F., EUGSTER, H.P. & RETTIG, S.L. (1977) Hydrochemistry of the Lake Magadi Basin, Kenya. *Geochim. Cosmochim. Acta* **41**, 53–72.

KAZMIN, V. (1980) Transform faults in the East African Rift System. In: *Geodynamic Evolution of the Afro-Arabic Rift System*, pp. 65–73. Accademia Nazionale dei Lincei (Atti dei Convegni Lincei, 47), Roma.

KELTS, K. (1988) Environments of deposition of lacustrine petroleum source rocks: an introduction. In: *Lacustrine Petroleum Source Rocks* (Eds Fleet, A., Kelts, K. & Talbot, M.), Geol. Soc. London Spec. Publ. **40**, 3–27.

KILHAM, P., KILHAM, S.S. & HECKY, R.E. (1986) Hypothesised resource relationships among African planktonic diatoms. *Limnol. Oceanogr.* **31**, 1169–1181.

KIRK-OTHMER (1985) *Concise Encyclopedia of Chemical Technology*, 3rd edn, c1978–1984. Wiley, New York.

KIRKPATRICK, I.M. (1969) The thermal springs of Malawi. *XXIII Int. Geol. Congr.* **19**, 111–120.

KLING, G.W., CLARK, M.A., COMPTON, H.R., DEVINE, J.D., EVANS, W.C., HUMPHREY, A.M., KOENIGSBERG, E.J., LOCKWOOD, J.P., TUTTLE, M.L. & WAGNER, G.N. (1987) The 1986 Lake Nyos gas disaster in Cameroon, West Africa. *Science* **235**, 169–175.

LANG, J., KOGBE, C., ALIDOU, S., ALZOUMA, K., DUBOIS, D., HOUESSOU, A. & TRICHET, J. (1986) Le Sidérolithique du Tertiaire ouest-africain et le concept de Continental terminal. *Bull. Soc. Géol. Fr.* **(8)2**, 605–622.

LANGGUTH, H.R. & POUCHAN, P. (1975) Caractères physiques et conditions de stabilité du lac Assal (T.F.A.I.). In: *Afar Depression of Ethiopia* (Eds Pilger, A. & Röesler, A.), pp. 250–258. E. Schweizerbart'sche, Stuttgart.

LE DOUARAN, S. (1986) Exemples de structure profonde des marges. Implications sur l'histoire de la subsidence et de la reconstitution des températures dans les sédiments. *Bull. Cent. Rech. Explor. Prod. Elf-Aquitaine* **10**, 165–166.

LE MUT, C. (1983) *Optimisation de la recherche pétrolière dans les rifts: le fossé du lac Tanganyika, Rift Est-africain; structure géologique et sédiments*. Rapport Société Nationale Elf-Aquitaine (Production), Réf. DEX/RAG. Lab. BSS n° 002/83, Boussens.

LOPOUKHINE, M. (1973) *Rôle de la géochimie dans la recherche d'énergie géothermique. Application au T.F.A.I.* Thèse 3ème cycle, Université Paris.

LOUPEKINE, I.S. (1971) *A catalogue of felt earthquakes in Kenya, 1892–1969.* Unpublished thesis, University of Nairobi, 125 pp.

MCCALL, G.J.H. (1967) Geology of the Nakuru–Thompson's Falls–Lake Hannington area. *Geol. Surv. Kenya Rep.* **78**, 122 pp.

MCCONNELL, R.B. (1967) The East African Rift system. *Nature* **215**, 578–581.

MCCONNELL, R.B. (1972) Geological development of the rift system of Eastern Africa. *Geol. Soc. Am. Bull.* **83**, 2549–2572.

MCKENZIE, D.P., DAVIES, D. & MOLNAR, P. (1970) Plate tectonics of the Red Sea and East Africa. *Nature* **226**, 243–248.

MELACK, J.M. (1979) Photosynthetic rates in four tropical African fresh waters. *Freshwater Biol.* **9**, 555–571.

MILBRINK, G. (1977) On the limnology of two alkaline lakes (Nakuru and Naivasha) in the East Rift Valley system in Kenya. *Int. Rev. Gesamten Hydrobiol.* **62**, 1–17.

MOHR, P.A. (1982) Musing on continental rifts. In: *Continental and Oceanic Rifts* (Ed. Palmason, G.), vol. 8, *Geodynamics Series*, pp. 293–309. American Geophysicists' Union, Washington, DC.

MOHR, P.A. & WOOD, C.A. (1976) Volcano spacings and lithospheric attenuation in the eastern rift of Africa. *Earth Planet. Sci. Lett.* **33**, 126–144.

MONDEGUER, A., RAVENNE, C., MASSE, P. & TIERCELIN, J.-J. (1989) Sedimentary basins in an extension and strike-slip background: the 'South Tanganyika troughs complex', East African Rift. *Bull. Soc. Géol. Fr.* **(8)5**, 501–522.

MORLEY, C.K., WESCOTT, W.A., CUNNINGHAM, S.M. & HARPER, R.M. (1989) Recent exploration in the Lake Rukwa Area, East African Rift (abstr.). *28th Int. Geol. Congr.* **2**, 462–463.

MOUGENOT, D., RECQ, M., VIRLOGEUX, P. & LEPVRIER, C. (1986) Seaward extension of the East African rift. *Nature* **321**, 599–603.

MOUGENOT, D., HERNANDEZ, J. & VIRLOGEUX, P. (1989) Structure et volcanisme d'un rift sous-marin: le fossé des Kérimbas (marge nord-mozambique). *Bull. Soc. Géol. Fr.* **(8)5**, 401–410.

MÜLLER, G. & FÖRSTNER, V. (1973) Recent iron ore formation in lake Malawi, Africa. *Mineral. Deposita* **8**, 278–290.

NAYLOR, I. (1972) *Geology of the Lake Hannington geothermal prospect.* United Nations/Kenya Power Company, Unpublished report.

PALACIOS, C., DE RENEVILLE, P. & ROBERT, C. (1987a) 3. 5. 1. 2. Sédimentation biogénique. e) Nature biologique des matières organiques dans les sédiments actuels. In: *Le demi-graben de Baringo–Bogoria, Rift Gregory, Kenya. 30000 ans d'histoire hydrologique et sédimentaire* (Eds Tiercelin, J.-J., Vincens, A. (coordinators), *et al.*), Bull. Cent. Rech. Explor. Prod. Elf-Aquitaine **11**, 341–349.

PALACIOS, C., RAYNAUD, J.F., ROBERT, C. & PENIGUEL, G. (1987b) 4.- La sédimentation Pléistocène supérieur–Holocene. 4.1.7. Evolution des faciès organiques. In: *Le demi-graben de Baringo–Bogoria, Rift Gregory, Kenya. 30000 ans d'histoire hydrologique et sédimentaire* (Eds Tiercelin, J.-J., Vincens, A. (coordinators), *et al.*), Bull. Cent. Rech. Explor. Prod. Elf-Aquitaine **11**, 449–468.

PEIRCE, J.W. & LIPKOV, L. (1988) Structural interpretation of the Rukwa Rift, Tanzania. *Geophysics* **53**, 824–836.

PENIGUEL, G., COUDERC, R. & SEYVE, C. (1989) Les microalgues actuelles et fossiles — Intérêts stratigraphique et pétrolier. *Bull. Cent. Rech. Explor. Prod. Elf-Aquitaine* **13**, 455–482.

PERINET, G., TIERCELIN, J.-J. & BARTON, C.E. (1982) Présence de kanemite dans les sédiments récents du lac Bogoria, Rift Gregory, Kenya. *Bull. Minéral.* **105**, 633–639.

PICKFORD, M., SENUT, B., SSEMMANDA, I., ELUPU, D. & OBWONA, P. (1988) Premiers résultats de la mission de l'Uganda Palaeontology Expedition à Nkondo (Pliocène du Bassin du Lac Albert, Ouganda). *C. R. Acad. Sci.* **306**(II), 315–320.

PICKFORD, M., SENUT, B., ROCHE, H., MEIN, P., NDAATI, G., OBWONA, P. & TUHUMWIRE, J. (1989) Uganda Palaeontology Expedition: résultats de la deuxième mission (1987) dans la région de Kisegi–Nyabusosi (Bassin du Lac Albert, Ouganda). *C. R. Acad. Sci.* **308**(II), 1751–1758.

PICKFORD, M., SENUT, B., TIERCELIN, J.-J., KASANDE, R. & OBWONA, P. (1990) Résultats de la troisième mission (1988) de l'Uganda Palaeontology Expedition: Régions de Nkondo–Sebugoro et de Hohwa (Bassin du Lac Albert, Ouganda). *C. R. Acad. Sci.* **311**(II), 737–744.

PORTER, K.G. & ROBBINS, E.I. (1981) Zooplankton fecal pellets link fossil fuel and phosphate deposits. *Science* **212**, 931–933.

RADIER, H. (1957) Le Précambrien au sud de l'Adras des

Iforas. Le bassin crétacé et tertiaire de Gao. Contribution á l'étude géologique du Soudan oriental. Thèse Sci. Strasbourg, and *Bull. Serv. Geol. Prosp. Min. A.O.F.* **26**, 556 pp.

RENAUT, R.W. (1982) *Late Quaternary geology of the Lake Bogoria fault-trough, Kenya Rift Valley.* Unpublished PhD thesis, University of London, 498 pp.

RENAUT, R.W., TIERCELIN, J.-J. & OWEN, R.B. (1986) Mineral precipitation and diagenesis in the sediments of the Lake Bogoria basin, Kenya Rift Valley. In: *Sedimentation in the African Rifts* (Eds Frostick, L.E., Renaut, R.W., Reid, I. & Tiercelin, J.-J.), Geol. Soc. London Spec. Publ. **25**, 159–175.

ROBBINS, E.I. (1983) Accumulation of fossil fuels and metallic minerals in active and ancient rift lakes. In: *Processes of Continental Rifting* (Eds Morgan, P. & Baker, B.H.), Tectonophysics **94**, 633–658.

RONA, P.A. (1984) Hydrothermal mineralization at seafloor spreading centers. *Earth Sci. Rev.* **20**, 1–104.

ROONEY, D. & HUTTON, V.R.S. (1977) A magnetotelluric and magnetovariational study of the Gregory Rift Valley, Kenya. *Geophys. J. R. Astron. Soc.* **51**, 91–119.

ROSENDAHL, B.R. (1987) Architecture of continental rifts with special reference to East Africa. *Ann. Rev. Earth Planet. Sci.* **15**, 445–503.

ROSENDAHL, B.R., REYNOLDS, D.J., LORBER, P.M., BURGESS, C.F., McGILL, J., SCOTT, D., LAMBIASE, J.-J. & DERKSEN, S.J. (1986) Structural expressions of rifting: lessons from Lake Tanganyika, Africa. In: *Sedimentation in the African Rifts* (Eds Frostick, L.E., Renaut, R.W., Reid, I. & Tiercelin, J.-J.), Geol. Soc. London Spec. Publ. **25**, 29–43.

RUDD, J.W.M. (1980) Methane oxidation in Lake Tanganyika (East Africa). *Limnol. Oceanogr.* **25**, 958–963.

RUEGG, J.C. & KASSER, M. (1987) Deformations across the Asal-Ghoubbet Rift, Djibouti, uplift and crustal extension 1979–1986. *Geophys. Res. Lett.* **14**, 745–748.

RUEGG, J.C., KASSER, M., LEPINE, J.C. & TARANTOLA, A. (1979) Geodesic measurements of rifting associated with a seismo-volcanic crisis in Afar. *Geophys. Res. Lett.* **6**, 817–820.

SCHLUTER, T. (1986a) A cross section through the lacustrine environment in the Pleistocene Lake Manyara Beds at Minjingu, northern Tanzania (abstr.). *Dakar INQUA Symposium 'Changements globaux en Afrique',* pp. 419–421.

SCHLUTER, T. (1986b) Eine neue Fundstelle pleistozäner Kormorane (*Phalacrocorax* sp.) in Nord-Tanzania. *J. Ornithol.* **127**, 85–91.

SCHULL, T.J. (1984) Oil exploration in the non-marine rift basins of interior Sudan. *Can. Soc. Petrol. Geol. Reservoir* **12**(8), 1–2.

SIBSON, R.H. (1981) Fluid flow accompanying faulting: field evidence and models. In: *Earthquake Prediction: An International Review* (Eds Simpson, D.W. & Richards, P.G.), Am. Geophys. Union, Maurice Ewing Ser. **4**, 593–603.

SIBSON, R.H. (1987) Earthquake rupting as a mineralizing agent in hydrothermal systems. *Geology* **15**, 701–704.

SIMONEIT, B.R.T. (1985) Hydrothermal petroleum: genesis, migration, and deposition in Guaymas Basin, Gulf of California. *Can. J. Earth Sci.* **22**, 1919–1929.

STIELJES, L. (1973) Evolution tectonique récente du rift d'Asal. *Rev. Géogr. Phys. Géol. Dyn.* **15**, 425–436.

STOFFERS, P. & HECKY, R.E. (1978) Late Pleistocene–Holocene evolution of the Kivu–Tanganyika basin. In: *Modern and Ancient Lake Sediments* (Eds Matter A. & Tucker, M.E.), Spec. Publ. Int. Assoc. Sedimentol. **2**, 43–55.

SURDAM, R.C. & EUGSTER, H.P. (1976) Mineral reactions in the sedimentary deposits of the Lake Magadi region, Kenya. *Geol. Soc. Am. Bull.* **87**, 1739–1752.

TALBOT, M.R. (1988) The origins of lacustrine oil source rocks: evidence from the lakes of tropical Africa. In: *Lacustrine Petroleum Source Rocks* (Eds Fleet, A.J., Kelts, K. & Talbot, M.R.), Geol. Soc. London Spec. Publ. **40**, 29–43.

TAZIEFF, H. (1984) Mt. Niragongo: renewed activity of the lava lake. *J. Volcanol. Geotherm. Res.* **20**, 267–280.

THOMPSON, A.O. & DODSON, R.G. (1963) Geology of the Naivasha area. *Geol. Surv. Kenya Rep.* **55**, 80 pp.

TIERCELIN, J.-J. (1981) *Rifts continentaux. Tectonique, climats, sédiments. Exemples: la sédimentation dans le nord du Rift Gregory (Kenya) et dans le Rift de l'Afar (Ethiopie) depuis le Miocène.* Thèse Doct. Etat, Université Aix-Marseille II, 260 pp.

TIERCELIN, J.-J. (1986) The Pliocene Hadar Formation, Afar Depression of Ethiopia. In: *Sedimentation in the African Rifts* (Eds Frostick, L.E., Renaut, R.W., Reid, I. & Tiercelin, J.-J.), Geol. Soc. London Spec. Publ. **25**, 253–265.

TIERCELIN, J.-J. (1990) Rift-basin sedimentation: responses to climate, tectonism and volcanism. Examples of the East African Rift. In: *African Continental Sediments* (Eds Lang, J. & Kogbe, C.A.). J. Afr. Earth Sci. **10**, 283–305.

TIERCELIN, J.-J. & LE FOURNIER, J. (1980) Un exemple de sédimentation récente dans un rift continental: le demi-graben de Baringo–Bogoria, Rift Gregory, Kenya. In: *Géodynamique et mécanismes sédimentaires et géochimiques dans les rifts continentaux* (Ed. Tiercelin, J.-J.), Rech. Géol. Afr. **5**, 133–140.

TIERCELIN, J.-J. & MONDEGUER, A. (1991) The geology of the Tanganyika Trough. In: *Lake Tanganyika and Its Life* (Ed. Coulter G.W.), Natural History Museum Publications, Oxford University Press, Oxford.

TIERCELIN, J.-J., MICHAUX, J. & BANDET, Y. (1979) Le Miocène supérieur du Sud de la dépression de l'Afar, Ethiopie: sédiments, faunes, âges isotopiques. *Bull. Soc. Géol. Fr.* (7)**21**, 255–258.

TIERCELIN, J.-J., LE FOURNIER, J., HERBIN, J.P. & RICHERT, J.P. (1980) Continental rifts: modern sedimentation, tectonic and volcanic control. Example from the Bogoria–Baringo graben, Gregory Rift, Kenya. In: *Geodynamic Evolution of the Afro-Arabic Rift System,* pp. 143–164. Accademia Nazionale dei Lincei (Atti dei convegni Lincei, 47), Roma.

TIERCELIN, J.-J., PERINET, G., LE FOURNIER, J., BIEDA, S. & ROBERT, P. (1982) Lacs du rift est-africain, exemples de transition eaux douces — eaux salées: le lac Bogoria, Rift Gregory, Kenya. *Mém. Soc. Géol. Fr.* **144**, 217–230.

TIERCELIN, J.-J., VINCENS, A. (coordinators), BARTON, C.E., CARBONEL, P., CASANOVA, J., DELIBRIAS, G., GASSE, F., GROSDIDIER, E., HERBIN, J.P., HUC, A.Y.,

JARDINE, S., LE FOURNIER, J., MELIERES, F., OWEN, R.B., PAGE, P., PALACIOS, C., PAQUET, H., PENIGUEL, G., PEYPOUQUET, J.P., RAYNAUD, J.F., RENAUT, R.W., DE RENEVILLE, P., RICHERT, J.P., RIFF, R., ROBERT, P., SEYVE, C., VANDENBROUCKE, M. & VIDAL, G. (1987) Le demi-graben de Baringo–Bogoria, Rift Gregory, Kenya. 30.000 ans d'histoire hydrologique et sédimentaire. *Bull. Cent. Rech. Explor. Prod. Elf-Aquitaine* **11**, 249–540.

TIERCELIN, J.-J., CHOROWICZ, J., BELLON, H., RICHERT, J.P., MWANBENE, J.T. & WALGENWITZ, F. (1988a) East African rift system: offset, age and tectonic significance of the Tanganyika–Rukwa–Malawi intracontinental transcurrent fault zone. *Tectonophysics* **148**, 241–252.

TIERCELIN, J.-J., MONDEGUER, A., GASSE, F., HILLAIRE-MARCEL, C., HOFFERT, M., LARQUE, P., LEDEE, V., MARESTANG, P., RAVENNE, C., RAYNAUD, J.F., THOUVENY, N., VINCENS, A. & WILLIAMSON, D. (1988b) 25.000 ans d'histoire hydrologique et sédimentaire du lac Tanganyika, Rift Est-africain. *C. R. Acad. Sci.* **307**(II), 1375–1382.

TIERCELIN, J.-J., VASLET, N., THOUIN, C., KALALA TSHIBANGU, CHARLOU, J.L., FOUQUET, Y. & MONDEGUER, A. (1988c) Découverte d'une activité hydrothermale à sulfures massifs en contexte d'ouverture intracontinentale. Les sites de Pemba et du Cap Banza, Fossé Nord-Tanganyika, Rift est-africain (abstr.). *Colloque National sur l'Hydrothermalisme Océanique*, Brest, 22–24 November 1988, pp. 138–140.

TIERCELIN, J.-J., THOUIN, C., KALALA TSHIBANGU & MONDEGUER, A. (1989a) Discovery of sublacustrine hydrothermal activity and associated massive sulfides and hydrocarbons in the north Tanganyika trough, East African Rift. *Geology* **17**, 1053–1056.

TIERCELIN, J.-J., SCHOLZ, C.A., MONDEGUER, A., ROSENDAHL, B.R. & RAVENNE, C. (1989b) Discontinuités sismiques et sédimentaires dans la série de remplissage du fossé du Tanganyika, Rift Est-africain. *C. R. Acad. Sci.* **309**(II), 1599–1606.

TIETZE, K., CEGH, M., MULLET, H., SCHRODER, L., STAHL, W. & WEHNER, H. (1980) The genesis of the methane in Lake Kivu (Central Africa). *Geol. Rundsch.* **69**, 452–472.

TISSOT, B.P. & WELTE, D.H. (1978) *Petroleum Formation and Occurrence.* Springer, New York, 538 pp.

VALETTE, J.N. (1975) Geochemical study of Lake Asal and Ghoubet el Kharab (T.F.A.I.). In: *Afar Depression of Ethiopia* (Eds Pilger, A. & Röesler, A.), pp. 239–250. E. Scheweizerbart'sche, Stuttgart.

VARESCHI, E. (1978) The ecology of lake Nakuru (Kenya). I. Abundance and feeding of the Lesser Flamingo. *Oecologia* **32**, 11–35.

VASLET, N. (1987) *Minéralisations hydrothermales en contexte d'ouverture intracontinentale. Fossé du Lac Tanganyika, Rift Est-africain (Zaire)*. Rapp. Diplôme d'Etudes Approfondies, Université Bretagne Occidentale, 52 pp.

VASLET, N., THOUIN, C., FOUQUET, Y., BRICHARD, P.J., KALALA TSHIBANGU, MONDEGUER, A. & TIERCELIN, J.-J. (1987a) Découverte de sulfures massifs d'origine hydrothermale dans le Rift Est-africain. Minéralisations sous-lacustres dans le fossé du Tanganyika. *C.R. Acad. Sci.* **305**(II), 885–891.

VASLET, N., TIERCELIN, J.-J., MONDEGUER, A. & THOUIN, C. (1987b) Activité hydrothermale, sédiments métallifères et hydrocarbures. Le cas des fossés Nord-Tanganyika et Kivu (Rift Est-africain) (abstr.). *1er Congrès Français de Sédimentologie*, Paris, 19–20 November 1987, pp. 323–324.

VERSFELT, J. & ROSENDAHL, B.R. (1989) Relationships between pre-rift structure and rift architecture in Lakes Tanganyika and Malawi, East Africa. *Nature* **337**, 354–357.

VIDAL, G. (1985) Analyse par télédétection et étude de terrain du Rift Est-africain au Kenya: le problème du linéament d'Assoua. *Univ. P. & M. Curie, Mém. Sci. Terre* **85**–40, 262 pp.

VON HERZEN, R.P. & VACQUIER, V. (1967) Terrestrial heat flow in lake Malawi. *J. Geophys. Res.* **72**, 4221–4226.

WELHAN, J.A. & LUPTON, J.E. (1987) Light hydrocarbon gases in Guaymas Basin hydrothermal fluids: thermogenic versus abiogenic origin. *Bull. Am. Assoc. Petrol. Geol.* **71**, 215–223.

WILLIAMS, L.A.J. (1972) Geology of the Amboseli area. *Geol. Surv. Kenya Rep.* **90**, 86 pp.

WILLIAMS, M.A.J., GETANEH ASSEFA & ADAMSON, D.A. (1986) Depositional context of Plio-Pleistocene hominid-bearing formations in the Middle Awash valley, southern Afar Rift, Ethiopia. In: *Sedimentation in the African Rifts* (Eds Frostick, L.E., Renaut, R.W., Reid, I. & Tiercelin, J.-J.), Geol. Soc. London Spec. Publ. **25**, 241–251.

WOHLENBERG, J. (1969) Remarks on the seismicity of East Africa between 4°N–12°S and 23°E–40°E. *Tectonophysics* **8**, 567–577.

WONG, H.-K. & VON HERZEN, R.P. (1974) A geophysical study of Lake Kivu, East Africa. *Geophys. J. R. Astron. Soc.* **37**, 371–389.

YEMANE, K., ROBERT, C. & BONNEFILLE, R. (1987) Pollen and clay mineral assemblages of a late Miocene lacustrine sequence from the northwestern Ethiopian highlands. *Palaeogeogr., Palaeoclimatol., Palaeoecol.* **60**, 123–141.

YURETICH, R.F. (1979) Modern sediments and sedimentary processes in Lake Rudolf (Lake Turkana), eastern Rift Valley, Kenya. *Sedimentology* **26**, 313–331.

YURETICH, R.F. (1982) Possible influences upon lake development in the East African Rift Valley. *J. Geol.* **90**, 329–337.

Spec. Publs Int. Ass. Sediment. (1991) **13**, 39–55

Sedimentology of sodium sulphate deposits and special clays from the Tertiary Madrid Basin (Spain)

S. ORDÓÑEZ*, J.P. CALVO*, M.A. GARCÍA DEL CURA[†], A.M. ALONSO-ZARZA[†] *and* M. HOYOS[‡]

**Departamento de Petrología y Geoquímica, Facultad de Ciencias Geológicas, Universidad Complutense, 28040-Madrid, Spain*
[†]*Instituto de Geología Económica, C.S.I.C. Facultad de C. Geológicas, Universidad Complutense, 28040-Madrid, Spain*
[‡]*Museo Nacional de Ciencias Naturales, C.S.I.C. C/José Gutiérrez Abascal, 2-28006-Madrid, Spain*

ABSTRACT

The Madrid Basin is a large Tertiary intra-cratonic depression, which contains some of the largest fossil sodium sulphate and sepiolite deposits in the world. These minerals, together with several types of Mg smectites ('bentonites'), occur in lacustrine sequences within the Neogene sedimentary record of the basin. Sodium sulphates (glauberite and thenardite) are restricted to the Lower Saline Unit, where they are associated with anhydrite, halite, magnesite, and minor clays. Glauberite and thenardite are thought to have been deposited in the most central part of a perennial saline lake. The accumulation of thenardite might have taken place during a stage of contraction of the lake system at the beginning of the middle Aragonian (middle Miocene). Polyhalite occurs as a diagenetic saline phase related both to calcium and sodium sulphates.

Both sepiolite and bentonite deposits occur widely distributed within distal fan and marginal lacustrine sequences in the so-called Intermediate Unit of the Miocene (middle to upper Aragonian). Thick beds of nearly pure sepiolite were deposited in ponds extended at the toes of arkosic alluviums. Sepiolite is also found within calcrete profiles in these environments. Minor amounts of sepiolite are commonly recognized along with palygorskite in open lacustrine areas. On the other hand, Mg bentonites characteristically occur associated with dolostones and fine micaceous sands in sequences that provide evidence of fluctuations in the lake level. The mineralogy of the bentonite deposits closely reflects the environmental variations in the lake margins.

INTRODUCTION

The Madrid Basin, also known as the Tajo Basin, is one of three large continental basins developed in the interior of the Iberian Peninsula during the Tertiary (Fig. 1). The Tertiary successions outcropping in the basin yield a rather large variety of industrial minerals and rocks, most of them occurring within lacustrine sequences. Not withstanding the enormous economic importance of other sedimentary rocks, the emphasis of this paper is on two groups of mineral deposits that are relevant both from an economic as well as a geological or sedimentological point of view. These two groups are sodium sulphate deposits and special clays (bentonites, sepiolite), both positioned at different levels within lacustrine or peripheral lake sequences in the basin.

The production of natural sodium sulphate in the Madrid Basin yearly exceeds 200 000 tons. Sodium sulphate is recovered from thenardite and glauberite deposits, which are exploited by underground mining and open pit solution mining, respectively. Reserves of sodium sulphate have been estimated to exceed 1000 million metric tons. They constitute, together with the western side of the Tertiary Ebro Basin in northern Spain, most of the Spanish sodium sulphate reserves and also make the Iberian continental Tertiary basins the most important deposits of fossil sodium sulphate known in the western countries.

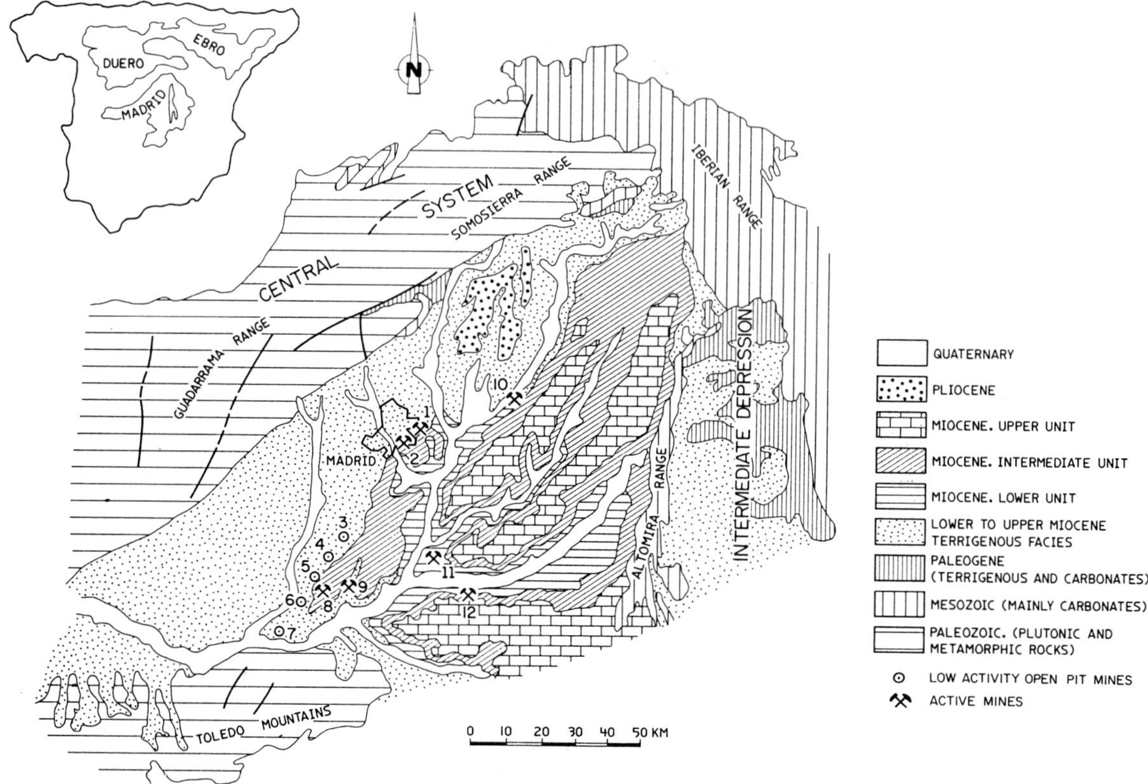

Fig. 1. Geological map of the Madrid Basin showing the location of the principal extractions of industrial minerals. 1 and 2, Vicálvaro and San Blas (sepiolite); 3–8, Yepes−Cabañas de la Sagra area (bentonites and sepiolite); 9 and 10, La Sagra and Alcalá (ceramic raw clays); 11, Villaconejos (glauberite); 12, Villarrubia (thenardite and glauberite).

Sodium sulphate in the Madrid Basin is extracted from the mine 'El Castellar' (which yields both thenardite and glauberite), and from nearby sites (Fig. 1), for example, Villaconejos, which contains only glauberite (Coope, 1983). Other occurrences of sodium sulphate deposits in the basin have also been known for a long time, the exploitation of these materials having started during the eighteenth century. Different aspects of these deposits, such as mining geology, mineral contents, and petrographic and geochemical characteristics of the sulphate deposits have been discussed by García del Cura (1979), Ortí *et al.* (1979), and Ordoñez *et al.* (1982,1987).

On the other hand, large volumes of both bentonite and sepiolite clays are extracted from an area extending from the outskirts of Madrid to Toledo (Fig. 1). Important reserves of ceramic raw clay materials, which have been estimated at nearly 500 millions tons (Menduiña, 1988), are also present in this area. The most important deposits of sepiolite are located at the southern limit of the city of Madrid, especially in Vicálvaro-Vallecas (Galán & Castillo, 1984). Total sepiolite resources in the region are estimated at 100 millions tons.

In this paper, the term 'bentonite' is used in a broad sense to indicate 'any clay composed dominantly of a smectite clay mineral, whose physical properties are dictated by this mineral' (Grim & Güven, 1978). This term is used in a commercial sense in the Madrid Basin for clays that possess suspension properties as well as high exchange and swelling capacities. These properties, which have been determined in a recent paper by Galán *et al.* (1986), make these bentonites appropriate for use as foundry sand, drilling muds, pelletizing, and other uses. The bentonite clays in the Madrid Basin are, in general, trioctahedral smectites (Doval *et al.*, 1985;

Galán *et al.*, 1986; IGME, 1989), similar to other types that are considered bentonites, although they are neither sodium nor calcium varieties (Patterson & Murray, 1975). The bentonite deposits are widespread, like those of sepiolite, along a narrow fringe between Madrid and Toledo (Fig. 1). Reserves of bentonite have not yet been estimated accurately.

This paper focuses on the lithostratigraphy of the Tertiary units that contain sodium sulphates and clays, as well as on the sedimentological analysis of these deposits. The sequential arrangement of the mineral occurrences is emphasized and models of formation are postulated in view of both mineralogical and sedimentological data.

GEOLOGICAL SETTING AND NEOGENE STRATIGRAPHY

The Madrid Basin is a triangular-shaped depression that exceeds 10 000 km^2. The formation of the basin began during the early Paleogene as a response to differential tectonic movements between the Hesperic Massif and areas to both the north and south, which were compressed during Alpine movements (Álvaro *et al.*, 1979; Vegas & Banda, 1982; Portero & Aznar, 1984). This resulted in a progressive uplift of rigid crystalline blocks (Central System, ·Toledo Mountains) in the north and south of the

basin (Fig. 1) and the folding of Mesozoic sedimentary successions (Iberian Ranges) in the east. During the late Oligocene to early Miocene, both the basin margins and the previously deposited Paleogene units experienced strong deformation that was accompanied by overthrusting of the Altomira Range towards the west, resulting in the formation of two differentiated depressions (Madrid Basin and Intermediate Depression) (Fig. 1) after the late Oligocene.

Moderate to high subsidence rates during the Tertiary accompanied the accumulation of more than 3500 m of continental sedimentary units in the Madrid Basin. These units are recognized by deep boreholes and seismic profiles (Megías *et al.*, 1983; Racero, 1988). The combined thickness of the Miocene units accounts for one-fourth of the Tertiary section in the basin, although those units do not exceed 300 m of thickness in outcrops. The Miocene units are well exposed in the central part of the basin, where they are comprised of several sequences of evaporites and carbonates distributed in the lower and upper parts of the sections, respectively. These lacustrine facies grade laterally into progressively coarser terrigenous deposits towards the basin margins.

Figure 2 shows the stratigraphic relationship between marginal terrigenous and chemical lacustrine facies throughout a W−E profile in the basin. The

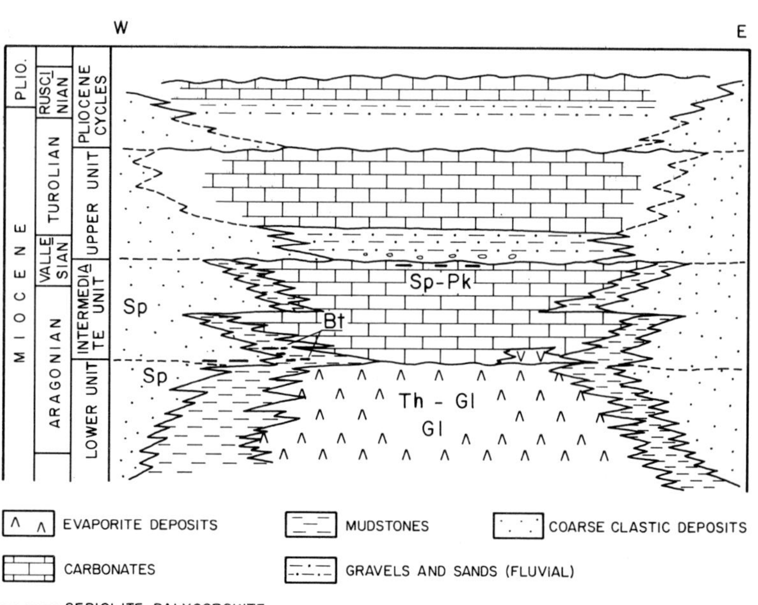

Fig. 2. Generalized Neogene stratigraphy of the Madrid Basin, showing the distribution of the major groups of facies in an idealized W−E profile. Symbols Sp, Bt, Pk, Th, Gl indicate sepiolite, bentonite, palygorskite, thenardite, and glauberite occurrences, respectively.

lithostratigraphy of the Neogene is divided into three major units, which are separated by regional disconformities: paleokarst surfaces, erosional surfaces related with the progradation of marginal terrigenous complexes, etc. (Alberdi *et al.*, 1983; Junco & Calvo, 1983; Megías *et al.*, 1983). The distinction of these three units is relevant in terms of the distribution of lacustrine mineral resources as their occurrence is controlled by the sedimentary changes that took place during the evolution of the basin in the late Tertiary.

The Miocene Lower or Saline Unit is a relatively thick (near 500 m) evaporite deposit, which is composed predominantly of anhydrite (commonly transformed to gypsum in outcrops), halite, clays, and non-skeletal carbonates (magnesite, dolomite), as well as important deposits of sodium sulphates. Both glauberite and thenardite are extracted by underground and open pit mining in the central parts of the basin where the Saline Unit is well exposed. The evaporite facies grade laterally to reddish-green mudstones containing anhydrite and/or gypsum nodules, and then to coarser clastics deposited in alluvial environments. A concentric arrangement of these facies is deduced from mapping of the lacustrine and alluvial sequences along the basin, which suggests a closed basin geomorphic and hydrological setting during the sedimentation of the Saline Unit (Fig. 3).

The Miocene Intermediate Unit shows, like the underlying unit, a concentric sedimentary pattern (Fig. 4), but lacks the characteristic saline deposits. In fact, open lacustrine facies within the Intermediate Unit consist exclusively of carbonate and gypsum (detrital and chemical). In addition, the sedimentation of the Intermediate Unit was accompanied by a net progradation of the marginal clastic complexes towards more central parts in the basin. This was probably the result of tectonic reactivation in the northwest basin margin (Central System), whereas the other margins, both to the south and the east, remained fairly stable during that epoch (Calvo *et al.*, 1989). The Miocene Intermediate Unit contains large deposits of economic clays, mainly of the bentonite and sepiolite types, which are interpreted as having been deposited in a number of lacustrine or peripheral lake subenvironments. Whatever its specific origin and accumulation process, the presence of bentonite and sepiolite is restricted to the northwest and south of the basin. Such areas were dominated by the influx of arkosic materials that were supplied from granitoid and high-grade

metamorphic terranes (Figs 3, 4). The close genetic relationships among source areas, depositional systems and the occurrence of sepiolite and bentonite in the Madrid Basin have been noted by Leguey *et al.* (1984), Doval *et al.* (1985), Calvo *et al.* (1989), and others. These aspects will be discussed later.

Finally, the Miocene Upper Unit (Fig. 2) is composed of a rather thin (0−50 m) sequence of clastic and carbonate deposits that, in general, can be interpreted as having been accumulated in a fluviolacustrine system. Both facies associations and their depositional arrangement within this unit strongly suggest a dramatic change in the palaeogeography of the basin during the late Miocene, as well as provide evidence of a fresher stage during the evolution of the lake system in this period.

SEDIMENTOLOGY OF THE SODIUM SULPHATE DEPOSITS

As mentioned above, the occurrence of sodium sulphate deposits within the Neogene units of the Madrid Basin is restricted to the Lower Saline Unit. Sodium sulphate deposits have been investigated in quarries, natural outcrops, and galleries, but the most relevant information on these facies derives from boreholes made for exploration of new reserves in previously exploited areas. A number of the studied boreholes were obtained from drilling in the vicinity of the mine 'El Castellar' and others from Villaconejos, San Martín de la Vega and the southern part of Madrid (García del Cura *et al.*, 1986).

The stratigraphy of the Lower Unit in the most central part of the basin is composed of a rather homogeneous sequence of massive, roughly laminated evaporite deposits with some fine terrigenous intercalations. The evaporite minerals in this sequence are anhydrite, glauberite, polyhalite, magnesite, and halite (Fig. 5) and, as discussed below, they occur in a variety of textures, sequences, and mineralogical assemblages. The top of this sequence is commonly weathered to secondary gypsum, which gives place to thick indurated beds throughout the area. To the south of the Tajo river valley, the uppermost part of the Saline Unit contains additionally a massive deposit of thenardite 7 m thick, which includes some glauberite in the form of scant idiomorphic crystals. This thenardite deposit is separated from the glauberite beds observed at depth by a conspicuous level of red clays, which, in turn, are overlain by massive halite and an irregular bed

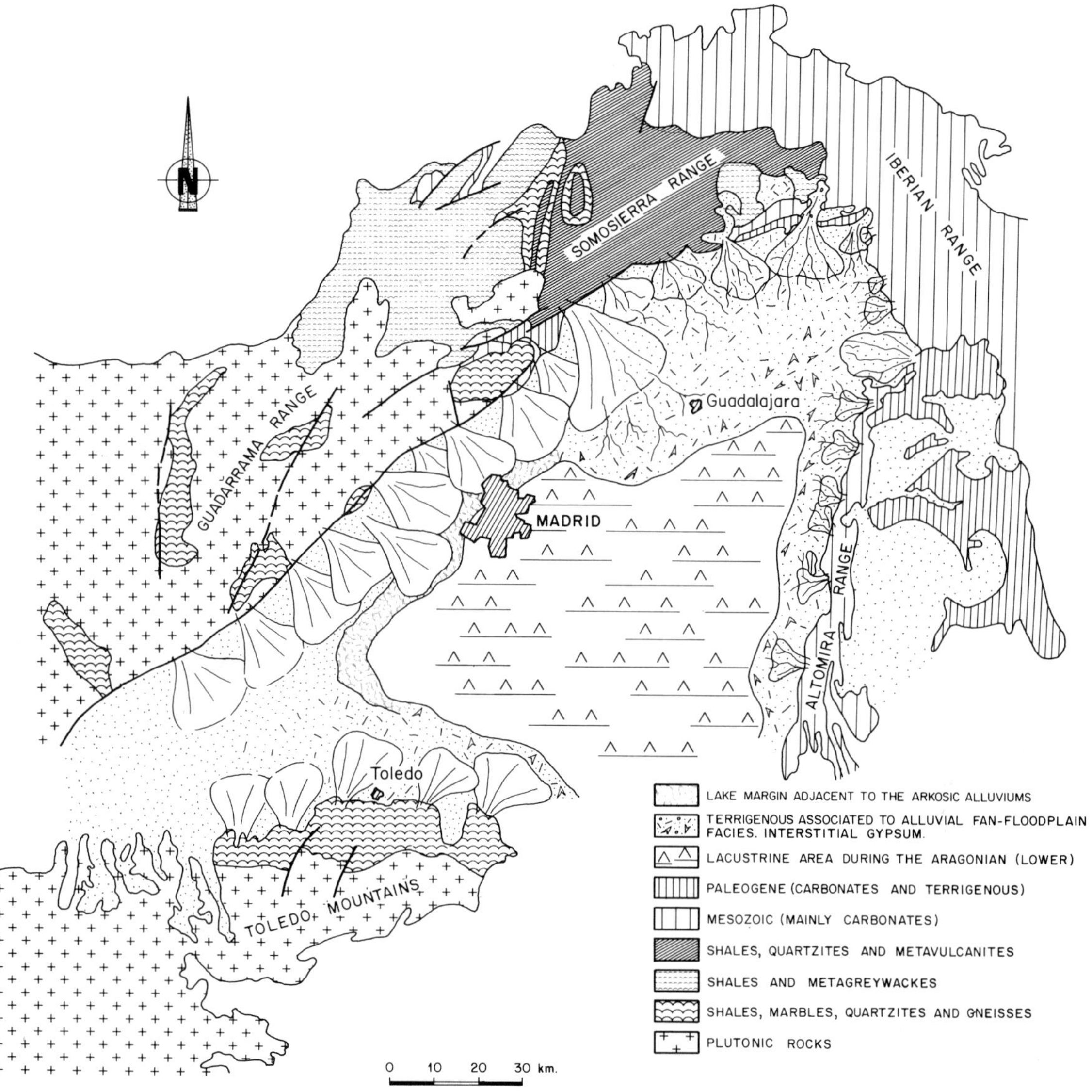

Fig. 3. Sketch of depositional systems in the Madrid Basin during the early Aragonian (Saline Unit).

of glauberite decimetres thick (so-called 'black rock' by the miners in the area). Thus, contrary to glauberite, the thenardite deposit is restricted to a relatively small area in the central part of the basin.

All of the evaporite deposits in the central part of the basin change laterally to mudstones, which can be massive or roughly laminated and which contain nodular anhydrite layers, and further change to red mudstones with a progressive increase of coarser clastic deposits. This trend is also observed in the eastern part of the basin. However, in this area, it is accompanied by a particular feature: the coarse terrigenous deposits of the alluvial fan facies (Fig. 4) are mainly composed of gypsum and dolomite clasts derived from Late Cretaceous formations in the Altomira Range. As discussed below, this is a significant fact, which provides evidence of the specific contribution of some pre-Tertiary evaporites to the formation of saline deposits during the Neogene.

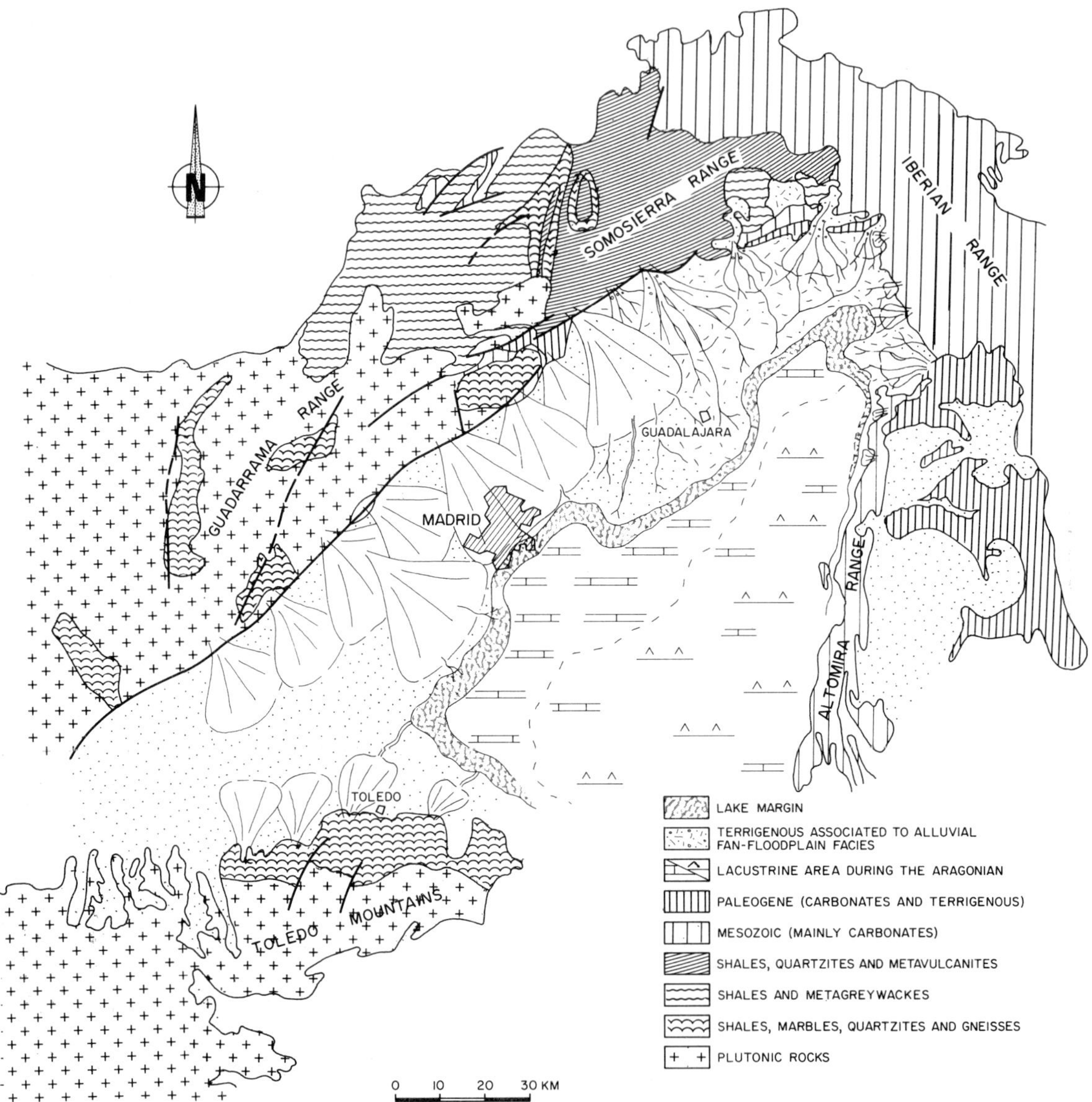

Fig. 4. Sketch of depositional systems in the Madrid Basin during the middle Aragonian (Intermediate Unit). Note the different length of the alluvial fan systems in the western part of the basin as compared with Fig. 3.

Textural characteristics of the evaporites

The evaporite deposits within the Saline Unit exhibit a variety of textures as well as sequential arrangements. These features can be summarized as follows:

Anhydrite and magnesite. Couplets of anhydrite and magnesite, usually arranged in rhythms, constitute a common facies observed in drill holes traversing the Saline Unit in the central part of the basin. The thickness of the individual couplets does not commonly exceed 10 mm. Magnesite and anhydrite laminae in the couplet have an average thickness of $4.3-0.3$ mm and $3.2-0.3$ mm, respectively. Anhydrite is commonly associated in the laminae with halite crystals that exhibit different morphol-

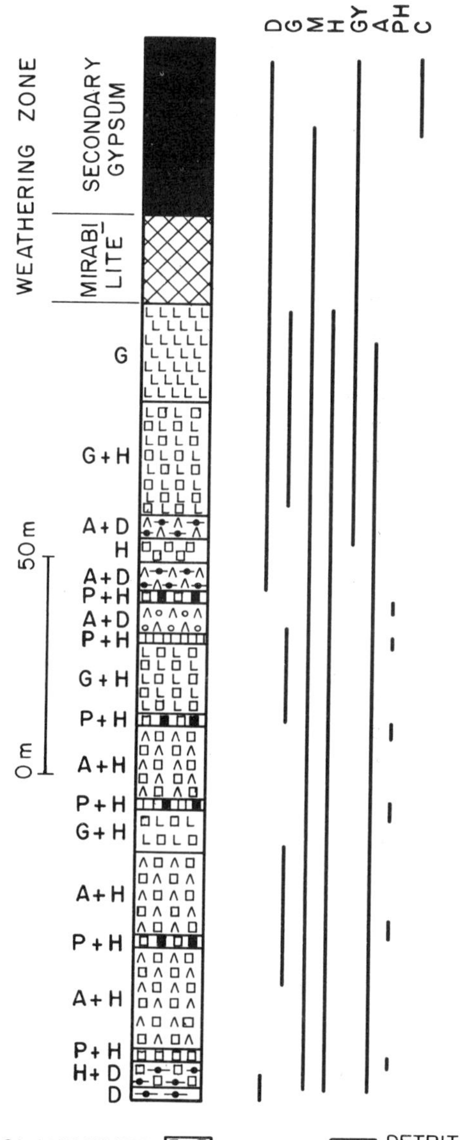

Fig. 5. Measured section from a borehole in the Saline Unit. G, glauberite; M, magnesite; H, halite; D, detrital minerals; A, anhydrite; P, polyhalite; C, calcite; Gy, gypsum.

ogies, such as fascicular, idioblastic, bundle, and bow-tie. Frequently, the halite crystals show traces of partial dissolution. Magnesite occurs in the couplets as micrite aggregates where the minute

crystals have an average diameter of 2 μm. This magnesite micrite is commonly mixed with fine clastic particles, mainly consisting of quartz, feldspar, and clays. Illite is the predominant clay mineral, whereas kaolinite, smectite, chlorite, and mixed-layer clays occur in minor amounts. The laminated structure defined by the couplets may be locally deformed by the diagenetic growth of some crystals, mainly glauberite and gypsum.

In addition to its occurrence in the aforementioned couplets, anhydrite forms centimetre-thick beds that alternate with the couplets and other evaporite facies.

Glauberite. Glauberite ($Na_2Ca(SO_4)_2$) occurs commonly as laminae 3–15 mm thick consisting of a dense aggregate of idiomorphic, lens-shaped, glauberite crystals that alternate with millimetre-thick laminae of anhydrite and/or magnesite. Occasionally, glauberite occurs mixed with anhydrite as large poikilotopic crystals. Layers of massive glauberite centimetres to decimetres thick are locally observed in some boreholes. Whatever the case, the glauberite crystals are free of inclusions and are positioned with their long axis parallel to the bedding.

On the other hand, the so-named 'black-rock' is a nearly pure glauberite deposit that consists of an aggregate of alotriomorphic, diffusely or wavy rimmed crystals that usually contain solid inclusions of magnesite, quartz, and micas.

Polyhalite. Polyhalite ($K_2Ca_2Mg(SO_4)_4.2H_2O$) has been observed as felty and spherulitic aggregates that alternate with centimetre-thick halite layers or millimetre-thick glauberite laminae. The polyhalite crystals are always associated with micritic magnesite (Fig. 6). In its turn, the felty polyhalite may be associated with skeletal glauberite crystals. The halite crystals commonly exhibit chevron-type morphologies. The thickness of the individual layers of halite ranges from 1 to 6 cm.

Thenardite. Thenardite (Na_2SO_4) has been found only within the uppermost part of the Saline Unit, where it occurs as aggregates of large anhedral thenardite crystals that include scant idiomorphic glauberite crystals (Fig. 7). The surfaces of the glauberite crystals usually exhibit reactive rims when they are in contact with the thenardite cement.

Weathering has affected the primary deposits of the Saline Unit and it is responsible for the formation

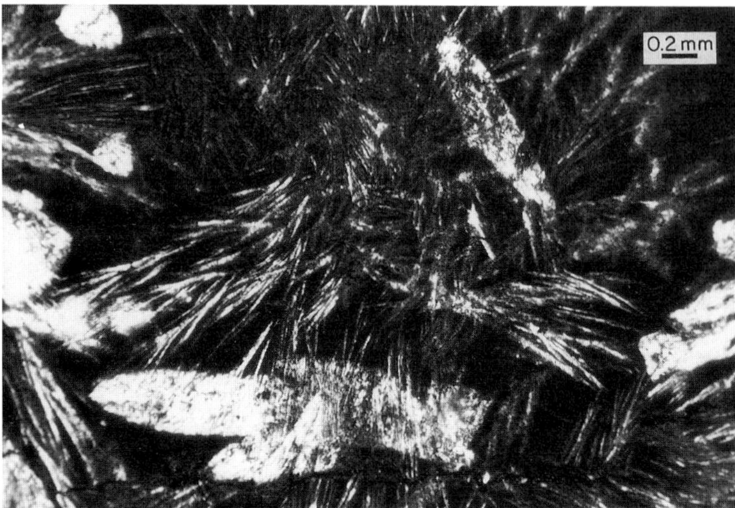

Fig. 6. Photomicrograph of stellate fibrous polyhalite crystals growing on magnesite (dark) and glauberite idiomorphs (crossed polars).

Fig. 7. Photomicrograph of idiomorphic glauberite crystals cemented by large, poikilotopic thenardite (crossed polars).

of large volumes of selenitic, secondary gypsum as well as the incongruent dissolution of glauberite. The latter process is made evident today by the development of mirabilite−bloedite efflorescences in outcrops and also by the presence of sodium sulphate brines in wells and springs located in the region. These brines are commonly collected for use by the sodium sulphate industry, as well as for medicinal waters.

Interpretation and source of the evaporites

The isotopic composition of the most common min-erals in the Saline Unit is shown in Fig. 8A,B. As previously noted by Ordoñez *et al.* (1983), it is quite unlikely that the evolution of waters generated by leaching of granitic and metamorphic rocks of the Central System could constitute the source waters for the Miocene evaporites of the Madrid Basin. Utrilla *et al.* (1987) argued that both $\delta^{34}S$ and $\delta^{18}O$ values obtained from sulphate minerals of the Saline Unit could be explained accurately through the re-cycling of Mesozoic evaporites of the Iberian System and La Mancha. This evidence, together with our structural and sedimentological observations, lead us to conclude that the main source of sulphates in

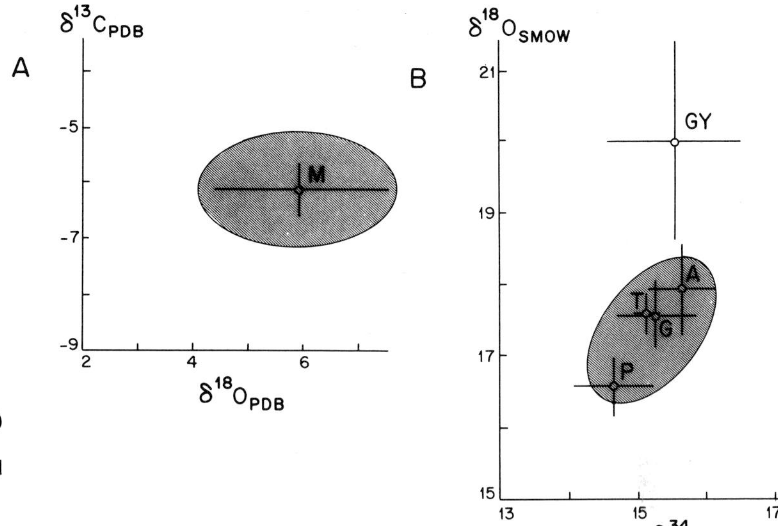

Fig. 8. Carbon, oxygen, and sulphur average isotopic values of evaporite minerals of the Saline Unit. (A) Carbon and oxygen isotopic values of magnesite (M) ($n = 5$). (B) Oxygen and sulphur isotopic values of gypsum (GY) ($n = 10$), anhydrite (A) ($n = 6$), glauberite (G) ($n = 5$), thenardite (T) ($n = 3$), polyhalite (P) ($n = 3$). Axes in ellipsoidal hatched area are defined by the standard deviation of data ($\sigma = 1$). (n = number of analysed samples.)

the Saline Unit corresponds to the recycling of Late Cretaceous evaporites through chemical and mechanical erosion. The contribution of other sources, such as Triassic and other Mesozoic evaporites, and possibly some sulphide deposits in Palaeozoic terranes of the Central System, must not be neglected, although their possible contribution is thought to be minor. Finally, the isotopic composition from magnesite (Fig. 8A) reflects the continental, freshwater origin of the carbonate ions and the derivation from a brine that has been submitted to strong evaporation.

The origin of the evaporite deposits in the Saline Unit can be determined by analysing the evolution of a brine within the system $Mg-K-Cl-SO_4-H_2O$ under conditions of saturation with respect to ClNa. Addition of calcium bicarbonate before saturation with ClNa and magnesium salts would result in the separation of double sulphate salts, especially glauberite and polyhalite. In the aforementioned five-component system, both gypsum and syngenite occupy large stability fields at 0°C, whereas polyhalite and glauberite are absent (Stewart, 1963). With increasing temperature, the stability fields of both polyhalite and glauberite become progressively larger. These minerals are the dominant phases above 55°C. Table 1 shows the chemical composition of a brine that, according Borchert (1940), would be in equilibrium with anhydrite−polyhalite−glauberite at the saturation point of 25°C. This table lacks precise values of Ca^{+2} and CO_3H^- ions, which could not be determined accurately by the author.

According to Strakhov (1970) and Valyashko (1972), the constant processes that change the chemical composition of brines in lakes result from the combined effect of environmental, that is geological and climatic factors. These constant processes, named 'metamorphosis of brines' by the Russian authors, act only in one direction. Because of this, only slightly soluble salts are present within the resulting saline sequence. Together with these constant processes, other processes of cyclic character, for example those determined by climatic changes throughout a year, can effect the chemical composition of brines. These cyclic processes are very probably the most important factor in the formation of couplets.

Table 1. Chemical composition of a brine at the glauberite−polyhalite−halite saturation point, at 25°C (after Borchert, 1940); used as a reference for the origin of the evaporite deposits of the Saline Unit

Ion	mol/kg H_2O
K^+	0.21
Mg^{2+}	3.49
SO_4^{2+}	1.14
Na^+	0.275
Cl^-	5.185

The formation of the alternations of magnesite with anhydrite from a brine with the composition noted in Table 1 could be achieved according to the following non-equilibrium equation (modified from Strakov, 1970):

$$2CO_3H^- + Ca^{2+} + SO_4^{2-} + 2Mg^{2+} + 2OH^- + 6H_2O \rightarrow SO_4Ca.2H_2O + 2(CO_3Mg.3H_2O).$$
$$\text{(gypsum)} \qquad \text{(nesquehonite)} \qquad (1)$$

The nesquehonite deduced from the equation might be the precursor of magnesite, which has been recognized as a common mineral in most of the evaporite sequences. Nesquehonite may be interpreted as the spring deposit in seasonally sedimented rhythms, whereas gypsum (the probable anhydrite precursor) may be considered a summer deposit. The commonly observed association of magnesite with fine detrital material and of anhydrite with halite in the rhythms contributes to this hypothesis. The suggested transformations of nesquehonite to magnesite and of gypsum to anhydrite could be conceived as early diagenetic, even penecontemporaneous, processes. Halite occurring in the rhythms usually displays solution features, which are thought to be caused by the freshening of the brine during the winter−spring period. The freshening, in short, influx of water supplied by the surrounding highlands, is also envisaged as a realistic mechanism for the increase in the amount of bicarbonate and calcium ions in the brine.

With regard to the alternations of glauberite with anhydrite and magnesite, these should be interpreted taking into account the following non-equilibrium equation (modified from Strakov, 1970):

$$2CO_3H^- + Ca^{2+} + 2SO_4^{2-} + 2Mg^{2+} + 2OH^- + 2Na^+ + 4H_2O \rightarrow (SO_4H)_2 CaNa_2 + 2(CO_3Mg.3H_2O).$$
$$\text{(glauberite)} \qquad \text{(nesquehonite)} \qquad (2)$$

Glauberite is envisaged as a primary precipitate. The glauberite crystals can be nucleated and then grow within the brine in a rather restricted range of physico-chemical conditions, in spite of the slow crystallization rate of this mineral. The formation of the deposit commonly would be realized through the sinking of idiomorphic glauberite crystals to the lake floor during the summer period. Subsequently, these crystals can either overgrow or be transformed to gypsum or anhydrite because of incongruent solution induced by seasonal freshening of the lake water. This may be the origin of the anhydrite that is associated with glauberite in the evaporite rhythms. On the other hand, nesquehonite may be considered

the spring deposit in the rhythms, although it is quickly transformed into magnesite.

Polyhalite is a rare mineral in continental saline deposits. This mineral has been recognized as an early replacement of gypsum in supratidal environments of the Ojo de Liebre area in Baja California (Pierre & Fritz, 1984). The process for the formation of polyhalite requires solutions rich in Mg^{2+}, K^+, and SO_4^{2-}. The textural features displayed by the polyhalite minerals of the Saline Unit in the Madrid Basin lead us to conclude that polyhalite was formed as an early diagenetic product from gypsum and/or anhydrite. This transformation can be expressed by the following equation:

$$4SO_4Ca.2H_2O + Mg^{2+} + 2K^+ \rightarrow$$
$$\text{(gypsum)}$$
$$(SO_4)_4.Ca_2.Mg.K_2.2H_2O + 2Ca^{2+} + 6H_2O.$$
$$\text{(polyhalite)} \qquad (3)$$

or by another equation (after Hardie & Eugster, 1980):

$$2SO_4Ca + 2K^+ + Mg^{2+} + 2SO_4^{2-} + 2H_2O$$
$$\text{(anhydrite)}$$
$$\rightarrow (SO_4)_4.Ca_2.Mg.K_2.2H_2O.$$
$$\text{(polyhalite)} \qquad (4)$$

An increase of salinity as well as a simultaneous decrease of water activity contribute to the growth of polyhalite. Accordingly, we suggest that equation (3) more consistently reflects the actual process of polyhalite formation. This process is strongly conditioned by the dehydration and/or dissolution of the primary precipitates. The crystallization of polyhalite in the gypsum−halite laminae would lead to the formation of polyhalite−halite and magnesite couplets. On the other hand, glauberite and anhydrite also may be replaced by polyhalite, as has been pointed out by Hardie & Eugster (1980). This replacement increases the amount of free sodium ions in the interstitial brine and leads to the precipitation of halite. This could explain the association of polyhalite either with halite or skeletal glauberite observed in our samples.

The occurrence of thenardite is thought to be related to a precursor mineral, probably mirabilite. Mirabilite has been recognized as the most common mineral in extremely shallow depressions (up to 25 cm in depth) in the Mormyshans Hoe Lake (Kulunda Steppe, USSR), whereas glauberite is formed in slightly deeper areas (up to 1 m in depth) (Strakhov, 1970). Grossman (1968) suggested that sodium sulphate deposits in Canada are formed by the influx of saline brines derived from Devonian evaporites. The saline brines are then carried through melt-water channels of glacial origin to lakes. Some of these lakes are deep enough to develop meromixis in the water column, thus favouring lower temperatures at the bottom and the sinking of mirabilite crystals. In addition to these models, the Great Salt Lake in Utah provides further evidence of the accumulation of sodium sulphate deposits (Spencer *et al.*, 1984). In short, the deposition of sodium sulphate in Great Salt Lake was closely related to a major drop in the water level that took place approximately 8000 yr BP. In these circumstances, beds of mirabilite were formed as a result of winter 'chill out' and were probably protected from back-dissolution owing to density induced stratification.

The models mentioned above can be used as a reference to explain the occurrence of thenardite and glauberite in the upper part of the Saline Unit. The absence of thenardite in the underlying levels could indicate a relative change in the sedimentary conditions towards the top of the unit. One interpretation that explains this situation is that during this period of the basin's evolution, large areas of the basin could have become more or less permanently exposed, so that the oldest glauberite-rich saline deposits were affected by seepage flux towards residual depressions generally located in the southern part of the basin. These depressions could then have been flooded by brines rich in sodium-sulphate–chloride that were generated by the incongruent dissolution of older glauberite deposits. The accumulation and preservation of mirabilite took place during meromictic stages of the lake. The supernatant water in the saline lakes could have reached the glauberite saturation level causing the glauberite crystals to sink to the bottom and mix with mirabilite. Further, early diagenetic reactions could have led to the transformation of mirabilite to thenardite to form thenardite–glauberite beds.

Both sedimentological characteristics of the deposits and the processes described above lead to the conclusion that most of the sedimentary sequences that form the Saline Unit were deposited in a closed lacustrine basin that was occupied by a large perennial saline lake. The vertical variations of the deposits that characterize the surrounding mudflats strongly suggest that the lake experienced strong oscillation in the water level. The composition of the brines in the lake was conditioned by the lithology of the drainage areas. Sulphate and magnesium ions were mainly derived from carbonate and evaporite rocks along the eastern margin, whereas chloride, sodium, magnesium, and potassium ions entered the basin through percolating waters from the granitic and metamorphic reliefs. In addition, Triassic salt deposits, as well as atmospheric precipitation, could be an additional source of chloride and sodium ions.

The presence of thenardite exclusively at the top of the Saline Unit leads to the conclusion that some changes took place in the lake system. We interpret these changes as a result of a drop in the water level, which in turn produced a drastic reduction in the lake area and the restriction of saline deposition to rather small ponds. This different facies pattern might have been accompanied by a change in the composition of the brines, thus leading to the precipitation of thenardite-rich deposits.

SEPIOLITE AND BENTONITE DEPOSITS

Within the lower and middle Miocene units of the Madrid Basin, both sepiolite and bentonite deposits are essentially restricted to a rather narrow, 4–5 km in width, fringe, which represents a 'transitional zone' (Galán, 1979) between the clastic wedges adjacent to the surrounding mountain fronts and the open lacustrine sequences, mainly dominated by carbonate and/or gypsum sedimentation. In this 'transitional zone' the two types of clay deposits occur in different lithofacies that are laterally equivalent. Figure 9 shows the idealized depositional relationship between sepiolite deposits and bentonites in an area near Madrid. The overlying of sepiolite facies on bentonites in vertical section is the result of the progradation of marginal clastic complexes during the middle Miocene.

The occurrence of sepiolite and bentonite deposits in the basin is closely related to arkosic materials that were derived from granitoid and high-grade metamorphic source areas. By contrast, they do not occur in the other basin margins, which are made up of carbonate or low-grade metamorphic rocks (Calvo *et al.*, 1989). Thus, differences in the lithology of the source areas exerted direct control on the distribution of clay minerals in the basin, this fact being significant for mining exploration as well as establishing sedimentary models for the Neogene lake systems of the basin.

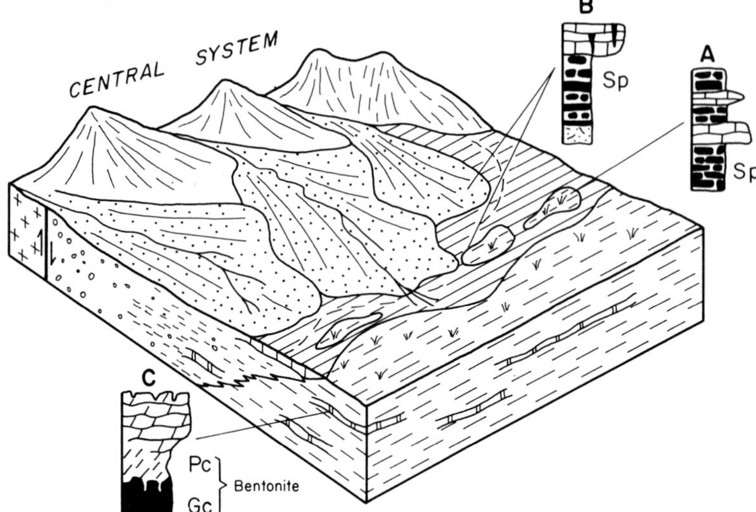

Fig. 9. Schematic block diagram sketching depositional environments at the western margin of the Madrid Basin. A indicates sepiolite accumulation in ponds. Sequence B corresponds to sepiolite within calcrete profiles. Sequence C is composed of greenish and pink clays (bentonites) that were deposited at the lake margin. Symbols: Sp, sepiolite; Gc, greenish clays (bentonite); Pc, pink clays (bentonite).

Sepiolite

The majority of the sepiolite deposits in the basin are located in the vicinities of Madrid and to the south in the province of Toledo (Galán & Castillo, 1984). A common feature of all these deposits is that the sepiolite levels occur in successions made up of fine arkosic sandstones and sandy mudstones. Sepiolite is extracted from a rather continuous level as much as 10 m thick, which also contains dolomite and limestone beds, chert, and minor amounts of fine sand and detrital clays (sequence A in Fig. 9). Carbonates form massive lenticular beds, a few decimetres thick, that are usually root-marked and frequently display horizontal and/or circular desiccation planes. Thin chert beds and nodules are irregularly distributed within the clay levels or replace carbonates; the mineralogy of the chert is mostly CT opal. Sepiolite occurs in two or three distinct beds, the thickest reaching 2 m. Pure sepiolite (95–100%) is generally massive to slightly brecciated and displays a variety of colour from whitish to pale brown, light greenish or pink; the lightest samples show an earthy luster, whereas denser sepiolite has a waxy appearance. An intraclastic texture (Khoury *et al.*, 1982) is commonly recognized in the sepiolite, probably resulting from fracturing induced either by dessication or root bioturbation. Thin channels and vugs filled by sepiolite are also found in the groundmass of the massive sepiolite beds and the carbonate material with which they are associated.

Sepiolite deposits in this area have been intensively investigated for industrial applications. Techno-logical properties of this sepiolite have been published by Galán (1987), who has also described the characteristic mineralogy of the deposits (Galán, 1979; Galán & Castillo, 1984). Typical X-ray diffraction patterns of the sepiolite samples correspond to fairly pure sepiolite, mixed to some extent with quartz, illite, feldspar, and carbonate (mainly calcite). Locally, the sepiolite deposits intercalate some thin levels of light brown or greenish massive clays composed of sepiolite, Mg smectites of stevensite type and variable amounts of Al smectites, which suggests that the deposit is a mixture of detrital and authigenic clays.

In addition to the described facies, sepiolite in the Madrid Basin is also found within calcrete profiles (sequence B in Fig. 9) that were developed in distal facies of the arkosic alluviums (Megías *et al.*, 1982; Calvo *et al.*, 1986). There, sepiolite occurs as a significant component, probably formed through precipitation from vadose solution in a similar way to that proposed by Hay & Wiggins (1980). Another occurrence of sepiolite is that observed in carbonate deposits that originated in an open lacustrine environment. In this situation the sepiolite is laminar and occurs in continuous beds 5–30 cm thick that alternate with bedded dolostones, which commonly exhibit nodular to spheroidal fracturing. Laminar sepiolite frequently displays deformation due to the load of the carbonates. This type of sepiolite deposit is usually referred to as 'mountain leather' (Galán & Castillo, 1984).

In addition to these common sepiolite occurrences,

minor amounts of sepiolite are often found associated with dolostones at the top of the Miocene Intermediate Unit in central parts of the basin. Both sepiolite and palygorskite occur in these deposits, which are thought to have been deposited in open areas of a shallow, brackish lake (Ordoñez *et al.*, 1985).

Bentonite

Bentonite deposits in the Madrid Basin are characterized by greenish to pale brown and pink clays that usually occur in laterally continuous levels 0.4–2 m thick. Bentonites are intercalated in the lacustrine sequences with fine micaceous sands, carbonates (mainly dolomite), some chert, and other mudstones. Several types of sedimentary sequences containing bentonite are recognized in sections of the 'transitional zone'. Such sedimentary sequences (Fig. 10) reflect different depositional settings along the lake margin zone.

Those areas under the influence of clastic influx derived from the arkosic alluviums received silt and fine-grained sands that entered the margins of the lake. Transport of the fine clastic material was either as non-confined flows or, more frequently, as wandering channels that finally built medium-scale fluvial–deltaic complexes into the lake (Fig. 10A). Bentonites in these types of sequences form thin beds that usually contain a high mixture of silty material.

Other sequences, which are laterally equivalent, show minor coarse clastic influx and are composed exclusively of clayey sediments and carbonates, with occasional chert nodules (Fig. 10B). These sequences are widespread in outcrops along the 'transitional zone' near Madrid and to the south. The thickness of the individual sequences is 2–4 m. Generally, they consist of greenish, massive to roughly laminated mudstones with a waxy luster, which grade upwards into massive, root-bioturbated, earthy pink clays. The occurrence of pink clays is sometimes accompanied by the development of chert nodules. These pink clays pass upwards to a whitish, irregularly dense or soft dolomitic carbonate bed, which usually shows prismatic structure due to root holes. The relative thickness of each lithology in the sequence is highly variable, but the clayey deposit, whether green or pink, is always considerably thicker than the carbonates.

The mineralogy of bentonite is essentially Mg smectites, mostly stevensite and saponite. The MgO contents determined from the whole sample in these bentonites usually range from 25 to 28% (Galán *et al*, 1986; IGME, 1989). The green clays are, in general, composed of nearly pure trioctahedral smectite, with variable amounts of illite, mixed-layer clays, and minor kaolinite (Doval *et al.*, 1985; IGME, 1989). The pink clays have received considerable attention by several authors in view of its particular mineralogy (Doval *et al.*, 1986,1987; Jones *et al.*, 1986; Martín de Vidales *et al.*, 1988,1989). According to some of these recent investigations, the pink clays consist largely of interstratified kerolite, most likely stevensite with interlayered kerolite (Martín de Vidales *et al.*, 1989), and some sepiolite. This sepiolite occurs as aggregates of fibres rimming the sheets of trioctahedral smectites.

Interpretation of sepiolite and bentonite deposits

The accumulation of sepiolite and bentonite deposits in the Madrid Basin is mainly related, in terms of sedimentary facies, to a clay-dominated narrow belt between arkosic alluvium and lacustrine successions in which gypsum and carbonate rocks are dominant.

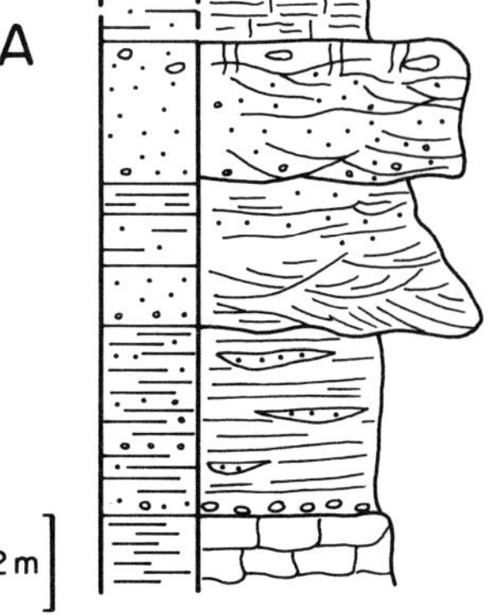
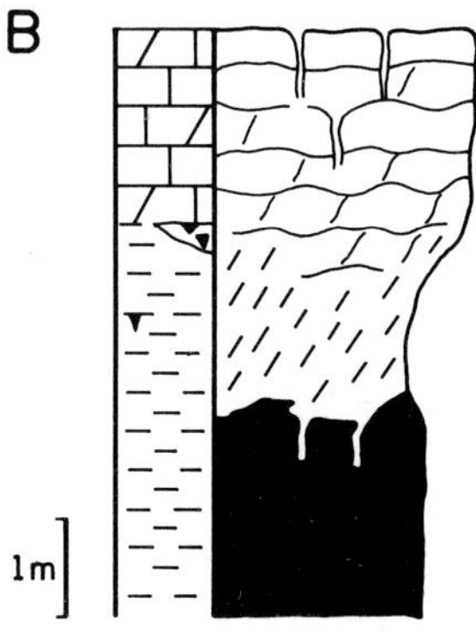

Fig. 10. Common sedimentary sequences containing Mg smectites in lake margin areas. (A) Coarsening and thickening upwards cycle composed of greenish clays and micaceous sandstones that correspond to a fluvio-deltaic complex. (B) Shallowing upwards cycle consisting of greenish and pink clays (interstratified kerolite–stevensite) and paludal carbonates.

This belt corresponds to the deposits that formed at the toes of arkosic alluvium fans as well as in the fringing mudflats and marginal facies of a shallow perennial lake.

The interbedding of sepiolite levels with very fine arkoses, bioturbated mudstones, and, in some places, intraclast-filled channels (Alonso *et al.*, 1986), strongly suggests that the accumulation of sepiolite took place in almost flat areas at the toes of the arkosic alluvial fans. These areas contained a mosaic of short-lived ponds and marshes that were intermittently fed by springs resulting from discharge of ground-water of the granitic terranes of the Central System. Sepiolite could be formed in these depressions by precipitation from silica-bearing waters that were mixed with brackish or saline Mg-contributor waters from the nearby lake. The extreme shallowness of the ponds is indicated by the abundance of desiccation features and root traces in the sepiolite deposit.

Similar sedimentary conditions to those that contributed to the accumulation of rather thick deposits of sepiolite have been described in the playa–marshland complex of the Amargosa Desert (Khoury *et al.*, 1982; Hay *et al.*, 1986) and in other closed basins in the western USA (Jones, 1986). In the Amargosa Basin, sepiolite was formed both in ponds and in caliche profiles, and it occurs associated with Mg smectite, carbonates, chert, and other clays. The study of sepiolite and Mg smectite deposits in Amargosa was helpful in revealing the principal physico-chemical and sedimentary controls governing the formation of these clay minerals (Khoury *et al.*, 1982). One of the most outstanding conclusions, further developed by Jones (1986), is that for typical natural waters, kerolite, rather than sepiolite, would precipitate with increasing pH and solute Mg and decreasing silica content.

The formation of sepiolite and Mg smectite (bentonite) deposits in the Madrid Basin is thought to have occurred under similar chemical constraints to those inferred from the Amargosa and other USA recent lake basins (Jones *et al.*, 1986). The distribution of the two types of clay deposits was controlled by specific chemical conditions (e.g. alkalinity, variation of solute contents, salinity) present in the different depositional subenvironments. Thus, the accumulation of sepiolite in ponds at the toe of the arkosic alluviums was favoured by relatively dilute, silica-rich waters under moderately alkaline conditions, the pH requirements for sepiolite precipitation ranging approximately from 8.0 to 8.5 (Siffert

& Wey, 1962). The high SiO_2 content of waters discharging from the crystalline terranes and arkoses could also account for extensive silicification (CT opal) of calcretes in low-lying areas adjacent to the sepiolite deposits (Hay *et al.*, 1986).

On the other hand, the formation of bentonites is chiefly restricted to the lake area and surrounding mudflats. In this environment, detrital clay consisting mainly of dioctahedral smectite (Doval *et al.*, 1986), which entered the lake either non-confined or within channelled flows, might be transformed into trioctahedral smectite by the addition of solute magnesium under increasing pH. This general process of clay transformation in lacustrine sediments has been documented properly from a number of closed lake basins (Stoessell & Hay, 1978; Jones & Weir, 1983; Jones, 1986; Hay *et al.*, 1986) as well as perimarine environments (Weaver & Beck, 1977). This mechanism for the formation of Mg smectite is consistent with the occurrence of bentonite in the lacustrine sequences of the Madrid Basin. Accordingly, greenish clays were accumulated in nearshore areas of a shallow, moderately alkaline perennial lake. Towards the central lake area the precipitation led to the deposition of massive to laminar dolomite and gypsum.

Periodic fluctuation of the lake level is indicated by rhythmically repeated sequences, some metres thick, of greenish clays passing upwards into pink clays and carbonates (Fig. 10B). The sedimentological features of each of these facies may be interpreted as the result of progressive shallowing of the lake culminating in a period of subaerial exposure in the margins. Mineralogical changes throughout the individual sequences also can be considered as indicators of the same trend. The transition from trioctahedral smectite in the base of these sequences to interstratified kerolite–smectite with some sepiolite (Jones *et al.*, 1986; Martín de Vidales *et al.*, 1988,1989) towards the top very likely reflects a progressive decrease in detrital input, the pink clays being formed at higher ratios of precipitate/detritus than the stevensitic and/or saponitic green clays. Both interstratified kerolite–smectite (usually kerolite–stevensite) and pure kerolite phases have been found in lacustrine deposits of recent closed basins, mainly in the western USA (Eberl *et al.*, 1982; Khoury *et al.*, 1982; Jones & Weir, 1983; Jones, 1986; Hay *et al.*, 1986) and eastern Africa (Stoessell & Hay, 1978). Experimental and sedimentological evidence lead to the conclusion that the formation of interstratified kerolite–

smectite is favoured by a rise in pH and salinity (Jones, 1986). This seems to fit well with the occurrence of these mineral phases in the bentonite sequences of Madrid. According to the observed sequential arrangement, the pink clays containing interstratified kerolite—stevensite were formed under more extreme evaporative conditions than the Mg smectite greenish clays and probably through their transformation in a vadose environment. This is in agreement with that suggested by Martín de Vidales *et al.* (1988), who also explain pink clays as a result of increasing evaporation subsequent to subaerial exposure. The origin of the sepiolite, which is found in minor amounts within the pink clays, is interpreted by the same authors as being caused by the dissolution of the interstratified kerolite—smectite and further precipitation of sepiolite. This process would be favoured by episodic input of meteoric water (i.e. rainfall) in the exposed marginal lake sediments.

The palaeogeographical distribution of clay minerals within the Miocene units of the Madrid Basin is in contrast with that established as a classic clay mineral chemical sequence in closed-basin deposits by Millot (1964). In the Millot model, the most aluminous clays commonly occur at the periphery and are succeeded by more magnesian smectites basinward. Sepiolite would be found at the centre of the basin. In contrast, major accumulation of sepiolite in the Madrid Basin does not occur at the centre, but in a belt between clastic deposits bearing detrital dioctahedral smectites (the arkosic alluvial fan systems) and lacustrine deposits mainly composed of trioctahedral smectites, carbonates, and gypsum. Paradoxically, some scarce traces of sepiolite in central positions in the basin are associated with carbonate deposits that were formed from brackish or slightly saline lake water.

CONCLUSIONS

During the early and middle Miocene, significant amounts of sodium sulphates and several types of clays were accumulated in lacustrine and peripheral lake environments of the Madrid Basin.

In the lower Miocene unit, sodium sulphate deposits, mainly glauberite and thenardite, occur associated with calcium sulphate, magnesite, and minor clays. Glauberite is widespread in the central part of the basin, where it characteristically alternates with anhydrite and/or magnesite. In addition,

couplets of magnesite mixed with fine clastics and anhydrite with halite constitute a common facies in this area. Such types of rhythmic sediments are interpreted to have been formed as a result of periodical (i.e. seasonal) changes in a perennial lake system.

Thenardite, which only occurs at the uppermost part of the Saline Unit, was formed during a separate stage of the evolution of the basin. The sedimentation of the precursor phase of thenardite (mirabilite) took place in more restricted saline lakes in which meromictic conditions were developed.

Polyhalite, which is not commonly found in continental environments, occurs within the lacustrine sequences of the Madrid Basin as a diagenetic phase, which probably replaced gypsum and perhaps glauberite and anhydrite. This is also in agreement with the observations made by Ortí & Pueyo (1980) who suggested that the polyhalite could replace some early diagenetic minerals.

Both sepiolite and bentonite deposits in the Madrid Basin are restricted to a narrow area that fringes the distal parts of the arkosic alluviums in the western part of the basin. The sedimentological analysis of the sepiolite deposits and associated facies leads to the conclusion that economic deposits of sepiolite were mainly accumulated by precipitation in ponded areas that extended to the toes of the alluvial fans. The stratigraphic relationships as well as the sequential arrangement of the bentonites indicate that these deposits were formed in a marginal lacustrine environment, probably by interaction of slightly alkaline lake waters with detrital fine clays. The resulting Mg smectites were submitted to episodic exposure that led to the formation of interstratified kerolite—stevensite.

The pattern of clay mineral distribution in the Madrid Basin differs with regard to previously proposed models of geochemical mineral sequences in closed lake basins. In our case, sepiolite does not represent clays that originated in the centre of a lake, but is a clay deposit that formed in rather marginal areas under moderately alkaline and saline conditions.

ACKNOWLEDGEMENTS

We thank J.M. Angulo for assistance with illustrations. This work is included in the Project P-84-0078-C02-02: 'Evolución geológica de la Cuenca Media y Alta del Tajo' (CSIC—CICYT).

REFERENCES

ALBERDI, M.T., HOYOS, M., JUNCO, F., LÓPEZ-MARTINEZ, N., MORALES, J., SESÉ, C. & SORIA, D. (1983) Biostratigraphie et évolution sédimentaire du Néogene continental de l'aire de Madrid. *RCMNS Interim-Coll.*, *Montpellier*, 15−18.

ALONSO, A.M., CALVO, J.P. & GARCÍA DEL CURA, M.A. (1986) Sedimentología y petrología de los abanicos aluviales y facies adyacentes en el Neógeno de Paracuellos de Jarama (Madrid). *Estud. Geol. (Madrid) 42*, 79−101.

ÁLVARO, M., CAPOTE, R. & VEGAS, R. (1979) Un modelo de evolución geotectónica para la Cadena Celtibérica. *Acta. Geol. Hisp.* **14**, 174−179.

BORCHERT, H. (1940) Die Salzlagerstätten des deutschen Zechsteins. *Archiv. Lagerstättenforsch.* **67**, 167 pp.

CALVO, J.P., ALONSO, A.M. & GARCÍA DEL CURA, M.A. (1986) Depositional sedimentary controls on sepiolite occurrence in Paracuellos de Jarama, Madrid Basin. *Geogaceta*, **1**, 25−28.

CALVO, J.P., ALONSO, A.M. & GARCÍA DEL CURA, M.A. (1989) Models of Miocene marginal lacustrine sedimentation in response to varied depositional regimes and source areas in the Madrid Basin (central Spain). *Palaeogeogr. Palaeoclimatol. Palaeoecol.* **70**, 199−214.

COOPE, B. (1983) Sulquisa − Spain's new producer of natural sodium sulphate. *Ind. Miner. (London)* **June 1983**, 79−83.

DOVAL, M., DOMÍNGUEZ DÍAZ, M.C., BRELL, J.M. & GARCIA ROMERO, E. (1985) Mineralogía y sedimentología de las facies distales del borde norte de la Cuenca del Tajo. *Bol. Soc. Esp. Mineral.* **8**, 257−269.

DOVAL, M., CALVO, J.P., BRELL, J.M. & JONES, B.F. (1986) Clay mineralogy of the Madrid Basin: comparison with other lacustrine closed basins (abstr.). *Symposium on Geochemistry of Earth Surface Processes and Mineral Formation, Granada*, pp. 188−189.

DOVAL, M., RAUTUREAU, M., BRELL, J.M. & FONTAINE, C. (1987) Crystal-chemistry of 'pink clays' from the Tertiary Madrid Basin. Its relationship with the sepiolite occurrence (abstr.). *Sixth Meeting of European Clay Groups, Sevilla*, pp. 203−204.

EBERL, D.D., JONES, B.F. & KHOURY, H.N. (1982) Mixed-layer kerolite/stevensite from the Amargosa Desert, Nevada. *Clays Clay Miner.* **30**, 321−326.

GALÁN, E. (1979) The fibrous clay minerals in Spain. *Proceedings of the Eighth Conference on Clay Mineralogy and Petrology*, Teplice, pp. 239−249.

GALÁN, E. (1987) Industrial applications of sepiolite from Vallecas−Vicálvaro, Spain: a review. *Proceedings of the International Clay Mineral Society*, pp. 400−404. Bloomington, Indiana.

GALÁN, E. & CASTILLO, E. (1984) Sepiolite−palygorskite in Spanish Tertiary basins: genetical patterns in continental environments. In: *Palygorskite −Sepiolite. Occurrences, Genesis and Uses* (Eds Singer, A. & Galán, E.), pp. 87−124. Elsevier, Amsterdam.

GALÁN, E., ALVAREZ, A. & ESTEBAN, M.A. (1986) Characterization and technical properties of a Mg-rich bentonite. *Appl. Clay Sci.* **1**, 259−309.

GARCÍA DEL CURA, M.A. (1979) *Las sales sódicas, calcosódicas y magnésicas de la Cuenca del Tajo*. Serie Universitaria n 109, Fundación March, Madrid, 93 pp.

GARCÍA DEL CURA, M.A., ORDÓÑEZ, S. & CALVO, J.P. (1986) La Unidad Salina (Mioceno) en el área de Madrid. Características petrológicas y mineralógicas. *Bol. Soc. Esp. Miner.* **9**, 329−338.

GRIM, R.E. & GÜVEN, N. (1978) *Bentonites. Geology, Mineralogy, Properties and Uses*. Elsevier, Amsterdam, 253 pp.

GROSSMAN, I.G. (1968) Origin of the sodium sulfate deposits of the Northern Great Plains of Canada and the United States. *U.S. Geol. Surv. Prof. Pap.* **600-B**, 104−109.

HARDIE, L.A. & EUGSTER, H.P. (1980) Evaporation of seawater: calculated mineral sequences. *Science* **208**, 498−500.

HAY, R.L. & WIGGINS, B. (1980) Pellets, ooids, sepiolite and silica in three calcretes of the southwestern United States. *Sedimentology* **27**, 559−576.

HAY, R.L., PEXTON, R.E., TEAGUE, T.T. & KYSER, T.K. (1986) Spring-related carbonate rocks, Mg clays, and associated minerals in Pliocene deposits of the Amargosa Desert, Nevada and California. *Geol. Soc. Am. Bull.* **97**, 1488−1503.

IGME (1989) *Mapa geológico de España 1/50 000, Sheet 19-22, Madrid*. Instituto Geológico Minero de España, Madrid, 71 pp.

JONES, B.F. (1986) Clay mineral diagenesis in lacustrine sediments. In: *Studies in Diagenesis* (Ed. Mumpton, F.A.), U.S. Geol. Surv. Bull. **1578**, 291−300.

JONES, B.F. & WEIR, A.H. (1983) Clay minerals of Lake Abert, an alkaline, saline lake. *Clays Clay Miner.* **31**, 161−172.

JONES, B.F., DOVAL, M., CALVO, J.P. & BRELL, J.M. (1986) Clay mineral authigenesis in lacustrine closed basins; comparison of the Madrid Basin with U.S. occurrences. *Soc. Econ. Paleontol. Mineral. Ann. Meet. Abstr.* p. 87.

JUNCO, F. & CALVO, J.P. (1983) Cuenca de Madrid. In: *Geología de España, II* (Ed. IGME), pp. 534−543. Instituto Geológico Minero de España, Madrid.

KHOURY, H.N., EBERL, D.D. & JONES, B.F. (1982) Origin of magnesium clays from the Amargosa Desert, Nevada. *Clays Clay Miner.* **30**, 327−336.

LEGUEY, S., ORDOÑEZ, S., GARCIA DEL CURA, M.A., & MEDINA, J.A. (1984) Estudio geoquímico y mineralógico de las facies arcósicas distales de la Cuenca de Madrid. *I Congreso Español de Geología*, **2**, 355−371.

MARTÍN DE VIDALES, J.L., POZO, M., MEDINA, J.A. & LEGUEY, S. (1988) Formación de sepiolita-paligorskita en litofacies lutítico-carbonáticas en el sector de Borox-Esquivias (Cuenca de Madrid). *Estud. Geol. (Madrid)* **44**, 7−18.

MARTÍN DE VIDALES, J.L., POZO, M. & LEGUEY, S. (1989) Kerolite−stevensite mixed layer from Neogene Madrid Basin. Genetic implications (abstr.). *International Clay Conference*, Strasbourg, p. 246.

MEGIAS, A.G., LEGUEY, S. & ORDÓÑEZ, S. (1982) Interpretación tectosedimentaria de la génesis de fibrosos de la arcilla en series detríticas continentales (Cuencas de Madrid y del Duero) España. *Quinto Congr. Latinoamericano Geol., Buenos Aires*, **2**, 427−439.

MEGIAS, A.G., ORDÓÑEZ, S. & CALVO, J.P. (1983) Nuevas aportaciones al conocimiento geológico de la Cuenca de Madrid. *Rev. Mat. Proc. Geol.* **1**, 163−192.

MENDUIÑA, J. (1988) *Geología y significado económico de las arcillas cerámicas de la Cuenca de Madrid*. Tesis Doctoral, Universidad Complutense, Madrid, 305 pp.

MILLOT, G. (1964) *Géologie des argiles*. Masson et Cie, Paris, 510 pp.

ORDÓÑEZ, S., MENDUIÑA, J. & GARCÍA DEL CURA, M.A. (1982) El sulfato sódico natural en España. *Tecniterrae* **46**, 16–33.

ORDÓÑEZ, S., FONTES, J.CH. & GARCÍA DEL CURA, M.A. (1983) Contribución al conocimiento de la sedimento-génesis evaporítica de las cuencas neógenas de Madrid y del Duero en base a datos de isótopos estables (abstr.). *Comunicaciones X Congr. Nac. Sedimentologia*, Menorca, pp. 49–52.

ORDÓÑEZ, S., GARCÍA DEL CURA, M.A., HOYOS, M. & CALVO, J.P. (1985) Middle Miocene paleokarst in the Madrid Basin (Spain). A complex karstic system (abstr.). *Sixth European Regional Meeting, International Association of Sedimentologists*, Lleida, pp. 624–627.

ORDÓÑEZ, S., FONTES, J.C. & GARCÍA DEL CURA, M.A. (1987) Estudio isotópico (^{18}O y ^{13}C) de la paragénesis sulfatada sódica, calcosódica y cálcica de la Unidad Salina de la Cuenca de Madrid (abstr.). *II Congreso Geoquím. España*, Soria, pp. 95–98.

ORTÍ, F. & PUEYO, J.J. (1980) Polihalita diagenética en una secuencia evaporítica continental (Mioceno, Cuenca de Tajo, España). *Rev. Inst. Invest. Geol.* **34**, 209–222.

ORTÍ, F., PUEYO, J.J. & SAN MIGUEL, A. (1979) Petro-genésis del yacimiento de sales sódicas de Villarrubia de Santiago, Toledo (Terciario continental de la Cuenca del Tajo). *Bol. Geol. Min.* **90**, 347–373.

PATTERSON, S.H. & MURRAY, H.H. (1975) Clays. In: *Industrial Minerals and Rocks* (Ed. Lefond, S.J.), pp. 519–586. American Institute of Mining and Metallurgy, Baltimore, Maryland.

PIERRE, C. & FRITZ, B. (1984) Remplacement précoce de gypse par la polyhalite: l'exemple de la bordure sud-orientale de la lagune d'Ojo de Liebre (Basse Californie, Mexique). *Rev. Géogr. Phys. Geol. Dynam.* **25**, 157–166.

PORTERO, J.M., & AZNAR, J.M. (1984) Evolución morfo-tectónica y sedimentación terciaria en el Sistema Central y cuencas limítrofes (Duero y Tajo). *I Congreso España de Geología* **3**, 253–263.

RACERO, A. (1988) Consideraciones acerca de la evolución geológica del margen NW de la Cuenca del Tajo durante el Terciario a partir de los datos del subsuelo. *II Congreso España de Geología*, Simposios, pp. 213–222.

SIFFERT, B. & WEY, R. (1962) Synthèse d'une sepiolite à temperature ordinaire. *C.R. Acad. Sci.* **254**, 1460–1463.

SPENCER, R.J., BAEDECKER, M.J., EUGSTER, H.P., FORESTER, R.M., GOLDHABER, M.B., JONES, B.F., KELTS, K., MCKENZIE, J., MADSEN, D.B., RETTIG, S.L., RUBIN, M. & BOWSER, C.J. (1984) Great Salt Lake, and precursors, Utah: the last 30000 years. *Contrib. Mineral. Petrol.* **68**, 321–334.

STEWART, F.H. (1963) Marine evaporites. *U.S. Geol. Surv. Prof. Pap.* **440-Y**, 53 pp.

STOESSELL, R.K. & HAY, R.L. (1978) The geochemical origin of sepiolite and kerolite at Amboseli, Kenya. *Contrib. Mineral. Petrol.* **65**, 255–267.

STRAKHOV, N.M. (1970) *Principles of Lithogenesis*, vol. 3. Plenum, New York, 577 pp.

UTRILLA, R., PIERRE, C., ORTÍ, F., ROSELL, L., INGLÉS, M. & PUEYO, J.J. (1987) Estudio isotópico de los sulfatos en formaciones evaporíticas mesozoicas marinas y terciarias continentales. Aplicación a la Cuenca del Tajo (abstr.). *II Congreso Geoquim. España*, Soria, pp. 91–94.

VALYASHKO, M.G. (1972) Scientific works in the field of geochemistry and the genesis of salt deposits in the USSR. In: *Geology of Saline Deposits*, pp. 289–311. Proceedings of the Hannover Symposium, 1968. Earth Sci. 7, Unesco, Paris.

VEGAS, R. & BANDA, E. (1982) Tectonic framework and alpine evolution of the Iberian Peninsula. *Earth Evol. Sci.* **4**, 320–343.

WEAVER, C.E. & BECK, K.C. (1977) Miocene of the S.E. United States: a model for chemical sedimentation in a peri-marine environment. *Sediment. Geol.* **17**, 1–234.

Spec. Publs Int. Ass. Sediment. (1991) **13**, 57–74

Lacustrine carbonates: facies models, facies distributions and hydrocarbon aspects

N.H. PLATT* *and* V.P. WRIGHT[†]

**Geologisches Institut, Universität Bern, Baltzerstrasse 1, CH-3012 Bern, Switzerland*
[†]Postgraduate Research Institute for Sedimentology, University of Reading, Reading RG6 2AB, UK

ABSTRACT

Lacustrine carbonates are dominantly biogenic or bio-induced precipitates. While lake basinal carbonates may be modelled in terms of hydrological factors, marginal lacustrine carbonate facies show great variability. At low-energy lake margins, bioturbated micrites dominate. At high-energy margins, lenticular carbonate sands and coated grains are developed. Lakes with low-gradient ('ramp'-type) margins show dominantly marginal lacustrine facies; lakes with high-gradient ('bench'-type) margins display greater development of basinal facies. Progradation of lake margins commonly leads to the deposition of regressive sequences, which may be modelled in four categories according to the morphology and energy of lake margins. These categories are: low-energy 'bench'; high-energy 'bench'; low-energy 'ramp'; high-energy 'ramp'.

Carbonate deposition in lakes is sensitive to climatic and tectonic influences and dependent upon lake hydrology and morphology. Climate controls the rate and nature of biogenic productivity, influences chemical weathering, erosion, and runoff rates in the catchment area, and thus determines carbonate supply.

Lake carbonates may form in a variety of structural settings, especially where the catchment area geology is dominated by carbonates or calcic basement rocks. Tectonic controls determine the rate of subsidence and thus influence sedimentation rates. High-gradient, bench-type lake margins commonly occur at faulted boundaries of rapidly subsiding rift basins where subsidence exceeds sedimentation. At more slowly subsiding rift borders, in larger strike-slip basins, in foreland settings, and in sag basins, low-gradient, ramp-type lake margins dominate. Tectonic factors also influence patterns of continental drainage and the location of clastic sediment input, each of which condition facies distributions within lake basins. Carbonate facies are deposited in areas of low alluvial clastic supply, either in central basin areas away from major bounding faults or prograding thrust fronts, or at basin margins starved of clastic input.

Carbonate lacustrine systems may have hydrocarbon potential. Good source rock prospects occur in deeper, stratified lakes, where anoxic bottom conditions and low detrital input permit the deposition of laminated organic-rich sediments. Reservoirs are more problematic; although reservoirs with primary and secondary porosity may occur in lake marginal bioherms and shoals, most reservoir potential is likely to occur in intercalated alluvial or sublacustrine clastic facies. The extreme lateral variability of lacustrine systems, particularly in rift and strike-slip basins and in high-energy settings, may make accurate prediction of source rock and reservoir facies distributions difficult.

INTRODUCTION

The literature on carbonate lake deposits is scattered, and, despite recent interest in lacustrine sediments, there has been relatively little attempt to synthesize these existing data. Descriptions in the literature of modern carbonate lakes and their deposits are commonly based on studies of small lakes with very short geological histories. In contrast, many ancient lacustrine deposits are thick and laterally extensive, recording deposition in very large, long-lived, tectonically controlled lake basins. Basic summaries include the reviews by Kelts & Hsü (1978), Dean (1981), Dean & Fouch (1983), Eugster

& Kelts (1983), and Allen & Collinson (1986), but no synthesis has been produced to date that includes facies models.

Idealized vertical facies models can be generated for lakes as they can for marine carbonates. For example, higher rates of production in shallow water than in deeper water result in progradation of lake shorelines, leading to the deposition of regressive sequences (a situation directly analogous to marine carbonate sedimentation). However, facies belts in lakes are commonly narrow and lake shorelines are more susceptible to rapid climatically and tectonically controlled fluctuation than marine shorelines. Thus ancient lacustrine deposits characteristically show abrupt and extremely complex lateral and vertical facies changes, making facies sequence analysis particularly difficult.

This paper aims to review lacustrine carbonate facies models, to consider the tectonic settings of carbonate lake systems, and to discuss the potential of lacustrine carbonate facies for hydrocarbon exploration.

HYDROLOGY AND LAKE DYNAMICS

Although carbonates may occur in lakes of different hydrological types, lake hydrology is an important control on mineralogy and facies development. Hydrological classification of modern lake systems permits recognition of two main categories (see e.g. Eugster & Kelts, 1983).

1 Hydrologically open lakes; these have an outlet (they are 'exorheic') and have relatively stable shorelines. Evaporation and outflow are balanced by precipitation and inflow. Lake chemistry is dilute and dominated by meteoric waters; low Mg calcite is the normal precipitate.

2 Hydrologically closed lakes; these have no outflow, and may be perennial or ephemeral, depending upon the climate, which controls the net-water budget (inflow versus evaporation).

Perennial lakes range from saline to dilute, and carbonates occur in high and low salinity examples. The circulation dynamics of perennial lakes are controlled by climate; many show thermal and chemical stratification (see e.g. Kelts & Hsü, 1978). Near the surface, in the epilimnion, solar heating and atmospheric exchange results in warm, oxygenated waters (Fig. 1). Deeper waters of the hypolimnion (below the thermocline) are cooler and denser and may become oxygen depleted. In cold and temperate climates, seasonal cooling and input of cold stream waters leads to turnover of the water column (holomixis). Tropical lakes experience smaller thermal variations and stratification is more stable (provided wind stress is low, see Talbot (1988)). Holomixis then occurs only rarely, owing to high winds or cold weather; such lakes are termed oligomictic. Lakes where turnover is incomplete are termed meromictic. In ancient sequences, analysis of the occurrence and distribution of laminated lacustrine facies may permit recognition of lake stratification and anoxic bottom conditions.

Ephemeral lakes are especially subject to repeated climatically controlled expansion and contraction; periods of low lake salinity after rain are followed by progressive evaporation and hypersalinity. Ancient ephemeral lake carbonates often occur interbedded with evaporites, stromatolites, and fenestral algal carbonates; subaerial exposure surfaces are common.

Interpretation of ancient lacustrine deposits requires consideration of possible changes in the hydrological balance (see e.g. Gore, 1989), since climate and drainage may fluctuate with time,

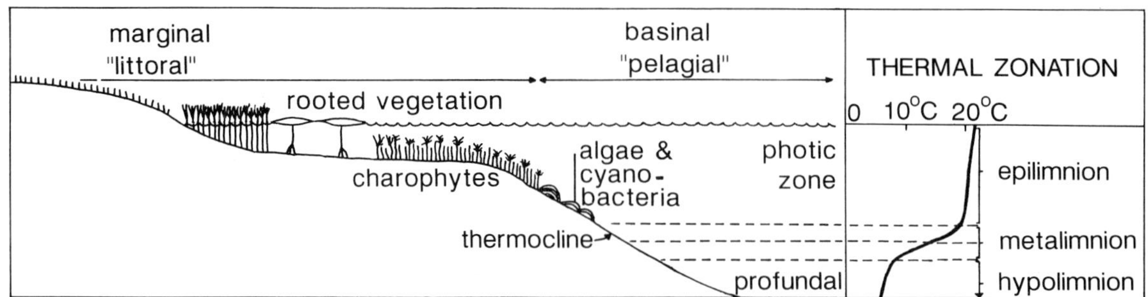

Fig. 1. Subdivision of lacustrine environments, showing two main facies groups: marginal and basinal. Thermal zonation shown for a temperate zone, thermally stratified lake (summer).

changing lake hydrology. This is especially true in the case of shallow lake systems, such as the Miocene Ries Crater of Germany (Füchtbauer *et al.*, 1977) and Paleocene–Eocene Lake Flagstaff of central Utah (Wells, 1983).

CARBONATE LAKE FACIES MODELS

Facies models derived from the scattered literature on recent and ancient carbonate lake systems identify two main facies groups (Fig. 1): lake margin ('littoral') and lake basin ('pelagial'). In ancient sequences, these facies may occur in close association, and are commonly arranged in regressive sequences (see Fig. 2).

Lake margin

Carbonate production is mostly biogenic or bio-induced in shallow water (< 10 m). Inorganic carbonate precipitation may also occur in shallow water owing to warming or wave agitation, or through mixing of Ca-rich stream inflow with carbonate-rich lake waters. Mudstone facies dominate in most shallow lake environments. Shoal or shoreline grainstone facies are poorly developed compared with marine analogues, reflecting the lower wave energies and negligible tidal action in lakes. Littoral carbonates may pass laterally into floodplain or fluvial clastic deposits. Minor evaporites, soils or peats may be deposited at low lake stands.

The shoreline zone is dominated by bioclastic and algal sediments. Littoral environments are commonly colonized by rooted aquatic plants. Charophyte algae may grow on low-energy substrates down to 15–20 m depth (Cohen & Thouin, 1987). Nearshore carbonate productivity occurs through photosynthetic uptake of CO_2, and leads to carbonate encrustation of reeds and charophyte stems and the calcification of charophyte female reproductive structures (oogonia) to form gyrogonites (see e.g. Murphy & Wilkinson, 1980; Dean & Fouch, 1983). Most encrusted material is rapidly broken down virtually *in situ* into carbonate mud. Carbonate is also produced as molluscan or ostracode shell material.

Relatively small bioherms built by green algae and cyanobacteria may occur at lake shorelines and carbonate may also be produced through calcification of shallow water cyanobacterial stromatolites or oncoids (see many examples in Dean & Fouch,

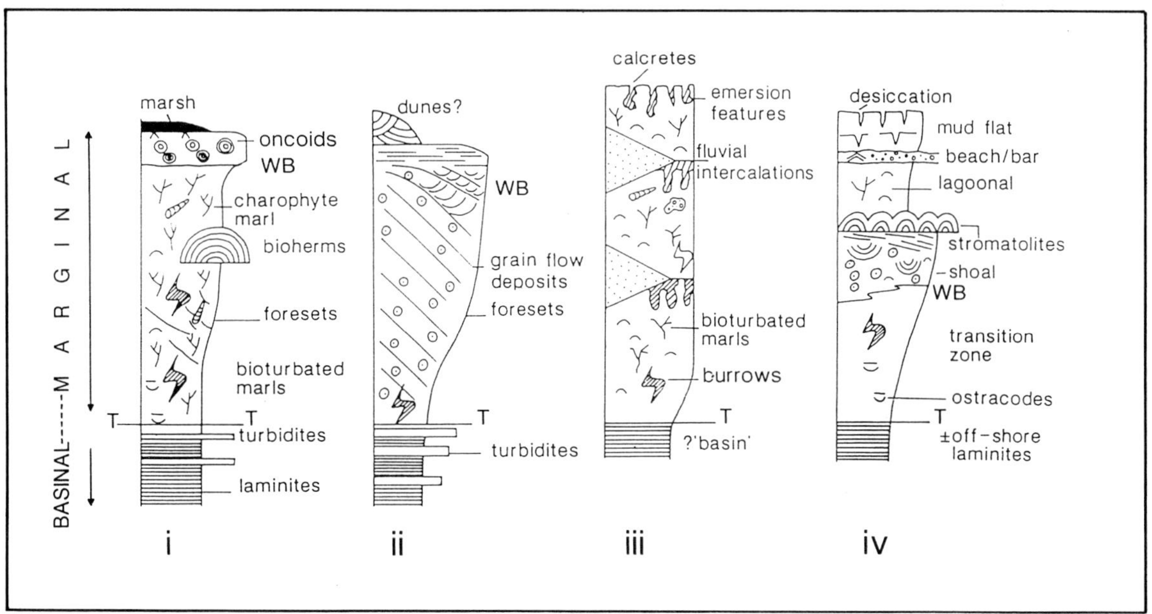

Fig. 2. Facies models for lacustrine carbonates showing typical shallowing upwards, progradational 'regressive' sequences for: (i) low-energy bench margin; (ii) high-energy bench margin; (iii) low-energy ramp margin; (iv) high-energy ramp margin. T, thermocline; WB, wave base.

1983). Recent lacustrine stromatolites show a variety of modern cyanobacteria genera, such as *Rivularia*, *Schizothrix*, and *Scytonema* (Osborne *et al.*, 1978; Pentecost, 1978; Schneider *et al.*, 1983).

The palaeoenvironmental significance of lacustrine stromatolites is unclear since their varying contexts suggest a range of possible lake salinities and chemistries. Fresh to brackish water stromatolites, oncoids, and algal travertines occur in the Plio-Pleistocene of the East African rift (Casanova, 1986) and the Triassic of Greenland (Clemmensen, 1978). Freshwater to mesohaline stromatolites with cyanobacterial layers and *Microcodium* were described recently from the Oligocene of southern France (Casanova & Nury, 1989). Stromatolites are also reported from saline and alkaline playa-lake deposits, e.g. in the Tertiary of southern France (Truc, 1978), the Eocene Green River Formation of the USA (Surdam & Wolfbauer, 1975; Surdam & Wray, 1976) and the Cambrian of South Australia (White & Youngs, 1980; Southgate *et al.*, 1989).

Stromatolite/tufa bioherms also commonly occur near slope breaks on prograding margins (Riding, 1979; Dean & Fouch, 1983). Thrombolitic 'reefs' have been reported from slightly deeper waters (15–50 m) in Lake Tanganyika by Cohen & Thouin (1987) and may be analogous to Mesozoic examples from South Atlantic marginal basins (Bertani & Carozzi, 1985). Marginal bioherms in the Miocene Ries Basin of southern Germany reach thicknesses of 7 m and diameters of 15 m (Riding, 1979). In fresh water, the bioherms are mostly composed of low Mg calcite. In many saline lakes (e.g. Great Salt Lake; Eardley, 1938,1966) the bioherms are aragonitic. Calcitic and aragonitic algal tufas may also occur at springs in alkaline and saline lakes; many examples are recorded, including Mono Lake, California and Pyramid Lake, Nevada (Scholl & Taft, 1964). One variety of tufa found in the Quaternary of the Lahontan Basin, Nevada, as well as in the Mono Lake Basin, is now composed of calcite but contains aggregates of elongate pseudomorphs. Shearman *et al.* (1989) suggested that these 'thinolitic' tufas may originally have precipitated during low temperature episodes as the mineral ikaite ($CaCO_3.6H_2O$).

Lake basin

Sediments of deeper, open water lake settings lack evidence of *in situ* vegetation. Although rates of carbonate precipitation are high, sedimentation rates are low. In the pelagial zone, carbonate is supplied mainly from bio-induced precipitation by phytoplankton, and from resedimentation, although inorganic precipitation may also occur during lake turnover. The fine siliciclastic content is commonly higher than in marginal lake deposits (see Fig. 2).

Basinal facies are strongly controlled by lake dynamics. Under permanent stratification, anoxia develops and organic-rich facies may be deposited. Modern Lake Tanganyika is anoxic below 150–250 m (Cohen, 1989) and bottom sediments contain between 7 and 11% organic carbon. Absence of bioturbating bottom fauna leads to the deposition and preservation of basinal 'laminites', with alternating organic-rich and carbonate-rich laminae reflecting seasonal variation in phytoplankton production. During algal blooms, thin layers of carbonate are precipitated as a result of biological uptake of CO_2. Later settling out of the planktonic organisms leads to sedimentation of an organic-rich layer (for further details see Kelts & Hsü, 1978). In shallow or polymictic lakes that are not permanently stratified, poorly or non-laminated facies predominate, since periodic bottom oxygenation results in bioturbation of bottom sediments, leading to oxidation of organic matter and destruction of laminite fabrics.

MORPHOLOGY AND ENERGY OF LACUSTRINE MARGINS

Lake depth and the morphology of lake margins control the development of lake basin and lake margin facies. Lakes with high-gradient, 'bench' type margins commonly show extensive development of lake basin facies. In contrast, lakes with low-gradient, 'ramp' type shorelines are dominated by marginal lacustrine facies. Very shallow lakes may not have a definable deeper water, basinal facies. Lake morphology may be controlled tectonically, and some deep rift lakes (e.g. Lake Tanganyika; Cohen, 1989) have a steeply inclined faulted bench margin and a ramp margin on the opposite side. High-energy lake margins are characterized by lenticular carbonate sand bodies, sometimes cross-bedded or with coated grains, while bioturbated micrites occur at low-energy margins.

This permits subdivision into four types of lake margin: (i) low-energy bench; (ii) high energy, wave dominated bench; (iii) low-energy ramp; (iv) high energy, wave dominated ramp.

Progradation of marginal facies results from higher rates of carbonate production in shallower water than in basinal areas and typically leads to the deposition of regressive sequences. This permits the construction of vertical facies models for carbonate lakes according to the morphology and energy of their margins. Figure 2 shows idealized sequences for each type of lake margin.

Steep-gradient 'bench' margins — low energy

Carbonate sediments at this type of lake margin are commonly fine grained and show low Mg calcite mineralogy. High productivity by shallow water benthic plants in the littoral zone results in build-up of the shore zone and its progradation into the lake. Deeper water, bench slope facies may show evidence of resedimentation and downslope transport.

Lake Littlefield, Michigan, is a small, temperate carbonate lake up to 20 m deep (Fig. 3) with shoreline benches (Murphy & Wilkinson, 1980). This dimictic, stratified lake shows prograding bench-type margins made up of a gently sloping (2°) bench platform in 1.5 m of water and a 30° dipping bench slope. Sediments coarsen upwards from micritic laminites (profundal and sublittoral zones) to bioclastic sands and gravels (littoral zone).

Profundal zone sediments comprise laminated silty ostracodal pelmicrites. Pyrite is present, but the occurrence of a benthic fauna indicates oxygenation during mixing. Thin turbidites or grain flow deposits contain resedimented littoral material. The lower part of the bench slope lies just above the hypo-limnion and shows finely laminated gastropod and ostracod micrites with bivalves. Slumps and resedimented units are evident in cores. The upper bench slope displays bioturbated sandy micrites with encrusted charophyte stem fragments. The shallow bench top is covered by cyanobacterial oncoid ('pisoid') gravels. The lake is bordered by a largely subaerial peat swamp.

A similar coarsening upward, prograding lake bench shoreline sequence occurs in Sucker Lake, Michigan (Treese & Wilkinson, 1982). Encrusted charophyte sands occur in the littoral zone. Bench slope sediments comprise finely laminated, resedimented charophyte−gastropod sands, which may be turbiditic, and laminated silts, which represent suspension deposits. Allochthonous blocks of bench carbonates are transported by slides into deeper water.

An ancient, large-scale example of a low energy, steep margin system comes from the Lower Cretaceous Peterson Limestone of Wyoming and Idaho, which covers $20\,000$ km^2 and is up to 60 m thick (Glass & Wilkinson, 1980). The abundance of micrites indicates mainly low depositional energies, but extensive resedimentation from steeper margins is suggested by the presence of graded, silty micrite beds, slumped horizons, and intra-formational conglomerates.

Steep-gradient 'bench' margins — high energy (wave dominated)

Lakes with this type of margin commonly show

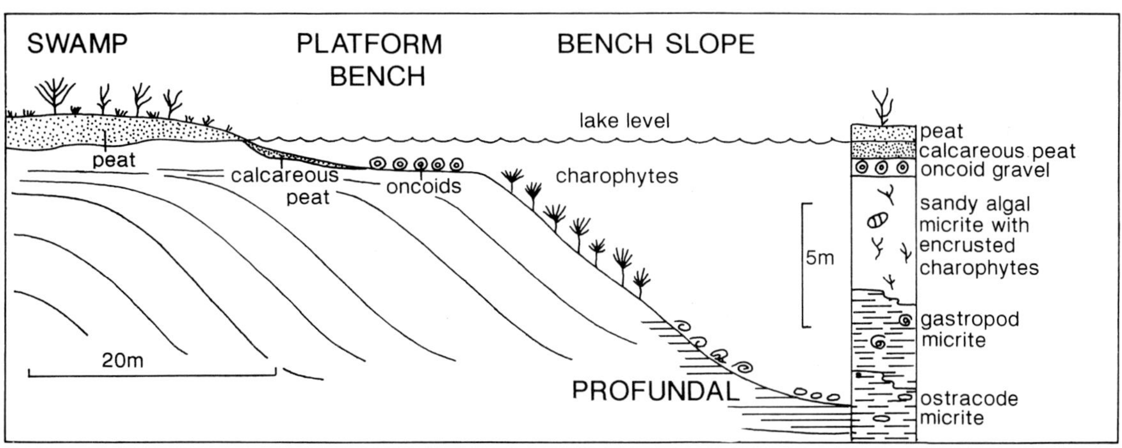

Fig. 3. Low energy, bench-type lake margin: modern example from Lake Littlefield, Michigan, USA (modified after Murphy & Wilkinson, 1980).

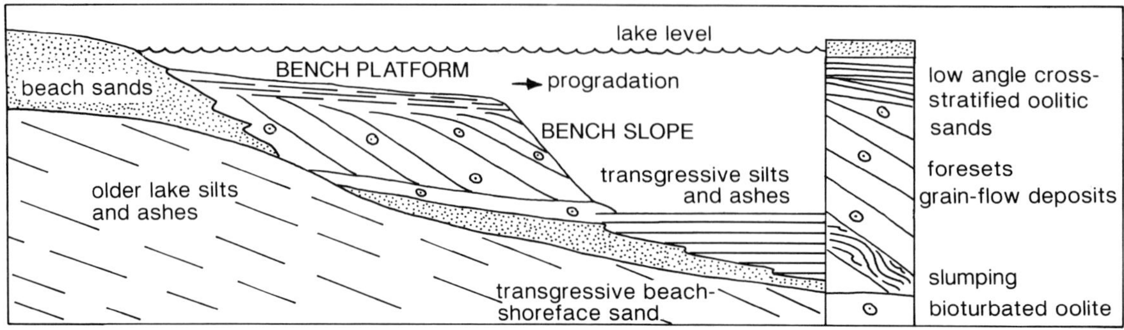

Fig. 4. High energy, bench-type lake margin: ancient example and idealized sequence from the Pliocene Shoofly Oolite, Glenns Ferry Formation, Snake River Plain, USA (after Swirydczuk *et al.*, 1980).

marked facies asymmetry owing to variable wave energy. Bioturbated, largely structureless, lime mudstones dominate at protected low-energy margins while packstones and grainstones occur in higher energy bench settings. Coarse-grained calcarenites, cross-bedded oolites, and shelly lags occur where margins are open to wave activity, and small-scale carbonate bioherms may also develop. In saline/alkaline lakes, ooids may be aragonitic, as found on wave-built benches around Lake Bonneville (Gilbert, 1885,1890).

An ancient example of a high energy, bench-type lake margin sequence comes from the Pliocene Glenns Ferry Formation Shoofly Oolite of the northeastern USA (Swirydczuk *et al.*, 1979,1980; Fig. 4). This 35-m-thick oolite deposit is traceable over an outcrop of 45 km. The oolite contains three transgressive sequences, which start with medium-grained oolitic carbonate bench deposits up to 12 m thick. The ooids formed on the wave-influenced bench surface, but were redeposited in large-scale foresets dipping lakeward at an average of 26°. These foresets record progradation of the bench slope. Massive units up to 1 m thick with thickly coated, sometimes phosphatized, ooids rest on a basal erosion surface, which truncates the topsets of the underlying bench platform facies. Reverse grading near the top of the unit indicates grain flow and avalanching; slumps and dish structures occur with normally graded laminae (turbidites) near the base.

Facies asymmetry is evident from the lacustrine carbonates of the mid-Tertiary Camp Davis Formation, northeastern Wyoming (Davis & Wilkinson, 1983). Facies vary *along* depositional strike, parallel to the palaeoshoreline. Wind

exposure was affected by shoreline relief and configuration, so that rooted charophyte marls formed in protected bays between alluvial fans and grainstone facies were deposited along wave-exposed headlands and at alluvial fan fronts.

Low-gradient 'ramp' margins — low energy

Lakes with low-gradient margins are commonly shallow and are rarely stratified. Shallow-water limestones (rich in charophytes, including stems) show low detrital contents, probably reflecting baffling and trapping of clastic material in lake marginal marsh zones. Owing to the low gradients, lake marginal facies may be extensive and small-scale water-level fluctuations cause exposure of large areas. Minor evaporites and intercalations of fine alluvial deposits are common. The tops of regressive sequences commonly show evidence of subaerial exposure produced on emergence of the lake deposits at low lake stands. Pedogenesis is evident from calcrete textures, brecciation, and 'pseudo-microkarst' (Plaziat & Freytet, 1978) — small irregular cavities and fillings produced by plant bioturbation and minor dissolution.

These facies are typical of many Mesozoic and Tertiary lacustrine sequences of southern Europe (e.g. Freytet & Plaziat, 1982; Cabrera *et al.*, 1985; Platt, 1989a). Similar deposits occur in the Paleogene of Utah (Lake Flagstaff; Stanley & Collinson, 1979; Wells, 1983), and in the Lower Cretaceous Draney Limestone of Wyoming and Idaho (Brown & Wilkinson, 1981). Other examples include the Oligocene of western Portugal (Azeredo & Galopim de Carvalho, 1986) and southern England (Marshall *et al.*, 1988).

The Upper Cretaceous to Paleogene of southern France (Languedoc; Freytet & Plaziat, 1982; Freytet, 1984) contains over 2 km of continental sediments, including intervals with shallow lacustrine, palustrine, and pedogenic carbonates. Laminated 'basinal' facies are scarce and evaporites are relatively rare, suggesting dominantly low salinities in holomictic or polymictic lakes. Typical regressive sequences start with resedimented pellet and intraclastic grainstones and packstones. These are overlain by open lake micrites, with gastropods, molluscs, and charophyte stems, and gyrogonites. Above these are brecciated limestones showing pedogenic fabrics, such as mottling, microkarst cavities, and *Microcodium*. Some sequences are incomplete or superposed; pedogenic features, such as root tubules, may cut several sequences.

The basal Cretaceous Rupelo Formation of the western Cameros Basin, northern Spain (Platt, 1989a), is up to 100 m thick and comprises a variety of palustrine and open lake carbonates (Fig. 5). Palustrine facies exhibit abundant pedogenic features, such as mottling, desiccation brecciation, microkarst cavities, and root structures. Open lake facies contain ostracods, charophyte stems and gyrogonites, gastropods, and sauropod bones. Laminated basinal facies are rare, suggesting mostly shallow, oxygenated conditions. Intraclast grainstones and conglomerates indicate reworking during storms and at low lake stands. A silicified evaporite horizon near the top of the sequence records a brief period of hypersalinity.

A Palaeozoic example of a low-energy shallow gradient system comes from the Mississippian

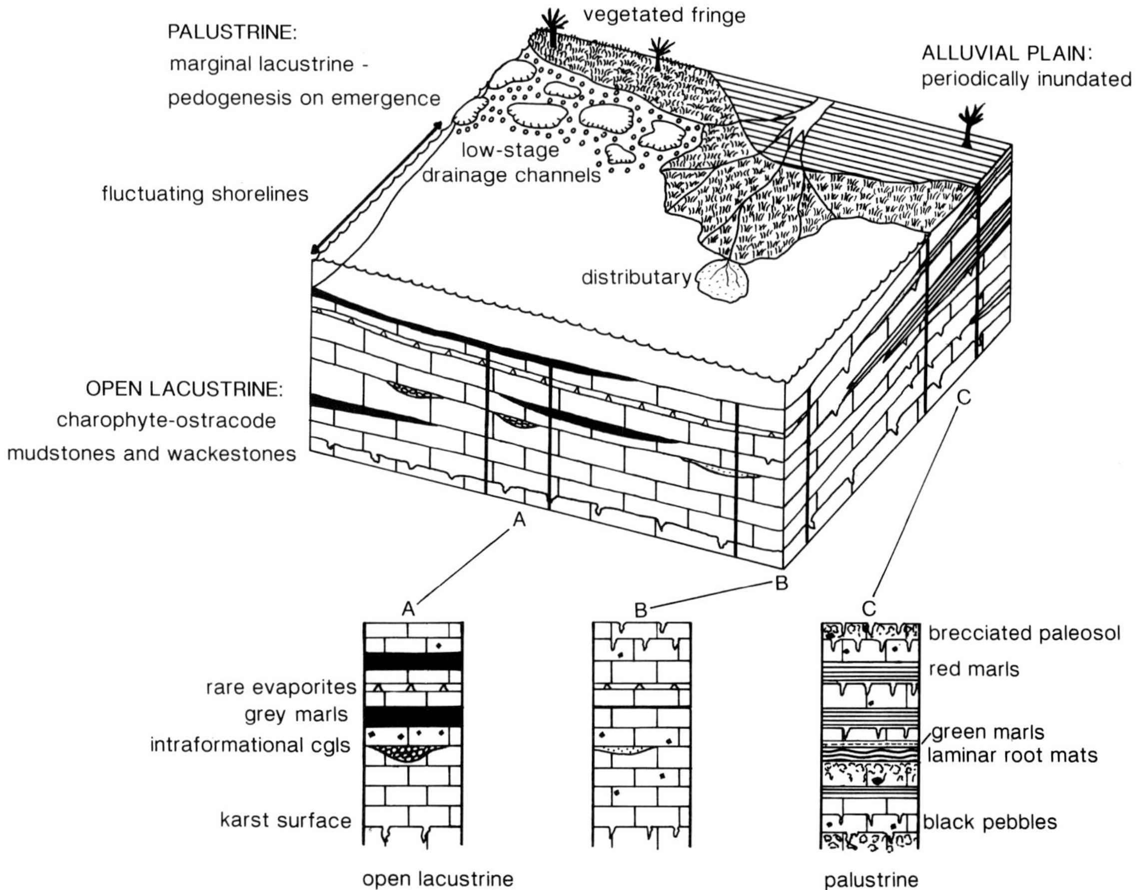

Fig. 5. Low energy, ramp-type lake margin: example from basal Cretaceous of central northern Spain with idealized ancient sequences (modified after Platt, 1989a).

Burdiehouse Limestone of Central Scotland (Loftus & Greensmith, 1988). The sequence is mud-dominated, with charophytes, small cyanobacterial bioherms, and impersistent shoreline oolitic pack-stones. The lake was deep enough for bottom anoxia to occur and a major oil shale unit is also present. However, evidence of slumping and resedimentation is rare, suggesting that lake margins were gently shelving.

Low-gradient 'ramp' margins — wave influenced

At strongly wave influenced, low-gradient margins, winnowed grainstones dominate and nearshore bars may develop. Modern examples include the Great Salt Lake of Utah, which has shoreline oolitic deposits with offshore bars and extensive bioherms (Eardley, 1938,1966; Sandberg, 1975,1980; Dean & Fouch, 1983). A well-documented ancient analogue occurs in Late Cretaceous to Eocene sediments of the Green River Formation and the Flagstaff Member of the Watsatch Formation in the Uinta Basin of northeastern Utah (Fig. 6). Facies deposited include biomicrites, calcarenites, sandstones, oil shales, evaporites (trona, halite), and stromatolitic limestones. Open lake facies up to 900 m thick consist of dark, organic-rich laminated carbonates and claystones (Ryder *et al.*, 1976). Low Mg calcite laminae are probably photosynthetically induced precipitates, and alternate with thin kerogen laminae, which represent microbial oozes. Organic matter also occurs as finely dispersed kerogen in unlaminated beds and as algal sapropels with *Botryococcus*. These are overlain by grey, horizontally bedded, fossiliferous limestones, and the shallowing upward cycle is completed by cross-bedded, bioclastic, oolitic–pisolitic grainstones with wave and current ripples, probably deposited in less than 10 m of water. Williamson & Picard (1974) assigned the shallow water carbonates of the Green River Formation to a complex system of barred shorelines and protected lagoons.

Another ancient carbonate-bearing stratified lake sequence is the 4-km-thick Devonian Caithness Flagstone Group (Donovan, 1975; Rogers & Astin, this volume) of the Orcadian Basin in northeastern Scotland. Donovan's facies model shows a ramp-type margin. Lake margin facies consist of fenestral limestones, stromatolites, and tufas. The offshore facies consists of thinly bedded laminites probably deposited below the thermocline. These are interpreted as the products of seasonal algal blooms. The presence of pyrite and well-preserved fish indicate anaerobic bottom conditions. Oil shales also occur in the offshore deposits.

FACIES DISTRIBUTIONS

Having discussed the processes of carbonate sedimentation in lakes, and having introduced some basic facies models for lake carbonate deposition, it is now instructive to consider the controls on facies distributions within lake basins. Facies distributions reflect carbonate and clastic supply, topography, and bedrock composition of the catchment area, as well as the tectonic setting of the lake system.

Carbonate and clastic supply to lake systems

Climate controls the rate and nature of organic productivity, and influences runoff, erosion, and

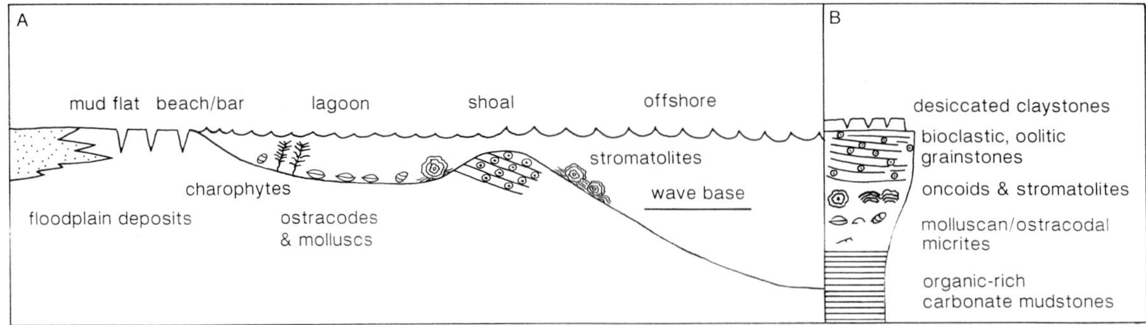

Fig. 6. High energy, ramp-type lake margin. (A) Idealized example from the Green River Formation (based on data in Williamson & Picard, 1974). (B) Ancient sequence from the Flagstaff Member of the Watsatch Formation, Uinta Basin, northeast Utah (based on data in Ryder *et al.*, 1976).

chemical weathering rates in the catchment area, thus determining the carbonate and clastic supply. Carbonate minerals may be supplied to lakes through detrital input of carbonate lithoclasts by rivers or shoreline erosion, but more commonly Ca^{2+}, Mg^{2+}, and HCO_3^- arrive in solution via surface or ground-water inflow. Drainage area geology and hydrology are thus critical controls on the sites of lacustrine carbonate sedimentation.

Areas of carbonate sedimentation in lakes are characterized by relatively low clastic input. Since carbonate sedimentation rates are commonly lower than those for clastic deposits, local or temporary increase in clastic supply (e.g. through changes in climate, in the intensity of tectonic activity or in the drainage pattern) leads to the deposition of clastic facies, or of impure carbonates with increased detrital content. Clastic intercalations are common within deeper perennial lakes and may represent seasonal varves (deposited in times of increased runoff) or turbidites (mostly in temperate and tropical lakes). Clastic horizons within shallow ephemeral lake sequences may record fluvial progradation, shoreline retreat owing to progressive basin infilling, or flash floods during storm rains.

Topography and bedrock composition of catchment area

Lacustrine deposits in tectonically active basins commonly occur above regional unconformities, which may be pronounced towards faulted basin margins. The bedrock geology of underlying and catchment areas strongly influences lake sedimentology and chemistry. Yuretich (1982) reported that this factor accounted for 50% of facies variation in Lake Turkana in the East African rift. Large catchments give greater runoff, and those with high relief may generate abundant coarse clastics. Weathered acidic igneous and metamorphic rocks and coarse clastics yield sand and mud. In contrast, limestones largely dissolve, yielding mud, and calcium carbonate in solution. Lakes in limestone terrains have larger areas of fine-grained sedimentation than those in granite terrains, but surface runoff is low owing to the porosity of the bedrock. Calcium carbonate may also be derived from calcic metamorphics and volcanics, although surface runoff over unweathered volcanics may likewise be limited, as at Lake Magadi (Eugster, 1986).

An example of the control of basement geology and palaeotopography on lacustrine carbonate development comes from the Triassic of South Wales, where marginal lacustrine carbonates and evaporites occur with alluvial fan conglomerates and playa-lake mudstones (Tucker, 1977,1978). Areas draining hills of Carboniferous Limestone received high dissolved and detrital carbonate supply but relatively low siliciclastic input, favouring deposition of lacustrine carbonates. In contrast, in areas near fans draining non-carbonate basement rocks, clastic supply was much higher, so that red dolomitic mudstone facies dominate and fewer limestones occur.

Another example of the control of catchment area geology on lake sedimentation comes from the Tertiary rift basins of China and the South China Shelf. Over 2 km of lake deposits are present within some of these extensional basins (Mitchell & Reading, 1986). However, development of carbonates is limited, particularly on the South China Shelf. This largely reflects the high clastic supply to the lake basins, which were mostly set in granitic terrain with few carbonates (Sladen, pers. comm., 1986, 1987).

Tectonic setting of carbonate lakes

The development of most larger long-lived lake basin sites is tectonically controlled. Carbonate lakes occur in a variety of tectonic environments: in rift basins, in strike-slip settings, in foreland basins, in internal drainage intra-continental 'sags', or on glaciated cratons and in mountain belts. Subsidence rates must be greater than the rate of deposition for the establishment of a permanent stable lake basin.

Faulting controls subsidence and the relative subsidence of different fault blocks plays an important role in determining both the most favoured site for a lake basin and the location of deeper or hydrologically isolated environments favourable for the deposition of organic-rich and evaporitic facies. While rapid tectonic subsidence may lead to the development of localized clastic traps, faulting may also control the formation of clastic-starved highs favourable for the development of freshwater carbonate build-ups, or of clastic-starved lows where isolated evaporitic or carbonate lake basins may form. The three-dimensional geometry or 'architecture' of the various lacustrine facies will depend primarily on the local pattern and intensity of tectonic activity and the location of stream and sediment input.

Rift basins

Longitudinal, basin-bounding normal faults and transverse wrench or 'transfer' faults compartmentalize rifts into small blocks (Gibbs, 1984; Rosendahl et al., 1986). Lake systems in rifts (e.g. East African rift; papers in Frostick *et al.*, 1986; Tiercelin, this volume) show complex facies distributions (Fig. 7A). Clastic input is strongly localized by tectonic factors, and adjacent rift segments may show different subsidence rates, volcanism, hydrology, water chemistry, and biological productivity. Neighbouring blocks may display different facies evolution, particularly during periods of evaporation or partial anoxia, as faulting may facilitate the formation of hydrological barriers, or 'sills', between the sub-basins.

Facies distributions in half-graben systems are commonly non-symmetrical, reflecting the pattern of tilted fault blocks (see, for example, Scholz, Rosendahl & Scott, 1990). Footwall uplift promotes the formation of coarse detrital belts at the base of fault-scarp slopes (Leeder & Gawthorpe, 1987). Lake carbonates may form only in areas where clastic supply is low. In rifts with major fluvial inflow from the footwall margin, this may be in the basin centre, away from boundary faults. However, in many tectonically active rift systems, major drainage is directed outwards from graben crests, so that clastic input is low and fine-grained deposits, including lacustrine carbonates and evaporites, may dominate over wider areas. In such cases, lacustrine facies may form directly above the locus of maximum subsidence and may thus extend to areas closer to faulted basin margins (Blair & Bilodeau, 1988; Mack & Seager, 1990).

Lakes developed in rift basins may be shallow or deep, depending upon the relative rates of subsidence and sedimentation. If subsidence exceeds sedimentation, then high-gradient, bench-type margins may develop, particularly where clastic input is low. A modern example is reported from Lake Tanganyika, which is up to 1500 m deep (Cohen, 1989) and displays laminated organic-rich deposits laid down in deep water, as well as a wide variety of shallow water carbonate facies (Cohen & Thouin, 1987). Deep ancient rift lakes also formed in the Proto-South Atlantic during the Early Cretaceous (Brice *et al.*, 1980).

In contrast, if sedimentation keeps pace with subsidence, then substantial relief is not generated and shallow lakes with ramp-type margins develop. Shallow lake carbonates occur in the Late Jurassic to Early Cretaceous of Iberia. Lacustrine deposits in the Upper Oxfordian to Lower Kimmeridgian of the Portuguese Lusitanian Basin (Wright & Wilson, 1985) and the Berriasian of the Cameros Basin in Spain (Platt, 1989a) formed during relatively slow early phases of rifting. Sedimentation broadly kept pace with subsidence so that low-gradient, ramp-type margins dominate. Both deposits lie unconformably upon thick marine Jurassic carbonate successions and show fine-grained, carbonate lake facies with low detrital content. Later increase in clastic supply (see Platt, 1989b; Hill, 1989) may reflect uplift and erosion of basement massifs. In both areas, relatively slow subsidence prevented the generation of significant depositional relief. Nevertheless, abrupt thickness and facies variations reflect strong tectonic controls on differential subsidence during these early rifting stages (see e.g. Platt, 1989c, 1990a).

Strike-slip basins

Lakes are common in strike-slip terrains, where high subsidence rates are characteristic (see examples in Ballance & Reading, 1980; Biddle & Christie-Blick, 1985). Complex fault patterns provide many enclosed, and sometimes isolated, potential lake basin sites. Basin shape may vary from wedge-shaped or elliptical in areas of complexly anastomosing or curved faults to simpler rectangular or rhomboidal geometries of pull-aparts associated with offset faults. Juxtaposition of uplifted sediment source terrains and rapidly subsident basins (e.g. Ridge Basin: Link & Osborne, 1978; Dead Sea Basin: Ten Brink & Ben-Avraham, 1989) commonly leads to high clastic sediment supply and consequently to elevated sedimentation rates close to bounding faults (Fig. 7B). However, lateral facies changes are abrupt, with potential for marked facies asymmetry and varying styles of sedimentation in adjacent basins.

Lacustrine deposition in small strike-slip basins is generally dominated by mudstones or muddy carbonates (e.g. in the Ridge Basin: Link & Osborne, 1978) and/or diatomitic facies (e.g. Cenajo Basin: Bellanca *et al.*, 1989; Cerdanya Basin: Cabrera *et al.*, 1988). Carbonate facies are relatively rare, although they may occur where climate is arid (e.g. saline lake, laminated carbonates occur associated with evaporites in the Dead Sea Rift: Zak & Freund, 1981; Manspeizer, 1985). However, lake carbonates are more common in the central areas of

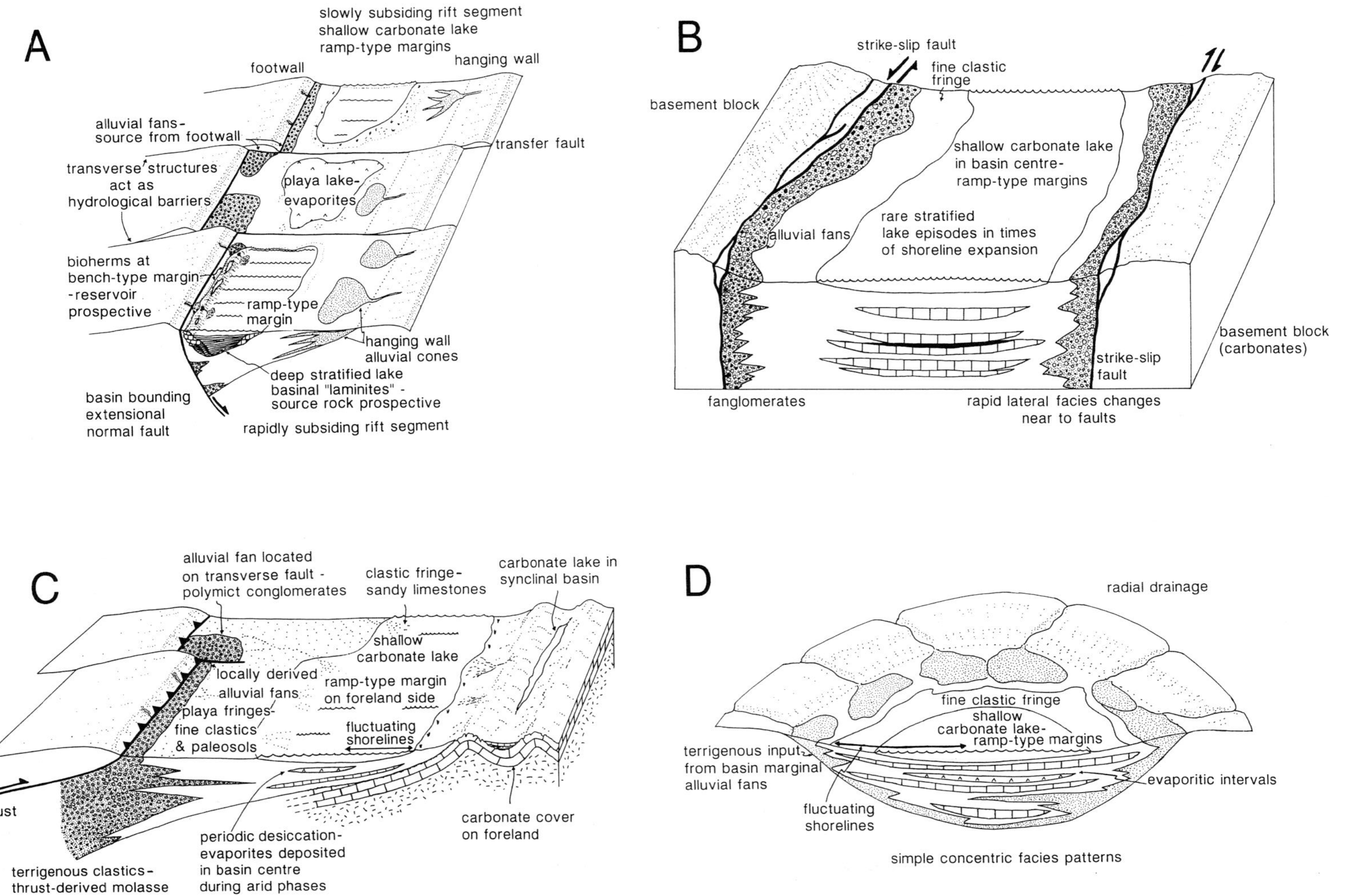

Fig. 7. Models showing tectonic settings of lacustrine carbonate facies. (A) Rift/half-graben basin complex; (B) large strike-slip basin; (C) foreland basin complex (also showing development of synclinal basin in carbonate foreland); (D) sag basin.

larger strike-slip basins, where low-gradient lakes with ramp-type margins may develop. Despite the low-gradient, 'basinal' laminated facies may occur at high lake stand in the centre of large basins.

Another example may come from the Orcadian Basin of northeastern Scotland. The Orcadian Basin was about 100 000 km^2 in size, and displays a continental sequence 4 km thick deposited within 10 Ma. Although classically considered as a strike-slip basin, recent opinion favours a rift origin for this Devonian Basin (Astin, 1990). For much of the period, alluvial clastic supply was high and lakes occupied less than 20 000 km^2 in the basin centre. Only in one high-stand episode were lacustrine conditions widely developed (Donovan, 1975; Janaway & Parnell, 1989). In this interval, lower clastic input and high carbonate supply from planktonic algae and seasonal precipitation led to the deposition of shallow water stromatolitic and fenestral limestones at low-gradient ramp-type lake margins. Laminated limestones and dolomites formed offshore. Organic-rich facies laid down in central basin areas form important hydrocarbon source rocks (Peters *et al.*, 1989). Detrital and dissolved carbonate was probably derived from erosion of Cambro-Ordovician carbonates and of metamorphic and igneous rocks in the Caledonian mountains bordering the basin.

Foreland basins

Lacustrine carbonate deposits are relatively common in foreland basins. Lake basin sites form in topographic depressions developed in advance of propagating thrust fronts, where clastic supply may be high. Proximal areas are generally dominated by coarse-grained alluvial sedimentation (Fig. 7C), with fine-grained sediments occurring in more distal areas. However, in times of thrust emplacement, subsidence is commonly very rapid, so that coarse-grained deposits may be laid down at depocentres adjacent to the thrust front. Fine-grained sedimentation may then dominate over the remainder of the basin (Heller *et al.*, 1988). Those distal parts of the basin where drainage is from carbonates in the foreland may be ideal sites for lacustrine carbonate deposition. Sedimentary gradients are generally low on the foreland side, where gently shelving, ramp-type lake margins dominate.

In the Narbonnais of the northeastern Pyrenees in southern France, Late Cretaceous to Paleogene, lacustrine/palustrine carbonates and evaporites (Freytet & Plaziat, 1982) record deposition in a closed shallow lake with low-gradient, ramp-type shorelines. Carbonates in the basin centre pass laterally into fluvial clastics at the basin margins, where alluvial fans drained the Pyrenees to the south and the Montagne Noire in the north. Abundant carbonate supply reflected erosion of the Mesozoic cover of the Pyrenees.

South of the Pyrenees in Spain, the Ebro Basin formed a large, complex, foreland basin during the Paleogene. Alluvial fan conglomerates at both borders of the basin (Anadón *et al.*, 1985; Hirst & Nichols, 1986; Cabrera & Sáez, 1987) reflect thrusting at the northern basin margin and thrusting and strike-slip faulting at the southern margin. In times of low clastic supply during the Oligocene to early Miocene, palustrine/lacustrine marls and limestones, with fine distal alluvial clastics, evaporites, and lignites, were deposited in the basin centre, and sometimes almost up to the southern basin margin. Calcium carbonate supply was from erosion of Mesozoic carbonates in the Catalanides and Iberian Chains and of calcic and carbonate rocks in the Pyrenees.

The Lower Freshwater Molasse of Switzerland is dominated by alluvial fan and fluvial deposits derived from the Alps (Heim, 1919; Trümpy, 1980; Homewood *et al.*, 1986). In distal areas, high rates of carbonate supply from erosion of the Jura Mountains in the foreland permitted deposition of marginal lacustrine carbonate mudstones, laminated open lake deposits, and evaporites (Jordi, 1955; Kissling, 1974; Platt, 1990b) in periods when clastic supply from the Alps was reduced. The carbonates probably reflect deposition in a closed-basin lake with a ramp-type northern margin. Nearer the Alps towards the south, the limestones show a progressively higher clastic component.

Minor lake carbonate deposition may also occur in small synclines in folded terrain where the substratum is dominated by carbonate lithologies (Fig. 7C). Such synclines are a feature of foreland fold and thrust belts, as in the Swiss Jura Mountains, where both the Oligocene and Miocene contain lacustrine carbonates (Trümpy, 1980). Lacustrine carbonate deposition coincided with erosion of marine limestones exposed in flanking anticlines and occurred intermittently during times of reduced clastic input.

Sag basins

Sag basins are extensive, commonly symmetrical, gently shelving crustal sags. Such sags generally show relatively low subsidence rates, favouring the development of fairly shallow, low-gradient lakes, commonly characterized by wide, climatically controlled shoreline fluctuations. Lakes developed in such sags are likely to show 'ramp'-type margins, with simple concentric facies patterns (Picard & High, 1972,1981). Lakes are surrounded by alluvial/aeolian deposits, which pass into shoreline clastics and/or carbonates and then into offshore deposits in central basin areas (Fig. 7D). Modern examples include Lake Chad (Burke, 1976; Reading, 1982) and Lake Victoria, which has an area of 70 000 km^2 and a maximum depth of less than 80 m (Kendall, 1969) but displays few carbonates (Talbot, pers. commun, 1990).

In the western USA, carbonate lakes developed in the Early Cretaceous (Glass & Wilkinson, 1980), and Paleocene–Eocene (Ryder *et al.*, 1976; Surdam & Stanley, 1979; Yuretich, 1989). Although this area forms part of the Rocky Mountains foreland basin complex (Cross, 1986), maximum subsidence was located in a large, subcircular crustal sag well to the east of the Rocky Mountains. In the Tertiary, clastic input was apparently controlled by local tectonics to a lesser degree and facies belts were nearly concentric (Picard & High, 1972,1981).

HYDROCARBON ASPECTS

Fouch & Dean (1982, p. 87) noted that although lacustrine deposits are common in many parts of the world, relatively few have been the focus for exploration for oil and/or gas. However, source rock prospects are commonly good (see Powell, 1986; papers in Fleet *et al.*, 1988). Many carbonate lakes have high organic productivity with good preservation of organic matter, particularly oil-prone kerogens of algal origin. Many lakes are thermally or chemically stratified; where stratification is permanent (as in many deeper or saline low-latitude lakes) anoxic bottom conditions allow accumulation of organic-rich sediments (see review by Kelts, 1988).

Lacustrine oil shales occur in many Chinese Tertiary basins (Powell, 1986; Watson *et al.*, 1987; Fu Jiamo *et al.*, 1988), but catchment geology constraints dictated high clastic supply (see above) so that carbonates occur only rarely (e.g. in Bohai Basin: Kelts, pers. comm., 1989). In contrast, stratified lakes developed in carbonate terrains (e.g. in Late Tertiary rifts of northeastern Spain: Anadón *et al.*, 1989) may show laminated oil shales interbedded with open lacustrine basinal and resedimented carbonates.

Localized strong subsidence in rift basins may exceed sedimentation rate, favouring the formation of deep, stratified lakes with basinal sediments of high source potential (Robbins, 1983; Anadón *et al.*, 1988, this volume). Lacustrine source rocks may be deposited in great thicknesses in rifts (up to 700 m in some Chinese and South Atlantic basins: Brice *et al.*, 1980; Powell, 1986; possibly several thousand metres beneath modern Lake Tanganyika; Cohen, 1989). Although deep water facies are generally thickest on the downdip sides of asymmetrical half-grabens (Cohen, 1989), major exploration problems are caused by lateral variability in rift basins where deep lake systems may be adjacent to ephemeral ones, and predictability is low.

Lakes in small, high-gradient strike-slip basins are likely to display mostly fine clastic facies (e.g. Ridge Basin: Link & Osborne, 1978). However, some larger lakes in large, low-gradient, strike-slip basins may develop stratification, permitting the deposition of organic-rich deposits in central basin areas. An example occurs in the Orcadian Basin, northeastern Scotland, where Devonian offshore carbonate and detrital laminites deposited below wave base in the basin centre are probably a source of oil in the North Sea Beatrice Field (Duncan & Hamilton, 1988; Peters *et al.*, 1989). In contrast, most lakes in foreland basins are shallow. However, potential gas-prone, terrestrial source rocks, such as coals or lignites, may occur in association with lacustrine carbonates, as in the Paleogene of the central Ebro Basin (Cabrera & Sáez, 1987). Laminated fine carbonate 'oil-shale' deposits may also occur associated with shallow lacustrine carbonates and evaporites, as in the Tertiary Green River Formation of the western USA, although there is debate as to whether these oil shales represent deposition in shallow, but anoxic, brackish–saline playa-lakes or record deeper lake conditions (see discussion in Talbot, 1988).

Only a few low-gradient lakes in sag basins are deep enough for the development of potential source facies (e.g. Lake Victoria, which is up to 80 m deep, Kendall (1969)).

The fine-grained nature of many lake carbonates favours low primary porosities, and (with the exception of molluscan bioclasts) most carbonate is low Mg calcite, providing little prospect of secondary porosity formation. However, in saline lakes, such as Great Salt Lake, much of the carbonate produced (ooids, bioherms, physico-chemically produced mud) is aragonitic (Sandberg, 1975,1980) and secondary (mouldic) porosity may develop on burial. Resedimented units are generally thin, so that porosity in lake carbonates themselves appears to be most common in marginal shoal deposits with solution-enhanced porosity, as in reservoirs noted in Barremian lacustrine shoal carbonates from the Cabinda Basin, Angola by Brice & Pardo (1980).

The 200–1500 m thick Lower Cretaceous Lagoa Feia Formation of the Campo Basin, offshore Brazil (Bertani & Carrozi, 1985) has reservoirs in packets of lacustrine molluscan (aragonitic) limestone with good primary and secondary porosity, the latter formed through meteoric diagenesis on exposure. Porous cyanobacterial or green algal bioherms of low-energy bench margins are also potential reservoirs. However, while shoal carbonates may be extensive laterally and show tabular geometries, bioherms tend to occur in discontinuous linear belts, commonly along faulted borders (Cohen, 1989).

Most potential reservoirs in lake deposits are likely to occur in intercalated clastic facies, including marginal lacustrine delta-channel sandstones, as well as shoreline, shoal, mouth-bar and sublacustrine fan sediments. Clastic reservoir distribution reflects tectonic control. In tectonically active rift, strike-slip and foreland basins, target clastic units are likely to occur mainly near to active faults and thrusts, passing sharply into carbonate facies in the basin centre. However, during tectonically quiescent phases, some clastic units may spread further out into the basin (see discussions by Blair & Bilodeau, 1988; Heller *et al.*, 1988; Mack & Seager, 1990). In the case of slowly subsiding sag basins, alluvial and aeolian deposits may be laterally extensive and pass gradationally into carbonate facies.

Trap styles depend largely upon the tectonic setting. In rifts, structural traps reflecting the pattern of extensional faulting are common, but basin inversion may be a problem in some cases and differential geothermal gradients in adjacent rift segments may be a factor in maturation. Peripheral structures and hanging wall anticlines may be important in the exploration of strike-slip and foreland basins. In sag basins especially, most traps are likely to be stratigraphic, reflecting complex lateral facies changes, and prediction requires detailed facies analysis.

CONCLUSIONS

Carbonate sedimentation in lakes is chiefly biogenic or bio-induced. Charophytes are the dominant carbonate producers in low-energy, shallow fresh-water settings, where carbonate muds are formed. Blue-green algal stromatolites, oncoids or small bioherms may occur in lakes of differing salinities, and oolitic carbonates characterize many higher energy lakes. Greater carbonate production in shallow water commonly leads to progradation of lake margins and to the deposition of regressive sequences.

Modern lakes have sharp environmental gradients and ancient lake deposits are characterized by abrupt areal and vertical facies changes. However, analysis of lacustrine carbonate sequences and recognition of the basic facies types permits palaeoenvironmental interpretation of ancient carbonate lake systems. Lake carbonate sediments may be grouped into 'marginal' and 'basinal' facies. Marginal lake carbonates show great sedimentological variability and facies development reflects shoreline energy. High-energy margins are characterized by lenticular bioclastic carbonate sands and coated grains (ooids, oncoids), while bioturbated micrites are developed at low-energy margins. Climate largely determines lake hydrology, circulation dynamics, and rates of organic productivity, which in turn control the development of lake stratification and govern deposition of basinal 'laminites'. In polymictic or shallow lakes that are not permanently stratified, poorly or non-laminated facies predominate.

Lake morphology determines the relative importance of these facies types. Lakes with higher gradient ('bench'-type) margins may show more extensive development of basinal facies. Lakes with low-gradient ('ramp'-type) margins are commonly dominated by marginal lake facies, with evidence of frequent subaerial exposure. Vertical facies models (summarized in Fig. 2) may be developed for carbonate lakes according to the morphology and energy of their margins.

Catchment area geology conditions the nature of detrital input and influences lake chemistry. Tectonics is a fundamental control on lake systems, influencing lake basin geometry, subsidence rates and clastic supply. In rifts and in some strike-slip basins, high-gradient, bench-type margins may develop if subsidence exceeds sedimentation, as may occur at faulted margins. Lakes with ramp-type margins are common in more slowly subsiding rift segments and in the central areas of larger strike-slip basins. Low-gradient, ramp-type margins are also likely to dominate in foreland and sag basins. A clear understanding of the climatic setting, regional geology, and tectonic regime of a target lacustrine system is thus of prime importance if lake facies and their distributions are to be modelled successfully.

Hydrocarbon source rock potential exists in many deeper, stratified carbonate lakes, where anoxic bottom conditions permit the deposition of organic-rich sediments. Potential source facies may be deposited in rapidly subsiding rift sectors or in the central areas of larger low-gradient lakes. Lake marginal bioherm and shoal deposits commonly develop at faulted, bench-type rift margins and show reasonable reservoir potential, but intercalated alluvial or sublacustrine clastic facies may be more promising reservoir targets. Abrupt facies and thickness changes in tectonically active lake basins render accurate prediction of facies distributions of critical importance in exploration.

ACKNOWLEDGEMENTS

We thank Pere Anadón, Lluís Cabrera, Kerry Kelts and Harold Reading for their helpful reviews. We also appreciate valuable discussions with Brian Williams and Albert Matter. We are grateful to AMOCO for financial support and especially thank Allen Ormiston for his interest. University of Reading P.R.I.S. Contribution 050.

REFERENCES

ALLEN, P.A. & COLLINSON, J.D. (1986) Lakes. In: *Sedimentary Environments and Facies* (Ed. Reading, H.G.), pp. 63–94. Blackwell Scientific Publications, Oxford.

ANADÓN, P., CABRERA, L., GUIMERA, J. & SANTANACH, P. (1985) Paleogene strike-slip deformation and sedimentation along the southeastern margin of the Ebro Basin. In: *Strike-slip Deformation, Basin Formation and Sedimentation* (Eds Biddle, K.T. & Christie-Blick, N.), Soc. Econ. Paleontol. Mineral. Spec. Publ. **37**, 303–318.

ANADÓN, P., CABRERA, L. & JULIA, R. (1988) Anoxic–oxic cyclical lacustrine sedimentation in the Miocene Rubielos de Mora Basin, Spain. In: *Lacustrine Petroleum Source Rocks* (Eds Fleet, A.J., Kelts, K. and Talbot, M.R.), Geol. Soc. London, Spec. Publ. **40**, 353–367.

ANADÓN, P., CABRERA, L., JULIA, R., ROCA, E. & ROSELL, L. (1989) Lacustrine oil-shale basins in Tertiary grabens from NE Spain (Western European Rift System). In: *The Phanerozoic Record of Lacustrine Basins and their Environmental Signals* (Ed. Talbot, M.R. & Kelts, K.), Palaeogeogr. Palaeoclimatol. Palaeoecol. **70**, 7–28.

ASTIN, T.R. (1990) The Devonian lacustrine sediments of Orkney, Scotland, implications for climate cyclicity, basin structure and maturation history. *J. Geol. Soc. London*, **147**, 141–151.

AZEREDO, A.C. & GALOPIM DE CARVALHO, A.M. (1986) Novos elementos sobre o 'Paleogénico' carbonatado dos arredores de Lisboa. *Comun. Serv. Geol. Portugal* **72**, 111–118.

BALLANCE, P.F. & READING, H.G. (Eds) (1980) *Sedimentation in Oblique-slip Mobile Zones*. Spec. Publ. Int. Ass. Sediment. **4**, 265 pp.

BELLANCA, A., CALVO, J.P., CENSI, P., ELIZAGA, E. & NERI, R. (1989) Evolution of diatomite carbonate cycles of Miocene age, southeastern Spain: petrology and isotope geochemistry. *J. Sediment. Petrol.* **59**, 45–52.

BERTANI, R.T. & CAROZZI, A.V. (1985) Lagoa Feia Formation (Lower Cretaceous) Campos Basin offshore Brazil: rift valley stage carbonate reservoirs, I & II. *J. Petrol. Geol.* **8**, 37–58 & 199–220.

BIDDLE, K.T. & CHRISTIE-BLICK, N. (Eds) (1985) *Strike-slip Deformation, Basin Formation and Sedimentation*. Soc. Econ. Paleontol. Mineral. Spec. Publ. **37**, 386 pp.

BLAIR, T.C. & BILODEAU, W.L. (1988) Development of tectonic cyclothems in rift, pull-apart and foreland basins: sedimentary response to episodic tectonism. *Geology* **16**, 517–520.

BRICE, S.E. & PARDO, G. (1980) Hydrocarbon occurrences in nonmarine, pre-salt sequences of Cabinda, Angola. *Bull., Am. Assoc. Petrol. Geol.* **64**, 681.

BRICE, S.E., KELTS, K.R. & ARTHUR, M.A. (1980) Lower Cretaceous lacustrine source beds from the early rifting phase of S. Atlantic. *Bull., Am. Assoc. Petrol. Geol.* **64**, 680–681.

BROWN, R.E. & WILKINSON, B.H. (1981) The Draney Limestone: Early Cretaceous lacustrine carbonate deposition in western Wyoming and southeastern Idaho. *Contrib. Geol., Univ. Wyoming* **20**, 23–31.

BURKE, K. (1976) Development of graben associated with the initial rupture of the Atlantic Ocean. In: *Sedimentary Basins of Continental Margins and Cratons* (Ed. Bott, M.H.P.), Tectonophysics **36**, 93–112.

CABRERA, L. & SAEZ, A. (1987) Coal deposition in carbonate-rich shallow lacustrine systems: the Calaf and Mequinenza sequences (Oligocene, eastern Ebro Basin, NE Spain). *J. Geol. Soc. London* **144**, 451–461.

CABRERA, L., COLOMBO, F. & ROBLES, S. (1985) Sedimentation and tectonics interrelationships in the Palaeogene marginal alluvial systems of the SE Ebro Basin; transition from alluvial to shallow lacustrine environments. In: *Excursion Guide, Sixth European Regional Meeting, International Association of Sedimentologists* (Eds Milá, M.D. and Rosell, J.), pp. 393–492.

CABRERA, L., ROCA, E. & SANTANACH, P. (1988) Basin formation at the end of a strike-slip fault: the Cerdanya Basin (eastern Pyrenees). *J. Geol. Soc. London* **145**, 261–268.

CASANOVA, J. (1986) East African Rift stromatolites. In: *Sedimentation in the African Rifts* (Eds Frostick, L.E., Renaut, R.W., Reid, I. and Tiercelin, J.-J.), Geol. Soc. London Spec. Publ. **25**, 201–210.

CASANOVA, J. & NURY, D. (1989) Biosédimentologie des stromatolites fluvio-lacustres du fossé oligocène de Marseille. *Bull. Soc. Géol. Fr.* **8**, 1173–1184.

CLEMMENSEN, L.B. (1978) Lacustrine facies and stromatolites from the Middle Triassic of East Greenland. *J. Sedim. Petrol.* **48**, 1111–1128.

COHEN, A.S. (1989) Facies relationships and sedimentation in large rift lakes and implications for hydrocarbon exploration: examples from Lakes Turkana and Tanganyika. In: *The Phanerozoic Record of Lacustrine Basins and their Environmental Signals* (Eds Talbot, M.R. & Kelts, K.), Palaeogeogr. Palaeoclimatol. Palaeoecol. **70**, 65−80.

COHEN, A.S. & THOUIN, C. (1987) Nearshore carbonate deposits in Lake Tanganyika. *Geology* **15**, 414−418.

CROSS, T.A. (1986) Tectonic controls of foreland basin subsidence and Laramide style deformation, western United States. In: *Foreland Basins* (Eds Allen, P.A. & Homewood, P.), Spec. Publ. Int. Assoc. Sedimentol. **8**, 15−39.

DAVIS, R.L. & WILKINSON, B.H. (1983) Sedimentology and petrology of freshwater lacustrine carbonate: mid-Tertiary Camp Davis Formation, northwestern Wyoming. *Contrib. Geol., Univ. Wyoming* **22**, 45−55.

DEAN, W.E. (1981) Carbonate minerals and organic matter in sediments of modern north-temperate hard-water lakes. In: *Recent and Ancient Nonmarine Depositional Environments: Models for Exploration* (Eds Ethridge, F.G. & Flores, R.M.), Soc. Econ. Paleontol. Mineral. Spec. Publ. **31**, 213−231.

DEAN, W.E. & FOUCH, T.D. (1983) Lacustrine environment. In: *Carbonate Depositional Environments* (Ed. Scholle, P.A. Bebout, D.G. & Moore, C.H.), Mem., Am. Assoc. Petrol. Geol. **33**, 96−130.

DONOVAN, R.N. (1975) Devonian lacustrine limestones at the margin of the Orcadian Basin, Scotland. *J. Geol. Soc. London* **131**, 489−510.

DUNCAN, A.D. & HAMILTON, R.F.M. (1988) Palae-olimnology and organic geochemistry of the Middle Devonian in the Orcadian Basin. In: *Lacustrine Petroleum Source Rocks* (Eds Fleet, A.J., Kelts, K. & Talbot, M.R.), Geol. Soc. London Spec. Publ. **40**, 173−201.

EARDLEY, A.J. (1938) Sediment of Great Salt Lake, Utah. *Bull., Am. Assoc. Petrol. Geol.* **22**, 1305−1411.

EARDLEY, A.J. (1966) Sediments of Great Salt Lake. *Utah Geol. Soc. Guidebook Geol. Utah* **20**, 105−120.

EUGSTER, H.P. (1986) Lake Magadi, Kenya: a model for rift valley hydrochemistry and sedimentation. In: *Sedimentation in the African Rifts* (Eds Frostick, L.E., Renaut, R.W., Reid, I. & Tiercelin, J.-J.), Geol. Soc. London, Spec. Publ. **25**, 177−189.

EUGSTER, H.P. & KELTS, K. (1983) Lacustrine chemical sediments. In: *Chemical Sediments and Geomorphology* (Eds Goudie, A.S. & Pye, K.), pp. 321−368. Academic Press, London.

FLEET, A.J., KELTS, K. & TALBOT, M.R. (Eds) (1988) *Lacustrine Petroleum Source Rocks*. Geol. Soc. London Spec. Publ. **40**, 391 pp.

FOUCH, T.D. & DEAN, W.E. (1982) Lacustrine and associated clastic depositional environments. In: *Sandstone Depositional Environments* (Eds Scholle, P.A. & Spearing, D.), Mem. Am. Assoc. Petrol. Geol. **32**, 87−114.

FREYTET, P. (1984) Carbonate lacustrine sediments and their transformations by emersion and pedogenesis. Importance of identifying them for paleogeographical reconstructions. *Bull. Cent. Rech. Explor. Prod. Elf-Aquitaine* **8**, 223−247.

FREYTET, P. & PLAZIAT, J.C. (1982) *Continental Carbonate Sedimentation and Pedogenesis — Late Cretaceous and Early Tertiary of Southern France* (Ed. Purser, B.H.), Contrib. Sedimentol. **12**, 213 pp.

FROSTICK, L.E., RENAUT, R.W., REID, I. & TIERCELIN, J.-J. (Eds) (1986) *Sedimentation in the African Rifts*. Geol. Soc. London Spec. Publ. **25**, 382 pp.

FÜCHTBAUER, H., VON DER BRELIE, G., DEHM, R., FÖRSTNER, U., GALL, H., HÖFLING, J., HOEFS, J., HOLLERBACH, A., HUFNAGEL, H., JANKOWSKI, B., JUNG, W., MALZ, H., MERTES, A., ROTHE, P., SALGER, M., WEHNER, H. & WOLF, M. (1977) Tertiary lake sediments of the Ries research borehole Nördlingen 1973 — a summary. *Geol. Bavarica* **75**, 13−19.

FU JIAMO, SHENG GUOYING, & LIU DEHAN (1988) Organic geochemical characteristics of major types of terrestrial petroleum source rocks in China. In: *Lacustrine Petroleum Source Rocks* (Eds Fleet, A.J. Kelts, K. & Talbot, M.R.), Geol. Soc. London Spec. Publ. **40**, 279−289.

GIBBS, A.D. (1984) Structural evolution of extensional basin margins. *J. Geol. Soc. London* **141**, 153−160.

GILBERT, G.K. (1885) The topographic features of lake shores. *U.S. Geol. Surv. Ann. Rep.* **5**, 75−123.

GILBERT, G.K. (1890) Lake Bonneville. *U.S. Geol. Surv. Monogr.* **1**, 438 pp.

GLASS, S.W. & WILKINSON, B.H. (1980) The Peterson Limestone — Early Cretaceous lacustrine carbonate sedimentation in western Wyoming and southeastern Idaho. *Sediment Geol.* **27**, 143−160.

GORE, P.J.W. (1989) Toward a model for open- and closed-basin deposition in ancient lacustrine sequences: the Newark Supergroup (Triassic−Jurassic), Eastern North America. In: *The Phanerozoic Record of Lacustrine Basins and their Environmental Signals* (Eds Talbot, M.R. & Kelts, K.), Palaeogeogr. Palaeoclimatol. Palaeoecol. **70**, 29−51.

HEIM, A. (1919) *Geologie der Schweiz*, Vols 1−3. Tauchnitz, Leipzig.

HELLER, P.L., ANGEVINE, C.L., WINSLOW, N.L. & PAOLA, C. (1988) Two-phase stratigraphic model of foreland-basin sequences. *Geology* **16**, 501−504.

HILL, G. (1989) Distal alluvial fan sediments from the Upper Jurassic of Portugal: controls on their cyclicity and channel formation. *J. Geol. Soc. London* **146**, 539−555.

HIRST, J.P.P. & NICHOLS, G.J. (1986) Thrust tectonic controls on Miocene alluvial distribution patterns, southern Pyrenees. In: *Foreland Basins* (Eds Allen, P.A. & Homewood, P.), Spec. Publ. Int. Assoc. Sedimentol. **8**, 247−258.

HOMEWOOD, P., ALLEN. P.A. & WILLIAMS, G.D. (1986) Dynamics of the Molasse basin of western Switzerland. In: *Foreland Basins* (Eds Allen, P.A. & Homewood, P.), Spec. Publ. Int. Assoc. Sedimentol. **8**, 199−217.

JANAWAY, T.M. & PARNELL, J. (1989) Carbonate production within the Orcadian Basin, Northern Scotland: a petrographic and geochemical study. In: *The Phanerozoic Record of Lacustrine Basins and their Environmental Signals* (Eds Talbot, M.R. & Kelts, K.), Palaeogeogr. Palaeoclimatol. Palaeoecol. **70**, 89−105.

JORDI, H.A. (1955) *Geologie der Umgebung von Yverdon (Jurafuss und mitteländische Molasse)*. Beitr. Geol. Karte der Schweiz, N.F. **99**, 84 pp.

KELTS, K. (1988) Environments of deposition of lacustrine petroleum source rocks: an introduction. In: *Lacustrine Petroleum Source Rocks* (Eds Fleet, A.J., Kelts, K. & Talbot, M.R.), Spec. Publ. Geol. Soc. London **40**, 3–26.

KELTS, K. & HSÜ, K.J. (1978) Freshwater carbonate sedimentation. In: *Lakes: Chemistry, Geology and Physics* (Ed. Lerman, A.), pp. 295–323. Springer Verlag, New York.

KENDALL, R.L. (1969) The ecological history of the Lake Victoria Basin. *Ecol. Monogr.* **39**, 121–176.

KISSLING, D. (1974) *L'oligocène de l' extrémité occidentale du bassin molassique suisse. Stratigraphie et aperçu sédimentologique.* Thèse No. 1648, Université de Genève, 94 pp.

LEEDER, M.R. & GAWTHORPE, R.L. (1987). Sedimentary models for extensional tilt-block/half-graben basins. In: *Continental Extensional Tectonics* (Eds Coward, M.P., Dewey, J.F. & Hancock, P.L.), Geol. Soc. London Spec. Publ. **28**, 139–152.

LINK, M.H. & OSBORNE, R.H. (1978) Lacustrine facies in the Pliocene Ridge Basin Group, Ridge Basin, California. In: *Modern and Ancient Lake Sediments* (Eds Matter, A. & Tucker, M.E.), Spec. Publ. Int. Assoc. Sedimentol. **2**, 169–187.

LOFTUS, G.W.F. & GREENSMITH, J.T. (1988) The lacustrine Burdiehouse Limestone Formation — a key to the deposition of the Dinantian Oil Shales of Scotland. In: *Lacustrine Petroleum Source Rocks* (Eds Fleet, A.J., Kelts, K. & Talbot, M.R.), Geol. Soc. London Spec. Publ. **40**, 219–234.

MACK, G.H. & SEAGER, W.R. (1990) Tectonic control on facies distribution of the Camp Rice and Palomas Formations (Pliocene–Pleistocene) in the southern Rio Grande rift. *Geol. Soc. Am. Bull.* **102**, 45–53.

MANSPEIZER, W. (1985) The Dead Sea Rift: impact of climate and tectonism on Pleistocene and Holocene sedimentation. In: *Strike-slip Deformation, Basin Formation and Sedimentation* (Eds Biddle, K.T. & Christie-Blick, N.). Soc. Econ. Paleontol. Mineral. Spec. Publ. **37**, 143–158.

MARSHALL, J.D., PAUL, C.R.C. & WRIGHT, V.P. (1988) Diagenesis in Tertiary palustrine paleosols. *Soc. Econ. Paleontol. Mineral. Ann. Midyear Meet. Abstr.* **5**, 34.

MITCHELL, A.H.G. & READING, H.G. (1986) Sedimentation and tectonics. In: *Sedimentary Environments and Facies* (Ed. Reading, H.G.), pp. 471–519. Blackwell Scientific Publications, Oxford.

MURPHY, D.H. & WILKINSON, B.H. (1980) Carbonate deposition and facies distribution in a central Michigan marl lake. *Sedimentology* **27**, 123–135.

OSBORNE, R.H., LICARI, G.R. & LINK, M.H. (1982) Modern lake stromatolites, Walker Lake, Nevada. *Sediment. Geol.* **32**, 39–61.

PENTECOST, A. (1978) Blue-green algae and freshwater carbonate deposits. *Proc. R. Soc. London Ser. B* **200**, 48–63.

PETERS, K.E., MOLDOWAN, J.M., DRISCOLE, A.R. & DEMAISON, G.J. (1989) Origin of Beatrice oil by co-sourcing from Devonian and Middle Jurassic source rocks, Inner Moray Firth, United Kingdom. *Bull., Am. Assoc. Petrol. Geol.* **73**, 454–471.

PICARD, M.D. & HIGH, L.R. (1972) Criteria for recognizing lacustrine rocks. In: *Recognition of Ancient Sedimentary Environments* (Eds Rigby, J.K. & Hamblin, W.K.), Soc. Econ. Paleontol. Mineral. Spec. Publ. **16**, 108–145.

PICARD, M.D. & HIGH, L.R. (1981) Physical stratigraphy of ancient lacustrine deposits. In: *Recent and Ancient Nonmarine Depositional Environments: Models for Exploration* (Eds Ethridge, F.G. & Flores, R.M.), Soc. Econ. Paleontol. Mineral. Spec. Publ. **31**, 233–259.

PLATT, N.H. (1989a) Lacustrine carbonates and pedogenesis: sedimentology and origin of palustrine deposits from the Early Cretaceous Rupelo Formation, W Cameros Basin, N Spain. *Sedimentology* **36**, 665–684.

PLATT, N.H. (1989b) Climatic and tectonic controls on sedimentation of a Mesozoic lacustrine sequence: the Purbeck of the Western Cameros Basin, Northern Spain. In: *The Phanerozoic Record of Lacustrine Basins and their Environmental Signals* (Eds Talbot, M.R. & Kelts, K.), Palaeogeogr. Palaeoclimatol. Palaeoecol. **70**, 187–197.

PLATT, N.H. (1989c) Continental sedimentation in an evolving rift basin: the Lower Cretaceous of the western Cameros Basin (Northern Spain). *Sediment. Geol.* **64**, 91–109.

PLATT, N.H. (1990a) Basin evolution and fault reactivation (western Cameros Basin, Northern Spain). *J. Geol. Soc. London* **147**, 165–175.

PLATT, N.H. (1990b) Lacustrine carbonates from the Swiss Molasse. *Abstracts and Papers, XII International Sedimentological Congress, Nottingham*, pp. 428–429.

PLAZIAT, J.-C. & FREYTET, P. (1978) Le pseudomicrokarst pédologique: un aspect particulier des paléo-pédogenéses developpées sur les dépots calcaires lacustres dans le tertiare du Languedoc. *C.R. Acad. Sci., Sér. D* **286**, 1661–1664.

POWELL, T.G. (1986) Petroleum geochemistry and depositional setting of lacustrine source rocks. *Mar. Petrol. Geol.* **3**, 200–219.

READING, H.G. (1982) Sedimentary basins and global tectonics. *Proc. Geol. Assoc.* **93**, 321–350.

RIDING, R. (1979) Origin and diagenesis of lacustrine algal bioherms at the margin of the Ries Crater, Upper Miocene, southern Germany. *Sedimentology* **26**, 645–680.

ROBBINS, E.I. (1983) Accumulation of fossil fuels in ancient rift lakes. *Tectonophysics* **94**, 633–658.

ROSENDAHL, B.R., REYNOLDS, D.J., LORBER, P.M., BURGESS, C.F., McGILL, J., SCOTT, D., LAMBIASE, J.J. & DERKSEN, S.J. (1986) Structural expressions of rifting: lessons from Lake Tanganyika, Africa. In: *Sedimentation in the African Rifts* (Eds Frostick, L.E., Renaut, R.W., Reid, I. & Tiercelin, J.-J.), Geol. Soc. London Spec. Publ. **25**, 29–43.

RYDER, R.T., FOUCH, T.D. & ELISON, J.D. (1976) Early Tertiary sedimentation in the western Uinta Basin, Utah. *Geol. Soc. Am. Bull.* **87**, 496–512.

SANDBERG, P.A. (1975) New interpretation of Great Salt Lake ooids and of ancient non-skeletal carbonate mineralogy. *Sedimentology* **22**, 497–538.

SANDBERG, P.A. (1980) The Pliocene Glenns Ferry Oolites: lake margin carbonate deposition in the southwestern Snake River Plain — discussion. *J. Sediment. Petrol.* **50** 997–998.

SCHNEIDER, J., SCHRÖDER, H.G. & LE CAMPION-ALSUMARD, TH. (1983) Algal micro-reefs — coated grains from freshwater environments. In: *Coated Grains* (Ed. Peryt, T.), pp. 284—298. Springer-Verlag, Berlin.

SCHOLL, D.W. & TAFT, W.H. (1964) Algae, contributors to the formation of calcareous tufa, Mono Lake, California. *J. Sediment. Petrol.* **34**, 309—319.

SCHOLZ, C.A., ROSENDAHL, B.R. & SCOTT, D.L. (1990) Development of coarse-grained facies in lacustrine rift basins: examples from East Africa. *Geology,* **18**, 140—144.

SHEARMAN, D.J., McGUGAN, A., STEIN, C. & SMITH, A.J. (1989) Ikaite, $CaCO_3.6H_2O$, precursor of the thinolites in the Quaternary tufas and tufa mounds of the Lahontan and Mono Lake Basins, western United States. *Geol. Soc. Am. Bull.* **101**, 913—917.

SOUTHGATE, P.N., LAMBERT, I.B., DONNELLY, T.H., HENRY, R., ETMINAN, H. & WESTE, G. (1989) Depositional environments and diagenesis in Lake Parakeelya: a Cambrian alkaline playa from the Officer Basin, South Australia. *Sedimentology* **36**, 1091—1112.

STANLEY, K.O. & COLLINSON, J.W. (1979) Depositional history of Paleocene—Lower Eocene Flagstaff Limestone and coeval rocks, Central Utah. *Bull. Am. Assoc. Petrol. Geol.* **63**, 311—323.

SURDAM, R.O. & STANLEY, K.O. (1979) Lacustrine sedimentation during the culminating phase of Lake Gosiute, Wyoming (Green River Formation). *Geol. Soc. Am. Bull.* **90**, 93—110.

SURDAM, R.C. & WOLFBAUER, C.A. (1975) Green River Formation, Wyoming: a playa-lake complex. *Geol. Soc. Am. Bull.* **86**, 335—345.

SURDAM, R.C. & WRAY, J.L. (1976) Lacustrine stromatolites, Green River Formation, Wyoming. In: *Stromatolites* (Ed. Walter, M.R.), Developments in Sedimentology, **20**, 535—541. Elsevier, Amsterdam.

SWIRYDCZUK, K., WILKINSON, B.H. & SMITH, G.R. (1979) The Pliocene Glenns Ferry Oolite: lake-margin carbonate deposition in the southwestern Snake River Plain. *J. Sediment. Petrol.* **49**, 995—1004.

SWIRYDCZUK, K., WILKINSON, B.H. & SMITH, G.R. (1980) The Pliocene Glenns Ferry Oolite — II — sedimentology of oolitic lacustrine terrace deposits. *J. Sediment. Petrol.* **50**, 1237—1247.

TALBOT, M.R. (1988) The deposition of lacustrine oil source rocks: evidence from the lakes of tropical Africa. In: *Lacustrine Petroleum Source Rocks* (Eds Fleet, A.J., Kelts, K. & Talbot, M.R.). Geol. Soc. London Spec. Publ. **40**, 29—43.

TEN BRINK, U.S. & BEN-AVRAHAM, Z. (1989) The anatomy of a pull-apart basin: seismic reflection observations of the Dead Sea Basin. *Tectonics* **8**, 333—350.

TREESE, K.L. & WILKINSON, B.H. (1982) Peat—marl deposition in a Holocene paludal lacustrine basin — Sucker Lake, Michigan. *Sedimentology* **29**, 375—390.

TRUC, G. (1978) Lacustrine sedimentation in an evaporitic environment: the Ludian (Palaeogene) of the Mormoiron Basin, SE France. In: *Modern and Ancient Lake Sediments* (Eds Matter, A. & Tucker, M.E.), Spec. Publ. Int. Assoc. Sedimentol. **2**, 187—202.

TRÜMPY, R. (1980) *Geology of Switzerland: a Guide Book. Part A: an Outline of the Geology of Switzerland.* Schweizerische Geologische Kommission, Wepf, Basel, 104 pp.

TUCKER, M.E. (1977) The marginal Triassic deposits of South Wales: continental facies and palaeogeography. *Geol. J.* **12**, 169—188.

TUCKER, M.E. (1978) Triassic lacustrine sediments from South Wales: shore zone clastics, evaporites and carbonates. In: *Modern and Ancient Lake Sediments* (Eds Matter, A. & Tucker, M.E.), Spec. Publ. Int. Assoc. Sedimentol. **2**, 205—224.

WATSON, M.P., HAYWARD, A.B., PARKINSON, D.N. & ZHANG, ZH.M. (1987) Plate tectonic history, basin development and petroleum source rock deposition onshore China. *Mar. Petrol. Geol.* **4**, 205—225.

WELLS, N. (1983) Carbonate deposition, physical limnology and environmentally-controlled chert formation in Palaeocene—Eocene Lake Flagstaff, central Utah. *Sediment. Geol.* **35**, 263—296.

WHITE, A.H. & YOUNGS, B.C. (1980) Cambrian alkali playa-lacustrine sequence in the northeastern Officer Basin, South Australia. *J. Sediment. Petrol.* **50**, 1279—1286.

WILLIAMSON, C.R. & PICARD, M.D. (1974) Petrology of carbonate rocks of the Green River Formation (Eocene). *J. Sediment. Petrol.* **44**, 738—759.

WRIGHT, V.P. & WILSON, R.C. (1985) Lacustrine carbonates and source rocks from the Upper Jurassic of Portugal. *Abstracts, sixth European Regional Meeting, International Association of Sedimentologists,* pp. 487—490.

YURETICH, R.F. (1982) Possible influences upon lake development in the East African rift valleys. *J. Geol.* **90**, 329—337.

YURETICH, R. (1989) Paleocene lakes of the central Rocky Mountains, western United States. In: *The Phanerozoic Record of Lacustrine Basins and their Environmental Signals* (Eds Talbot, M.R. & Kelts, K.), Palaeogeogr. Palaeoclimatol. Palaeoecol. **70**, 53—63.

ZAK, I. & FREUND, R. (1981) Asymmetry and basin migration in the Dead Sea rift. In: *The Dead Sea Rift* (Eds Freund, R. & Garfunkel, Z.), Tectonophysics **80**, 27—38.

Spec. Publs Int. Ass. Sediment. (1991) **13**, 75–92

Structural and depositional patterns of the Tertiary Baise Basin, Guang Xi Autonomous Region (southeastern China): a predictive model for fossil fuel exploration

LIN CHANGSONG, YANG QI *and* LI SITIAN

China University of Geosciences, Beijing, China

ABSTRACT

Baise Basin is a Paleogene (Eocene–Oligocene) coal and oil-bearing intermontane half-graben (90 × 16 km). It was formed during transtensional movements along the You Jiang Fault, which defines the northern basin margin, and infilled with over 3000 m thickness of alluvial–fluvial and lacustrine sediments. Five genetic stratigraphic units characteristic of distinct facies configurations and basin infilling settings have been identified in the basin. They show that the basin evolved from initial alluvial–fluvial episode (unit 1), to a shallow and deep lacustrine stage (units 2–4), and finally to alluvial–fluvial conditions (unit 5). This infilling evolution is attributed to tectonic change from initial sinistral transtension to final transpressive movement along the You Jiang Fault.

Several major facies associations have been identified in the basin fill: (i) marginal fan and (ii) fan-delta complexes, which developed along the northern boundary fault and include alluvial fan, fan–braided-stream delta, and deep-lake marginal fan-delta or subaqueous fan deposits; (iii) fluvial deltas, comprising the fluvial deltas fed from the southern side of the basin and the axial fluvial system impinging from the western and northern basin margin; (iv) terrigenous and carbonate shallow and deep lacustrine facies, consisting dominantly of fossiliferous mudstones and micrites mostly deposited in central basin areas. Peat swamp deposits have been recorded associated mainly with the marginal fan-delta and the fluvial delta complexes. The asymmetric facies development in the basin was likely controlled by its half-graben structure. Moreover, the generation and distribution of economic deposits in the Baise Basin were strongly controlled by structural and depositional patterns. The two major coal-bearing and oil-producing units in the basin underlie and overlie the Upper Natu Formation, a lacustrine unit 300–700 m thick that consists mainly of mudstones rich in organic matter. The coal-rich zones coincide with the basin subdepressions and are usually associated with fluvial delta, shallow lacustrine deposits and fan–braided-stream deltaic sandstone deposits along the northern basin margin. Moreover these sandstones are the major oil reservoirs found in the basin.

INTRODUCTION

The Baise Basin is the largest (90 × 16 km) among more than 30 Tertiary intermontane fault-bounded basins in the Guang Xi Autonomous Region, southeastern China. This basin contains important mineable coal seams, oil reservoirs, and other economic deposits. Its basin fill records various evolutionary stages and was strongly controlled by the strike-slip movement of the You Jiang Fault, which bounds the northern basin margin. The present Baise Basin pattern can be considered as original because of only minor later deformation. The nearly continuous out-crop of the basin fill along the basin margin, more than 600 drilled boreholes, and a dense network of available seismic profiles, provide excellent conditions for the study of the structural and sedimentary evolution of the basin.

The purpose of this paper is to concisely describe and discuss the structural framework and depositional pattern of the Baise Basin. This paper suggests that the palaeostructure and the spatial configuration of the major facies, particularly the northern basin margin alluvial fan and fan-delta complexes, were the main factors controlling the distribution of the coal-rich zones and oil reservoirs developed in the basin.

GEOLOGICAL SETTING AND MAJOR BASIN FEATURES

The Paleogene Baise Basin was formed along the southwest margin of the You Jiang fault system (Fig. 1).

The basin is underlain and surrounded by folded Triassic turbiditic rocks. It was mainly infilled by siliciclastic sediments fed from northern and north-western source areas, although a restricted source area located to the south also shed sediments into southern basin sectors. The basin was tectonically active from early Eocene to late Oligocene and infilled with extremely thick (up to 3000 m) composite alluvial, fluvial, and lacustrine clastic deposits with fossil molluscs, ostracods, vertebrate remains, and other non-marine fossil taxa. Carbonate and coal deposits also developed in relation to the lacustrine units (Fig. 2).

There are two major groups of syndepositional faults that affect the basin basement. These have been identified from the analysis of depositional sections and seismic profiles.

1 The northwest striking fault system defines subordinate fault blocks or grabens and includes the northern boundary fault. This northern fault is split into a series of side-stepping subordinate faults that controlled the axis of maximum basin subsidence (Fig. 3).

2 The northeast striking fault system is arranged as a series of faults that divided the basin into several subordinate uplifts and depressions. These resulted in significant variation of basin infill thickness (Figs 3, 4).

Most of these faults are inferred to be developed along previously weak fractures, based on a comparison of the intra-basin faults with the old fractures observed outside the basin (Fig. 3). The overall setting and structural features in the basin indicates that its origin may be related to sinistral transtensional displacement along the You Jiang Fault during Eocene to early Oligocene time. Up to the late Oligocene, the basin was mildly compressed and uplifted, a process that may be attributed to transpressive displacement along the You Jiang fault system.

The northern boundary fault is a normal one dipping toward the central basin, as is clearly shown in seismic sections (Fig. 5) and in outcrops. Very thick marginal fan complexes developed along this fault, indicating that sedimentation was contemporaneous with the faulting. This faulting caused basement tilting and led to an asymmetric structural and stratigraphic framework (Figs 3, 4). The faulting patterns of the boundary fault changed from west to east, and this fact exerted a great influence on the basin geometry and the development of the marginal fan systems (Fig. 3). It has been argued (Heward, 1978; Reading, 1980) that persistent faulting patterns

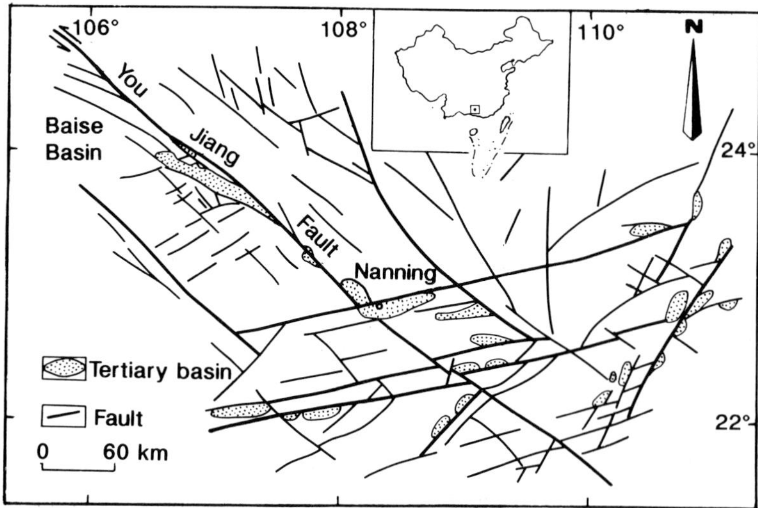

Fig. 1. Tertiary basins and their major controlling faults in the Guang Xi Autonomous Region. Baise Basin is located at the southwestern side of the You Jiang Fault.

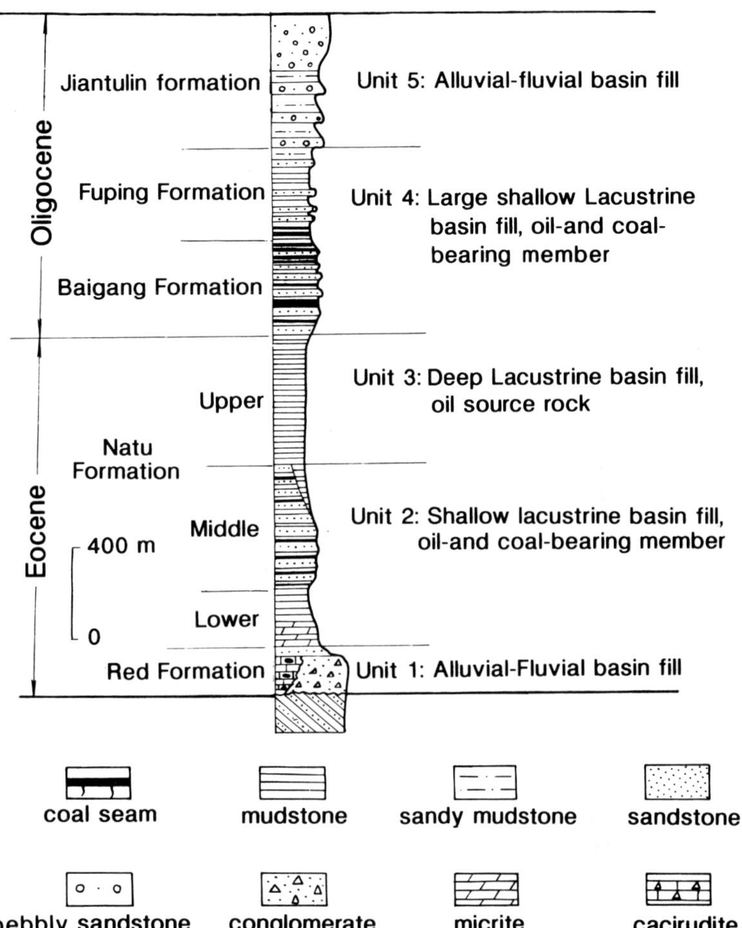

Fig. 2. Infill sequence of the Baise Basin, showing the lithostratigraphic and genetic units, and the basin infilling evolution from early alluvial–fluvial deposits (unit 1), through shallow and deep lacustrine units (units 2–4), and finally to alluvial–fluvial sequences (unit 5).

are presumably associated with major strike-slip zones.

STRATIGRAPHIC FRAMEWORK

Sedimentary infill of the Baise Basin consists of alluvial–fluvial and lacustrine deposits, ranging from 1000 to 3200 m in thickness (Fig. 2). The lowermost sedimentary unit in the basin is the Red Formation, which is mainly composed of alluvial–fluvial red or brown coloured breccias, pebbly sandstones, and mudstones containing gypsum and calcareous concretions. Fossil ostracods (*Limnocythere* sp. and *Eucypris* sp.) and vertebrate remains (Lacertidae) have been found in this unit. An early–middle Eocene age has been suggested for it (Hu Yiankuin, 1979).

The Red Formation is overlain by the Natu Formation, which is about 300–1500 m thick and can be divided into three members (Fig. 2). The lower and middle parts of the formation are coal and oil-bearing intervals composed of micrites or muddy limestones (Lower Natu Formation), mudstones, sandstones, and conglomerates (Middle Natu Formation). The Upper Natu Formation is a widely developed mudstone unit, about 300–700 m thick, comprising the major source rocks in the basin (Figs 2, 4, 5).

The Baigang Formation, which overlies the Upper Natu mudstone member, is another major coal-bearing and oil-producing interval, consisting mainly of sandstones and mudstones, and characterized by well-developed coarsening upward deltaic sequences. The lithofacies of the overlying Fuping Formation is similar to that of the Baigang Forma-

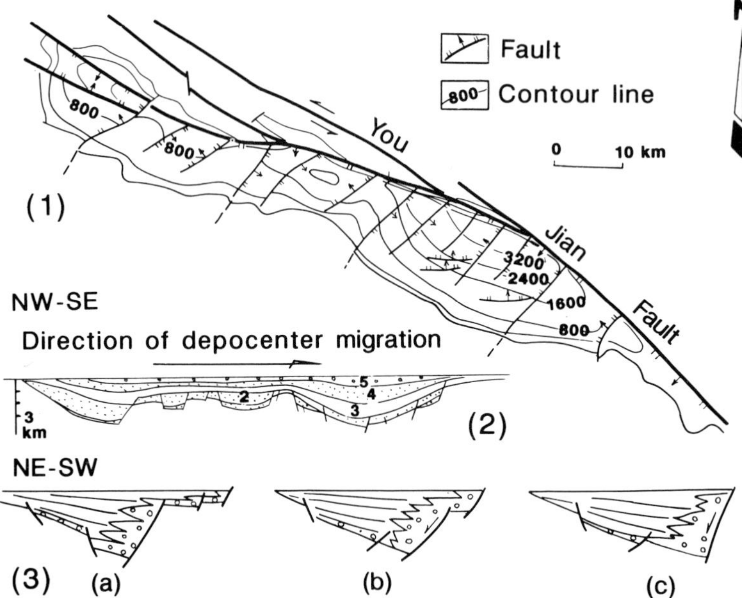

Fig. 3. Structural pattern of the Baise Basin: (1) contour map of the basin basement and the arrangement of the major basement faults that were active during basin infilling. (2) Longitudinal section of the Baise Basin showing the distribution of subdepressions and uplifts, which are defined by NNE–SSW oriented basement faults. Note the shifting of the depocentres from west to east. (3) Cross-sections showing the half-graben structure and the changing basin geometry from east (c) to west (a) in relation to the varied arrangement of boundary faults.

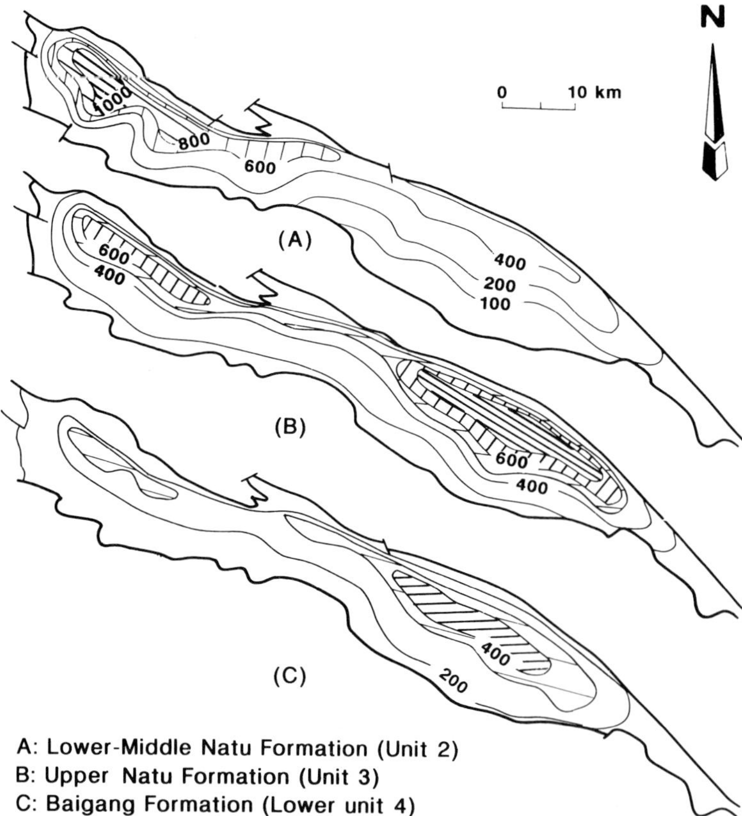

Fig. 4. Isopach maps of the lower–middle Natu Formation (A), upper Natu Formation (B) and Baigang Formation (C). Note the thickness variation influenced by syndepositional faults and the depocentre migration from west to east (see Fig. 3).

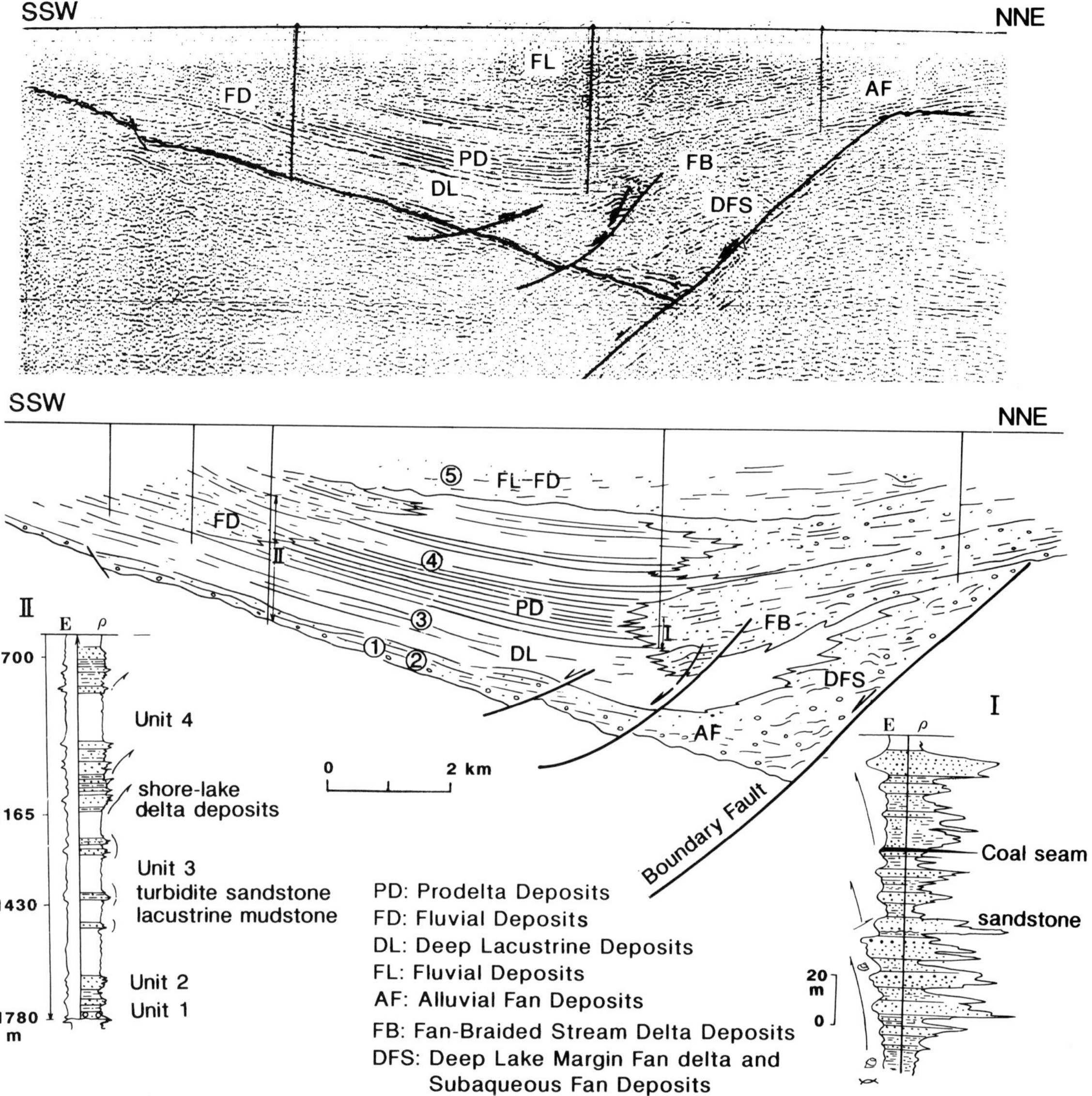

Fig. 5. Seismic profile and its facies explanation showing the major facies configuration in the basin. Note the evolution of basin infill from unit 1 to 5 and the asymmetric facies development (units 2–4). Alluvial fan and fan-delta systems (coarsening upward sequences in profile I) formed along the northern boundary fault, meanwhile deep–shallow lacustrine and fluvial delta deposits developed in central and southern basin areas (sequences in profile II). (Seismic profile provided by Guang Xi Oil Exploration Company.)

tion, although the Fuping Formation mudstones are obviously green coloured and do not contain economic coal seams. The thickness of both the Baigang and the Fuping Formations ranges from 500 to 800 m (Figs 2, 4, 5). The Jiang Tulin Formation, which tops the basin fill section (Fig. 2), is mainly composed of conglomerates, pebbly sandstones, and purple coloured gypsum-bearing mudstones. Fining upward sequences about 10–30 m thick, including pebbly sandstones with scoured bases that pass vertically into interbedded sands and muds, characterize this formation. These sequences are interpreted as fluvial deposits (Chen Baiyue & Xie Yongsen, 1987).

Based on the integrative analysis of facies associations, seismic sequences, and tectonic palaeostructures, five genetic stratigraphic units have been recognized. Each one of them represents a distinct depositional episode or basin infill stage (Fig. 2). It is apparent that the basin fill evolved from initial alluvial−fluvial basin fill to shallow and deep lacustrine basin fills, and finally to alluvial−fluvial basin fill (Figs 2, 5). This evolutionary history may be related to the changing basin tectonics.

The tectonic control on Paleogene sedimentation in the Baise Basin is clearly shown by the geometry and spacial arrangement of the above-mentioned stratigraphic units, which are strongly dependent on the structural framework of the basin. It is clearly shown in cross-sections that the thickness of these units increases northward and decreases southward owing to the differential subsidence caused by the major boundary fault (Figs 3, 4, 5). Meanwhile, the successive units intermittently overlapped the northern margin as the back-faulting took place, and also overlapped gradually on to the southern margin as a consequence of the southward basement tilting (Fig. 3). The overlapping also occurred in a longitudinal axial direction. The earlier units were overlapped by younger ones within the subordinate depressions. The basin depocentres moved gradually from west to east (Figs 3, 4). Peat accumulating centres shifted in the same direction through time. This process is presumably attributed to the sinistral transtension along the You Jiang fault system that generated a series of faults developing gradually from west to east (Fig. 3).

MAJOR FACIES ASSEMBLAGE ANALYSIS

A series of depositional systems or facies associations, which comprise the major segments of the basin fill, have been identified (Fig. 5).
1 Alluvial fan deposits.
2 Fan-delta assemblages, including fan−braided-stream delta, deep-lake marginal fan-delta, and subaqueous fan deposits.
3 Fluvial and fluvial−lacustrine delta deposits.
4 Terrigenous and carbonate shallow and deep lacustrine deposits.

The following discussion will focus on these basin-fill major facies, most of which were recorded in the Natu, Baigang, and Fuping Formations.

Alluvial fan, fan-delta or fan−braided-stream delta deposits, recognized along the northern basin margin, differ significantly from each other with respect to sedimentary structures and facies associations. They represent the marginal facies deposited during different episodes of the basin infill and will be discussed separately. Sedimentary structures, vertical sequences, facies geometry, electric log and seismic reflection patterns characterizing the above-mentioned deposits are integrated and discussed below.

Alluvial fan deposits

Alluvial fan deposits adjacent to the northern boundary fault scarp were widely developed in the basin during the initial and uplift stages of the basin infill (units 1 and 5). Most of these facies were formed, and commonly coalesced to form wedge-shape *bajada* deposit (1−3 km wide), along the northern side of the basin (Fig. 5). The alluvial fan deposits of the Lower−Middle Natu Formation, Baigang Formation and Fuping Formation are mainly composed of debris flow and water-laid sediments, and generally show downfan reduction in grain size, interfingering basinward with fluvial floodplain or shallow lacustrine deposits. The alluvial fan seismic facies display varied amplitude, low continuity, low to middle frequency, and chaotic reflection configuration (Fig. 5; Lin Changsong, 1987). This reflects the high-energy sedimentation of the alluvial fan palaeoenvironment (Brown & Fisher, 1977; Sangree & Widmier, 1977).

Debris flow conglomerates and breccias are usually matrix supported and poorly sorted. They display unordered fabric and lack any bedding structures (Fig. 6A). These deposits generally contain boulders (ranging from 0.5 to 1 m in diameter) supported by mixed sands and muds, and vertically oriented elongate gravels (Lin Changsong, 1987). These features indicate a high matrix strength and lack of clast to clast movement during flow (Steel & Gloppen, 1980). Some of these conglomerates, displaying flat erosional bases and normal graded bedding, are considered as low-viscosity debris flow deposits (Steel & Gloppen, 1980). Pebbly mudstones (pebbly content < 20%) associated with the conglomerates are interpreted as mud flow sediments. Consistent with this interpretation is a lack of stratification, poor sorting, and randomly scattered angular pebbles. Very thick accumulations of pebbly mudstones along the northern basin margin is plausibly related to the mud-dominated source rock lithology of the folded Triassic rocks.

Fig. 6. Alluvial fan and fan-delta facies in the Baigang and Natu Formations: (A) coarsening upward alluvial fan sequence changing from fine-grained sheetflood deposits to overlying debris flow and coarse-grained channel fills; (B) coarsening upward sequence of braided stream delta front deposit, evolving from basal, thinly interbedded sandstones and mudstones up to thick-bedded medium to coarse-grained sandstones with large-scale cross-bedding and erosional surfaces; (C) subaqueous base-scoured channel sandstones incised into fine prodelta deposits.

Water-laid deposits are obviously better sorted and better stratified than the debris flow deposits, and are mostly composed of horizontally stratified conglomerates, interbedded cross-bedded pebbly sandstones and sandy mudstones. The conglomerates, which display imbricated gravels and low-angle cross-stratifications, are analogous to longitudinal, channel fill, gravel bar deposits on modern proximal fan zones (Collinson & Thompson, 1982). These units have markedly eroded basal surfaces and lenticular geometry, and are associated with debris flow or coarse-grained sheet-flood deposits. Thinly interbedded conglomeratic sandstones or coarse-grained sandstones and sandy siltstones characterize the sheet-flood sediments. Individual pebbly sand beds usually have a thickness ranging from 0.4 to 0.1 m, and show sheet-eroded bases, massive or graded beddings, and fining-upward, occasionally with horizontal lamination and ripple cross-bedding, in the upper part of the units. The siltstones or sandy mudstones, 0.1~0.2 m thick, commonly contain wavy bedding, horizontal lamination and rootlets (Lin Changsong, 1987). These couplets of sand and mud represent the intermittent sheet-flood events. The alternating sheet beds become finer and thinner in the distal fan zones, where they are commonly eroded by braided channel infill sandstones displaying tabular and trough cross-bedding.

Relatively regular downfan changes in grain size, sedimentary style, and facies association generally permit ready identification of proximal, mid- and distal marginal fan deposits. The proximal fan facies are dominated by debris flow conglomerates, mud flow pebbly mudstones, and associated coarse-grained sheet-flood deposits. Coarsening upward sequences, 20−40 m thick, are commonly observed in mid-fan deposits (Fig. 6A), which may represent individual progradation events of alluvial fans (Heward, 1978). The distal fan facies vary considerably in colour and component, recording changes in palaeoclimatic conditions and in the basin setting. In the coal-bearing formations (Natu, Baigang, and Fuping Formations) these deposits are relatively dark, and occur associated with coal seams or swamp deposits. On the other hand, those in the Red Formation are typically brown or red coloured and are associated with thin evaporitic deposits, such as gypsum. Distal fan facies in the Middle Natu Formation, Baigang Formation and Fuping Formation, in fact, comprise the braided-stream delta plain deposits (Figs 5, 7).

Fan-delta deposits

Fan-deltas are defined as the alluvial fans that prograde directly into a standing body of water from an

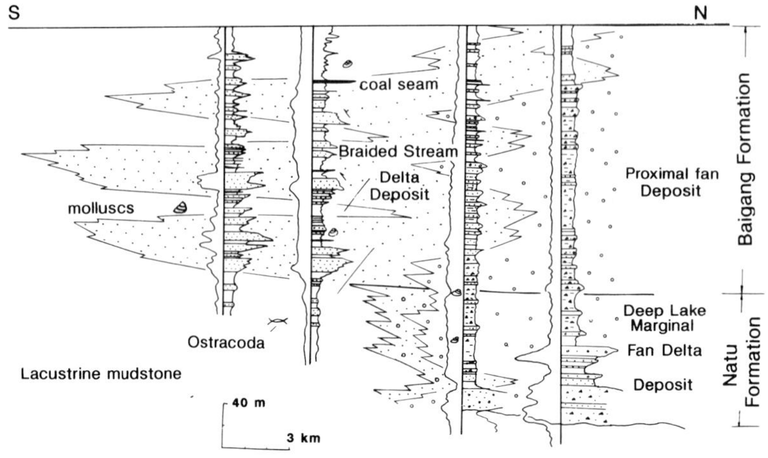

Fig. 7. Depositional section of fan-delta complexes in the Baigang and Natu Formations. Facies assemblages and electric-log patterns of the braided-stream delta and deep-lake marginal fan deposits, which developed along the northern boundary fault, are shown.

adjacent highland (Holmes, 1965; McGowen, 1970). With the detailed investigation carried out on modern and ancient fan-deltas, many workers have recognized that they vary in style (Nemec & Steel, 1988). Wescott & Ethridge (1980) recognized two types of fan-delta depositional sequences based on the investigation of modern marine fan-deltas. McPherson *et al.* (1987,1988) have distinguished two types of coarse-grained deltas: fan-deltas and braid-deltas, the latter having been previously classified as fan-deltas (Galloway & Hobday, 1983).

The study in the Baise Basin has shown that the lacustrine fan deltas also can be subdivided into two categories in terms of sedimentary features and processing characteristics: fan−braided-stream delta and deep-lake marginal fan-delta. They have been widely identified in genetic stratigraphic units 2−4 (Fig. 5).

Fan−braided-stream delta deposits

This facies assemblage is well developed along the northern basin margin in the Middle Natu, Baigang, and Fuping Formations, and resulted from the progradation of stream-dominated alluvial fans into relatively shallow and open lacustrine zones. It is suggested that the sedimentary processes of the system were dominated by aggradation of proximal alluvial fans and progradation of braided-stream deltas, which gave rise to a series of wide, lobate depositional bodies extending far into the central basin zones (Fig. 7). The seismic reflection characters of these bodies display better continuity and stronger amplitude than that of the alluvial fans, and

commonly show a divergent foreset or oblique progradation reflection configuration (Lin Changsong, 1987). These facies grade basinward into high continuity, high frequency, mid-amplitude, and sheet-like parallel seismic facies that represent the prodelta and shallow and deep lake deposits formed in low-energy basin environments (Fig. 5). The fan−braided-stream delta system is characterized by forming a relatively large deltaic plain and can be divided clearly into four depositional facies from fault scarp to the central basin.

Fanglomerate facies. These are adjacent to the marginal fault, consisting mostly of breccias, unstratified conglomerates, pebbly mudstones, and interbedded sandstones and siltstones. They mainly represent subaerial debris flow, channel fill, and sheet-flood deposits similar to the proximal to mid-fan alluvial facies discussed above (Fig. 7).

Braided-stream delta plain facies. These generally include cross-bedded sandstones, horizontally laminated conglomerates deposited by braided stream channels, and thinly interbedded sandstones and mudstones formed as overflood sediments. These units are commonly associated with coal seams, bioturbated fine sandstones, lenticular bioclastics, and mollusc-bearing mudstones, owing to frequent fluctuation of lake level and/or abandonment of the delta system (Fig. 7).

Subaqueous delta-plain sandy facies. These mainly include fluvial mouth bar and subaqueous channel fill sandy deposits (Figs 6B,C, 7). The fluvial mouth bar sandstones show a coarsening and thickening

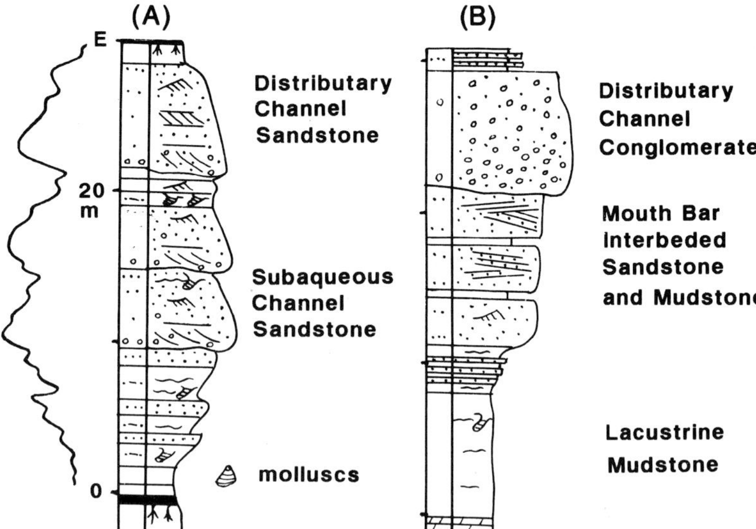

Fig. 8. Two types of vertical sequences in the fan—braided-stream delta deposits of the Baigang Formation: (A) coarsening upward sequence interrupted by subaqueous channel sandstones; (B) coarsening upward sequence with well-developed mouth bar sandstones observed in outcrops.

upward sequence from basal lacustrine wavy bedded and burrowed, sandy mudstones, which contain ostracods, to alternating fine-grained sandstones and siltstones (distal bar), and finally up to the thick-bedded, medium to coarse-grained sandstone showing large-scale cross-bedding and intra-erosive surfaces representing proximal mouth bar or channel deposits (Figs 6B,C, 8). Interbedded sandstones in the distal bar are usually composed of massive or parallel laminated sand with flat erosional surfaces, which reflect that underflow frequently occurred in the delta front area. These units are commonly burrowed at the top and covered with wave marks, and bioturbated siltstones or mudstones formed in sub-aqueous environments. Low-angle cross-bedding in the upper part of the proximal bars indicate wave reworking along the lake shoreline (Fig. 8B). Large-scale foresets are also observed occasionally, and are commonly eroded by distributary and sub-aqueous channels. The latter contain a lot of intra-clastic mud pebbles and are commonly eroded into more distal prodelta deposits, surrounded by lacus-trine mudstones (Fig. 6C). These units are generally capped by shallow lake and shoreline sandstones and sandy mudstones, which contain fossil molluscs, and display a bioturbated structure. Stacked channel sandstone sequences from basal subaqueous channel to upper distributary channel deposits have been occasionally examined from successive borehole cores (Fig. 8A). Some of the high-angle foresets and

the gradually coarsening upward sequences could plausibly have been produced by perennial fan streams, which produced isopycnal flows and laid sediments in the river mouth. Nevertheless the extensively developed subaqueous channel fills and sheet-like sandstones with a flat scoured base and parallel bedding observed in the delta front deposits indicate that hypopycnal flow often occurred at the river mouth. The density of flood discharges from alluvial fans was commonly higher than that of fresh water in a lake.

Prodelta muddy facies. These are composed of silt-stones and mudstones intercalated with thin turbidite sandstones, deformed beds and slump-generated breccias. These sediments are commonly laminated, burrowed, and contain limnic fauna, such as ostra-cods. Some large-scale slumping structures are also identified from seismic profiles (Fig. 5).

Progradation of braided streams of 'wet' fans into the standing body of lacustrine water that developed in the central basin zones, generated coarsening upward deltaic sequences consisting of distal pro-delta, subaqueous delta plain, and braided-stream delta plain facies (Figs 5I, 8). Individual lobate deltas representing progradational events are usually 30—60 m thick, and have an area of 16—20 km^2 (Lin Changsong, 1987). The alternation of constructive and destructive processes generated very thick poly-cyclic deltaic sections along the northern basin mar-

                                    Lin Changsong et al.

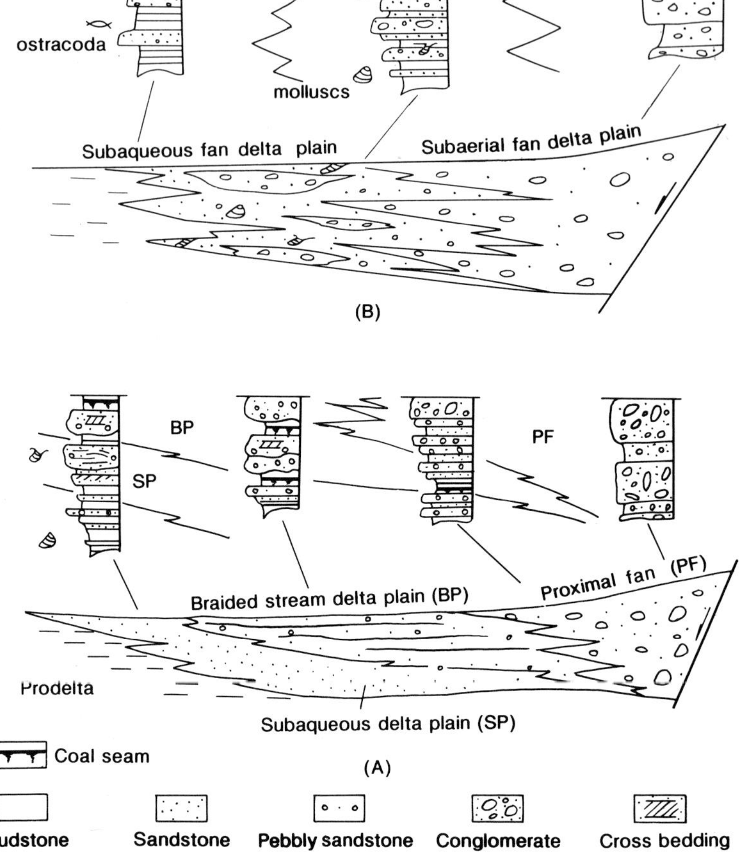

Fig. 9. Facies architecture of fan−braided-stream delta deposits (A) formed in a relatively large and shallow basin margin (Baigang Formation), and of deep-lake marginal fan-delta successions (B) developed during a deep lacustrine basin infilling episode (Upper Natu Formation).

gin in response to the intermittent activity of the boundary fault and the fluctuations in lake level. As a consequence a complex depositional architecture resulted for the braided-stream delta systems (Fig. 9A).

It has been recorded (Lin Changsong, 1987) that mineable coal seams in the Baigang Formation mainly formed during the abandonment period of the braided-stream deltas, thus providing extensive platforms suitable for peat accumulation. Moreover the braided stream channel and delta front sandstones, reworked by stream flows and wave action, are the most important oil reservoirs developed in the basin.

Deep-lake marginal fan-delta deposits

This type of fan-delta is mainly recognized in Middle−Upper Natu and Lower Baigang Formations, and was formed by the direct entering of proximal alluvial fan zones into the relatively steep and deep lake margins that had developed along the northern basin margins. These fan-deltas often developed during rapid lake-level rises or high lake-level periods (Figs 5, 7). Debris flows and flood flows from the proximal fan zones poured into the lake and did not build up relatively large subaqueous fan-deltas, but small subaerial fan-deltas near the boundary fault scarp. Two types of facies assemblages have been distinguished in these fan-delta deposits (Fig. 9B).

1 Subaerial proximal fan deposits — similar to the alluvial fan or fan-delta facies assemblages discussed above. They are light-grey coloured or grey-brownish and devoid of lacustrine fossils.

2 Subaqueous facies assemblage — dominantly

made up of massive or normally graded conglomerates and gravelly mudstones deposited by subaqueous debris flow and parallel laminated and graded sandstones that formed as sheet-flood deposits. Wave-generated cross-bedding, ripple marks, ripples, and trace fossils occur at the top of these units. Bioturbated sandy mudstones capping the surface of the gravel beds, and their association with fossiliferous lacustrine mudstones, indicate the subaqueous origin of these facies (Figs 5, 7, 9B).

Deep-lake marginal fan-deltas formed a relatively narrow facies zone (Figs 5, 7). It is postulated that the sedimentary accumulation was dominated by aggradation for lack of a typical delta sequence. The coarse clastics were distributed over the whole system (Fig. 9B). This feature distinguishes the assemblage from the fan—braided-stream delta system (Fig. 9A).

The analysis of seismic facies and electric-log patterns shows that some of the marginal fans in the Upper Natu Formation are enclosed spatially by thick lacustrine mudstone. Thus, they have been interpreted as subaqueous fan deposits that formed along a previous boundary fault scarp, which was located below lake level during rapid lake-level rise or high period (Fig. 5). These deposits mainly consist of subaqueous debris flow conglomerates and turbidite sandstones. It may be difficult to distinguish the deep-lake marginal fan-delta deposits from subaqueous fan assemblages. The sedimentary features of this facies remain for further, more detailed, examination.

Fluvial—lacustrine delta deposits

Apart from the marginal alluvial-fan and fan-delta complexes along the northern basin margin, a great deal of fluvial and fluvial—lacustrine delta deposits have been identified in the remaining basin areas. Although widely developed in all the basin during the early and final periods of the basin infill (Red and Jiatulin Formations), fluvial deposits were mainly formed along the southern and southwestern basin margins during lacustrine basin infill stages (Units 2—4). Thus fluvial delta deposits dominated the marginal lacustrine facies association developed in these basin zones during the lake development periods (Fig. 5).

These deposits are dominantly present in the lower Natu Formation, Baigang Formation, and Fuping Formation in southern and central parts of the basin. They were mainly formed by the fluvial systems that impinged from southern and western basin margins into the central lake basin. They consist of relatively thinner and finer sandstones than the fan deltas, and are commonly organized as coarsening upward deltaic sequences, 5—10 m thick, consisting of three major sedimentary types (bottomsets, foresets, and topsets) whose vertical stacking succession records the deltaic progradation (Fig. 10). The deltaic sequences pass laterally into interdeltaic lacustrine facies, such as bioclastic deposits, biomicrites, sheetlike bioturbated sandstones, and mollusc-bearing mudstones. Distal bars and prodelta deposits are relatively thin, and composed of thinly laminated sands and muds displaying wavy bedding and burrows. River mouth bar sandbodies show a coarsening upward sequence, ranging from 3 to 5 m in thickness, containing small-scale cross-bedding or ripple lamination and burrows in the lower part and large-scale cross-bedding in the upper part. These units are overlain by scour-based distributary channel sandstones, or capped by rootlet mudstones, coal seams, bioturbated sandstones, bioclastic facies, and other deposits formed during destructive periods. Subaqueous channel sandstones also occurred in the delta front and prodelta deposits, but they are generally thinner than those of the braided-stream deltas along the northern basin margin. These sandstones are usually 2—4 m thick and show eroded bases. They were often modified by burrowing and wave action or directly covered by lacustrine mudstones. Mud pebbles eroded from prior deposited lacustrine mudstones often occur in these sandstone facies, this being the main distinctive feature.

In the western marginal basin zones (Baise Depression), fluvial delta deposits were formed by sediment contributions from an easterly flowing river system. As a result of the syndepositional fault activity, which continuously controlled deltaic depocentres, deltaic deposits are stacked vertically, producing very thick sections about 500 m in thickness (Middle Natu Formation).

These deltaic deposits, which contain important economic coal seams and whose sandstones are sealed by several hundred metres of lacustrine mudstones, may be the major potential oil reservoirs in this area.

Lacustrine deposits

Terrigenous and carbonate sediments representing shallow and deep lacustrine facies, excluding the delta deposits, are well developed in the Natu,

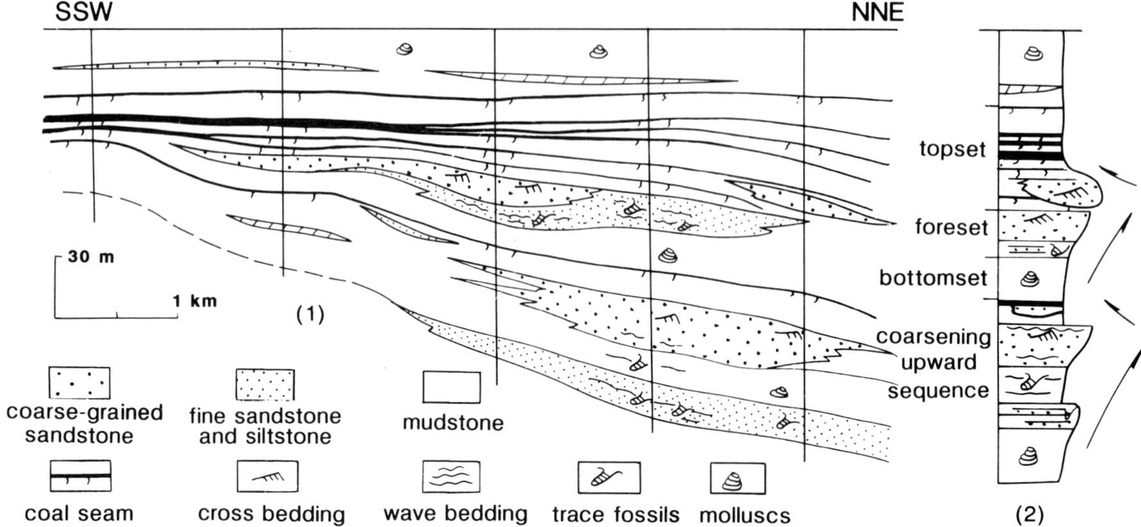

Fig. 10. Depositional section of fluvial delta deposits in Baigang Formation along the southern basin margin. Note the coarsening upward deltaic sequences, the lenticular sand body geometry, and the thinning of the coal seams towards the basin centre.

Baigang, and Fuping Formations. Shoreline and shallow lacustrine deposits formed mainly in interdeltaic areas, while deep lacustrine facies were deposited in central, inner lake basin zones.

Terrigenous shoreline and shallow lacustrine facies

These facies are relatively thin but widely spread in the basin, resulting from frequent fluctuations in lake level. Distinct lithofacies in these facies assemblages display well-developed, wave and current generated structures and horizontal lamination often affected by bioturbation. They typically contain numerous burrows, mollusc remains, and other bioclasts, as well as plant debris.

Common lithofacies, characteristic of terrigenous shoreline and shallow lake deposits include:

1 brown or grey mudstones containing a large number of well-preserved molluscs and siderite concretions;

2 siltstones and very fine sandstones with intensively bioturbated lenticular bedding;

3 well sorted, fine-grained sandstones with low-angle cross-bedding;

4 grey and brownish siltstone or fine sandstones displaying well developed, small-scale cross-bedding, wavy bedding, and trace fossils (Fig. 11A).

Their sedimentological features indicate that these facies were formed in shallow lake palaeoenvironments above the wave base. These facies are usually organized into coarsening upwards or shallowing upward sequences 1–3 m thick, apparently thinner than that of deltaic deposits. Open, shallow lacustrine mudstones (facies 1) are overlain by sandy mudstones or siltstones (facies 2) with bioturbated structures and lenticular beddings. These grade upward to siltstone or fine-grained sandstones (facies 3 and 4) which display cross-bedding, wavy bedding, low-angle cross-bedding, and erosive surfaces.

Well sorted, fine-grained and sheet-like sandstones with low-angle cross-bedding and gradually coarsening upward sequences characterize the longshore bar sandstones (Fig. 11c). Associated ripple marks, 'U' shaped or vertical trace fossils, and mollusc remains are commonly preserved in their upper part. The sandstone bodies are commonly 2–4 m thick, range at least from 30 to 100 m long, and were likely deposited parallel to the shoreline. Their upper parts were occasionally eroded by channelized, storm-generated channel flows (Fig. 11C). Mudcracks, rootlets, and lenticular bioclastic accumulations (5–10 cm thick), which formed as shoreline shell banks, are characteristic features in this assemblage. Moreover vertebrate remains are often present in the lakeshore mudflat mudstone facies.

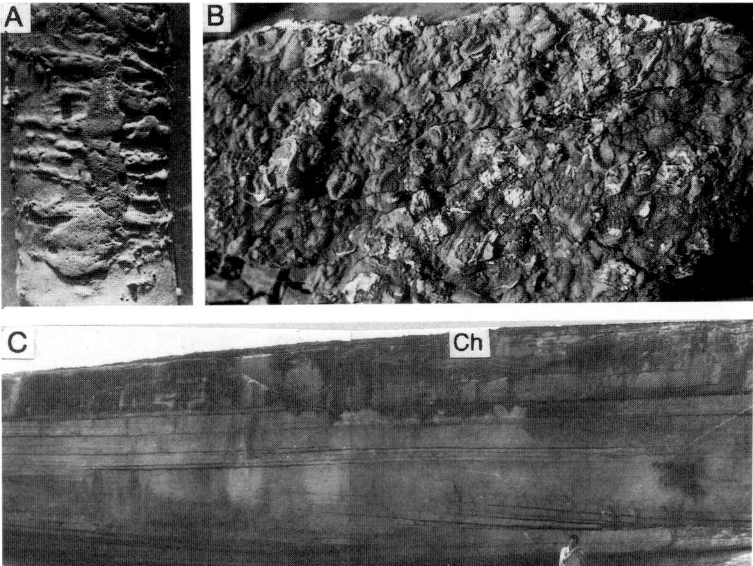

Fig. 11. Lacustrine facies: (A) trace fossil in shallow-lacustrine sandstone deposits (B) shelly limestone facies formed in shallow lacustrine environment, Lower Natu Formation (C) sheet-like, fine-grained coarsening upward sandstone bodies displaying low-angle cross-bedding, wavy bedding, horizontal lamination, ripple marks, and trace fossils. These deposits may record longshore bar sedimentation. Note incision in the upper part by lenticular channel fills (Ch), which may be storm-generated channel deposits (Baigang Formation).

Carbonate shoreline and shallow lacustrine facies

These deposits mainly constitute the Lower Natu Formation in the western part of the basin (Fig. 12). They are dominantly composed of bioclastic facies, oolitic limestones, biomicrites, shelly limestones, and bioclastic limestones. Wavy bedding and clastic-oriented fabric in bioclastics, bioclastic limestones or biomicrites indicate wave reworking during their deposition along the nearshore zone. Shelly limestones are usually 0.5–2 m thick, mainly composed of usually well-preserved gastropods and other molluscs (Fig. 11B). They were likely formed as shell accumulations in low-energy embayments or in the interdeltaic areas (Lin Changsong, 1987). Lenticular bioclastic accumulations consisting mostly of fine to coarse-grained bioclasts were likely deposited following the shoreline trend. They are interpreted as shoreline shell bank deposits similar to those often observed along modern shorelines. These deposits were usually formed in the basin margin zones and are occasionally found in coal seams. The shoreline and shallow carbonate lacustrine facies are mainly developed along the basin marginal areas, and they grade basinward into thick micritic limestone sections formed in open lacustrine environments (Fig. 12).

Shallowing upward sequences have been recog-

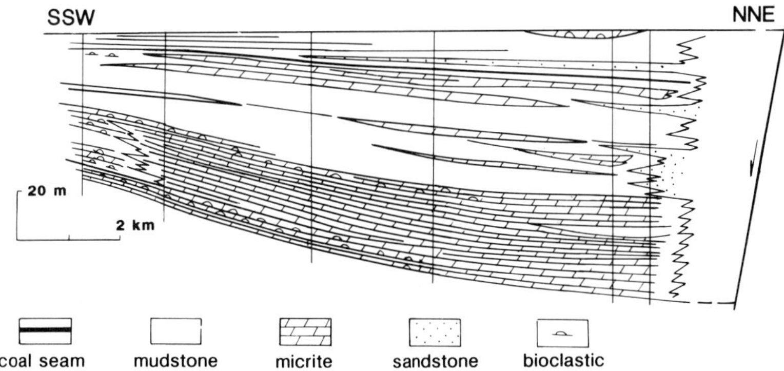

Fig. 12. Carbonate shallow and deep-lacustrine deposits of the Lower Natu Formation. The marginal bioclastic limestones and bioclastic facies grade into central massive or laminated micrites formed in open lacustrine environments.

nized in carbonate lacustrine deposits of Lower Natu Formation and they usually include several facies elements from bottom to top: (i) massive or horizontally laminated micrites of relatively deep lacustrine environment; (ii) biomicritic, oolitic, and wavy bedded bioclastic limestones deposited in wave-working zone, and (iii) rootlet swamp micrites or biomicrites, overlain by coal seams. These sequences are repeated in the sections and indicate frequent lake level fluctuations.

Deep lacustrine deposits

These deposits are commonly present in the Natu, Baigang, and Fuping Formations, and comprise dark mudstones, micrites, thin turbidite sandstones and slump breccias (Figs 5, 13). These lacustrine deposits are either massive or show horizontal laminations displayed by fine-grained graded bedding, striped side-rites, and so on. These features indicate that they were formed in deep lake environments below wave base. Associated fossils include fish remains, small Gastropoda, Ostracoda, and insects, but lacking any bivalve molluscs.

The Upper Natu Formation, an extremely thick mud member ranging from 400 to 600 m in thickness, is dominated by dark and massive mudstones with high organic carbon (1.5–3.0%), which have been considered as typical deep lacustrine deposits. This mud member is the major oil source rock in the basin. In its turn the overlying Baigang Formation and the underlying Middle Natu Formations are, respectively, the major coal-bearing and oil-producing units in the basin (Figs 2, 10).

Relatively thick lacustrine mudstones can be identified by seismic facies analysis. They are often represented by weak reflections and low-velocity zones with high continuity, weak or mid-amplitude, and low frequency (Fig. 5). These mudstone members with typical reflection characteristics, particularly the Upper Natu Formation, are the major markers for the correction of time-stratigraphic units throughout the whole basin.

DEPOSITIONAL BASIN MODEL FOR PREDICTIVE EXPLORATION

The depositional pattern of the Baise Basin infill (Figs 14, 15, 16) shows clearly that the infilling of the basin experienced various phases represented by five distinct genetic stratigraphic units or depositional episodes. The spatial facies configuration and distribution in each one of these genetic units are characteristically asymmetrical owing to the half-graben basin structure. These features markedly controlled the forming period and distribution of major economic deposits in the basin. Each one of the depositional episodes is characterized by a typical facies association of the basin fill, as briefly discussed below.

Unit 1: early alluvial–fluvial stage. Initial accumulation in the early evolutionary stage of the basin development mainly resulted in alluvial fan and fluvial deposits (Red Formation and lowest part of Natu Formation). These are composed of poorly sorted red or brown-coloured breccias, conglomerates, pebbly sandstones, and mudstones with calcareous concretions, possibly formed under relatively arid palaeoclimatic conditions. Alluvial fan deposits dominantly developed along the northern basin margin and some parts of the southern side where small-scale early boundary faults occurred. Fluvial and local shallow-lake deposits, containing gypsum and salt pseudomorphs, are dominantly distributed over the central and southern basin areas (Fig. 14A). The thickness of unit 1 ranges from 1 to 300 m and varies considerably in different places as a result of high relief topography in the basin.

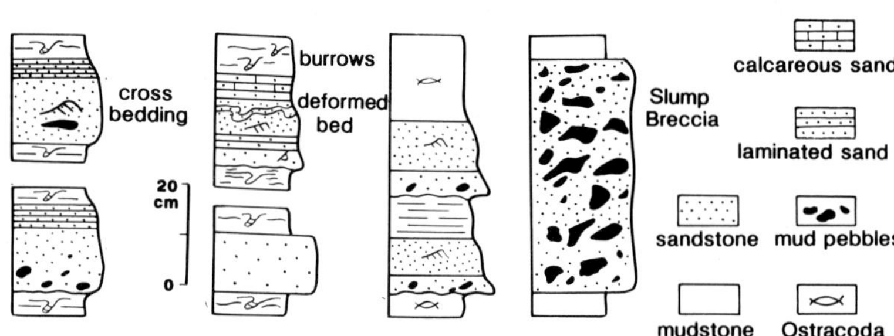

Fig. 13. Vertical sequences of thinly bedded turbidite sandstone in Upper Natu and Lower Baigang Formations.

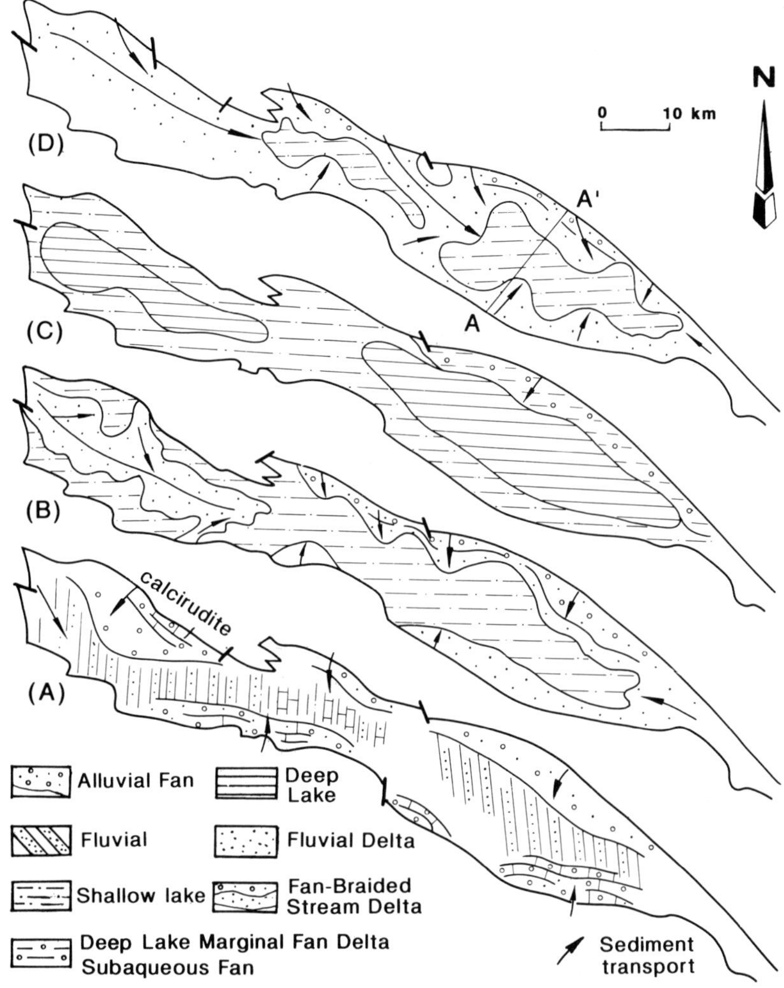

Fig. 14. Depositional patterns of the Baise Basin in different episodes of the basin infilling evolution: (A) alluvial−fluvial basin fill (unit 1); (B) shallow-lacustrine basin fill (unit 2); (C) deep-lacustrine basin fill (unit 3); (D) large shallow-lacustrine basin fill (unit 4). Note that each basin infill episode shows a distinctive facies configuration and a noticeable asymmetry in the facies development.

Unit 2: shallow lacustrine stage. This unit, including mainly the Lower and Middle Natu Formation, is characterized by the development of shallow lacustrine and delta deposits containing major workable coal seams. The lower part of the unit was mainly formed in partly isolated small lakes. The Lower Natu Formation was deposited in basin subdepressions and shows different facies associations. Thus in the western part of the basin (Baise Depression), major carbonate lacustrine deposits developed (Fig. 12), while the dominant facies of the Lower Natu Formation in the eastern part (Tian Tong Depression)

were fluvial sandstones and flood or shallow lacustrine fine-grained deposits. During the formation of the Middle Natu Formation the early small lakes gradually extended and connected to become a large shallow lacustrine basin. The basin depocentre during this time was in the western part. Along the northern basin margin, alluvial fan, braided-stream delta, and longshore sand bar deposits developed. Terrigenous shallow lacustrine and local small delta facies were mostly deposited in the central basin and along the southern margin. Relatively thick deltaic sections in the western central part of the basin were formed by

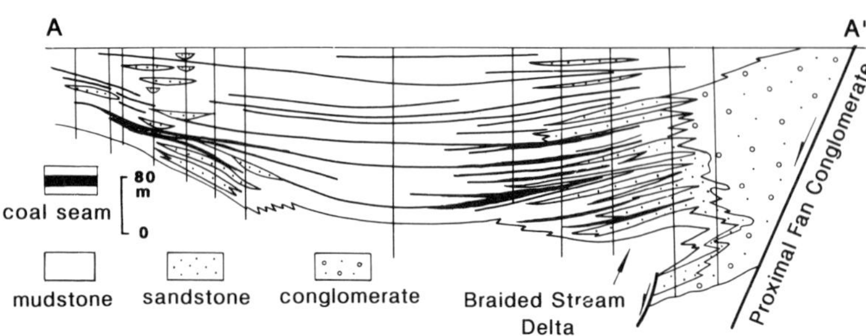

Fig. 15. Depositional cross-section (A−A′ in Fig. 14) of the Baigang Formation. Three facies zones can be distinguished across the basin: (1) northern marginal fan−braided-stream delta; (2) central shallow−deep lacustrine, and (3) southern fluvial delta and shallow lacustrine deposits. The major coal seams are mainly associated with the braided-stream delta and fluvial delta−shallow lacustrine deposits along both sides of the basin.

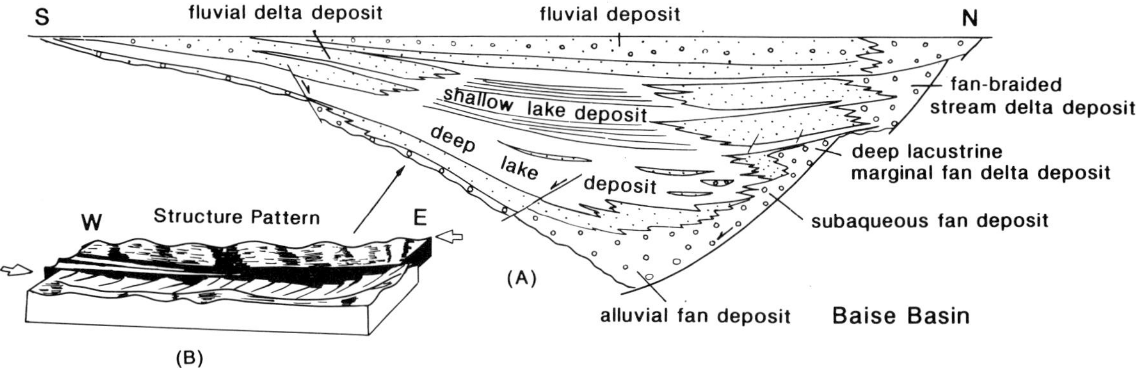

Fig. 16. Schematic depositional model (A) and structure pattern (B) of the Baise Basin. The asymmetric facies development is related to the half-graben structural framework, while the facies vertical changes in the basin fill are attributed to the tectonic evolution of the basin, from initial sinistral transtension to final dextral transpression movement along the You Jiang fault system.

a west to east progradation of an axial delta system (Fig. 14B).

The Lower Natu Formation contains mineable coal seams that accumulated in a small lake environment and are associated with micritic carbonate lacustrine facies, which were preserved in subdepressions. The coal-forming deposits of the Middle Natu Formation are mainly related to the axial deltaic and to the northern marginal fan−braided-stream delta systems. Their sequences include most of the major coal-bearing members in the basin fill sequence. The sandstones of these delta deposits, in particular in the fan−braided-stream delta facies along the northern margin, are predicted to be the potential major oil reservoirs.

Unit 3: deep lacustrine stage. During this period the basin extended and developed into a larger and deeper lake basin in response to the subsidence of the whole basin or rising lake levels. Deep and shallow lacustrine deposits more than 600 m thick, comprising the major source rock and dominated by mudstones, are distributed over most of the basin areal extent (Upper Natu Formation). Along the northern margin, subaqueous fan and deep-lake marginal fan-deltas formed a narrow coarse-grained marginal facies belt, in front of which a turbidite sandy complex, associated with deep lacustrine mudstones, developed (Fig. 14C).

Unit 4: large shallow lacustrine stage. During this period, the basin depocentre shifted to the eastern part of the basin and after a prior extensive overlapping along the basin margins, the lake basin became shallow and was infilled dominantly by deltaic deposits (Fig. 14D).

Widely developed deltaic systems (Baigang and Fuping Formations) characterize this part of the basin fill, which includes fan−braided-stream deltas along the northern margin, small shore-lake deltas on the southern side, and an axial delta system prograding eastward into the central basin part. The depositional facies of the Baigang Formation in this unit can be divided into three zones (Fig. 15, A–A' in Fig. 14; Lin Changsong, 1987):

1 fan–braided-stream delta deposits along the northern basin margin;

2 prodelta, shallow and deep lacustrine deposits in the central part of the basin;

3 shoreline and small-scale deltaic deposits in the southern basin zones.

Well-developed peat swamp deposits occur on the abandoned delta plain or interdeltaic areas, particularly in the braided-stream delta plains along the northern margin (Fig. 15). These sandstones are affected by various structural unconformities and make up stratigraphic oil traps in the basin. Thus, these sandstone facies are another major coal-bearing and oil-producing member that overlies the thick organic-rich lacustrine mudstones.

Unit 5: alluvial–fluvial stage. This unit mainly comprises alluvial deposits along the northern margin and fluvial deposits in the central and southern side. Some minor lacustrine fine-grained deposits formed in parts of the central basin area. Because the hydraulic balance was likely negative, peat accumulation occurred rarely. Finally red-coloured sediments with some gypsum formed. These deposits record the late infilling of the basin during a period of compression and uplift.

It is suggested here that the basin infilling evolution from unit 1 to unit 5 is linked to a change in tectonic setting. Integrative analysis of structural and depositional features shows that the initial formation of the basin and following alluvial–fluvial basin fill resulted from the cessation of sinistral transtensional movement along the You Jiang fault system. The sinistral strike-slip displacement of the regional northwest-west trending faults is in fact a regional tectonic event which affected southeastern China during the Paleogene. As a consequence of transtension and sediment loading, the basin sank rapidly and became a large shallow and deep lacustrine basin. During this stage sedimentation extended beyond the original basin margins (units 2–4). When extension ceased, the transpressive shearing displacement along the You Jiang fault system started and the lacustrine basin gradually filled up and became an alluvial–fluvial basin (unit 5), This structural and depositional basin evolution is analogous to that recorded in the Mesozoic, coal-bearing, fault-bounded lacustrine basins in northeast China (Li Sitian *et al.*, 1984).

CONCLUDING REMARKS

The Baise Basin is an elongate half-graben (90 × 16 km). It formed during the Eocene as a result of transtensional movement along the You Jiang Fault, which defines its northern basin margin. The basin was infilled with alluvial–fluvial and lacustrine deposits more than 3000 m thick, which bear coal, oil, and other economic deposits. Baise Basin shows an asymmetric stratigraphic framework and facies development that is likely attributed to the half-graben structure, which caused differential subsidence along the northwest-west boundary and its subsidiary faults on the northern margin. Meanwhile, another major group of northeast striking syndepositional faults divided the basin into several subordinate depressions and uplifts, which also exerted a great influence on the variation in facies and thickness.

Five genetic stratigraphic units have been identified on the basis of facies association and basin setting analysis. They show that the basin experienced various infilling episodes, generally from the initial alluvial–fluvial basin fill (unit 1), to shallow and deep lacustrine basin fill (units 2–4), and finally to alluvial and fluvial basin fill (unit 5), which may be attributed to the tectonic evolution of the You Jiang fault system from early sinistral transtension, through intensive subsidence, and finally to transpressive movements.

Several major facies associations in the alluvial–fluvial to lacustrine basin fill (units 2–4) have been identified: (i) alluvial fan and (ii) fan-delta complexes developed along the northern boundary fault, including fan–braided-stream delta, deep-lake marginal fan-delta, and/or subaqueous fan deposits; (iii) fluvial and fluvial lacustrine deltas comprising the small-scale fluvial deltas from the southern side of the basin and the axial fluvial and delta system fed from the western basin margin; (iv) terrigenous and carbonate shallow and deep lacustrine deposits, which are mostly present in central basin areas and consist dominantly of fossiliferous mudstones, bioclastics, bioturbated sandy mudstones, and thinly bedded turbidite sandstones. The asymmetric development of depositional facies is attributed to the half-graben structure.

The formation and distribution of economic deposits in the Baise Basin were strongly controlled by the palaeostructures and depositional pattern. Peat swamp deposits developed in relation to some of the fan-delta and fluvial deltaic assemblages, as well as to the shallow lacustrine assemblages. Deep-lacustrine mudstones (Upper Natu Formation) more than 600 m thick, which formed during a deep-lake basin infilling period, are major oil source rocks. The two major coal-bearing and oil-producing members in the basin underlie and overlie this mudstone member. On a basinal scale, the thickest coal-bearing sections and potential reservoir sandstones are confined to subdepressions. The coal-rich zones in subdepressions are usually associated with fluvial–lacustrine delta, shallow-lacustrine deposits and braided-stream delta deposits along the northern basin margin. The braided stream and fluvial delta sandstones comprise the major potential oil reservoirs. The migration of the depocentre also caused the shifting of peat accumulation from west to east. The overlapping of alluvial fan and fan-delta systems caused by back-faulting at the northern boundary fault produced the most important structure and unconformity oil traps found in the basin.

ACKNOWLEDGEMENTS

We thank Drs Ll. Cabrera, M. Esteban, K. Kelts and Professors Li Baofang and Liu Hepu, as well as an anonymous reviewer, for thier helpful comments and suggestions.

REFERENCES

BROWN, L.F.JR. & FISHER, W.L. (1977) Seismic-stratigraphic interpretation of depositional systems: examples from Brazilian rift and pull-apart basins. In: *Seismic Stratigraphy — Applications to Hydrocarbon Exploration* (Ed. Payton, C.E.), Mem. Am. Assoc. Petrol. Geol. **26**, 213–248.

CHEN BAIYUE & XIE YONGSEN (1987) On the depositional system of Baise basin, Guang Xi, and the accumulation of coal and oil. *Earth Sci. J. Wuhuan Coll. Geol.* **12**(1), 57–65.

COLLINSON, J.D. & THOMPSON, D.B. (1982) *Sedimentary Structures*. George Allen & Unwin, London, 194 pp.

GALLOWAY, W.E. & HOBDAY, D.K. (1983) *Terrigenous Clastic Depositional Systems*. Springer Verlag, New York, 423 pp.

HEWARD, A. (1978) Alluvial fan sequence and megasequence models: with examples from Westphalian B Coalfields, Northern Spain. In: *Fluvial Sedimentology* (Ed. Miall, A.D.), Can. Soc. Petrol. Geol. Mem. **5**, 669–802.

HOLMES, A. (1965) *Principles of Physical Geology*, 2nd edn. Ronald Press, New York, 1288 pp.

HU YIANKUIN (1979) Classification and correction of Tertiary strata in Guang Xi Autonomous Region. *Spec. Publ. Guang Xi Oil Geol.* **1969–1979**, 77–94.

LI SITIAN, LI BAOFANG, YANG SHIGONG, HUANG JIAFU & LI ZHEN (1984) Sedimentation and tectonic evolution of Mesozoic faulted coal basins in northeastern China. In: *Sedimentology of Coal-bearing Sequences* (Eds Rahman, R.A. & Flores, R.M.), Spec. Publ. Int. Assoc. Sediment. **7**, 387–406.

LIN CHANGSONG (1987) Coal accumulation model of the Baigang Formation in Baise Basin and its predicting significance. *Acta Sediment. Sin.* **5**, 113–123.

McGOWEN, J.H. (1970) Gum Hollow fan delta, Nueces Bay, Texas. *Tex. Univ. Bur. Econ. Geol. Rep. Inv.* **69**, 91 pp.

McPHERSON, J.G., SHANMUGAM, G. & MOIOLA, R.J. (1987) Fan-deltas and braid deltas: varieties of coarse-grained deltas. *Geol. Soc. Am. Bull.* **99**, 331–340.

McPHERSON, J.G., SHANMUGAM, G. & MOIOLA, R.J. (1988) Fan deltas and braid deltas: conceptual problems. In: *Fan Deltas. Sedimentology and Tectonic Settings* (Eds Nemec, W. & Steel, R.J.), pp. 14–22. Blackie & Son, Glasgow.

NEMEC, W. & STEEL, R.J. (1988) What is a fan delta and how do we recognize it?. In: *Fan Deltas. Sedimentology and Tectonic Settings* (Eds Nemec, W. & Steel, R.J.), pp. 3–13. Blackie & Son, Glasgow.

READING, H.G. (1980) Characteristics and recognition of strike-slip fault systems. In: *Sedimentation in Oblique-slip Mobile Zones* (Eds Ballance, P.F. & Reading, H.G.), Spec. Publ. Int. Assoc. Sediment. **4**, 7–26.

SANGREE, J.B. & WIDMIER, J.M. (1977) Seismic stratigraphy and global changes of sea level, part 9: seismic interpretation of clastic depositional facies. In: *Seismic Stratigraphy — Applications to Hydrocarbon Exploration* (Ed. Payton, C.E.), Mem., Am. Assoc. Petrol. Geol. **26**, 165–184.

STEEL, R. & GLOPPEN, T.G. (1980) Late Caledonian (Devonian) basin formation, western Norway: signs of strike-slip tectonics during infilling. In: *Oblique-slip Mobile Zones* (Eds Ballance, P.F. & Reading, H.G.), Spec. Publ. Int. Assoc. Sediment. **4**, 79–103.

WESCOTT, W.A. & ETHRIDGE, F.G. (1980) Fan-delta sedimentology and tectonic setting — Yallahs Fan-delta, Southeast Jamaica. *Bull. Am. Assoc. Petrol. Geol.* **64**, 374–399.

Spec. Publs Int. Ass. Sediment. (1991) **13**, 93–107

Tertiary lignite-bearing lacustrine facies of the Zittau Basin: Ohře rift system (Poland, Germany and Czechoslovakia)

J.R. KASIŃSKI

Geological Survey of Poland, Rakowiecka, 4, 00-975 Warszawa, Poland

ABSTRACT

Polyphasic faulting related to Tertiary rift systems in central Europe led to the development of fluvial, alluvial, and lacustrine basins favourable for the deposition of lignite. The continental Ohře rift system (Ohře is a river which in German is named Eger) along the Erzgebirge (mountain range in the borderlands of Germany and Czechoslovakia) shows this to be true. The Zittau Basin in the northwest part of the Ohře rift system contains Oligocene to middle Miocene sediments with mineable lignite, and in its lower parts volcanic and pyroclastic intercalations. These volcanic layers reflect local tectonic phases that influenced the development of the coal-bearing sequences. The sediments of the Zittau Basin appear to be connected closely with the structural developments controlling the distribution of fluviatile and lacustrine basins. Thick coal seams of lignite and sub-bituminous coal were deposited in the lacustrine systems in the central basin zones. Four major lacustrine environments developed, characterized by fan-delta, peat bog, transitional–marginal, and open lacustrine facies.

INTRODUCTION

Deep tectonic-rift troughs are commonly characterized by lacustrine sediment infill. Suitable palaeoclimatic conditions lead to the deposition of coal-bearing sequences within them. The significance of different palaeoenvironmental settings for the deposition of coal has been discussed extensively. Several authors (e.g. Frazier & Osanik, 1968; Baganz *et al.*, 1975) have suggested deltaic environments as the most suitable for the accumulation of plant remains. Recent work on non-marine basins, however (Galloway & Hobday, 1983; McCabe, 1984), opened new perspectives, as illustrated by mineable lignites in alluvial–lacustrine depositional systems in the central European rift basins.

The Ohře rift is a part of the central European rift system and is located between the Erzgebirge and the Bohemian massif, situated in the territories of Czechoslovakia, Poland, and Germany (Fig. 1). It records the developments of an intra-plate continental thermal rift (Kopecký, 1978,1979; Ziegler, 1982,1988; Kanasiewicz, 1983). Rifting processes were accompanied by volcanic activity and formed fault-bounded basins, most of which contain coal-bearing sediments (Kanasiewicz, 1983; Kasiński, 1984,1989a; Kasiński & Panasiuk, 1987; Brause, 1988).

This paper describes the major sequential organization and diverse lithofacies characterizing the sedimentary infill of some basins of the Ohře rift complex, with special emphasis on the Zittau Basin. The influence of tectonics on the sediments of the basins and the significance of this setting for coal accumulation in the Tertiary intra-plate continental rift are documented.

GEOLOGICAL SETTING: THE OHŘE RIFT

Several rift systems developed during the Tertiary on the foreland platform zones of the Alpine and Carpathian thrust–fold belts (Pozarisky & Brochwicz-Lewinsky, 1978; Malkovsky, 1980; Ziegler, 1987,1988). These rift systems are related to earlier fracture lineaments, which were reactivated during the post-Laramian extensional

                    J.R. Kasiński

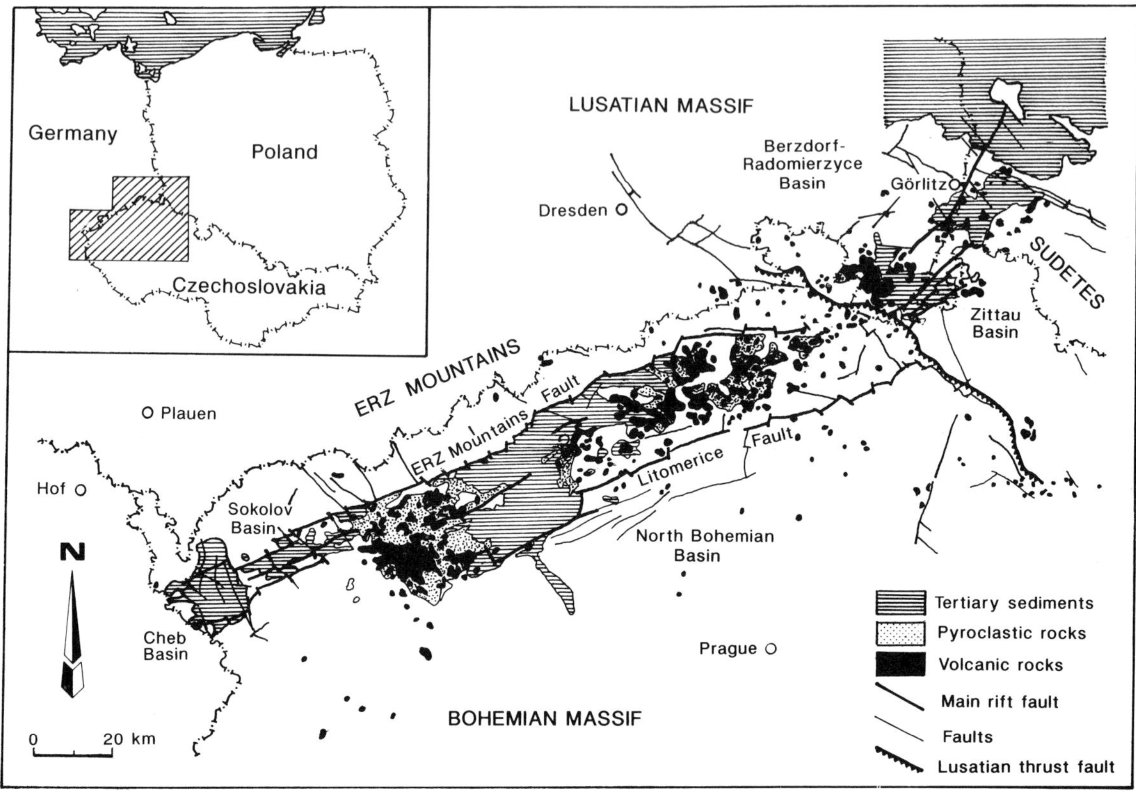

Fig. 1. Geological sketch of the Ohře rift showing its position between the Bohemian and Lusatian massifs (Erzgebirge) (slightly modified from Kasiński & Panasiuk, 1987).

episodes and affected the foreland platform zones.

The Ohře rift system (Erzgebirge Graben) is one of the main rift complexes of central Europe, and according to Kopecky (1978) is part of the rift system that includes the Lower Rhine Graben and the Lusatia−Elbe tectono-volcanic belt at the north-eastern margin of the Bohemian massif (Fig. 1).

The WSW−ENE striking longitudinal Ohře rift is over 240 km long and 30 km wide, and developed on the Proterozoic crystalline basement. The rift extends between the Erzgebirge in the north-northeast and the Bohemian Massif in the south-southeast (Fig. 1). Most of it is located in Czechoslovakia, though its north-northeast end stretches as far as Poland and Germany. Two major basement faults bound the Ohře rift; the Erzgebirge Fault to the north-northwest and the Litomerice Fault to the south-southeast

Until recently the Ohře rift was generally interpreted as a graben between the northwest part of the

Bohemian Massif and Erzgebirge horst. More recent studies (Kopecký, 1979; Malkowsky, 1980) showed that the uplift of this horst is much younger than the basin fill.

Various geophysical, geological, and geochemical evidence has made it possible to classify this structure as a continental palaeorift *sensu lato* (Kopecký, 1978,1979; Zeman, 1979; Kanasiewicz, 1983,1988; Brause, 1988). These are: extensive alkaline volcanism with carbonatite intrusions (Kanasiewicz, 1983), the presence of polzenite dykes (Kopecký, 1978), the occurrence of deep faults along basin boundaries and high-value heat flow and extensive mineralization on fault surfaces recorded in the central rift zone (Čermak *et al.*, 1968). This is contrary to Malkovský's theory (1980) who regarded it as a complex of individual gravitational tectonic depressions formed as spatial equivalents of great volcanic eruptions in the neighbouring areas.

The earliest volcanic activity recorded in the Ohře

rift began in the Late Cretaceous during a Laramide compressional phase. The oldest volcanic rock (syenitoid) is dated at 84.1 Ma (Panasiuk, 1986). This episode probably correlates to the initial rifting phase, proposed by Kopecký (1978) as ranging between 60.0 and 64.7 Ma. Later volcanic activity phases, combined with successive rifting episodes, all took place within the Tertiary (Figs 2, 3).

The last volcanic episode in the region (6.6–0.2 Ma; after Sibrava & Havlíček, 1980) was not related to the relaxation phase, which post-dates the rift activity. It shows a markedly compressional character and determined the late uplift of the Ohře Mountains block (Wolf & Hirschmann, 1964; Malkowský, 1980) and a few internal areas of the rift and the Lusatian overthrust. Volcanic activity has been recorded in recent times: an earthquake in 1908 in the Cheb Basin, a geyser in the town of Karlovy Vary (close to the Sokolov Basin), and a warm-water eruption at the floor of the 'Turow II' open-pit mine in the Zittau Basin, observed by the author in 1982.

BASIN INFILL OF THE OHŘE RIFT BASINS

Five Cenozoic sedimentary basins are located along the axis of the Ohře rift system. They are, from west-southwest to east-northeast, the Cheb, Sokolov, North Bohemian, Zittau, and Berzdorf-Radomiercyze Basins (Fig. 2). These basins are separated by horsts and outcrops of volcanic rocks. Their geological framework is well known from numerous lignite open-pits, deep subsurface mines, and hundreds of boreholes.

The beginning of Tertiary sedimentation along the Ohře rift zones was heterochronous. The earliest, late Eocene to earliest Oligocene, sequences have been recorded only in the southwestern basins (Cheb, Sokolov, and North Bohemian). From late Oligocene to the early Miocene, sedimentation spread all over the rift basins. Later it was again restricted to the northeastern rift basins (Zittau and Berzdorf-Radomierzyce Basins) during the middle–late Miocene (?) to Pliocene.

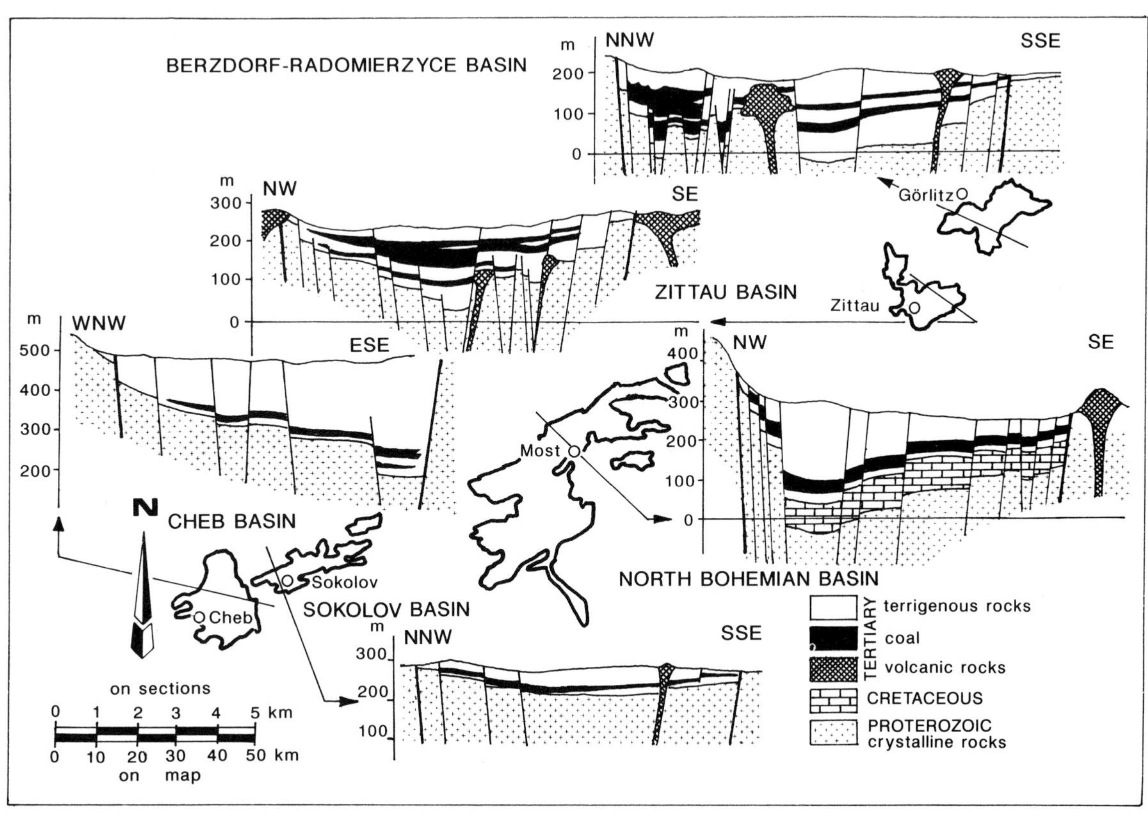

Fig. 2. Cross-sections of the Tertiary sedimentary basins related to the Ohře rift system.

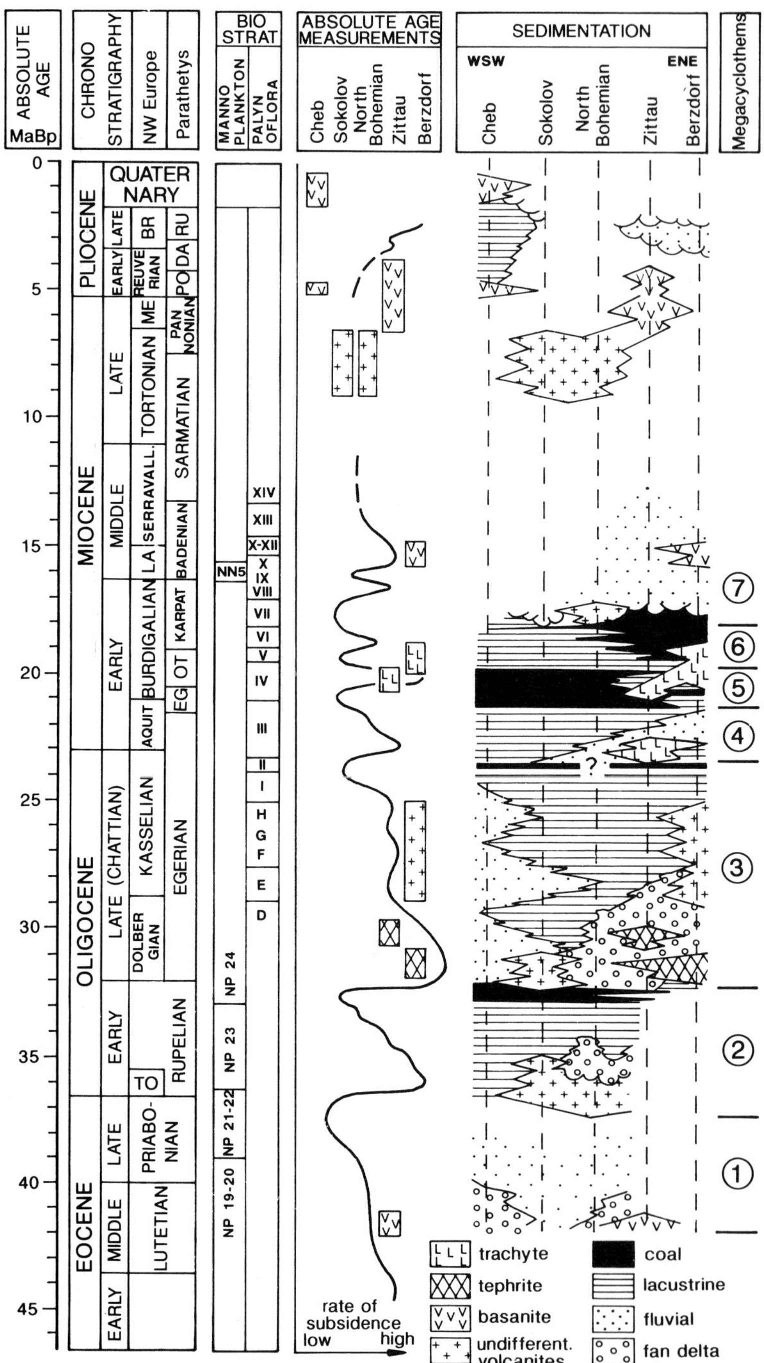

Fig. 3. Sedimentary facies succession in the Ohře rift basins related to stages of local tectonic and volcanic activities. Note the subdivision of the basin infills into major stratigraphic units (megacyclothems) recorded in the diverse basins. BR, Brussumian; DA, Dacian; EG, Eggenburgian; LA, Langhian; ME, Messinian; OT, Ottangian; PO, Pontian, RU, Rumanian; TO; Tongrian.

The sedimentary records of the Ohře rift basins, as well as in some neighbouring areas, display common features. The sedimentary basin infills are up to several hundred metres thick and consist mostly of terrigenous (conglomerates, sandstones, and mudstones) and organogenic (lignites) facies (Fig. 2).

Minor biomicritic limestones or marls, displaying diverse contents of freshwater fauna and vegetal remains have been recorded only in the North Bohemian and Sokolov Basins. Although the lowermost basin fill deposits in most of the basins consist of some tens of metres of conglomeratic and sandy facies, fine-grained, mudstone-dominated successions intercalated with changing amounts of sandstone facies are dominant in most of the overlying basin fill (Fig. 2). Lignite seams ranging from a few decimetres up to 90 m in thickness are included in these sedimentary successions.

The basin fills in the Ohře rift basins can be split into several major sedimentary sequences (megacyclothems, *sensu* Sakamoto, 1957) ranging from some tens to a few hundred metres in thickness. These first-order, major sequences consist of a lower coarse-grained unit, which is overlain by fine-grained dominated terrigenous facies assemblages overlain by a relatively thick lignite seam, sometimes capped by a fluvial dominated succession (Fig. 3).

The sedimentary evolution in the Ohře rift basins was punctuated by regional and/or local tectonic pulses accompanied by volcanic activity (Malkovský, 1980; Kasiński, 1984,1989a,b; Kasinski & Panasiuk, 1987; Brause, 1988). Each volcanic episode was probably related to a renewal of activity along the rift faults, this fact being recorded in turn by the basin fill megasequential organization.

The first Cenozoic volcanic episode in the rift system occurred in the Zittau Basin zone. There, basanites and nephelinites 41.4 Ma old have been found (Panasiuk, 1986). The correlative upper Eocene to lower Oligocene deposits of the first megacyclothem related to this eruptive episode, exist only in the Cheb, Sokolov, and North Bohemian Basins. The sedimentation of the second megacyclothem was also related to an extensive volcanic episode and also remained restricted to the same basins. The sedimentary sequences of the third and fourth megacyclothems, punctuated by successive volcanic episodes in the rift system, spread over all the rift basins and gave rise to some of the most important coal-bearing successions. The deposition of the latest megacyclothem directly related to the rift system was restricted to the east-northeast rift basins (Fig. 3). From the stratigraphic and palaeontological data currently available, it is possible to suggest that some of the major depositional processes recorded in the rift basins (i.e. the beginning and ending of the megacyclothems) were synchronous or just slightly diachronous. This fact aids the possibility

of correlating some of the episodes that generated the thickest coal seam over the entire Ohře rift system.

The estimated lignite resources in the Ohře rift basins and the major average chemical and technological parameters (ash content, heating value, sulphur content) are summarized in Table 1.

THE ZITTAU BASIN

Major structural and stratigraphic features

The Zittau Basin is located at the crossing of the rift axis and the Lusatia—Elbe tectono-volcanic belt. This basin displays most of the structural and stratigraphic features of the Ohře rift (Fig. 4). The position was controlled by extensive subsidence related to volcanic events (Kasiński, 1984,1989a). Numerous small tectonic blocks of the basin floor have been displaced at different rates during the structural development of this basin.

The sedimentary evolution of the Zittau Basin started with a post-Laramide tectono-volcanic episode (Fig. 3) (Kasinski, 1984). Phonotephrite eruptions preceded sedimentation of the lowermost members of the third megacyclothem, the first of which was deposited in this basin. Volcanic activity persisted during the first depositionary stage. Volcanic rocks and tufaceous layers alternate with coarse-grained sediments in the deepest part of the depression (Kanasiewicz, 1988). Several faults, often with veins of volcanic rocks, cut across Tertiary sediments and displaced them during the later Miocene and Pliocene tectonic phases. Clayey—silty units overlay volcanic, pyroclastic, and coarse-grained sedimentary rocks and are covered again by a thin lignite seam. Deposition of the third megacyclothem was interrupted by the next tectono-volcanic episode.

Deposition of each successive overlying megacycle was also preceded by new tectono-volcanic episodes. The fourth to sixth megacyclothems display a lithological succession similar to that of the third megacyclothem. Petrography and absolute ages of the volcanic rocks at the beginning of every megacyclothem are shown in Fig. 3.

Sedimentation of the seventh megacyclothem (well developed in the Zittau Basin) took place mainly in the east-northeast rift zones, but it has also been recorded in extra-rift Tertiary basins at the northwest Sudetes Mountains margin (Kasiński,

Table 1. Resources and major average chemical and technological parameters of coal in the Ohře rift basins (data after Piwocki, 1987; Pések, 1988)

Coal-bearing basin	Coal rank	Resources[a] (mln Mg)	Ash content, (d.b.) (%)	Calorific value, (w.b.) (kJ × kg^{-1})	Sulphur content, (d.b.) (%)
Bersdorf-Radomierzyce (Polish part only)	Lignite	180	31.61	7880	1.30
Zittau (Polish part)	Lignite	990	18.30	9752	1.04
Zittau (Czechoslovakian part)	Lignite	?	20.00	12100	0.30
North Bohemian	Sub-bituminous	6000	35.00	13300	1.75
Sokolov	Sub-bituminous	640	20.50	12320	1.95
Cheb	Lignite	1000	19.60	8900	2.20
Average		8810	30.25	12560	1.73

[a] Without German part of the basins and Czechoslovakian part of the Zittau Basin.
(d.b.) = dry basis;
(w.b.) = wet basis.

1989a) where sedimentation was strongly influenced by fluvial processes. This megacyclothem might not have been related directly to the rift evolution.

Lithofacies description and palaeoenvironmental interpretation

The study of the Zittau Basin fill is restricted to

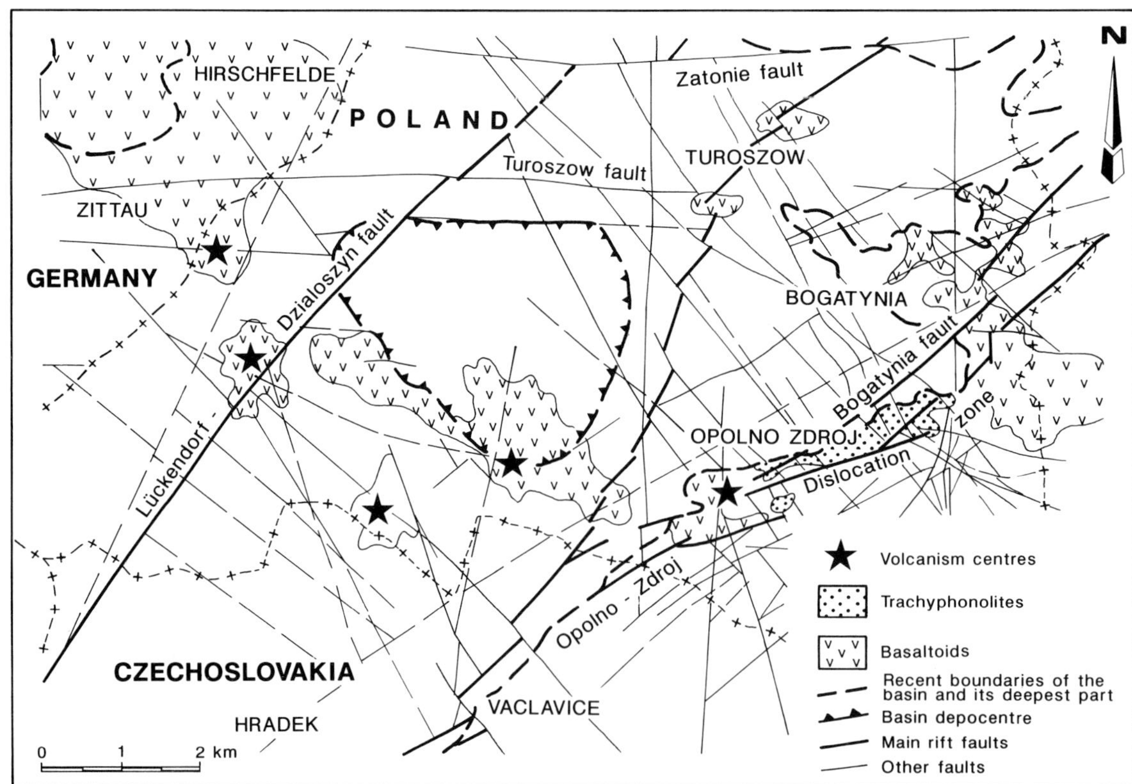

Fig. 4. Structural map of the Zittau Basin (based on Kanasiewicz, 1983; Kasiński & Panasiuk, 1987).

short-lived active open-pit trenches and boreholes, which precludes carrying out very detailed sedimentological studies. The available data were gathered from the study of more than 30 trench outcrop profiles in open pits and a great number of drill cores. The basin infill was deposited in an alluvial—fluvial to lacustrine depositional framework, which evolved from the late Oligocene until the late Miocene to Pliocene. A wide variety of facies make up the studied sequences. In broad terms, the lignite bearing units are composed of a variety of terrigenous deposits (mudstone, silt, sand, gravel) and lignite. Lithological, sedimentological, and palaeontological features were used to define the following major lithofacies.

Conglomeratic lithofacies (a)

This lithofacies consists of fine-grained conglomerates (a1) and sandy gravels (a2). The conglomerates are poorly sorted and consist of poorly rounded granitoid, granito-gneiss, volcanic, and quartz clasts. The sandy gravels consist of a quartz and feldspar sandy matrix, which includes matrix supported, poorly sorted quartz clasts (Figs 5a, 7b,c).

The conglomeratic lithofacies has not been observed in surface outcrops and is known only from borehole cores. No detailed information on fabrics and major sedimentary structures is available.

Sandy—gravelly lithofacies (b)

This lithofacies consists mainly of grey and yellow quartz and feldspar sands, and of white, grey, and yellow quartz grits (Figs 5, 7b). The sands usually display trough and planar cross-bedding, as well as small-scale cross-lamination. Sands and gravels are poorly sorted and the clasts are not well rounded. Basal scouring and erosive surfaces as well as load casts are commonly observed at the bottom of the beds formed by the lithofacies. High concentrations of plant seeds (*Xilomastixia* sp., *Mastixia amygdaleformis* and *Mastixiocaryum lusaticum*) often occur at the planar cross-bedding surfaces. The sandy beds are sometimes cemented by iron oxides, clay, or silica.

Silt—clay lithofacies (c)

This lithofacies is composed of monotonous, fine grained, and clay rich deposits, including silts (c1), silty clays (c2), and minor sandy streaks (Figs 5,

7b,c). Very fine mica blasts occur in this facies, together with the quartz grains. Small-scale cross-lamination (current ripples and ripple marks up to 15 mm in amplitude) are scarce. Load casts have been observed at the top of the beds.

Kaolinitic clay lithofacies (d)

This facies consists of pure kaolinitic clay (d1), which sometimes includes or alternates with lignitic clay intercalations (d2). Reed stalks, strongly flattened by compaction, sometimes have been observed in this facies (Figs 6, 7). The lignitic clay intercalations (d2) are rich in organic matter of lignitic dust, pollen and spores, plant detritus, and, occasionally, concentrations of seeds. Seeds of *Nyssa* sp., *Mastixia amygdaleformis*, *Mastixiocaryum lusaticum*, *Vitis lusatica*, and *Granitocera* sp. dominate this taphocoenosis. Syderitic concretions are common in both species types and were usually formed in extensive layers (see Figs 5b, 6). Palaeosoil horizons with rootmarks also have been commonly found in these facies.

Lignite lithofacies (e)

This lithofacies includes several types of lignite (see Figs 5b, 6, 7). Simplified, a threefold division may be established for this facies: (i) detrital, light (commonly yellow), low-density lignite — pyropissite (e1) — occasionally as very fine horizontal laminations; (ii) detrital dark (brown or black) denser lignite (e2); and (iii) xylitic yellow to dark brown lignite (e3). All these types show different microscopic features, particularly in pollen-spore spectra, the composition of plant detritus, and in microlithotype assemblages (Table 2). Rootmark horizons also have been observed in these facies.

Tree trunks (f)

Tree trunks are very common in many facies of the lignite-bearing association (Figs 6b, 7). A few specimen from the genera *Athrotaxis*, *Chamaeocypris*, *Cryptomeris*, *Dacrydium*, *Glyptostrobus*, *Juniperus*, *Libocedrus*, *Phyllocladus*, *Podocarpus*, *Sequoia*, *Taxodium*, and *Widdringtonia* have been identified in the basin (Zalewska, 1953,1955a,b; Greguss, 1955). *Sequoia*, *Taxodium*, and *Glyptostrobus* are the most frequent genera in the basin sedimentary fill.

They occur either as: (i) more or less continuous

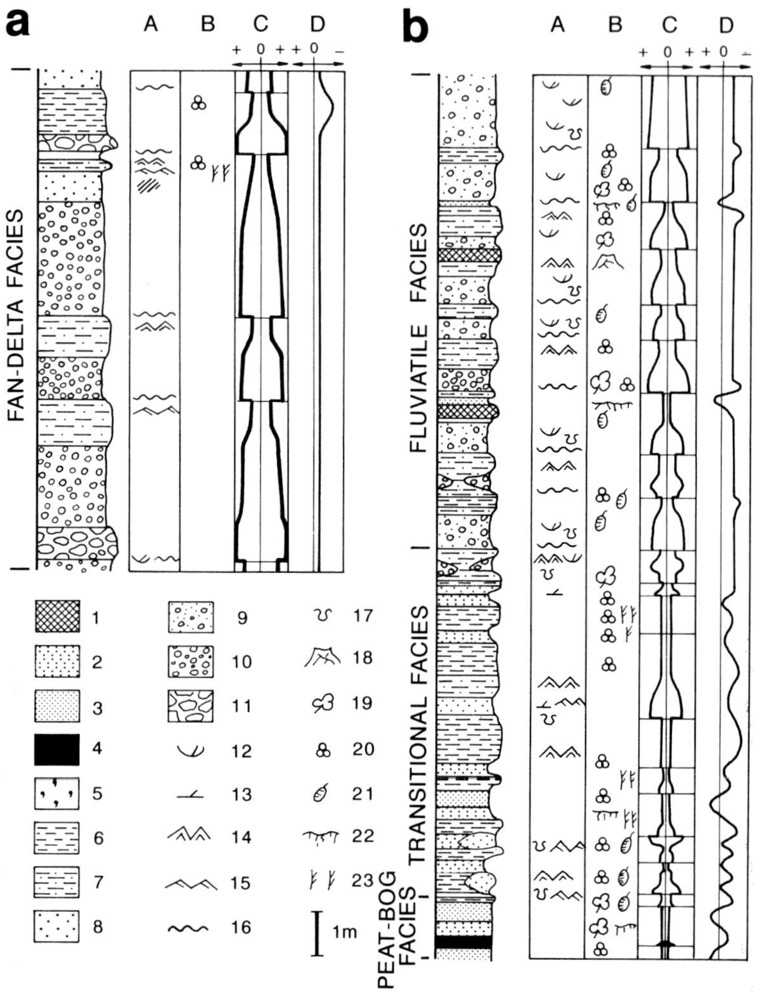

Fig. 5. Characteristic logs of the alluvial–lacustrine facies assemblage of sedimentary infill in the Zittau Depression. (a) Fan-delta facies log from the southeastern part of the basin near Opolno Zdroj, Late Oligocene, middle part of the megacyclothem 1. (b) Lacustrine to fluviatile facies, southern slope of the 'Turów II' open-cast mine, early to middle Miocene, megacyclothems 4 and 5. Facies and environmental trends; A, sedimentary structures; B, biogenic remains and structures; C, high-order sedimentary cycles (grain size variability); D, water-level oscillation. Facies and sedimentary structures: 1, siderite; 2, xylitic lignite (e3 facies); 3, detritic lignite (e2 facies); 4, pyropissite (e1 facies); 5, lignitic clay (d2 facies); 6, clay (d1); 7, silt (c facies); 8, sand (b facies); 9, sand with gravel (a facies); 10, gravel (a facies); 11, debris (a facies); 12, trough cross-lamination; 13, planar cross-lamination; 14, ripple marks; 15, current ripples; 16, erosional surface; 17, loadcasts; 18, tree trunk; 19, plant remains; 20, pollen grains (recorded in samples); 21, fruits; 22, rootlets; 23, reed stalks.

horizons, where numerous trunks are emplaced along the same stratum, their roots extending into lignite clays (d2), dark detritic lignite (e2) and palaeosoils; or (ii) irregularly deposited trunks at different levels, mostly in contact with the clastic facies (b, c1, and c2); their roots do not extend into rocks rich in organic matter or palaeosoils.

Facies assemblages and palaeoenvironmental interpretation

According to the relative development of the above-mentioned facies, their thickness relationships, geometry, mode of occurrence, and sequential organization, four main facies assemblages can be distinguished, each one corresponding to a major palaeoenvironment.

Coarse-grained dominated facies assemblage (alluvial fan and/or fan delta)

Coarse-grained dominated sequences, made up of the lithofacies 'a', 'b', and 'c' have been recorded from boreholes drilled in the basin. These lithofacies are organized into fining upward sequences up to several metres thickness, dominated by conglomeratic facies (Fig. 7b,c). The sequences are stacked up forming successions several tens of metres thick (Fig. 5a). In some southern basin sectors they can even attain up to 200 m thickness. Minor debris deposits occur intercalated with conglomerate beds. These coarse-grained sequences are more common at the marginal basin zones, its importance decreasing rather quickly towards the inner basin sectors (Fig. 8). Thin clay beds and lignite seams

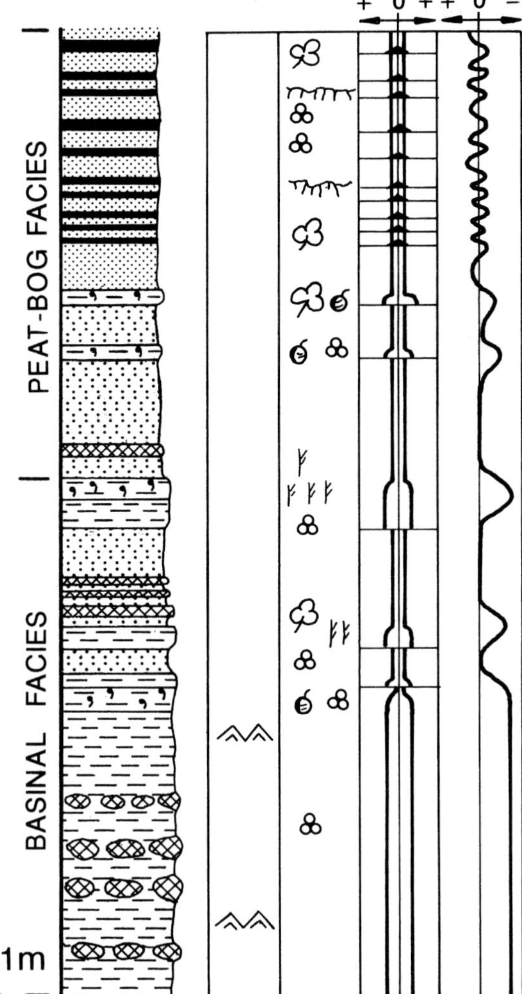

Fig. 6. Characteristic log of the open and marginal lacustrine facies assemblage of sedimentary infill in the Zittau Basin. Southeast slope of the 'Turow I' open-cast mine, early Miocene, lower part of megacyclothem 4. See Fig. 5 for explanation of symbols.

up to a few decimetres thick may occur in these successions.

Despite the scarcity of data, the sedimentological features and the recorded marginal distribution of this facies assemblage suggest that it was formed on alluvial fan or fan-delta systems spreading from the basin margins. These systems developed at the foot of highlands that surrounded the zones of deposition, locally framing the basin.

Sandstone dominated facies assemblage (braided and/ or meandering fluvial systems)

Sandstone dominated sequences, which also include minor conglomerates, mudstones, and lignites, have been recorded widely in the basin. Lithofacies 'b' and 'c' occur in greater numbers, while the mudstone and the lignite facies ('d' and 'e', respectively) are clearly subordinate (Fig. 5b). The lithofacies that make up these successions are organized mainly in few metre-thick fining upward sequences, which may be capped by lignitic horizons (Fig. 7b). The fining upward sequences are stacked to make up thick sedimentary piles, ranging from several tens to a few hundred metres in thickness. However, minor coarsening upward sequences also have been recorded in relation to these sequences; they are capped by rather thin, up to 0.6 m thick, sandy layers with sand/mud rates ranging from 0.1 to 1.0.

The most complete fining upward sequences have been observed in the upper part of diverse megacyclothems in basin zones where they are laterally and/or vertically related to coarse-grained lithosomes. Thin coarsening upward sequences also occur in these basin zones, but they are not as well developed.

The geometric relationships between this facies assemblage and the marginal coarse-grained sequences, as well as its spatial distribution in the basin (Figs 3, 8), suggest that it was deposited on distributive fluvial plains developed ahead of the proximal alluvial zones. The described sequences are interpreted as the result of sedimentation on fluvial, braided and/or meandering plains, including channel and crevasse splay deposits. This interpretation is supported by the observation in some open-pit trenches of diverse kinds of channel infill, which occur as isolated sandstone bodies included in the mud dominated floodplain deposits. The thin lignite seams that occasionally occur in this facies assemblage can be interpreted as minor peat-bog deposits developed in the interchannel zones.

Lignite dominated assemblage (peat bog to marginal lacustrine)

Thick lignite seams ranging from tens to a hundred metres thick have been recorded in the Zittau Basin. These seams show diverse lignite lithofacies with minor mudstone beds intercalated within them (Fig. 6). They are well developed in the inner basin zones (Fig. 8) and usually cap the basin mega-

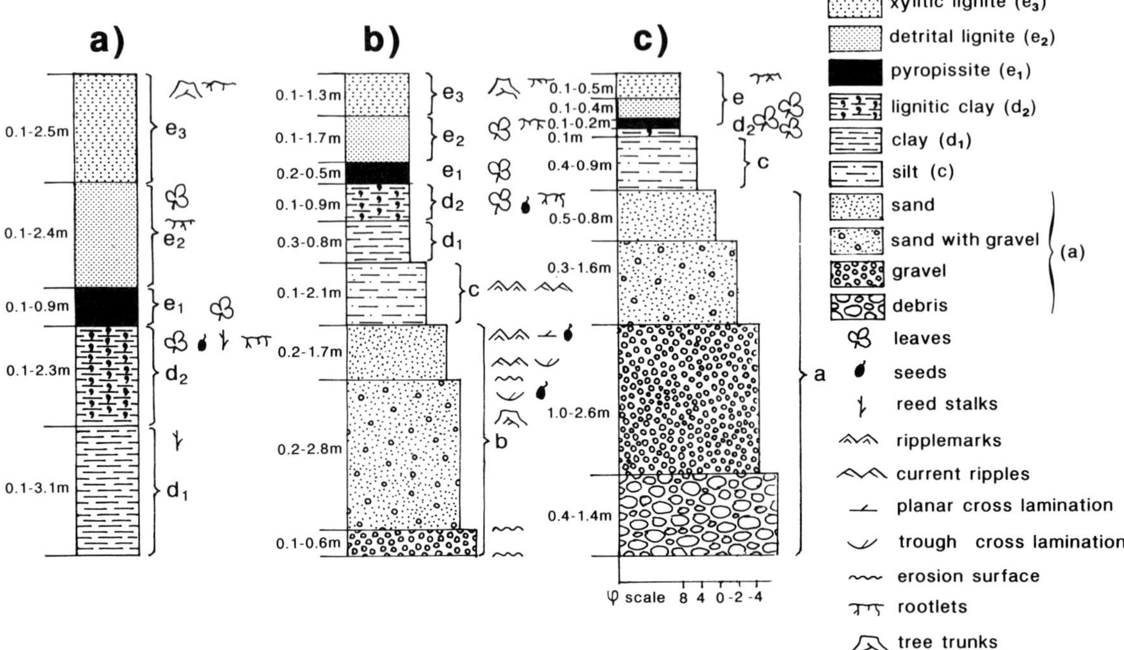

Fig. 7. Ideal lignite-bearing sequences in the Zittau Basin infill. The range of thickness of the diverse facies is indicated. See text for further explanation.

cyclothems (Fig. 3). Allochthonous pyropissite layers (e1) probably accumulated in subaqueous lacustrine environments under reducing conditions, as suggested by the presence of strongly corroded pollen grains. The remaining lignite lithofacies were deposited as autochthonous peats in marsh and

Table 2. Major petrological types of the Zittau Basin lignite

Petrological type	Pollen-spore spectrum	Phytodetritus content	Micropetrography
e1		Highest pollen-grain frequency (above 80%) Corrosional structures on a pollen grain surface	High humosapropelite content High sporinite content in maceral spectrum Clay minerals addition Strong bituminization Decay traces
e2	High frequency of the water-plant pollen grains (Sparganium and Butomus) and algal cells	High frequency of pollen grains (about 60%), resin grains (about 20%), and fungi spores (about 10%) Dispersed cuticulae of Marcoduria	Distinct admixture of humosapropelite (10−20%) Higher inertinite content Small addition of clay minerals
e3	High frequency of spores of Myricaceae−Cyrillaceae− Alnus (up to 40%) and Taxodiaceae− Cupressaceae−Nyssaceae (up to 20%) associations Polypodiaceae spores present	High wood-tissue frequency Relatively high frequency of resin grains (up to 20%) and fungi spores (up to 10%)	High content of humotelinite and humocolinite Relatively high content of humodetrinite High content of gelinite and corpohuminite and significant reenite addition in maceral spectrum

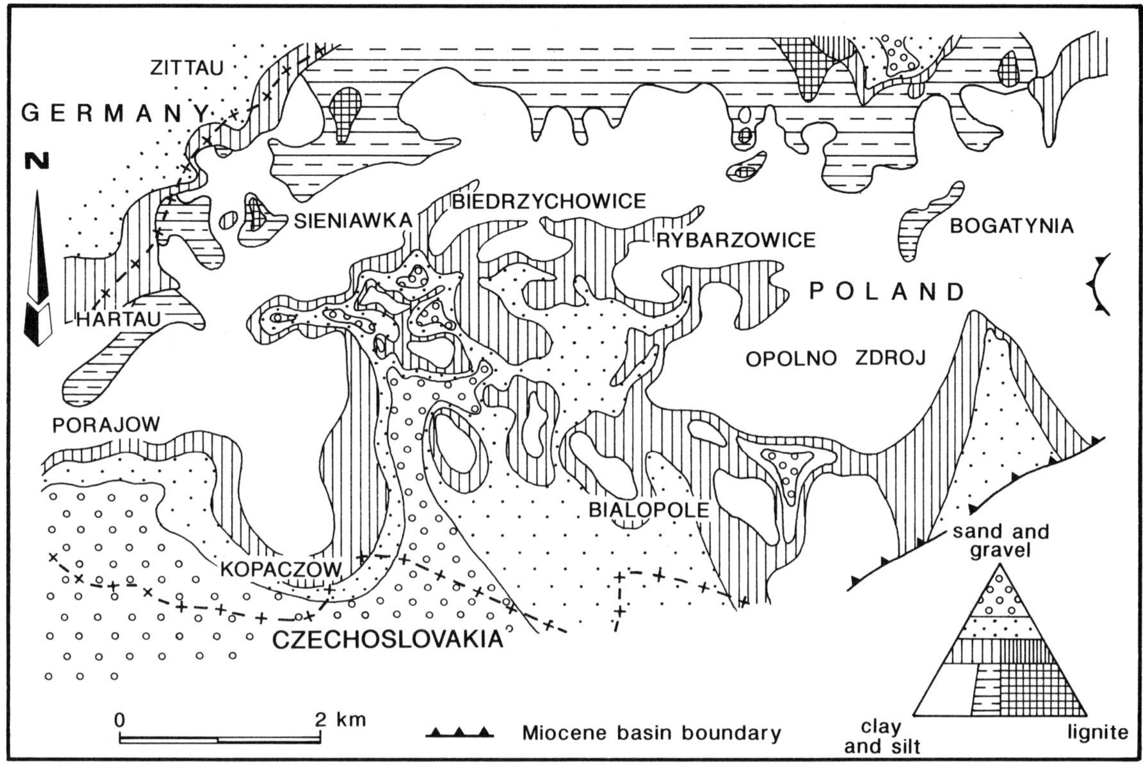

Fig. 8. Facies distribution map of the seventh megacyclothem (middle Miocene) in the southwestern part of the Zittau Basin.

swamp zones developed ahead of the terminal fluvial zones and surrounding inner lacustrine basin zones.

Peat-bog dominated environments developed during the early stages in nearshore marginal lacustrine zones during some evolutionary phases, but later spread over extensive marsh zones on to former inner lacustrine areas.

Clay dominated assemblage (open lacustrine facies)

Homogeneous sequences consisting of silty clays and kaolinitic clay dominated facies, including minor, centimetre-thick lignite beds, are well developed in inner basin zones protected from coarse terrigenous input from the terminal fluvial axis (Figs 6, 7a). At least part of these lithofacies (c, d) were deposited in open lacustrine environments dominated by deposition of fine-grained mineral and organic matter.

LIGNITE DEPOSITION AND BASIN EVOLUTION

Lignite deposition depends on the preservation of an approximate dynamic equilibrium between basin floor subsidence and accumulation of phytogenic material (Bouroz, 1960). Nevertheless, a persistent equilibrium is extremely unlikely because we are usually dealing with conditions far from the ideal.

The first palaeogeographical–palaeoecological model of Cenozoic peat-bog areas that led to lignite accumulations was proposed by Jurasky (1928) for the Lower Rhine Basin. His idea was the basis for a more precise and detailed model of the same area (Teichmüller, 1958). This model suggests a fivefold palaeoecological succession of vegetation types in a progradational peat-bog sequence. Starting from an open lacustrine zone, the sequential trend would consist of:

 J.R. Kasiński

1 deposition of clays with allochthonous phytogenic matter in marginal lacustrine zones;

2 accumulation of mixed autochthonous and allochthonous vegetal remains in reed marginal lacustrine zones;

3 autochthonous vegetal remains accumulating on a Taxodium−Nissa forest swamp;

4 autochthonous vegetal accumulation on a Miricaceae−Cyrilliciae swamp forest;

5 accumulation in a Sequoia forest.

The lignite facies accumulated in each palaeo-environment differ from each other in their palynological spectrum, phytodetritus content, and chemical composition (Teichmüller & Thompson, 1958).

This palaeoenvironmental trend of lignite generation was later discussed by Jux (1966), who suggested another position for the previous swamp forest zones, and added a leaf-tree swamp forest zone. Today, this model is not considered able to explain lignite deposition in the open, paralic, and shallow lacustrine to swamp basins of the Lower Rhine Basin (Hiltmann, 1976). Nevertheless, it may be useful for understanding lignite deposition in sedimentary settings similar to those observed in the Zittau Basin, that is, closed, distinctly limited, inter-montanc basins (Kasiński & Klimck, 1985).

An ideal lignite-bearing sequence for the Zittau Basin includes (Fig. 7):

1 clays and lignitic clays (d1 and d2 facies);

2 detrital, light, low-density yellow pyropissite (e1 facies);

3 detrital dark, brown or black lignite (e2 facies);

4 yellow to dark brown xylitic lignite (e3 facies).

Although this ideal sequence fits quite well with that previously proposed for the Lower Rhine Basin, some differences have been recorded in the Zittau sequences which make them distinct from those proposed by Teichmüller & Thompson (1958) and Jux (1966). The lower part of this ideal sequence (d1, d2, and e1) are interpreted as being related to subaqueous lacustrine sedimentation. The interpretation of the overlying sequential sections (detrital e1, e2, and xylitic e3 lignite facies) is more difficult because they are linked to the gradual shallowing of the lacustrine basin and its overgrowth by swamp plants. Thus some uncertainty arises regarding the origin of part of the dark detrital lignite horizons (e2) recognized in the Zittau Basin, as some of them could have formed in subaqueous conditions (von der Brelie & Wolf, 1981a,b). The dark, bright lignite horizons suggested as their equivalents in the Lower Rhine Basin could be related either to deposition on ombrogenous moors or to peat remains that underwent aerobic decomposition. Finally, the analysis of the lignite facies corresponding to the so-called Sequoia forest have confirmed the opinion that this phytocenosis records the climax of the peat-bog overgrowth, but it also has shown that peat decomposition was the dominant process in this palaeo-environment. Therefore the Sequoia forest cannot be related to any ecological zone of a functional living peat bog with a positive organic matter accumulation balance.

A similar lignite-bearing sequence was proposed to explain lignite accumulation in neighbouring Lower Lusatia (Schneider, 1980,1986). This sequence differs from that proposed in the Zittau Basin in terms of the significance of the light pyropissite (e1 facies), which is considered as capping each sequence. However, the pyropissite petrological and palaeontological features suggest its more direct relationship with a lignitic clay and a lower detrital dark lignite facies (deposited in nearshore and reed marginal lacustrine zones) than with sub-aerially influenced lignite facies.

The generation of a lignite-bearing sequence would be the consequence of the combination of the divcrsc subsidcncc mcchanisms that caused sinking of the depositional surface, infilling and shallowing of the lacustrine zones, and plant overgrowth. As a consequence of the changes in the relative influence and intensity of these processes, the above described ideal sequence is rarely completed owing to interruptions (e.g. a minor phase of basin-floor subsidence).

The sedimentary development of the Zittau Basin displays so-called allocyclic behaviour (Kasiński, 1983,1984). Tectonic subsidence acted as a major controlling factor. The megacyclothems (Fig. 3) may be regarded as a direct response to major tectonic events, and the facies distribution of any sequence reflects the scale and rate of subsidence. Rapid basin-floor subsidence started accumulation of alluvial fan and fan-delta sediments fringing the basin margins, while lacustrine environments developed in the inner basin zones (Fig. 9).

During earlier megacyclothem stages, related to noticeable subsidence rates, basinal lacustrine facies were restricted to the central parts of the basin. Later the lacustrine environments spread laterally as subsidence decreased. In a subsequent evolutionary stage, lacustrine environments covered most of the basin area (Fig. 3). Finally, after an initial partial infilling of the inner lacustrine areas, prograding

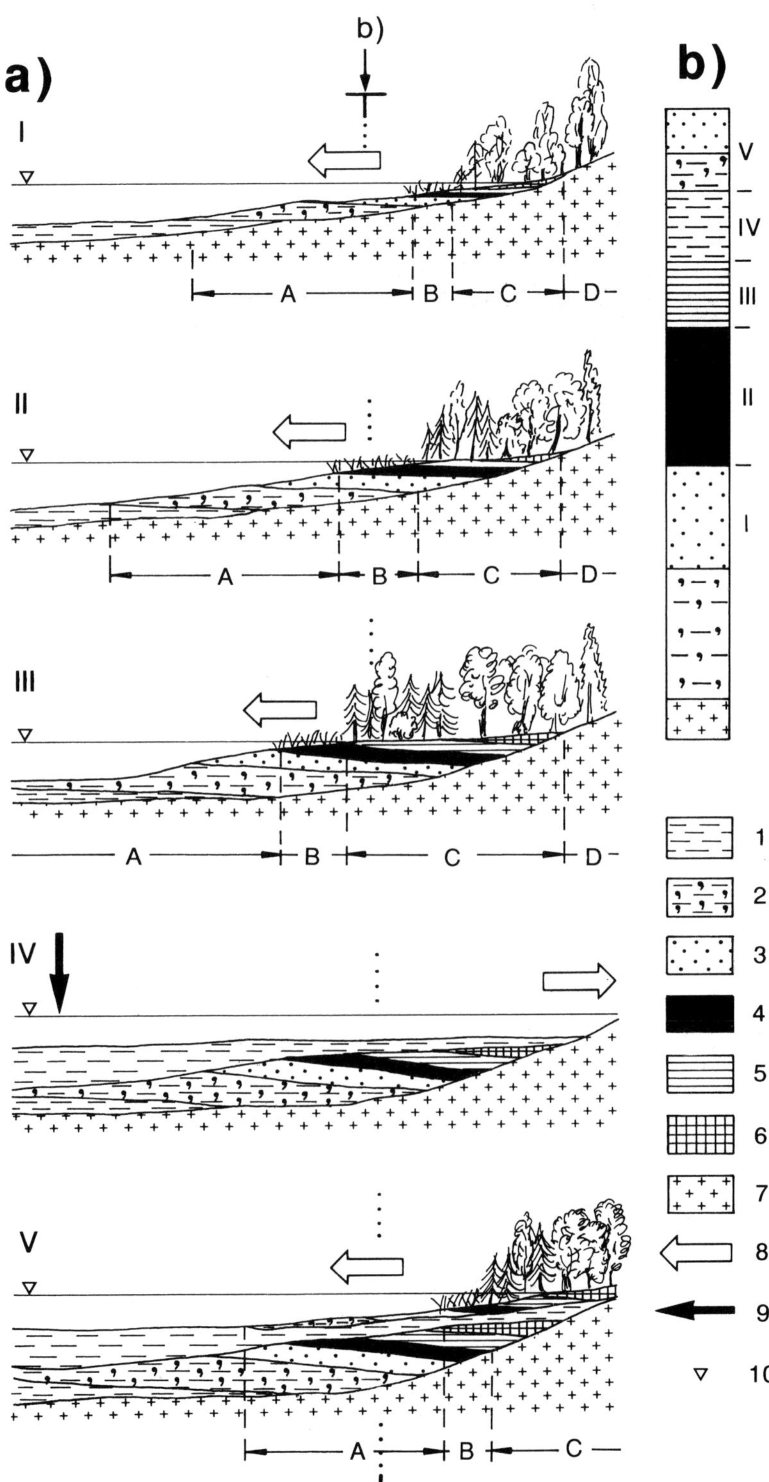

Fig. 9. Scheme showing the nearshore lacustrine peat-bog progradation and development stages (a) and the resulting ideal sedimentary sequence (b) (modified from Kasiński, 1983). Peat-bog palaeoenvironmental subdivision: A, nearshore open lake; B, reed area; C, swampy forest; D, humid basin-margin forest. Lithology and evolutionary features: 1, clay; 2, peaty clay; 3, detrital gyttja; 4, reed peat; 5, swamp-forest peat; 6, drier peat-bog margin phytogenic accumulate; 7, substratum; 8, direction of the shoreline migration; 9, subsidence; 10, lacustrine base level.

peat-bog facies spread from nearshore lacustrine zones towards the central zones. As a consequence, lignite-producing environments spread over the zones formerly occupied by subaqueous lacustrine facies.

During the last megacyclothem stage, shallow lacustrine to palustrine conditions were dominant and the influence of terminal fluvial plains increased.

As a consequence of this evolution, the thickest lignite deposits in the Zittau Basin are linked to the last stages of megacyclothem development, which are interpreted here as being associated with decreasing subsidence rates. In the Zittau Basin the main lignite accumulations developed during the sedimentation of the fourth and sixth megacyclothems, which bear almost all the economic lignite resources of the basin. It is suggested here that the occurrence of thick coal seams in both megacyclothems, which can be traced across the entire Ohře rift system (Fig. 3), derived from a particularly favourable period for coal accumulation, with both ideal palaeohydrology and tectonics.

In addition to tectonic activity, compaction processes may play a role in lowering the depositional surface in the basin. Compaction or degradation of thick peat layers, under sedimentary loading leads to a compaction coefficient estimated to be 1.7–4.0. Even the lowermost estimates could result in lowering of the depositional surface by about 50–100 m (Kasiński, 1984). Compactional subsidence could have been rather rapid.

CONCLUDING REMARKS

1 Some central European rift systems, including the Ohře rift, were especially favourable for thick, economic coal seams accumulating in alluvial lacustrine settings.

2 Faulting episodes with consequent widespread volcanic activity are well recorded in the rift system, where they gave rise to a series of major megacyclothems, which can be traced throughout most of the rift basins.

3 Each major stratigraphic unit (megacyclothem) displays a similar sequential organization, starting with terrigenous dominated sequences and finishing with lignite-bearing successions, including coal seams up to 100 m thick. The lower megacyclothem horizons are interpreted as being related to active tectonic episodes, while the lignite-bearing upper sequences seem to be linked to tectonic quiescence.

4 Some of the thickest lignite seams in the Ohře rift basins may be correlated directly, since they record generalized episodes of peat accumulation in the rift system.

5 Tectonics seem to be the major controlling factor that influenced coal accumulation, assuming an appropriate palaeoclimate. Compaction processes may be significant.

ACKNOWLEDGEMENTS

I am grateful to Drs Romeo Flores and Th. Kreuser for their valuable suggestions in the revision of the earlier version of this paper. Special acknowledgements are due to Drs P. Anadón, Ll. Cabrera, and K. Kelts for their critical and constructive comments on the paper contents and organization.

REFERENCES

BAGANZ, B.P., HORNE, J.C. & FERM, J.C. (1975) Carboniferous and recent Mississippi lower delta plains; a comparison. *Trans. Gulf Coast Assoc. Geol. Soc.* **26**: 183–191.

BOUROZ, A. (1960) La sédimentation des séries houillères dans leur contexte paléogéographique. In: *Congress pour l'Avancement des Études de Stratigraphie Carbonifère* (Ed. van der Waals, L.), Compte Rendu **4**, 65–78, van Aelst, Heerlen.

BRAUSE, H. (1988) *Die Wirkung endogener und glazigener Tektonik bei der Bildung von Braunkohlenlagerstatten.* Archiv VEB Geologische Forschung und Erkundung, Freiberg, 32 pp. (unpublished).

ČERMAK, V., JETEL, J. & KRČMAŘ, B. (1968) Terrestial heat-flow in the Bohemian Massif and its relation to the deep structure. *SBOR. Geol. Ved.* **7**, 25–41.

FRAZIER, D.E. & OSANIK, A. (1968) Recent peat deposits — Louisiana coastal plain. In: *Environments of Coal Deposition* (Eds Dapples, E.C. & Hopkins, M.E.). *Geol. Soc. Am. Spec. Pap.* **144**, 63–65.

GALLOWAY, W.E. & HOBDAY D.K. (1983) *Terrigenous Clastic Depositional Systems.* Springer, Berlin, 423 pp.

GREGUSS, P. (1955) Oznaczenie dolno-miocenskiego pnia drzewa z Turowa nad Nysa Zuzycka. *Acta Geol. Pol.* **5**, 273–275.

HILTMANN, W. (1976) *Pollenanalytische Untersuchungen im Rheinische Hauptbraunkohlenflös der Tagebauen Frechen und Fortuna unter besonderer Berucksichtigung der Makroppetrographischer Ausbildung der Kohle.* Unpublished Ph. thesis, Rheinisch-Westfalische Technische Hochschule, Aachen, 162 pp.

JURASKY, K.A. (1928) Aufgaben und Ausblicke fur die paleobotanische Erforschung der niederrhrinischen Braunkohle. *Braunkohle* **27**, 436–443.

JUX, U. (1966) Torfe des rheinischen Tertiär im Vergleich mit heutigen Bildungen an der amerikanischem

Ostküste. In: *Bericht über de Frühjahrs-tagung der Deutschen Geologischen Gesellschaft im Köln vom 18–21, Mai 1966; Thema: Die Rheinische Braunkohle* (Ed. Quitzow, H.W.), Z. Dtsch. Geol. Ges. **118**, 69–101.

KANASIEWICZ, J. (1983) O mozliwoci wystepowania karbonatytów facji subwulkanicznej w rejonie niecki zytawskiej, Sudety Zachodnie. *Kwart. Geol.* **27**, 755–762.

KANASIEWICZ, J. (1988) Magmy zasadowe jako zródzo uranu dla zlozuranowych. *Przegl. Geol* **36**, 482–484.

KASIŃSKI, J.R. (1983) Mechanizmy sedymentacji cyklicznej osadów trzeciorzedowych w zapadliskach tektonicznych przedola Sudetów. *Przegl. Geol.* **31**, 237–243.

KASIŃSKI, J.R. (1984) Synsedimentary tectonics as a factor controlling sedimentation of brown-coal formation in tectonic depressions in Western Poland. In: *Solid Fuel Mineral Deposits* (Ed. Borisov, V.S.) *Proceedings of the 27th International Geological Congress, Moscow, 1984*, pp. 247–279. VNU Science Press, Utrecht, 288 pp.

KASIŃSKI, J.R. (1989a) Tectonic control of Tertiary terrigenous formations in the intermontane depression along the NW margin of the Sudetes Mts., SW Poland. In: *Proceedings of International Symposium on Intermontane Basins: Geology and Resources* (Eds. Thanasuthipitak, T. & Ounchanum, P.), pp. 353–373. Chiang Mai University, Chiang Mai.

KASIŃSKI, J.R. (1989b) Lacustrine sedimentary sequences in the Polish Miocene lignite-bearing basins — facies distribution and sedimentary development. *Palaeogeogr. Palaeoclimatol. Palaeoecol.* **70**, 287–304.

KASIŃSKI, J.R. & KLIMEK, W. (1985) Analiza mikrofacjalna wegla brunatnego jako metoda rekonstrukcji srodowiska sedymentacji na przykladzie wybranego profilu z dolnego pokladu zloza Turów. *Biul. Inst. Geol., Warsaw* **349**, 93–131.

KASIŃSKI, J.R. & PANASIUK, M. (1987) Geneza i ewolucja strukturalna niecki zytawskiej. *Biul. Inst. Geol., Warsaw* **357**, 5–35.

KOPECKÝ, L. (1978) Neoidic taphrogenic evolution and young alkaline volcanism of the Bohemian Massif. *SBOR. Geol. Ved.* **31**, 91–140.

KOPECKÝ, L. (1979) Magmatism of the Ohře Rift in the Bohemian Massif, its relationship to the deep fault tectonics and to the geologic evolution, and its mineralization. In: *Czechoslovak Geology and Global Tectonics* (Eds Mahel, M. & Reichwalder, P.), pp. 167–181. Veda, Bratislava.

MALKOVSKÝ, M. (1980) Model of the origin of the Tertiary basins at the foot of the Krusne Hory Mts.: volcano-tectonic subsidence. *Vestn. Ustred. Ustavu Geol.* **55**, 141–150.

McCABE, P.J. (1984) Depositional environments of coal and coal-bearing strata. In: *Sedimentology of Coal and Coal-bearing Sequences* (Eds Rahmani, R.A. & Flores, R.M.), Spec. Publ. Int. Assoc. Sediment. **7**, 13–42.

PANASIUK, M. (1986) Wyniki datowania wieku bezwzgled-nego law wulkanicznych rejonu Bogatyni metoda potasowo-argonowa K–Ar. *Przegl. Geol.* **34**, 149–152.

PÉSEK, J. (1988) Brown-coal and lignite deposits in Czechoslovakia. *Folia Mus. Rer. Nat. Bohemise Occ. Geol. Brno* **27**, 56 pp.

PIWOCKI, M. (1987) Charakterystyka chemiczno-technologiczna glównych grup trzeciorzędowych węgli brunatnych w Polsce. *Biul. Inst. Geol., Warsaw* **357**, 41–60.

POZARISKY, W. & BROCHWICZ-LEWINSKY, W. (1978) On the Polish Trough. *Geol. Mijnbouw* **57**, 545–558.

SAKAMOTO, T. (1957) Cycle of sedimentation in the Late Paleozoic coal-bearing formation in the Kailan Coalfield, north China. *XX Int. Geol. Congr.* **5**, 375–387.

SCHNEIDER, W. (1980) Zu einigen Gesetzmässigkeiten der faziellen Entwicklung im 2. Lausitzer Flöz. *Z. Angew. Geol.* **24**, 125–130.

SCHNEIDER, W. (1986) Cryptomeria Don (Taxodiaceae) — ein Kohlenbildner im mitteleuropäischen Tertiär. *Z. Geol. Wiss.* **14**, 735–744.

ŠIBRAVA, V. & Havlíček, P. (1980) Radiometric age of Plio-Pleistocene volcanic rocks of the Bohemian Massif. *Vestn. Ustred. Ustavu Geol.* **55**, 129–139.

TEICHMÜLLER, M.-L. (1958) Rekonstruktionen verschie-dener Moortypen des Hauptflözes der niederrheinischen Braunkohle. In: *Die Niederrheinische Braunkohlenfor-mation* (Eds Teichmüller, R. & van der Brelie, G.), Fortschr. Geol. Rheinl. Westfalen **2**, 599–612.

TEICHMÜLLER, M.-L. & THOMSON, P.W. (1958) Vergleichende mikroskopische und chemische Untersuchungen der wichtigsten Faziestypen im Hauptflöz der niederrheinischen Braunkohle. In: *Die Niederrheinische Braunkohlenformation* (Eds Teichmüller, R. & van der Brelie, G.), Fortschr. Geol. Rheinl. Westfalen. **2**, 573–598.

VON DER BRELIE, G. & WOLF M. (1981a) Zur Petrographie und Palynologie heller und dunkel Schichten im Rheinischen Hauptbraunkohlenflöz. In: *Geologie und Lagerstättenerkundung im Rheinischen Braunkohlen-revier* (Ed. Hilden, H.D.), Fortschr. Geol. Rheinl. Westfalen **29**, 95–163.

VON DER BRELIE, G. & WOLF, M. (1981b) 'Sequoia' und Sciadopitys ins der Braunkohlenmooren der Niederrhei-nischen Bucht. In: *Geologie und Lagerstättenerkundung im Rheinischen Braunkohlenrevier* (Ed. Hilden, H.D.) Fortschr. Geol. Rheinl. Westfalen **29**, 177–191.

WOLF, L. & HIRSCHMANN, G. (1964) Posthume pleistozäne Bruchtektonik in der Östlichen Oberlausitz. *Geologie* **13**, 1229–1234.

ZALEWSKA, Z. (1953) Trzeciorzędowe szczątki drewna z Turowa nad Nysą łuzycką, cz. 1. *Acta Geol. Pol.* **3**, 481–543.

ZALEWSKA, Z. (1955a) Trzeciorzedowe szczątki drewna z Turowa nad Nysą łuzycką, cz. 2. *Acta Geol. Pol.* **5**, 277–304.

ZALEWSKA, Z. (1955b) Trzeciorzedowe szczątki drewna z Turowa nad Nysą łuzycka, cz. 3. *Acta Geol. Pol.* **5**, 517–537.

ZÉMAN, J. (1979) The influence of paleorifts on the devel-opment of continental crust in the Bohemian Massif. In: *Czechoslovak Geology and Global Tectonics* (Eds Mahel, M. & Reichwalder, P.), pp. 167–181. Veda, Bratislava.

ZIEGLER, P.A. (1982) *Geological Atlas of Western and Central Europe.* Elsevier, Amsterdam, 130 pp.

ZIEGLER, P.A. (1987) Late Cretaceous and Cenozoic intra-plate compressional deformation in the Alpine foreland — a geodynamic model. *Tectonophysics* **137**, 389–420.

ZIEGLER, P.A. (1988) Evolution of the Arctic–North Atlantic and Western Tethys. *Am. Assoc. Petrol. Geol. Mem.*, **43**, 198 pp.

Spec. Publs Int. Ass. Sediment. (1991) **13**, 109–125

Carbonate and coal deposition in an alluvial–lacustrine setting: Lower Cretaceous (Weald) in the Iberian Range (east-central Spain)

E. GIERLOWSKI-KORDESCH*, J.C. GÓMEZ
FERNÁNDEZ[†] *and* N. MELÉNDEZ[†]

**Department of Geological Sciences, Ohio University, Athens, Ohio, 45701, USA*
[†]*Departamento Estratigrafía, Facultad C.C. Geológicas, Universidad Complutense, E-28040 Madrid, Spain*

ABSTRACT

Lower Cretaceous (Upper Hauterivian? to Lower Barremian) carbonate and clastic rocks interbedded with coal from the Weald of the Serranía de Cuenca (Cuenca Province, Iberian Range, Spain) were analysed. The study area centred around Uña (north of the capital of Cuenca), where a lignite deposit exists; it is completely enclosed in carbonates and contains a vertebrate fauna including mammals. From six measured sections and other data, four facies associations containing 14 separate facies were identified: coal–limestones (association A), charophyte-micrites–silty-marls (association B), clayey-marls–limestones–sandstones (association C), and coal–clastics (association D). Based on evidence from tectonics, fossils, and regional geology, a continental palaeoenvironment comprising a carbonate–clastic alluvial plain transitional to a marginal lacustrine setting adjacent to a large lacustrine delta was envisaged.

The sediments of the carbonate floodplain were deposited and constantly recycled as bedload and as physico-chemical and biochemical precipitates. The continual recycling of the carbonates was attributed to early diagenetic processes including karstification, calichification, marmorization, micritization, nodularization, and illuviation. Algal morphologies within oncolites, as well as diagenetic patterns, suggest alternating wet and dry periods on the floodplain.

The limnic coal deposit (lignite) is interpreted to have developed within a prograding lake delta. Evidence includes facies geometry and lack of exposure features. From climatological and sedimentological arguments, a seasonal, warm and semi-arid climate was established for the Weald at Uña. Subsidence, sediment supply, and hydrology, not climate, were named as important factors in coal preservation. The Jurassic karst substratum was deemed sufficient to control the hydrology (water levels) and carbonate sedimentation in the Uña Basin for maintenance of constant water cover for coal formation and preservation.

INTRODUCTION

Descriptions of coal localities and their associated depositional models are usually provided from siliciclastic systems (e.g. McCabe, 1984; Fielding, 1987). Coal formation in carbonate systems of a continental nature is rarely described (Durand, 1980; Cabrera & Sáez, 1987; Loftus & Greensmith, 1988), and is never extensive since the conditions for preservation are limited (Teichmüller & Teichmüller, 1982).

The study area of this paper contains an extensive alluvial–lacustrine carbonate system centring around Uña, which is about 30 km northeast of the provincial capital of Cuenca in east-central Spain (Fig. 1a). To the west of Uña, a sub-economic lignite mine was exploited by residents in the earlier part of this century. This coal is found within a thick sequence of carbonates and contains a vertebrate fauna (Kühne & Crusafont, 1968; Henkel & Krebs, 1969; Henkel, 1970; Fey, 1988). The sedimentological model for coal deposition is explored in this paper. Some contemporaneous, deep lacustrine deposits south of Uña are dealt with in Gómez Fernández & Meléndez (this volume).

E. Gierlowski-Kordesch et al.

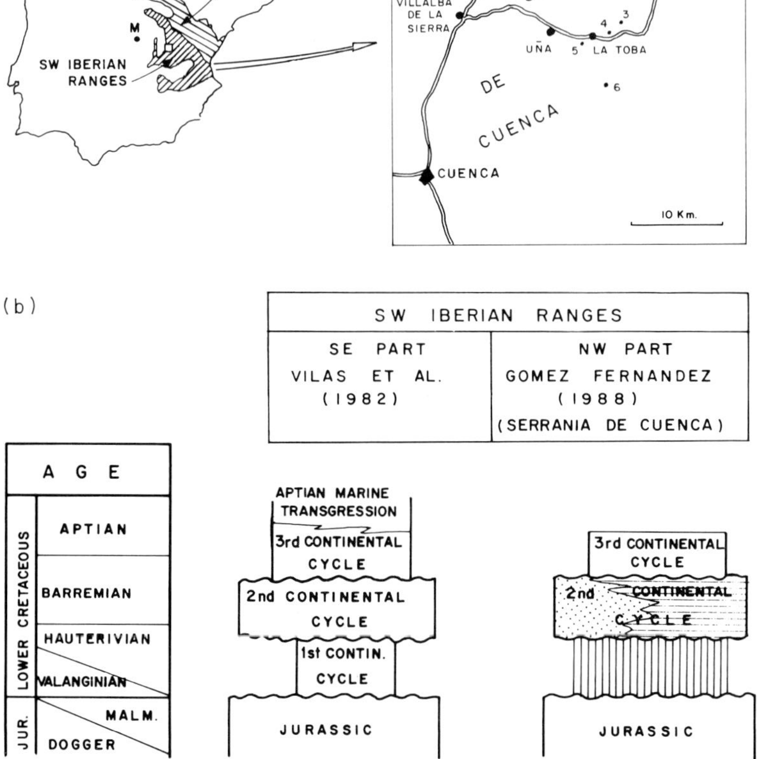

Fig. 1. (a) Location of the study area in the southwestern Iberian Range, which is east of Madrid (M). Measured sections in the Serranía de Cuenca, northeast of Cuenca, are numbered: 1, Las Majadas I; 2, Las Majadas II; 3, Los Lagunillos; 4, La Toba; 5, Uña; 6, Cabeza Gorda. (b) General stratigraphy for the Upper Jurassic to Lower Cretaceous of the southwest Iberian Range. Jurassic marine limestones are overlain by three sedimentary cycles, continental in origin, followed by a marine transgression in the Aptian. Locally in the Serranía de Cuenca, only the second and third sedimentary cycles of the Lower Cretaceous are present. Rocks described in this paper form the second continental cycle.

GEOLOGICAL SETTING

The Iberian Range of eastern Spain (Fig. 1a) is mainly composed of Permian to Upper Cretaceous rocks with a slightly metamorphosed Palaeozoic basement exposed in places along the range axis. These mountains are interpreted to be developed in an aulacogen in which the initial graben stage began in the Early Triassic (Alvaro *et al.*, 1979). After a transitional stage of volcanic activity in the Late Triassic, the flexure stage lasted from the Jurassic through to the end of the Late Cretaceous, with a folding stage occurring in the Oligocene. During the Early Cretaceous, perturbations due to the rotation of the Iberian Plate caused more internal rifting and the creation of many small basins. Vilas *et al.* (1983) believed that these perturbations extended through-

out the Early Cretaceous and marked the beginning of the next aulacogen-forming stage in the Cretaceous.

Within the Early Cretaceous basins of the southwest Iberian Range (see Fig. 1a), continental deposition occurred towards the northwest (Serranía de Cuenca) while marine conditions dominated towards the southeast (toward Valencia). These conditions lasted until the middle Aptian, within the so-called Iberian Basin (Mas *et al.*, 1982; Vilas *et al.*, 1983). Three sedimentary cycles characterize the Lower Cretaceous Weald (Fig. 1b); however, only the second and third sedimentary cycles are represented in the Serranía de Cuenca (Meléndez, 1983; Vilas *et al.*, 1983; Gómez Fernández, 1988). Only the second Lower Cretaceous cycle is discussed in this paper (see 2 in Fig. 1b).

In the Uña study area of the southeast Iberian Basin (Serranía de Cuenca), continental sedimentation began in the Upper Hauterivian? to Lower Barremian upon the karstic limestones of the Jurassic (Dogger) after a depositional hiatus and unconformity (Meléndez, 1982,1983; Mas *et al.*, 1982; Vilas *et al.*, 1982,1983). This sedimentary cycle (second continental cycle — see 2 in Fig. 1b) is composed of two general units: the El Collado and La Huérguina Formations. The El Collado represents alluvial deposits, such as sandstones and oncolitic grainstones, whereas the La Huérguina Formation comprises lacustrine to alluvial deposits, such as marls, micrites, and shales (Meléndez, 1983; Gómez Fernández, 1988).

METHODOLOGY

The stratigraphy and sedimentology of the Cretaceous of the northwestern part of the Iberian Basin (Serranía de Cuenca) are dealt with in detail in other works (Ramírez del Pozo & Meléndez Hevia, 1972; Meléndez Hevia *et al.*, 1975; Ramírez del Pozo *et al.*, 1975; Meléndez, 1982,1983; Mas *et al.*, 1982; Vilas *et al.*, 1982; Gómez Fernández, 1988; Gierlowski-Kordesch & Janofske, 1989; and others). In this paper, six detailed lithological sections from the area around the coal deposit at Uña are presented, although sedimentological data from various other outcrops are also utilized to complement this study. These sections (Uña, La Toba, Los Lagunillos, Las Majadas I and II, and Cabeza Gorda, Figs 2, 3) were measured using lithology, grain size, fossils, sedimentary structures, and diagenetic features. Correlation (based on lithology) is presented in Fig. 4. Representative samples were slabbed and thin-sections were made. Marls were sieved for charophytes and ostracods. The coal at Uña was sent out for petrographic analysis.

FACIES ASSOCIATIONS

Four facies associations containing 14 different facies are recognized within the palaeoenvironmental setting of the Weald around Uña (see Table 1 for summary). Two facies assemblages are mainly carbonate: coal—limestones (association A) and charophyte-micrites—silty-marls (association B); one is mixed carbonate and clastics: clayey-marls—limestones—sandstones (association C); and one contains clastics: coal—clastics (association D), as shown in Fig. 5. Associations A and B are interpreted as lacustrine deltaic to marginal lacustrine palaeoenvironments and associations C and D are interpreted as the adjacent alluvial plain deposits. Detailed descriptions and interpretations of the associations and their constituent facies follow.

Coal—limestones (association A)

This association is composed of marls and micrites interbedded with lignitic coal and is only found at Uña (Meter 31.5−38.7, total thickness is 7.2 m) (Figs 3, 4). Here the vertical sequence of facies is $A_1 \rightarrow A_2 \rightarrow A_3 \rightarrow A_2$ (Fig. 5).

Facies A_1 (marls)

Defined as mostly carbonate in composition, these marls are grey in colour and are finely laminated to massive with dispersed oncolites. Fossils include ostracods, gastropods, charophytes, and carbonized macrophyte debris. Early diagenetic gypsum crystal aggregates replaced charophyte fragments.

Facies A_2 (micrites with coalified material)

Well-cemented micrites (wackestones) are massively bedded, with units from 30 cm to 1 m in thickness (average 50 cm). In a few cases, coalified macrophytic debris appears to define relict cross-stratification. This debris is also dispersed throughout the units, but is mainly aligned horizontally with concentration decreasing upward within units. Units are composed of lenses at an outcrop scale and can be up to tens of metres in length. Coalified or carbonized *in situ* rhizoliths are common. Single specimens and accumulations of oncolites are dispersed throughout the units. Fossils include ostracods, charophytes, gastropods, and colonies of *Rivularia* and *Phormidium* (blue-green algae). No evidence for subaerial exposure was found.

Facies A_3 (coal)

Identified as lignite, this coal is mainly composed of macrophyte debris, allochthonous in origin. Micrite lenses are present in the coal (Fig. 6). Petrographic analysis from a six-sample average resulted in 83.8% vitrinite, 1.5% liptinite, and 14.7% inertinite. Other parameters include reflectance of 0.21, volatiles (dry, ash-free) at 67.1% (average), moisture content at

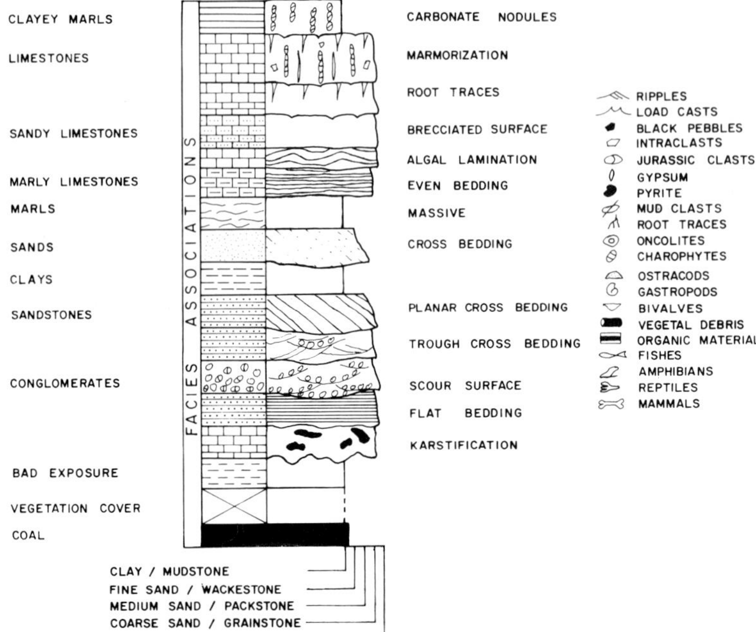

Fig. 2. Key for Fig. 3.

Table 1. Summary of the palaeoenvironmental analysis of the Weald around Uña in the Cuenca Province of east-central Spain

Association	Facies	Environment	Diagenesis
A: coal– limestones		**Carbonate lake delta**	
	A_1 — marls	Prodelta	
	A_2 — micrites with coalified material	Delta	Sparmicritization
	A_3 — coal	Delta	Recrystallization
B: charophyte- micrites–silty- marls		**Carbonate lake margin**	
	B_1 — silty marls	Shallow lakes	Sparmicritization
	B_2 — charophyte micrites	Ponds	Pedogenesis Recrystallization
C: clayey-marls– limestones– sandstones		**Mixed alluvial plain**	
	C_1 — carbonate conglomerates– oncolitic sandstones	Carbonate–Clastic channels	Brecciation Dissolution
	C_2 — grainstones	Carbonate channels	Karstification
	C_3 — packstones–wackestones	Carbonate channels	Marmorization Recrystallization
	C_4 — clayey marls	Paleosols	Pedogenesis
	C_5 — carbonaceous micrites	Marshes	Nodularization
	C_6 — laminated pisolitic wackestones	Caliche	Calichification Sparmicritization
D: coal–clastics		**Clastic alluvial plain**	
	D_1 — sandstones–lignites	Extra-channel	
	D_2 — conglomeratic sandstones	Channel	
	D_3 — claystones–lignites	Extra-channel	

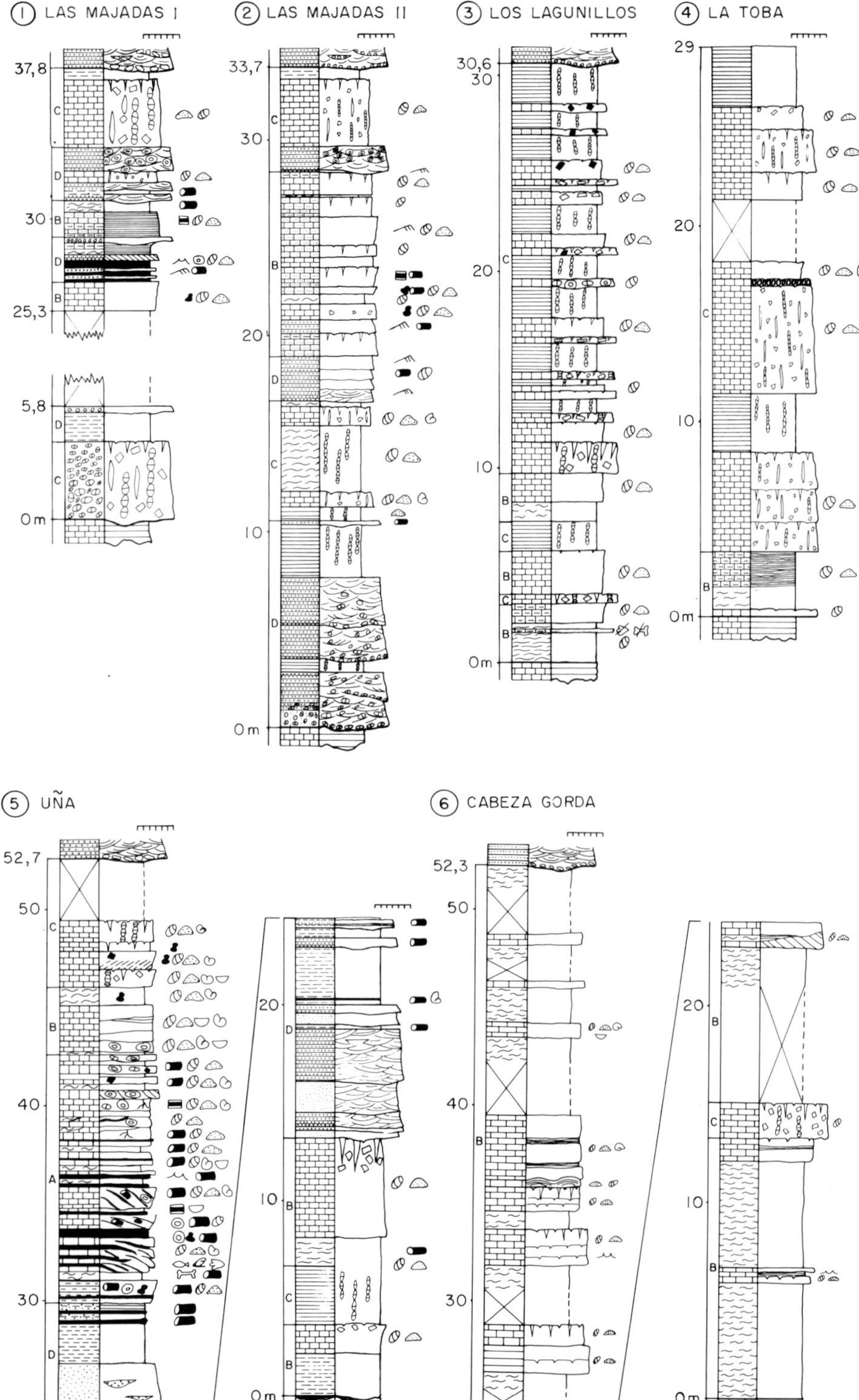

Fig. 3. Six measured sections from the Serranía de Cuenca of the southwestern Iberian Range. See Fig. 1 for location of sections.

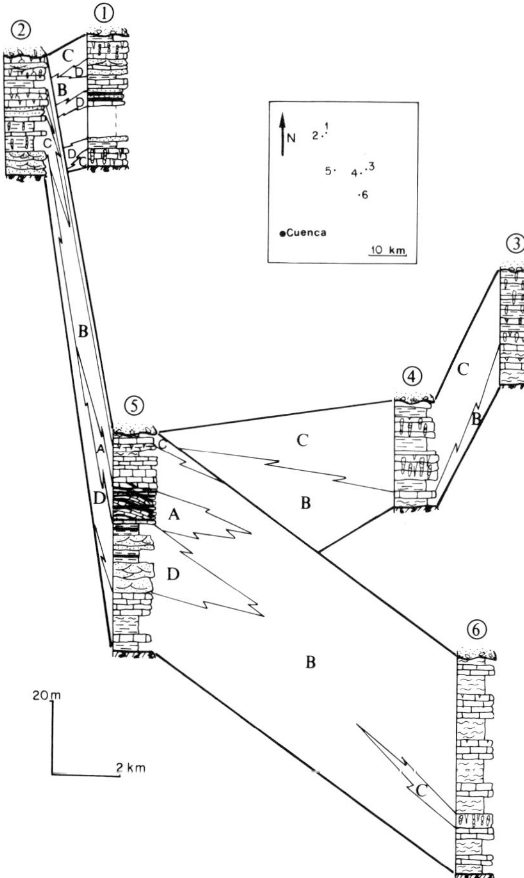

Fig. 4. Correlation among the sections based on facies associations (see Fig. 5). Location of sections are schematically shown within inset; compare with Fig. 1. 1, Las Majadas I; 2, Las Majadas II; 3, Los Lagunillos; 4, La Toba; 5, Uña; 6, Cabeza Gorda.

9.9% (average) with the ash content ranging from 4.7–37.6% and the sulphur content (dry, ash-free) ranging from 1.3–4.1%. Fossils include mammals (*Crusafontia cuencana* (Dryolestidae)) (Henkel & Krebs, 1969), frogs (*Wealdenbatrachus jucarensis* (Discoglossidae)) (Fey, 1988), crocodiles, lizards, other unnamed amphibians and reptiles (Kühne & Crusafont, 1968; Henkel, 1970), and pollen and spores dated as early Barremian (Mohr, 1987).

This coal is intercalated exclusively with carbonates. The thickest coal bed at Uña is 50 cm thick (Fig. 3; Uña section, 31.5 m), but above and below it, the thickness of micrite to marl layers increases and the amount of coal material decreases. Tracing the exposure from the neighbouring mine, the lateral extent of the main coal body is about 700 m, with its centre about 300 m southwest of the Uña road section. Laterally in the northeast direction, carbonates dominate.

Interpretation

The coal–limestones association represents lower to upper portions of a small lake delta with a vertical sequence of marls (A_1) → micrites (A_2) → coal (A_3) → micrites (A_2) (Fig. 3; Uña section, 30–43 m). The lack of ubiquitous exposure features and the presence of coal and coalified rhizoliths in the deltaic micrites indicate that conditions remained subaqueous throughout sedimentation. Autochthonous carbonate was generated through precipitation (both physico-chemical and biochemical (see Murphy & Wilkinson, 1980)). Allochthonous carbonate was deposited through traction deposition of carbonate bioclasts and rock fragments. The micritic texture of these carbonates is both diagenetic and authigenic. Evidence of bedload transport of carbonates includes relict cross-stratification vaguely defined by coalified macrophyte debris at Uña. Subaqueous synsedimentary processes, such as calcite precipitation, dissolution, and sparmicritization, mostly initiated by blue-green algae (Kahle, 1977), would tend to homogenize the sedimentary features of the deltaic carbonates.

Progradation of the Uña deltaic facies northeastward into the lake is indicated by the geometry and stratigraphic succession of this association at Uña. A modern analogue to this deltaic environment is found in Sucker Lake, a temperate marl lake in North America. There, allochthonous macrophyte fragments (phytoclasts) are accumulating as topset delta deposits at the mouth of a creek draining a vegetated marsh. Peat preservation is linked to burial as a result of delta progradation (Treese & Wilkinson, 1982).

Charophyte-micrites–silty-marls (association B)

This association comprises of up to 3-m sequences of silty marls grading upward into micrites, with tops characterized by diagenetic alteration and rhizoliths (Fig. 5). Association B dominates the section at Cabeza Gorda (charophyte micrites dominate) and is interbedded with associations C and D at Las Majadas I and II, Los Lagunillos, and La Toba (Figs 3, 4).

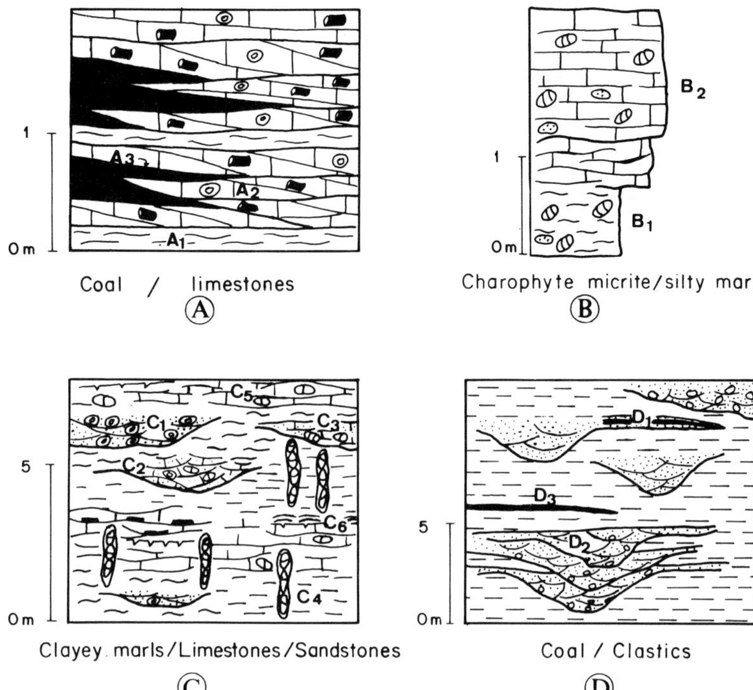

Fig. 5. Pictorial representation of the four facies associations (A–D) described in the area surrounding Uña. Component facies are described in the text and summarized in Table 1.

Facies B_1 (silty marls)

Units are only a few metres in lateral extent and some decimetres in thickness. Grey-green in colour, these marls are shaly to massive in appearance. Units are clay-rich at the bottom and grade upward to pure carbonate. Pyrite concretions are dispersed throughout the sedimentary unit. Rare occurrences of matrix- to grain-supported carbonate lenses contain quartz grains, carbonate clasts, and reptile bone fragments; early diagenetic cementation of these

Fig. 6. Outcrop photograph of main coal seam at Uña showing a lens of micrite 20 cm thick. Note coalified macrophyte debris scattered within lens (black pieces floating in matrix). This micrite lens is surrounded by lignite intercalated with smaller micrite lenses. A hammer head is present at lower right for scale (facies A_2 and A_3).

lenses is evidenced by differential compaction in relation to surrounding marls.

The transition layer to the upper charophyte micrites is c. 15−30 cm thick and comprises increasingly thicker lenses of micrites within the silty marls — the initial thickness of micrite units is 1−2 cm. Macrophyte debris and pyrite concretions are present in the marls between micrite lenses in the transition zone.

Facies B_2 (charophyte micrites)

This facies can be classed as a wackestone containing charophytes, either dispersed throughout units or in discrete accumulations. Units are from 100 to 200 m long with a maximum thickness of 1 m. Massive bedding predominates; rare smaller lenses appear to contain relict cross-stratification. Rare oncolites are dispersed through units. Fossils include gastropods, bivalves, ostracods, and rhizoliths. Rhizoliths occur at the top of units and appear as vertical cavities with or without a recrystallized calcitic cement lining. Small algal mounds have an irregular, draping morphology (10−15 cm high and 20−30 cm long) and are associated with charophyte accumulations. Algal mounds show irregular dark laminae alternating with clear sparite laminae and charophytes and ostracods 'caught' under algal layers.

Interpretation

The charophyte-micrites−silty-marls association is interpreted as being the deposits of shallow lakes (facies B_1) and ponds (facies B_2) distributed adjacent to the lake. These sediments were deposited in quiet conditions where charophyte/algal mounds grew and invertebrate remains, macrophytic debris, and bones could accumulate. Carbonate precipitation dominated, promoted to a large extent by blue-green algae and charophytes (Murphy & Wilkinson, 1980).

The sequence defined by the transition from silty marls to charophyte micrites is topped by diagenetically altered limestones containing rhizoliths and represents a shallowing upward sequence. The shallow lakes and ponds were slowly filled as channels switched across the alluvial plain. The final stage of subaerial exposure promoted pedogenic processes. Association B transitions repeat over most of the exposure at Cabeza Gorda (Fig. 3), indicating pulses of subsidence followed by lake infilling, and ending with soil development.

Clayey-marls−limestones−sandstones (association C)

This association dominates the section at Los Lagunillos and La Toba and is interbedded with association B at Las Majadas I and II (Figs 3, 4). There is no characteristic sequence of facies within this association (Fig. 5). Diagenetic overprint in this association can obliterate all primary structures.

Facies C_1 (carbonate conglomerates−oncolitic sandstones)

Transitional in nature and occurrence with facies C_2 and C_3, this facies also has a wide variety in the proportion of matrix and clast components. The carbonate conglomerate is grain-supported. Clast types include carbonate rock fragments (mostly Wealdian), carbonate black pebbles, oncolites, and clastic rock fragments, with clast sizes ranging from cobbles to granules. It is transitional in composition to oncolitic sandstones. Sedimentary units are decimetres thick and metres long, with some decimetre-scale, low-angle cross-stratification visible. Diagenetic overprint, in some cases, eliminates all internal fabric, leaving behind a micrite with only a weathered surface showing vague clast outlines.

The oncolitic sandstone units are only a few metres wide and up to 1.5 m thick. Interbedded with the limestone facies of association C, its distribution is random. The sandstone component alters along palaeocurrent direction, i.e. from north to south. The sandstone grains are quartzose and coarse to very coarse in size (more granular and pebbly to the north), with the sandstone/oncolite ratio decreasing from north to south. Units have erosive bases cut into the underlying limestones and are decimetres in thickness. Low-angle trough cross-stratification is visible with cross-sets about 30 cm thick. The component oncolites (including those of the carbonate conglomerates) range in size from 3−4 to 10−12 cm in diameter. They are round to ovoid in shape with cores composed of fragments of clastic rocks, carbonate rocks (Wealdian and Jurassic), biogenic fragments (ostracods and gastropods), or a combination thereof (Fig. 7). The oncolitic structures are singular to composite with erosive surfaces. Their internal structures comprise various types of alternations between algal populations of *Phormidium* and *Schizothrix* (type A, B, C, and D morphologies, see Monty & Mas, 1981).

Fig. 7. Rock slab of a carbonate conglomerate (facies C_1) from Las Majadas I. It is mainly composed of oncolite clasts, oncolites, and rock fragments. The cores of many oncolites are Jurassic and Weald rock fragments and many oncolites are composite in structure. Scale in centimetres.

Facies C_2 (grainstones)

This facies contains carbonate clasts (mostly Wealdian) interspersed with quartz grains, carbonate black pebbles, oncolites, biogenic fragments (ostracod and gastropod shells), and rare plant debris in varying proportions. Grain sizes are relatively smaller than facies C_1 (pebbles to sand). Generally trough cross-stratified, cross-sets are decimetres thick. Sedimentary units are lensoid and only metres long. This facies is continuous in composition with facies C_1. Diagenetic overprint includes brecciation, marmorization, and calichification.

Facies C_3 (packstones to wackestones)

Similar to charophyte micrites (facies B_2), this facies contains more 'detrital' material, including clastic fragment accumulations (lag deposits?). The percentage of clastic grains can be as high as 50% or more, as in one case where a rippled sandstone–micrite sequence (Fig. 8) is evidence for tractional (bedload) deposition of the mixed carbonate and sandstone rock. This shows that diagenetic overprint in the carbonate portion is especially strong; this facies contains brecciation fabrics, karst features, calichification, and marmorization.

Facies C_4 (clayey marls)

Defined as carbonate marls with varying amounts of clay, this facies is extremely friable and is red and yellow in colour. Units can be up to metres in thickness, with lengths beyond outcrop scale. Lateral gradation with limestones (mostly with facies in association C) appears as a successive degradation of the limestones into the clayey marls. These sections across the transition between clayey marls and surrounding limestones exhibit soil features, such as peds, clay cutans (Retallack, 1988), and tiny, branching, tube-like voids (see Fig. 9). Rare charophytes can be found in this facies and centimetre-scale pure carbonate nodularization is commonly developed within units.

Facies C_5 (carbonaceous micrites)

These are dark grey to brown carbon-rich micrites in lenses of 20–50 cm thickness and metres long. Fine lamination is defined by clastic material, organic detritus, and fine-grained carbonate. Bioturbation and root traces are rare. It is mainly found in the area between Uña and Los Lagunillos.

Fig. 8. Rippled sandy micrite from an outcrop along the road to the northeast of Los Lagunillos. Oversteepened ripple forms are defined by quartz sand grains, but featureless micrite comprises the remainder of the ripple forms. This supports the interpretation of bedload transport for the micrites; original carbonate structures and grains have been micritized by diagenesis (facies C_3). Scale: 3 cm. White splotches are lichens.

Facies C_6 (laminated pisolitic wackestone)

This facies comprises a crudely laminated wackestone (Fig. 10) found to date only in one layer sandwiched between a micrite below (facies C_3) and a clayey marl above (facies C_4) in the La Toba section. This layer is only 15 cm thick with a lateral extent difficult to measure owing to erosion and/or diagenesis. The centimetre-scale lamination is both undulatory and uneven, with laminae components comprising millimetre-scale pisolites and carbonate rock fragments bound by sparitic and micritic cement. Vertical microcracks and microscopic bird's eye texture are common. Diagenetic overprint includes brecciation.

Interpretation

The clayey-marls—limestones—sandstones association represents carbonate-dominated alluvial plain

Fig. 9. Thin-section of transition zone between clayey marl (facies C_4) and limestone lens (facies C_3) at Los Lagunillos. Dark ovoids are relict oncolites. Sparry calcite and microcrystalline micrite cements fill the elongated, tube-like voids interpreted as rhizoliths. Clay cutans (arrows) lined the root traces (pedogenic process: clay illuviation) before they were cemented. Scale: 0.5 mm.

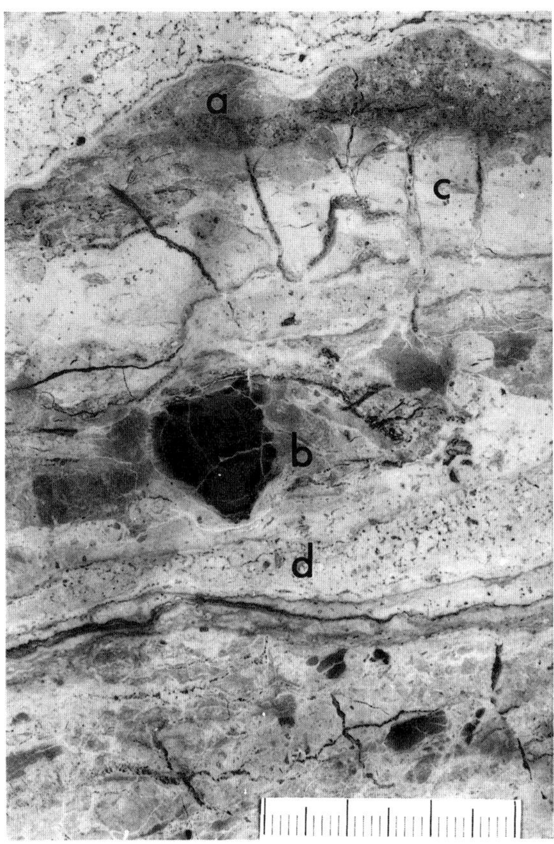

Fig. 10. Rock slab of laminated pisolitic wackestone (facies C_6) from La Toba. Lamination is uneven and undulatory, and contains pisolites (a). Note the floating rock fragment (b) and vertical microcracks (c). Microscopic bird's eye texture (d) suggests an originally porous texture for this interpreted caliche. Scale in millimetres.

deposits. Facies C_1, C_2, and C_3 are interpreted as channel deposits, facies C_4 as flood basin paleosols, facies C_5 as marsh, and facies C_6 as caliche. Early synsedimentary diagenetic processes affected the carbonates to varying degrees owing to subaerial exposure and re-immersion at surface or near-surface conditions.

The oncolitic sandstones of facies C_1 represent the minor clastic input into the carbonate-dominated portions of the Uña Basin. Although a few carbonate rocks (facies C_2 and C_3) contain up to 40% clastic components, the separation of clastics and carbonates is normally quite distinct and probably the result of density differences in hydraulic sedimentary processes. Fortunately, sand grains within the car-

bonates preserve the primary structures despite diagenetic overprinting of the carbonates (such as cross-lamination within a micrite (facies C_3), which is defined by quartz grains, east of Los Lagunillos (Fig. 8)) and confirm the traction deposition of the carbonates as bedload.

The existence of large numbers of oncolites in the sandstones indicates a high concentration of dissolved calcite in the Weald waters. The growth of oncolites is common in river channels, marshes, lakes, and floodplains, past and present (e.g. Głazek, 1965; Anadón & Zamarreño, 1981; Monty & Mas, 1981; Ordóñez & Garcia del Cura, 1983). The morphology of the algal components of the Wealdian oncolites suggest seasonal growth.

The clayey marls (facies C_4) represent soils that developed on the drier portions of the flood basin and/or where sedimentation rates were lower (Bown & Kraus, 1987; Kraus, 1987). The presence of cutans (Retallack, 1988), rhizoliths (Klappa, 1980) (Fig. 9), and nodularization (Gile *et al.*, 1965,1966), as well as the indistinct boundaries between the clayey marls and surrounding limestones, indicate that pedogenic processes altered carbonates on the floodplain. Other synsedimentary diagenetic processes at work on the Weald alluvial plain included caliche or calcrete development (facies C_6: Fig. 11) (Goudie, 1983; Vogt, 1984a,b; Wright *et al.*, 1988), calichification (cf. Esteban & Klappa, 1983), karstification (cf. James & Choquette, 1988), marmorization (hydromorphic soils — gleying; cf. Freytet, 1973,1984; Freytet & Plaziat, 1982), and micritization by algae (Kahle, 1977).

Freshwater diagenesis in carbonates is mainly early or synsedimentary (cf. Arribas, 1982,1986). Owing to its instability, calcite is subject to erosion, dissolution, and precipitation by physico-chemical and biochemical processes. In the Uña Basin, within an environment affected by oscillating water levels and changes in the level of the water table, the resultant deposits are products of carbonates repeatedly recycled and transported as well as altered *in situ* (see also Platt, 1989).

A modern analogue of a carbonate-dominated alluvial plain has been described from the semi-arid to subarid limestone plains of Bahia in northeast Brazil (Branner, 1911). This floodplain contained 'soft and marly fresh deposits' and 'rock hard' older deposits. Depressions and holes on the Bahia plains were distributed along the surface as a result of dissolution, whereas temporary lake deposits were strewn with molluscan shells. Although Branner

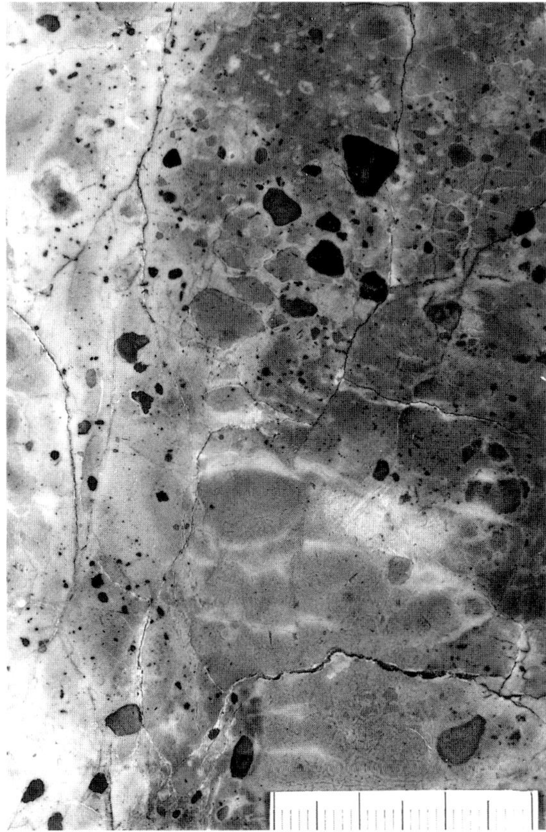

Fig. 11. Rock slab of calichified breccia (associated with facies C$_3$) from Los Lagunillos. Note angular to rounded, carbonate rock fragments (both dark and light) floating in micritic matrix. Colours include pink, yellow, and red; marmorization also affected this breccia. Scale in millimetres.

noted the ubiquitous precipitation of calcite, the presence of 'angular fragments and water-worn boulders of all kinds of rocks' indicated bedload transport of carbonate and clastic components alike.

Coal−clastics (association D)

This association (Fig. 5) is found at Uña and Las Majadas (Figs 3, 4). The coal is lignite (no detailed analysis has been done) and is mixed with sand to clay components. The lignite lenses are up to 30−40 cm thick. Facies D$_1$ is found singly at Las Majadas, whereas facies D$_2$ and D$_3$ are stacked one above the other in fining upward sequences at Uña for a total thickness of 16 m.

Facies D$_1$ (sandstones and lignites)

Medium to fine-grained sandstone is intercalated with lignite. Units are decimetres thick lenses and can be metres long. Ripple cross-lamination and low-angle trough cross-stratification is present.

Facies D$_2$ (conglomeratic sandstones)

This sandstone is especially well exposed at Uña, where it comprises a multi-storey channel fill with wings pinching out into the surrounding finer clastics (facies D$_2$). This amalgamated sandstone body complex is about 25 m wide and has four component sand bodies about 1.5 m thick containing trough cross-stratification. Cross-sets are about 30 cm thick; erosive scours and fining-upward sequences occur within the cross-sets. The grain size ranges from coarse to granular and grains are quartzose in composition. Palaeocurrent direction is N−NW to S−SE.

Facies D$_3$ (claystones and lignites)

Lignites are interbedded with siltstones and claystones in this facies. Exposure is very poor.

Interpretation

This coal−clastics association is interpreted as the clastic-dominated alluvial system within the Uña area. Facies D$_1$ may represent crevasse-splay deposits and facies D$_2$ and D$_3$ are interpreted as channel and overbank deposits. The changes between clastic and carbonate-dominated deposition could be the result of changing topography and source area and/or widening of the basin resulting from successive block faulting, but this is merely conjectural. A similar model was utilized to explain facies distribution patterns in the southeastern part of the southwest Iberian Range, in the province of Valencia (Monty & Mas, 1981). Exposure is not sufficient around Uña to confirm that proximal deposits comprised a clastic alluvial plain and the carbonate alluvial plain was exclusively distal in nature.

PALAEOENVIRONMENTAL SUMMARY

The Lower Cretaceous Weald in the study area around Uña is interpreted to have been deposited in a freshwater, continental basin containing hard waters. Fossil evidence and regional geology indicate a non-marine setting (Meléndez, 1983; Gómez Fernández, 1988). The provenance area surrounding the basin comprised Palaeozoic and Triassic rocks providing the clastics, whereas the basement constituted a Jurassic karst system delivering bedload and dissolved carbonates into the Uña Basin. Biochemically deposited carbonates (cf. Eugster & Kelts, 1983) were contributed by charophytes, blue-green algae, macrophytes, molluscs, ostracods, and gastropods.

The depositional environment around Uña is interpreted as an alluvial−lacustrine system containing a carbonate−clastic alluvial plain adjoining a carbonate, marginal lacustrine area, with ponds and small lakes, which is transitional to a carbonate lacustrine delta (Fig. 12; Table 1). Association A represents the shallow, marshy (facies A_2) to profundal (facies A_1) portions across a lake delta. Coal formation (facies A_3) occurred as allochthonous material collected on the delta plain and surrounding area. Association B is interpreted as representing shallow lakes (facies B_1) and ponds (facies B_2) dispersed over a carbonate alluvial plain surrounding the lake at Uña. Association C represents the deposits of a dominantly carbonate alluvial plain with channel deposits (facies C_1, C_2, and C_3) and extra-channel deposits, including palaeosols (facies C_4), carbonate marshes (facies C_5), and caliche (facies C_6). Association D is interpreted as the deposits of a clastic alluvial system containing minor coal; clastic deposition only occurred in portions of the basin. Early cementation and other diagenetic processes, and pedogenesis, erosion, repeated dissolution, and re-precipitation through physical and biological influences, masked many of the primary sedimentary structures in the carbonates.

This diagenetic overprint on the continental carbonates of the Uña area can cover sequences up to 5 m thick with great lateral and vertical variation between and among facies. Diagenesis can completely eliminate the original primary structures; the limestones of associations B and C are affected (carbonate alluvial plain) and this makes identifi-

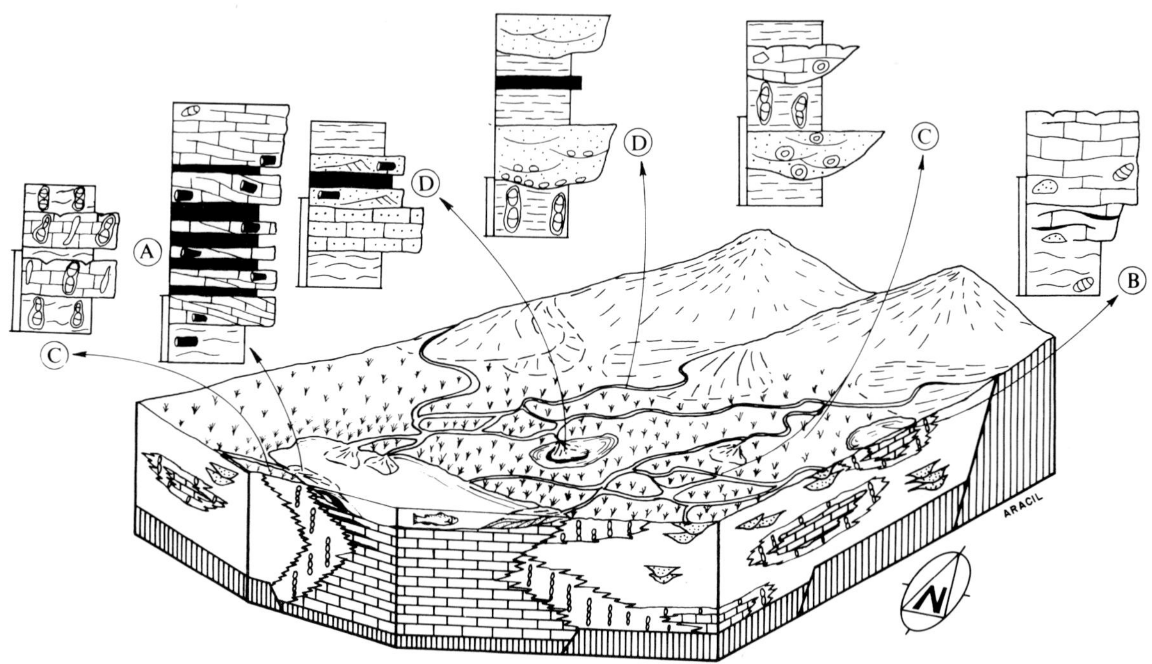

Fig. 12. Depositional model and characteristic lithological sequences of the facies associations for the lake−alluvial-plain palaeoenvironment around Uña. From left to right: association C (surrounding carbonate alluvial plain) and association A (lake delta) from Uña. Association D (clastic alluvial system) from La Majadas I and II. Association C (carbonate alluvial plain) from La Toba and Los Lagunillos. Association B (small lakes and ponds) from Cabeza Gorda. Key from Fig. 2. Scale bar on each lithological column represents 2 m.

cation of relationships between facies very difficult. The diagenetic processes appear to have been early; contemporaneous to syndepositional, since carbonates are very reactive, especially in continental environments (Golubic, 1973; Murphy & Wilkinson, 1980; Goudie, 1983). Processes included recrystallization, micritization, marmorization, karstification, nodularization, brecciation, calichification (facies C_6) (Arribas *et al.*, 1989), and clay illuviation (pedogenesis, facies C_4) (Goudie, 1983; Retallack, 1988).

Brecciated limestones comprise angular to rounded, carbonate rock fragments of Wealdian origin that are still *in situ* or floating in a micritic matrix (Fig. 11) (cf. Esteban & Klappa, 1983). At La Toba, over 1.5 m of brecciated limestones are exposed.

Evidence for karstification includes irregularly shaped cavities up to 1 m long to centimetre-scale fenestrae; these cavities are filled with clastic or carbonate lag material, or facies C_4.

Marmorized limestones are colourfully mottled carbonates; colours include yellow, brown, blue, red, pink, and tan. Separate grains are not readily discernable in many samples, though a spectrum from mottled to brecciated types occurs (Fig. 11). Thicknesses of up to 6 m can be marmorized.

COAL DEPOSITIONAL MODEL

In order to construct a depositional model for the coal at Uña, pertinent data from the facies associations about the hydrology, sediment supply, subsidence, and climate must first be examined. The hydrology of the Wealdian depositional system was stable and maintained lake levels despite periodically dry periods in the surrounding areas. The absence of subaerial exposure features in the lacustrine limestones at Uña indicates subaqueous conditions throughout deposition, while the diagenetic overprinting in the surrounding floodplain carbonates demonstrates varying water input over time. The Jurassic karst substratum probably sustained a constant water supply to the lake at Uña.

The sediment supply within the Uña Basin was continual. This is shown by the repeated infilling of the small lakes and ponds of association B; the Jurassic karst substratum supplied a steady input of carbonates. Evidence for subsidence in the Uña Basin lies in the succession of shallowing upward sequences (association B) at Cabeza Gorda; pulses of subsidence created lakes and ponds that were

subsequently infilled.

The *in situ* carbonized rhizoliths found at Uña may indicate rapid burial, perhaps as a result of subsidence. According to Teichmüller & Teichmüller (1982), low acidity or basic conditions, as found in carbonate-rich waters (hard waters), enhances bacterial activity and leads to extreme plant decomposition at surface to near-surface temperatures. Rapid burial under subaqueous conditions would be especially important in preserving coal in carbonates. This is significant because the coal at Uña is mainly allochthonous in origin and would require rapid burial/subsidence for preservation.

The interpretation of the coal–limestones association at Uña (association A) as a prograding lake delta fits the above constraints; hydrology (water levels), subsidence (rapid burial), and sediment supply were important for the preservation of plant material for coal formation.

Climate is more difficult to assess from the given data. The synsedimentary diagenetic features of the floodplain carbonates point to alternating wet and dry periods. The algal morphologies of the oncolites are interpreted as influenced by a seasonal climate. Seasonality may be indicated, but general climatic patterns cannot be inferred directly from the sediments. Theories on the climate of the Early Cretaceous vary.

Rat (1982) believed, according to palaeogeographic and sedimentary evidence, that the climate in the western Tethys during the Weald was drier and more contrasting than the hot and humid norm throughout the rest of the Cretaceous. Kemper (1987) theorized that the Cretaceous in the northern latitudes consisted of cold and warm cycles, with warm cycle postulated from the Hauterivian to Lower Aptian. Barron & Washington (1982), based on a numerical model of the Mid-Cretaceous, proposed no great warming trends on the continents or upper latitudes, or poleward displacement of circulation features; however, a subtropical high (hot and dry) did shift equatorward in association with the Tethys Ocean.

Ziegler *et al.* (1983) placed central Spain at about 30°N during the Early Cretaceous; this is in the dry, divergent subtropical zone, using today's climatic patterns. However, Ziegler *et al.* (1987), based on palaeobotanical evidence, postulated that the climatic zones were different owing to weaker polar fronts, so that seasonal climate probably occurred in tropical and subtropical zones. Their model would predict a seasonal, generally warm and semi-arid climate for central Spain in the subtropical zone of the Early Cretaceous. This prediction fits well with the presented depositional model for the Weald of the Uña Basin.

Coal depositional models invoke many variables to explain and predict coal location. For example, Cabrera & Sáez (1987) linked extensive coal formation to tectono-sedimentary and climatic factors; these basically determine groundwater levels and their intersection with the surface — defining lacustrine or paludal conditions. Li Sitian *et al.* (1984) stressed the importance of rapid deposition and subsidence in coal preservation. Fielding (1987) proposed that the subsidence regime and, less importantly, sediment supply are the main factors in coal distribution. These two latter models are applicable to the thin coal layer at Uña. Sediment supply and subsidence (rapid burial within a prograding lake delta) were important in preserving plant material within carbonates in a seasonal, semi-arid climate; hydrological conditions in the Uña Basin (ultimately based on tectono-sedimentary influences) also contributed to coal preservation. Climate itself, however, was *not* an important factor in coal formation.

CONCLUDING REMARKS

The Weald (Lower Cretaceous) in the Serranía de Cuenca of east-central Spain is interpreted as basinal fill of one of a series of rift basins in the southwest Iberian Range. The unusual nature of the basin deposits is the wide variety of facies and the predominance of carbonates in this continental depositional system: a carbonate-dominated, alluvial-plain—lacustrine system (see Table 1) containing limestone channels, ponds, and soils. With the Jurassic marine limestones as source rocks, carbonate transport was accomplished by both bedload and suspension transport. The preservation of primary sedimentary structures was rare because of the high reactivity of calcium carbonate, at surface conditions, in the presence of water. Carbonates were continually being recycled, both physico-chemically and biologically, within the depositional system; various diagenetic processes, including pedogenesis, sparmicritization, and brecciation, masked the original evidence of transport mode. More petrological research is needed to identify the numerous stages of synsedimentary and post-sedimentary diagenesis in a carbonate freshwater environment.

Climate in the Barremian of central Spain, as deduced from patterns in the carbonate deposits as well as from palaeogeographic studies, was seasonal with wet and dry periods. However, the hydrology of the Serranía de Cuenca Basin, controlled by the Jurassic karst substratum, permitted the maintenance of a prograding lacustrine delta in a perennial lake (as well as a larger, deeper lake to the south (see Gómez Fernández & Meléndez, this volume). Despite the high alkalinity of a carbonate depositional system, coal (lignite) was indeed preserved within the interpreted delta deposits at Uña. A modern analogue from a temperate, marl lake shows that peat can accumulate in a prograding delta system with high sedimentation rates. Sediment supply and subsidence were deemed the important physical variables controlling coal deposition at the Wealdian lake near Uña.

ACKNOWLEDGEMENTS

E.G.K. wishes to thank Erna Kordesch and Dorothea Janofske for field assistance and Joachim Reitner for discussion. The Deutsche Forschungsgemeinschaft supported her research through their post-doctoral stipend program. J.C.G.F. and N.M. wish to thank Maria Eugenía Arríbas for assistance with identification of diagenetic features and acknowledge financial support from Project ID-452, CAICYT CSIC. The coal at Uña was analysed by Dr J.G. Prado of the Instituto Nacional del Carbon in Oviedo, Spain. Helpful comments, which improved the manuscript considerably, were supplied by our reviewers, Lluís Cabrera and Thomas Kreuser.

REFERENCES

ÁLVARO, M., CAPOTE, R. & VEGAS, R. (1979) Un modelo de evolucion geotectónica para la Cadena Celtibérica. In: *Homenaje a Lluis Solé i Sabarís*, (Eds Riba, O., Santanach, P. & Solé, L.) Acta Geol. Hisp. **14**, 172–177.

ANADÓN, P. & ZAMARREÑO, I. (1981) Paleogene nonmarine algal deposits of the Ebro Basin, northeastern Spain. In: *Phanerozoic Stromatolites* (Ed. Monty, C.), pp. 86–120. Springer-Verlag, Berlin.

ARRIBAS, M.E. (1982) Petrología y sedimentología de las facies carbonáticas del Paleogeno de La Alcarria (Sector NW). *Estud. Geol.* **38**, 27–41.

ARRIBAS, M.E. (1986) Pedogenetic facies in a lacustrine sedimentation model: Paleogene of the Tertiary Tajo Basin. In: *Geochemistry and Mineral Formation at the Earth's Surface* (Eds Rodríguez Clemente, R. & Tardy,

Y.), pp. 339–350. *Proceedings of International Meeting on the Geochemistry of the Earth Surface and Processes of Mineral Formation*, Granada. Consejo Superior de Investigaciónes Científicas, Barcelona.

ARRIBAS, M.E., GÓMEZ FERNÁNDEZ, J.C., & MELÉNDEZ, N. (1989) Early diagenetic processes in lacustrine and in related floodplain sediments from the Lower Cretaceous (central Spain). *International Association of Sedimentologists, Abstracts of Tenth European Regional Meeting* (Ed. Kázmér, M.), pp. 3–4.

BARRON, E.J. & WASHINGTON, W.M. (1982) Cretaceous climate; a comparison of atmospheric simulations with the geologic record. *Palaeogeogr. Palaeoclimatol. Palaeoecol.* **40**, 103–133.

BOWN, T.M. & KRAUS, M.J. (1987) Integration of channel and floodplain suites. I. Developmental sequence and lateral relations of alluvial paleosols. *J. Sediment. Petrol.* **57**, 587–601.

BRANNER, J.C. (1911) Aggraded limestone plains of the interior of Bahia and the climatic changes suggested by them. *Geol. Soc. Am. Bull.* **22**, 187–206.

CABRERA, LL. & SÁEZ, A. (1987) Coal deposition in carbonate-rich shallow lacustrine systems: the Calaf and Mequinenza sequences (Oligocene, eastern Ebro Basin, NE Spain). *J. Geol. Soc. London* **144**, 451–461.

DURAND, J.-P. (1980) Les sédiments fuvéliens du synclinal de L'Arc (Provence). *Ind. Min. (Suppl.)* **62**(6), 13–25.

ESTEBAN, M. & KLAPPA, C.F. (1983) Subaerial exposure environment. In: *Carbonate Depositional Environments* (Eds Scholle, P.A., Bebout, D.G. & Moore, C.H.). Mem., Am. Assoc. Petrol. Geol. **33**, 1–54.

EUGSTER, H.P. & KELTS, K. (1983) Lacustrine chemical sediments. In: *Chemical Sediments and Geomorphology* (Eds Goudie, A.S. & Pye, K.), pp. 321–368. Academic Press, London.

FEY, B. (1988) *Die Anurenfauna aus der Unterkreide von Uña (Ostspanien)*. Dissertation, Freie Universität Berlin, 168 pp.

FIELDING, C.R. (1987) Coal depositional models for deltaic and alluvial plain sequences. *Geology* **15**, 661–664.

FREYTET, P. (1973) Petrography and paleoenvironments of continental carbonate deposits with a particular reference to Upper Cretaceous and Lower Eocene of Languedoc, southern France. *Sediment. Geol.* **10**, 25–60.

FREYTET, P. (1984) Les sediments lacustres carbonatés et leurs transformations par émersion et pédogenèse. Importance de leur identification pour les reconstructions paléogéographiques. *Bull. Cent. Rech. Exp. Elf. Aquitaine* **8**, 223–247.

FREYTET, P. & PLAZIAT, J.C. (1982) Continental carbonate sedimentation and pedogenesis — Late Cretaceous and Early Tertiary of southern France. *Contrib. Sedimentol.* **12**, 1–213.

GIERLOWSKI-KORDESCH, E. & JANOFSKE, D. (1989) Paleoenvironmental reconstruction of the Weald around Uña (Serranía de Cuenca, Cuenca Province, Spain). In: *Cretaceous of the Western Tethys* (Ed. Weidmann, J.), pp. 239–264. *Proceedings of Third International Cretaceous Symposium,* Tübingen, 1987. E. Schwiezerbart'sche Verlagsbuchhandlung, Stuttgart.

GILE, L.H., PETERSON, F.F. & GROSSMAN, R.B. (1965) The K horizon: a master soil horizon. *Soil Sci.* **99**, 74–82.

GILE, L.H., PETERSON, F.F., & GROSSMAN, R.B. (1966) Morphological and genetic sequences of carbonate accumulation. *Soil Sci.* **101**, 347–360.

GŁAZEK, J. (1965) Recent oncolites in streams of North Vietnam and of the Polish Tatra Mountains. *Rocz. Pol. Tow. Geol.* **35**, 221–242.

GOLUBIC, S. (1973) The relationship between blue-green algae and carbonate deposits. In: *The Biology of Blue-green Algae* (Eds Carr, N.G. & Whitton, B.A.), Bot. Monogr. **9**, 435–472.

GÓMEZ FERNÁNDEZ, J.C. (1988) *Estratigrafía y sedimentologia del Cretácico Inferior en 'Facies Weald' de la Región Meridional de la Serranía de Cuenca.* Masters Thesis, Universidad Complutense, Madrid, 228 pp.

GOUDIE, A.S. (1983) Calcrete. In: *Chemical Sediments and Geomorphology* (Eds Goudie, A.S. & Pye, K.), pp. 93–131. Academic Press, London.

HENKEL, S. (1970) Eine neue Fossillagerstätte in Ostspanien und ihre Bedeutung für die Stammgeschichte der Wirbeltiere. *Umsch. Wiss. Tech.* **8**, 247–248.

HENKEL, S. & KREBS, B. (1969) Zwei Säugetier-Unterkiefer aus der unteren Kreide von Uña (Prov. Cuenca, Spanien). *Neues Jahrb. Geol. Paläeontol. Monatsh.* 449–463.

JAMES, N.P. & CHOQUETTE, P.W. (Eds) (1988) *Paleokarst.* Springer-Verlag, New York.

KAHLE, C.F. (1977) Origin of subaerial Holocene calcareous crusts: role of algae, fungi, and sparmicritization. *Sedimentology* **24**, 413–435.

KEMPÈR, E. (1987) Das Klima der Kreide-Zeit. *Geol. Jahrb. Reihe A* **96**, 5–186.

KLAPPA, C.F. (1980) Rhizoliths in terrestrial carbonates: classification, recognition, genesis, and significance. *Sedimentology* **27**, 613–629.

KRAUS, M.J. (1987) Integration of channel and floodplain suites. II. Vertical relations of alluvial paleosols. *J. Sediment. Petrol.* **57**, 602–612.

KÜHNE, W.G. & CRUSAFONT, M. (1968) Mamíferos del Wealdense de Uña cerca de Cuenca. *Acta. Geol. Hisp.* **3**, 133–134.

LI SITIAN, LI BAOFONG, YANG SHIGONG, HUANG JIAFU, & LI ZHEN (1984) Sedimentation and tectonic evolution of Late Mesozoic faulted coal basins in northeastern China. In: *Sedimentology of Coal and Coal-bearing Sequences* (Eds Rahmani, R.A. & Flores, R.M.), Spec. Publ. Int. Assoc. Sediment. **7**, 387–406.

LOFTUS, G.W.F. & GREENSMITH, J.T. (1988) The lacustrine Burdiehouse Limestone Formation — a key to the deposition of the Dinantian oil shales of Scotland. In: *Lacustrine Petroleum Source Rocks* (Eds Fleet, A.J., Kelts, K. & Talbot, M.R.), Geol. Soc. London Spec. Publ. **40**, 219–234.

MAS, J.R., ALONSO, A., & MELÉNDEZ, N. (1982) El Cretácico basal Weald de la Cordillera Ibérica (NW de la provincia de Valencia y E de la provincia de Cuenca). *Cuad. Geol. Ibérica* **8**, 309–335.

McCABE, P.J. (1984) Depositional environments of coal and coal-bearing strata. In: *Sedimentology of Coal and Coal-bearing Sequences* (Eds Rahmani, R.A. & Flores, R.M.), Spec. Publ. Int. Assoc. Sediment. **7**, 13–42.

MELÉNDEZ, N. (1982) Presencia de una discordancia cartográfica intrabarremiense de la Cordillera Ibérica Occidental (provincia de Cuenca). *Estud. Geol.* **38**, 51–54.

MELÉNDEZ, N. (1983) El Cretácico de la Región de Cañete-

Rincón de Ademuz (provincias de Cuenca y Valencia). *Semin. Estrat. Ser. Monograf. Univ. Complutense, Madrid*, **9**, 242 pp.

MELÉNDEZ HEVIA, F., VILLENA MORALES, J., RAMÍREZ DEL POZO, J., PORTERO-GARCIA, J.M., OLIVÉ DAVÓ, A., ASSENS CAPARROS, J. & SÁNCHEZ-SORIA, P. (1975) Sintesis del Cretácico de la Zona Sur de las 'Rama Castellana' de la Cordillera Ibérica. *Actas Ier Symposium Cretácico Cordillera Ibérica*, Cuenca, pp. 241−252.

MOHR, B.A.R. (1987) Mikrofloren aus Vertebraten-Führenden Unterkreide-Schichten bei Galve und Uña (Ostspanien). *Berliner Geowiss. Abh.* **86A**, 69−85.

MONTY, C.L. & MAS, J.R. (1981) Lower Cretaceous (Wealdian) blue-green algal deposits of the province of Valencia, eastern Spain. In: *Phanerozoic Stromatolites* (Ed. Monty, C.), pp. 86−120. Springer-Verlag, Berlin.

MURPHY, D.H. & WILKINSON, B.H. (1980) Carbonate deposition and facies distribution in a central Michigan marl lake. *Sedimentology* **27**, 123−135.

ORDÓÑEZ, S. & GARCÍA DEL CURA, M.A. (1983) Recent and Tertiary fluvial carbonates in central Spain. In: *Modern and Ancient Fluvial Systems* (Eds Collinson, J.D. & Lewin, J.). Spec. Publ. Int. Assoc Sediment. **6**, 485−497.

PLATT, N.H. (1989) Lacustrine carbonates and pedogenesis: sedimentology and origin of palustrine deposits from the Early Cretaceous Rupelo Formation, W. Cameros Basin, N. Spain. *Sedimentology* **36**, 665−684.

RAMÍREZ DEL POZO, J. & MELÉNDEZ-HEVIA, F. (1972) Nuevos datos sobre del Cretácico inferior en facies 'Weald' de la Serranía de Cuenca. *Bol. Geol. Min.* **83**(6), 569−581.

RAMÍREZ DEL POZO, J., PORTERO GARCIA, J.M., OLIVÉ DAVÓ, A. & MELÉNDEZ HEVIA, F. (1975) El Cretácico de la Serranía de Cuenca y de la región Fuentes-Vilar del Humo: correlación y cambios de facies. *Actas Ier Symposium Cretácico Cordillera Ibérica*, Cuenca, pp. 189−205.

RAT, P. (1982) Factores condicionantes en el Cretácico de España. *Cuad. Geol. Ibérica* **8**, 1059−1076.

RETALLACK, G.J. (1988) Field recognition of paleosols. In: *Paleosols and Weathering through Geologic Time: Principles and Applications* (Eds Reinhardt, J. & Sigelo, W.R.), Geol. Soc. Am. Spec. Pap. 216, 1−20.

TEICHMÜLLER, M. & TEICHMÜLLER, R. (1982) The geological basis of coal formation. In: *Stach's Textbook of Coal Petrology* (Eds Stach, E., Chandra, D., Mackowsky, M.T., Taylor, G.H., Teichmüller, M. & Teichmüller, R.), pp. 5−86. Gebrüder Bornträger, Stuttgart.

TREESE, K.L. & WILKINSON, B.H. (1982) Peat−marl deposition in a Holocene paludal−lacustrine basin — Sucker Lake Michigan. *Sedimentology* **29**, 375−390.

VILAS, L., MAS, J.R., GARCÍA, A., ARIAS, A., ALONSO, A., MELÉNDEZ, N. & RINCÓN, R. (1982) Ibérica Suroccidental. In: *El Cretácico de España*, pp. 457−509. Universidad Complutense, Madrid.

VILAS, L., ALONSO, A., ARIAS, C., GARCÍA, A., MAS, J.R., RINCÓN, R. & MELÉNDEZ, N. (1983) The Cretaceous of the southwestern Iberian Ranges (Spain). *Zitteliana* **10**, 245−254.

VOGT, T. (1984a) *Croûtes calcaires: types et genèse. Exemples d'Afrique du Nord et de France Méditerranénne.* Dissertation, Université Louis Pasteur Strasbourg, 239 pp.

VOGT, T. (1984b) Problèmes de genèse des croûtes calcaires quaternaires. *Bull. Cent. Rech. Explor. Prod. Elf-Aquitaine* **8**, 209−221.

WRIGHT, V.P., PLATT, N.H., & WIMBLEDON, W.A. (1988) Biogenic laminar calcretes: evidence of calcified root-mat horizons in paleosols. *Sedimentology* **35**, 603−620.

ZIEGLER, A.M., BARRETT, S.F. & SCOTESE, C.R. (1983) Mesozoic and Cenozoic paleogeographic maps. In: *Tidal Friction and the Earth's Rotation, II* (Eds Brosche, P. & Sundermann, J.), pp. 240−252. Springer-Verlag, Berlin.

ZIEGLER, A.M., RAYMOND, A.L., GIERLOWSKI, T.C., HORRELL, M.A., ROWLEY, D.B. & LOTTES, A.L. (1987) Coal, climate and terrestrial productivity: the present and Early Cretaceous compared. In: *Coal and Coal-bearing Strata: Recent Advances* (Ed. Scott, A.C.), Geol. Soc. London Spec. Publ. 32, 25−49.

Modern Processes in
East African Rift Lakes

Spec. Publs Int. Ass. Sediment. (1991) **13**, 129–145

High-resolution acoustic character of Lake Malawi (Nyasa), East Africa and its relationship to sedimentary processes

D.L. SCOTT[*,1], P. NG'ANG'A[†], T.C. JOHNSON[‡] *and* B.R. ROSENDAHL[§]

[*]206 Old Chemistry, Department of Geology, Duke University, Durham, NC, USA
[†]P. O. Box 40658, National Museums of Kenya, Nairobi, Kenya
[‡]Duke University Marine Laboratory, Beaufort, NC, USA
[§]RSMAS, University of Miami, 4600 Rickenbacker Causeway, Miami, FL, USA

ABSTRACT

A comprehensive geophysical survey of Lake Malawi, located in the southern portion of the East African rift system (EAR), has provided new insights to modern geological processes in the synrift lacustrine environment. In particular, depositional processes have been identified by detailed interpretation of over 12 000 km of 28 kHz echosounder data, 1000 km of 1 kHz seismic reflection profiles, and 650 km of side-scan sonar (SSS) profiles. Separate examination of 19 gravity and 33 piston cores, and study of 3300 km of multifold seismic (MFS) data, presented elsewhere, are also used to constrain and enhance these interpretations. Present-day bathymetry suggests that there are three modern depositional provinces: the northern Livingstone, central Usisya–Mbamba and southern Metangula provinces. Each province has distinctive sediment types and process response features, which can be recognized by their acoustic expression. These differences can be attributed directly to source and degree of tectonic activity variations. In all data sets, there is an apparent increase in tectonic activity to the north. This is reflected by increasingly complex, but systematic patterns of acoustic response. Acoustic character correlates with either sediment type and/or depositional process features. Correlation of structural activity and depositional processes is evident. It is concluded that subaqueous mass-wasting and gravity induced creep processes account for the bulk of sediment transport in this synrift lacustrine environment.

INTRODUCTION

Lake Malawi (Nyasa) (Fig. 1) occupies a southern rift zone (Rosendahl, 1987) just south of the Rungwe Volcanic Centre in the East African rift system (EAR). The lake is approximately 570 km long, with an average width of 50–60 km. A maximum width of 90 km is measured across the dog-leg that divides the lake structurally into northern and southern segments at approximately 12°S (Fig. 2B). The northern segment is composed of two well defined, opposite polarity half-graben whose border faults exhibit substantial relief. The underlying structure south of the dog-leg is more subtle and

appears to be a mosaic of smaller blocks that are not as easily grouped into discrete half-graben. This results in the definition of three depositional provinces: the northern Livingstone, the central Usisya–Mbamba and southern Metangula provinces (Fig. 1).

The lake reaches a maximum depth of about 730 m in the Usisya deep (Fig. 2A). In terms of African lakes, only Tanganyika is deeper or longer. Bathymetric closure around this deep defines one large basin occupying the southern three-fourths of the lake. The gross bathymetry of this southern basin is that of a gentle, north sloping ramp with a gradient of about 1.5 m/1000 m, merging into a strongly asymmetric, steep sided, nearly rectangular deep. Closure around the Livingstone deep to the north

---

[1] Present address: Research School of Earth Sciences, G.P.O. 4, Canberra ACT 2601, Australia.

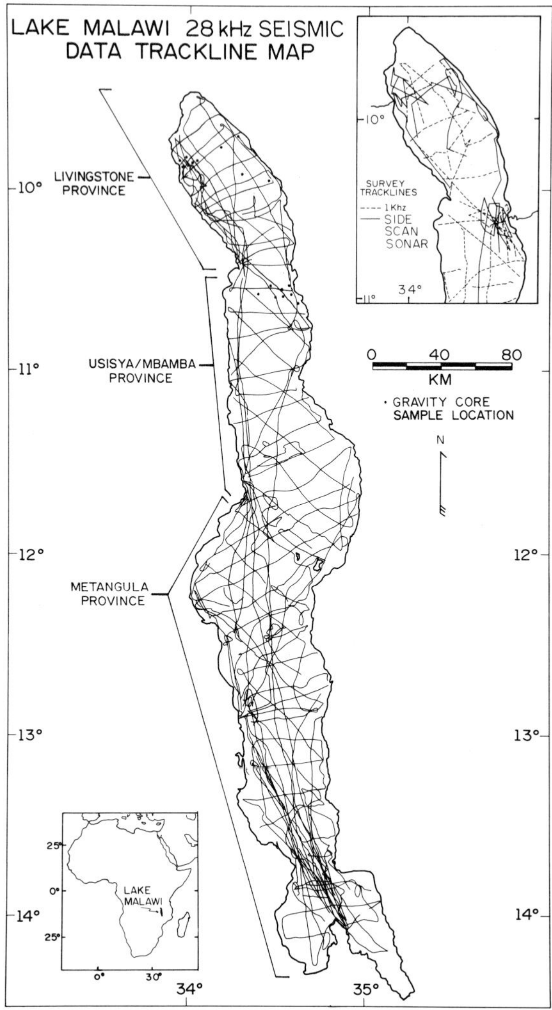

Fig. 1. Track-line map for 28 kHz and 1 kHz data, side-scan sonar (inset) data, and gravity core locations.

defines a smaller basin separated from the southern deep by a gentle saddle. The northern deep reaches maximum depths of just over 500 m. As in Lake Tanganyika (Lorber, 1984; Burgess, 1985; Burgess *et al.*, 1988), the series of depocentres appear to be defined by underlying half-graben units (Specht, 1987; Flannery, 1988). The size, morphology, and water chemistry of Lake Malawi and Lake Tanganyika set them apart from the other, smaller and/or alkaline lakes of the EAR (e.g. Swain *et al.*, 1981; Frostick *et al.*, 1986; Tiercelin *et al.*, 1987).

Lake Malawi is situated in a mobile zone of me-

dium to high-grade, banded, sillimanite—biotite gneiss and banded, cordierite gneiss of the Mozambique, Ubendian, and Irumide Belts. The bulk of the catchment area is composed of high-grade feldspathic gneisses, granulites, and schists of Precambrian age (Bloomfield *et al.*, 1966). Cretaceous N—S trending rifts and NE—SW Permo-Triassic rifts intersect the lake at high angles. These rifts strongly influence drainage patterns, which erode sandstones, calcareous mudstones, and carbonaceous shales. At the northern end of the lake the Pleistocene to Recent Rungwe Volcanics outcrop. These young volcanics range in composition from basalts to phonolitic trachytes (Harkin, 1960). Plio-Pleistocene sediments exposed along the western shore consist of lake-margin sands and gravels.

Figure 2B presents the general basement structure (Versfelt, 1988) underlying Lake Malawi, as interpreted from multifold seismic (MFS) data. Present-day bathymetry is the result of overprinting of the deep-seated structure by sedimentation and erosion within the lake. Sublacustrine structure (Specht, 1987; Versfelt, 1988) is dominated by three main border fault systems (BFS): the Livingstone, Usisya, and Metangula BFS. These fault systems alternate in polarity along the axis of the lake, without much overlap. In general, infrabasinal faulting is synthetic to the major BFS trend.

The Livingstone depositional province in the north has a well-defined depocentre. Located in the southeast quadrant of the province this deep is bounded to the east and southeast by a steep gradient and to the west and northwest by a shoaling ramp (Fig. 2A). This morphology corresponds to the northernmost structural unit, the Livingstone half-graben. This relatively simple half-graben is deepest in the east-northeast according to depth to acoustic basement in MFS data (Flannery, 1988), documenting a migration of the depocentre to the south through time. The eastern shoreline is defined by the border fault complex, whose footwall rises to about 2100 m above lake level. Depth to acoustic basement in MFS data exceeds 4 s two-way travel time in this province, indicating that total throw on the fault probably exceeds 6000 m.

The central Usisya—Mbamba depositional province encompasses the deepest point in the lake. This deep is bounded to the west by a steep bathymetric gradient and to the east by a shoaling ramp (Fig. 2A). Thus, in the general sense of depocentre asymmetry, the Livingstone and Usisya—Mbamba provinces are morphological mirror images. Again, the depo-

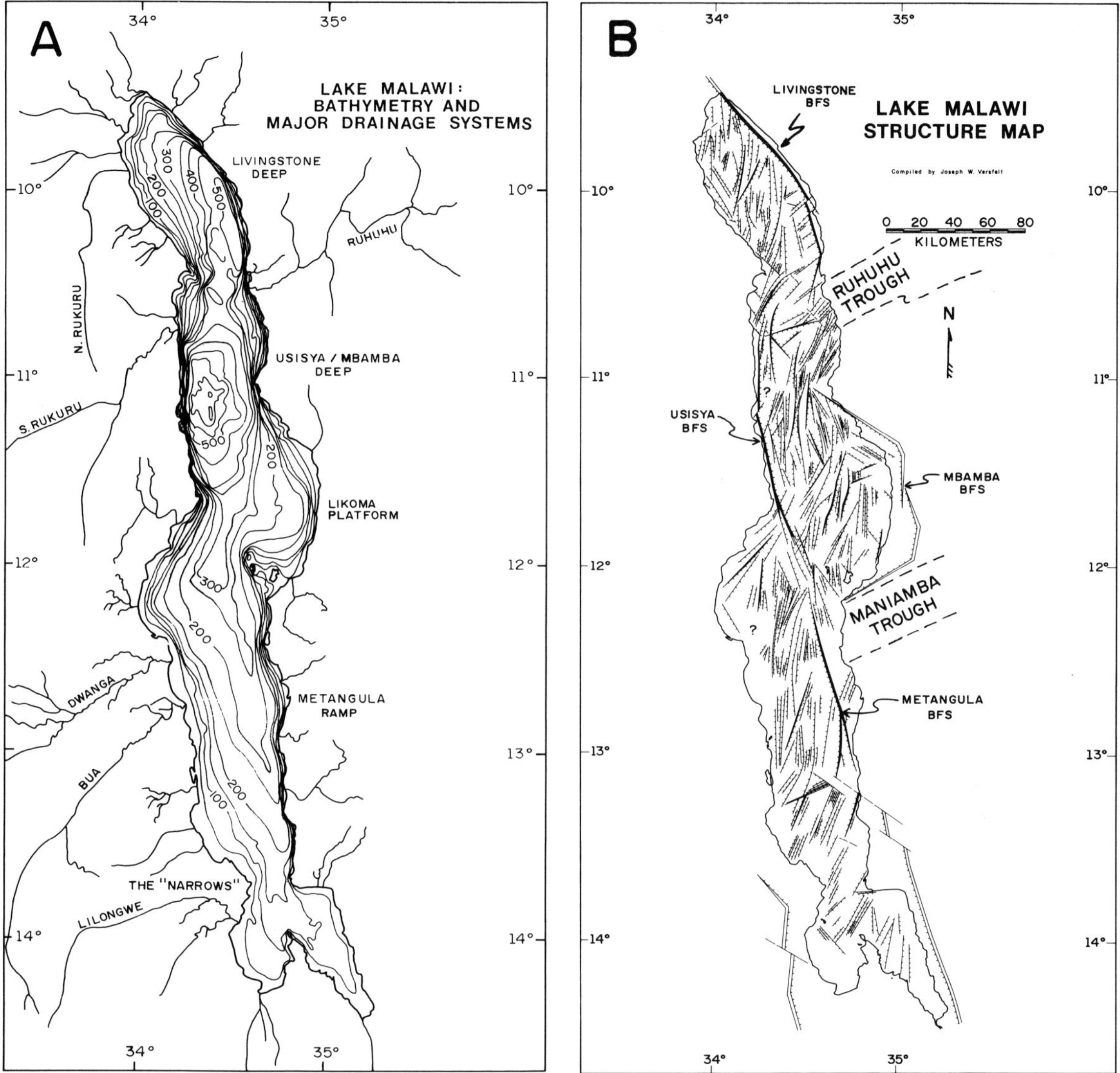

Fig. 2. (A) General bathymetry (Scott, 1988) and (B) structure (Versfelt, 1988). Compare with Fig. 4 for correlation of acoustic character with bathymetry and lake-floor structure.

sitional province corresponds directly to a well-defined structural unit, the Usisya half-graben, which has the opposite polarity from its northern neighbour. The border fault is on the west and it curls into the lake at its northern end. The rift shoulder rises to an elevation of about 1700 m above lake level, which combined with the depth to acoustic (MFS) basement, produces a total throw of 5000 m on the border fault. To the southeast, the steep bathymetric closure changes to the gentle gradient of the Likoma

platform. The MFS data indicate that this region is underlain by shallow acoustic basement at less than 1.0 s two-way travel time.

The down-to-the-west asymmetry of the Usisya half-graben changes in the dog-leg near 12°S to a down-to-the-east asymmetry in the Metangula half-graben (Fig. 2B). The bathymetric transition between depositional provinces is not as precise as between the Usisya−Mbamba and Livingstone provinces. In the southern Metangula province, the lake floor forms a gently northward sloping ramp without a well-defined depocentre. The gross bathymetry belies the structural complexity of the area (compare Figs 2A, 2B). The Metangula border fault and its occurrence well lakeward of the shoreline is a somewhat anomalous feature. It separates the basin from an adjacent continuous submerged platform rather than from greatly uplifted emergent shoulders. There are no significant rift mountains associated with any part of this BFS. Throws on the fault in the MFS data average 1000 to 1500 m.

The southern arms of the lake show very little internal faulting. However, very sparse data coverage in the arms preclude any definitive statements about the structure or sedimentation. Presumably, some type of mechanical accommodation zone must occur in the vicinity of the 'Narrows' (Fig. 2A), where rift structures change from a down-to-the-east half-graben to ill-defined, broadly spread faulting. The eastern shore is intersected by two NE−SW trending, Permo-Triassic Karroo troughs: the Ruhuhu and Maniamba (Fig. 2B). Both troughs deliver significant terrigenous sediments to the lake, although the sediment contribution of the northern Ruhuhu trough seems to be volumetrically more significant (Ng'ang'a, 1988).

Seasonal increases in rainfall correspond to greatly increased river flow and significant periodic increases in fluviatile sediment input, leading to a prevalence in the deep basins of laminated sediments with alternating layers of diatom ooze and terrigenous silts and clays (Johnson *et al.*, 1988; Johnson & Davis, 1989). Precipitation is significantly higher in the northern portion of the lake, so sediment input volume may be skewed to the north. Several axial rivers enter the lake at its northern end (Fig. 2A). Small fan-deltas are formed at the mouths of the axial rivers. The Shire River drains Lake Malawi at its southeastern end. Lake Malawi is permanently, but weakly, stratified in the thermal/chemical sense (Eccles, 1974), and anoxic below water depths of about 200 m. Local upwelling and turnover occur, but there are no historical records of main basin turnover. Surface temperatures of the lake tend to be higher than those of river waters, so the latter should sink to some equilibrium depth as they enter the lake.

In this study, selected interpretations and data from the analysis of 12 000 km of 28 kHz data (Scott, 1988) and 100 km of 1 kHz and 650 km of side-scan sonar data (Ng'ang'a, 1988; Johnson & Davis, 1989) from Lake Malawi are presented. Interpretations are constrained by sedimentological analyses of 19 gravity cores (Ng'ang'a, 1988). Results from the analysis of 33 piston cores (Johnson *et al.*, 1988) and multifold seismic data (Specht, 1987; Flannery, 1988) are considered also. Development of qualitative depositional process models is emphasized using correlation of acoustic response with sedimentological data and structural information presented elsewhere (Specht, 1987; Ng'ang'a, 1988; Johnson *et al.*, 1988; Flannery, 1988; Versfelt, 1988; Johnson & Davis, 1989; Scholz *et al.*, 1989).

RESULTS

The 28 kHz acoustic response of the lake sediments may be classified into three primary categories (Fig. 3): I is a distinct, prolonged echo, II is a semi-transparent, multiple reflection echo, and III is hyperbolic, irregular echo (after Damuth, 1975,1980; Damuth & Hayes, 1977). The distribution of acoustic character type is shown in Fig. 4. The opaque nature of the class I response limits the information in these data to surface sediments. Classes II and III show variations within the overall reflection character described by their class. That is, class II encompasses a spectrum of multiple reflection responses that grade from a very transparent water bottom (class IIA; Fig. 5A) to a fairly distinct water bottom (class IIB; Figs 5B, 6B). A local gradation can be seen in the profile in Fig. 5A. This subdivision is mapped as the prevalence of transparent versus distinct water bottom in Fig. 4. Class III can be divided into those hyperbola or hummocks that are regular and tangent to the lake floor gradient (class IIIB, Figs 5B, 7B, 8B) or those that are very irregular (class IIIA, Figs 3B, 9, 10).

Correlation of independent mapping of the 28 kHz data and the 1 kHz and SSS data has resulted in the recognition of general acoustic response categories. These categories can be correlated with sediment type or depositional process as follows.

Class I. This acoustic character is described in the 1 kHz and 28 kHz data sets by a distinct, prolonged echo, with no sub-bottom reflections. The side-scan sonar (SSS) records are generally smooth or streaky indicating a relatively planar lake floor. Darker and lighter tones on the SSS profiles depict coarser versus finer grained sediments exposed on the lake floor, respectively. This response is distributed primarily along the nearshore rim of the lake. Few cores have

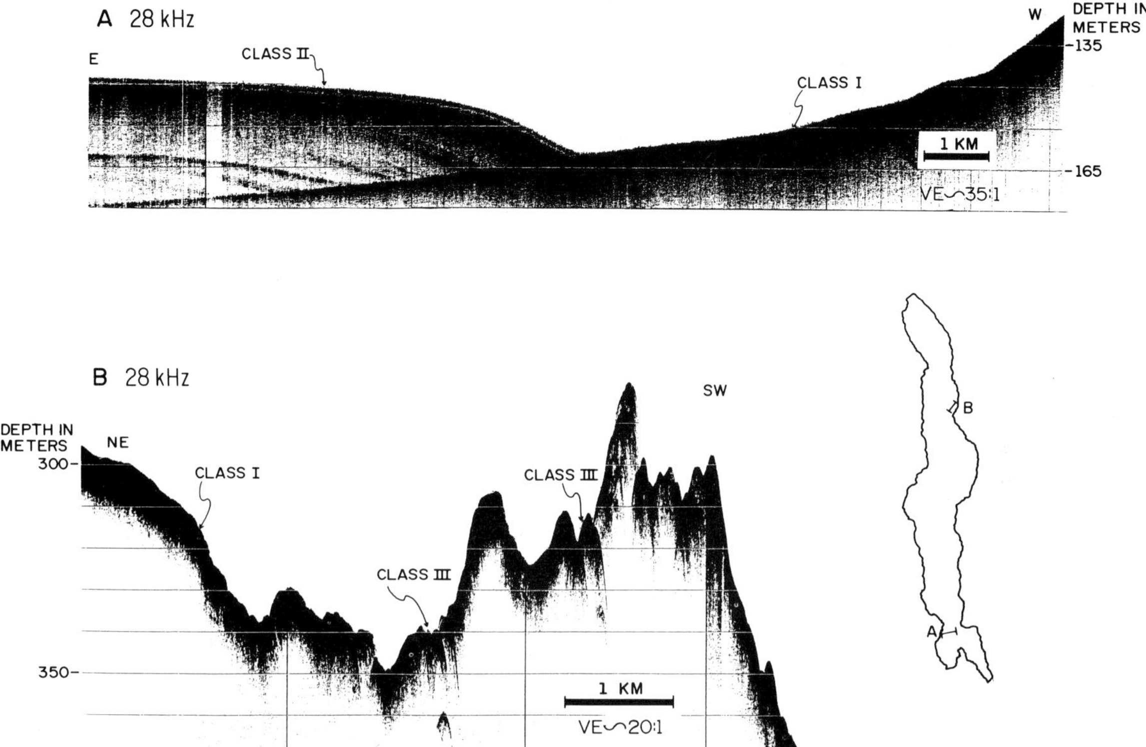

Fig. 3. Examples of the three main acoustic responses. (A) Class I — distinct, prolonged (on right-hand side of profile); and class II — multiple reflector (on left-hand side). (B) Class III — very irregular, hummocky or hyperbolic.

been obtained successfully from areas of this response because of the predominance of sand. These sediments are good reflectors of acoustic energy, allowing little or no energy to penetrate to reveal buried sediment interfaces. In the 28 kHz data this response extends into mid-lake positions in the northern third of the lake (Fig. 4). In these deeper environments, class I response tends to be less distinct and is coupled with class III as discussed below.

Class II. This acoustic character is by far the most widespread response, covering nearly all of the mid-lake environment (Fig. 4). This response varies from a distinct to a virtually transparent lake-floor reflection, overlying subparallel to parallel, sub-bottom reflections in 28 and 1 kHz profiles. The SSS records exhibit smooth, light monotone responses. Cores show that in the south the more prevalent, nearly transparent lake floor (class IIA, Fig. 5A) is indicative of a relatively pure diatom ooze (Johnson & Davis, 1989; Johnson, Hecky & Engstrom, unpub-

lished data). In the north, a higher prevalence of a more distinct lake-floor return (class IIB; Fig. 6B) represents a diatom ooze mixed with more terrigenous mud and sands (Ng'ang'a, 1988). The relative transparency of the water-bottom reflection appears to correlate to the ratio of biogenic to detrital components of the diatom ooze.

Class III. This class encompasses all other acoustic responses. Class III is much more common north of the dog-leg. In the 1 kHz and 28 kHz data the response is hyperbolic, hummocky, or very irregular. The SSS data in these regions is mottled and/or distinctly patterned. These responses correspond to a corrugated or irregular lake-floor morphology. Two large regions of very irregular or hyperbolic responses (class IIIA) are found from approximately 11°S northward (Fig. 4). In the 28 kHz data these distributions are intermixed by the aforementioned mid-lake Class I distinct echo returns (Figs 6, 10). These fields are located in areas of very gentle

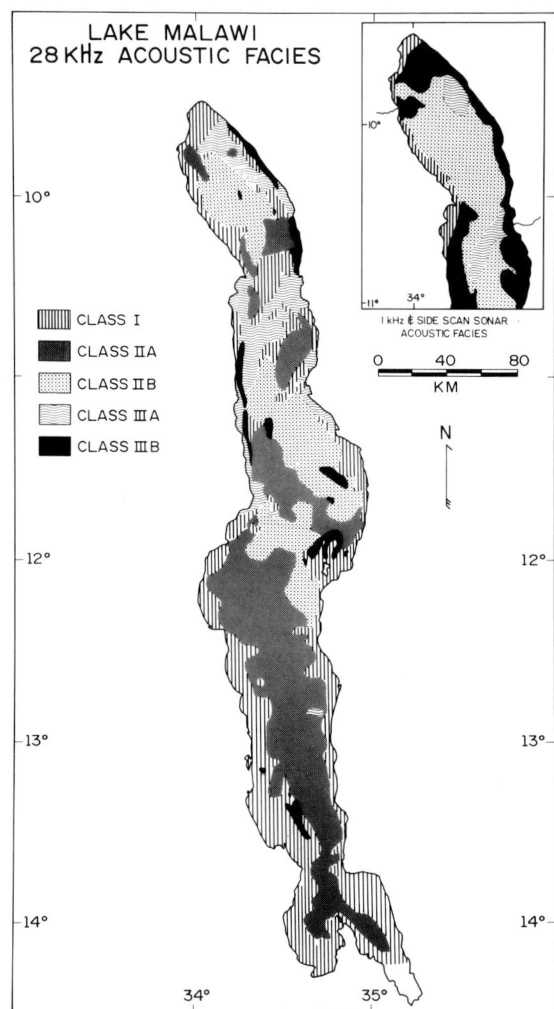

Fig. 4. Independently determined acoustic character distribution from 28 kHz and 1kHz data, and side-scan sonar (inset) data sets. See text for discussion.

gradient and extend from very shallow to the deepest water in the lake. Other regions exhibit very regular hyperbolic echoes whose vertices are tangent to the lake-floor slope (class IIIB). In the south, the regular hyperbola are limited to small restricted fields that are located on gentle bathymetric gradients (less than 1%; Fig. 5B). In the north this response is concentrated on steeper slopes corresponding to border fault systems or delta systems (Figs 7B, 8B, 11).

Cores retrieved from areas of Class III response primarily appear to record differences in depositional process rather than sediment type. The sediments recovered from northern Lake Malawi consist primarily of laminated diatom silts. These are interspersed with silty sands and sandy silts, which are locally slightly oxidized. Plant debris are often abundant in sandy units. Total organic carbon (TOC) values around deltas and border faults range from 0.35 to 4.44 wt.% (Ng'ang'a, 1988), as compared with 1.0–6.0 wt.% in open lake sediments in the north province (Johnson *et al.*, 1988). Carbon/nitrogen ratios range between 9.28 and 26.09.

DISCUSSION

Depositional processes

The Metangula province in the south includes the arms south of the Narrows and the main ramp between approximately 14°S and 12°S. The acoustic response in this province is predominated by class II 'layer cake' reflection geometries (Fig. 5) in the mid-lake position. Penetration averages about 30 m, but exceeds 50 m locally. The class II water–sediment interface in this province is commonly very transparent. This response correlates to a very fine grained, uncemented, relatively uncontaminated diatom ooze, suggesting that the primary depositional mechanism in this province is biogenic 'raining'.

Class II response is rimmed by distinct, prolonged echoes of Class I that represent nearshore sands and erosional surfaces. The boundary between these two responses commonly exhibits downlapping reflection geometries, as shown in Figs 3A, 5A, possibly indicative of a low-stand erosional surface or enhanced current activity. The profiles in Fig. 5 record the influence of bottom currents on sedimentation. At the acoustic facies boundary in Fig. 5A these data show definitive channel scouring. This is not surprising given the location of the profile across a significant constriction of the lake shore at about 13°40'S. At other boundaries, sub-bottom reflections simply disappear or merge into a class I prolonged homogeneous echo train.

Just to the north-northwest of the profile in Fig. 5A, at approximately 13°30'S (Fig. 4), a small patch of class IIIB hyperbolic response extends subparallel to bathymetric contours. These small, regular hyperbola, shown in Fig. 5B, are very slightly elongate along the NW–SE trend of the field (average wavelengths are 300 m with a deviation of up to

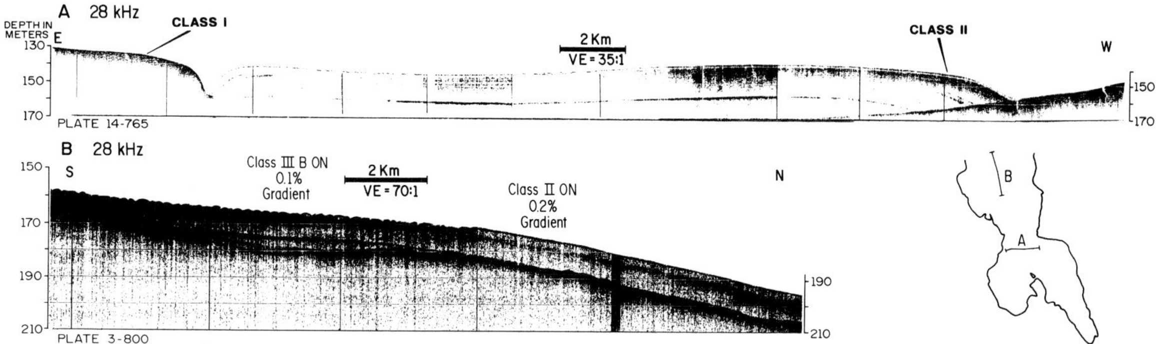

Fig. 5. Current-induced bedforms. (A) Downlapping reflectors at either end of class II field record a relative rise in lake level. Note the variation in reflectivity of water—sediment interface. Scoured channels at acoustic facies boundaries and (B) sediment waves distributed along bathymetric contours indicate significant bottom-current activity.

50 m in all profile directions). Figure 5B also indicates that these phenomena terminate abruptly at a change in slope, from 0.1 to 0.2% gradient. The correlation with bathymetry and proximity to other current-produced phenomena (e.g. Fig. 5A) suggest that these represent sediment waves or possibly current-induced furrows (Flood, 1980,1981; Flood & Hollister, 1980). The lack of evidence for underlying structural activity in MFS data supports the interpretation that this bedform is not tectonically induced.

Two small fields of similar class IIIB acoustic response are mapped on the Likoma Platform at about 11°40′S and 12°S, in the Usisya—Mbamba province (Fig. 4). The northern field follows bathymetric contours. The southern field, however, owes its distinctive distribution, which cross-cuts gross bathymetric contours, to subtle changes in the local bathymetry that mimic underlying horst and graben basement structure. Currents are essentially funnelled into these slight bathymetric lows so that the sediment waves are contained therein. Here there is a spatial correlation of bedform to underlying structure and these beds may represent a different depositional process (e.g. turbidites). However, morphological differences with other clear examples of tectonically induced turbidity deposition, lack of evidence for major recent activity on the underlying structure, and equivalent acoustic signature argue for a current-related process for these bedforms. The overall acoustic response on the Likoma Platform follows the mid-lake multiple reflection, rimmed by the nearshore prolonged response pattern of the Metangula province.

To the northwest, in the deep of the Usisya—

Mbamba province, acoustic responses are among the most spectacular in the lake. This apparently correlates with an increase in bathymetric gradient and structural activity leading to a predominance of gravitational, episodic deposition mechanisms. The Permo-Triassic Ruhuhu Trough funnels a large drainage system, draining approximately 10 000 km^2, into the northeast corner of the province (Fig. 2). A large delta and sublacustrine fan system is well defined by the shoreline protuberance and lobate form of the bathymetric contours, and covers an area of about 400 km^2 where the Ruhuhu River debouches into the lake. The lake-floor slope changes from 15% at the lake shore to less than 1% 20 km offshore (Figs 7, 8B). The MFS stratigraphy (Flannery, 1988; Scholz *et al.*, 1990) indicates that this delta package is at least a kilometre thick.

Deep distributary channels with well-developed levees splay radially outward from the river mouth. These well-defined channels (Figs 8A, 9) transport sediments up to 20 km into the basin. Two channels trend northward, suggesting possible spillover across the bathymetric divide into the Livingstone province (Figs 4, 8B). All other channels trend southwest into the Usisya—Mbamba province. The upper fan is dominated by slumps. Acoustically, the slump geometry has a staircase appearance in profile view (Fig. 7A). The rotational nature of the downthrown blocks is recognized by the reverse slope left hanging on the updip side creating slump scars. Close examination of the 1 kHz profiles reveals that the multiple concave-upward shears merge, at a depth of approximately 15—30 m, into a basal shear surface that is inclined parallel to the sediment surface.

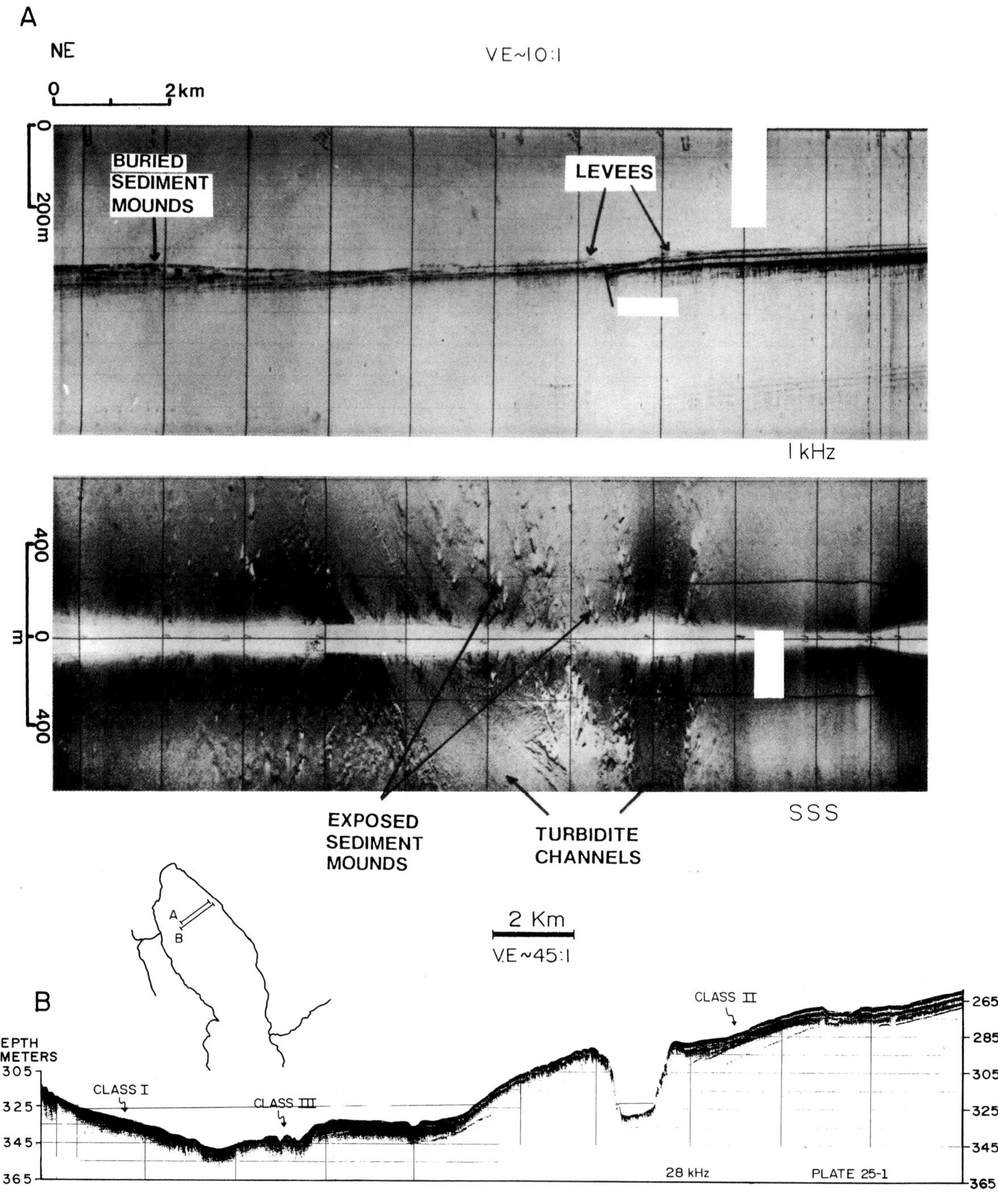

Fig. 6. (A) Profiles across the Livingstone basin document that the location and trend of channels in this basin are often controlled by active faulting. (B) A profile from the dense 28 kHz data coverage in the Livingstone basin describes a large, anastomosing channel−levee complex mapped as intermixed class I and class III. This complex extends along the entire trend of the border fault and at least 15 km lakeward of it. These channels trend nearly perpendicular to the turbidity channels observed along the border fault face, but are probably connected with the slope apron system described by Figs 11, 12A.

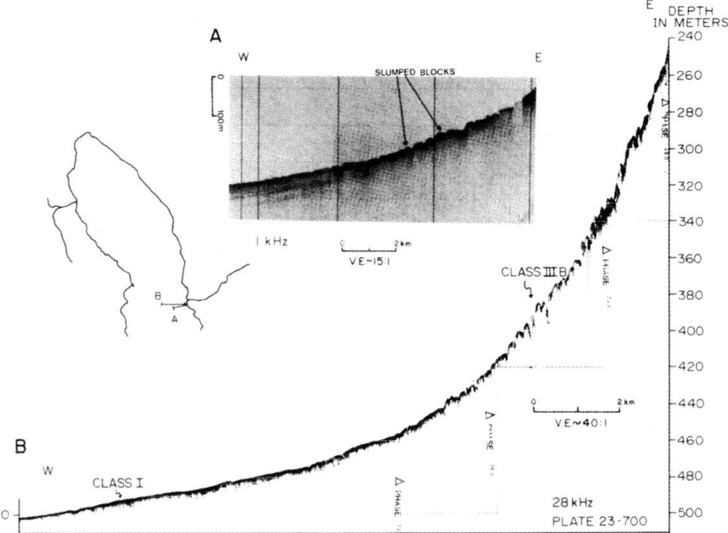

Fig. 7. (A) 1 kHz and side-scan sonar and (B) 28 kHz profiles down the front of the Ruhuhu delta—fan complex record large zones of slumping. Compare slump geometry with that in Fig. 13 from the North Rukuru delta.

Slumps extend 10 km into the province. Further lakeward they are replaced by an irregular, wavy SSS response (Figs 8A, 9A). This response is interpreted to be analogous to those reported by Damuth (1979) from the northern China Basin, that is, gravity induced creep structures, unlike the sediment waves recognized in the Metangula province, which are current produced. The respective acoustic responses faithfully record these process variations. This response covers an area of about 250 km². Deltaic sands are not restricted to the nearshore regions. Most of the upper fan sequence of slumps, creep features, and turbidity channels is covered with sands extending on average 20 km offshore. The estimated volume of sediments in the upper fan is 12.5×10^7 m³ (Ng'ang'a, 1988).

Five cores from the inner fan sequence and three distal cores were taken from this delta—fan package (Fig. 1). Ng'ang'a (1988) reports coarsening upward cycles of sandy silts and clays in the nearshore sediments. The thickness of the beds typically range from 10 to 20 cm. Channel and interchannel facies can be recognized in the textural analysis. Fining upward turbidites intermix with coarsening and fining upward debris flows. The acoustic response that correlates to the turbidity sequences extends well into the deepest part of this province in both the 28 and 1 kHz data (Figs 4, 10). How much of the most distal fan/turbidity character is due to the contribution of the Usisya BFS (see discussion on Livingstone BFS) is still undetermined. However,

documentation of a significant delta, a large sublacustrine fan, and turbidity sequences attest to the predominance of gravity flow and traction transport mechanisms in this province.

Figure 11 shows 1 kHz, 28 kHz, and SSS profiles from the border fault environment of the Livingstone province. The border fault complex on the east has been sampled by three gravity cores (Ng'ang'a, 1988). In addition, nine piston cores (Johnson *et al.*, 1988) were taken in the province. Where the SSS data changes from mottled to smooth appearance, depositional mechanisms associated with the border fault, such as debris flows, turbidity channels, and sediment waves, change to a generally pelagic depositional mode (Figs 6, 11). The hyperbolic acoustic response directly overlying the border fault is attributed to small sublacustrine fans formed by (subaerial?) mass-wasting processes (Fig. 12A). The lobate form of these debris flow fans is apparent in Fig. 11. The occurrence of a Class I distinct, prolonged response mapped by the 28 kHz data (Fig. 4) just lakeward of the hyperbolic response north of 10°S suggests distal fan sheets. Such sheets would be deposited relatively quickly and episodically, with intervening periods of biogenic ooze deposition. The continuous longitudinal extent of these facies suggests that the recent fans forming along the Livingstone BFS have coalesced to form a slope apron (Figs 4, 12).

Data also suggests that the border fault sediment pathways are connected and contribute to a large

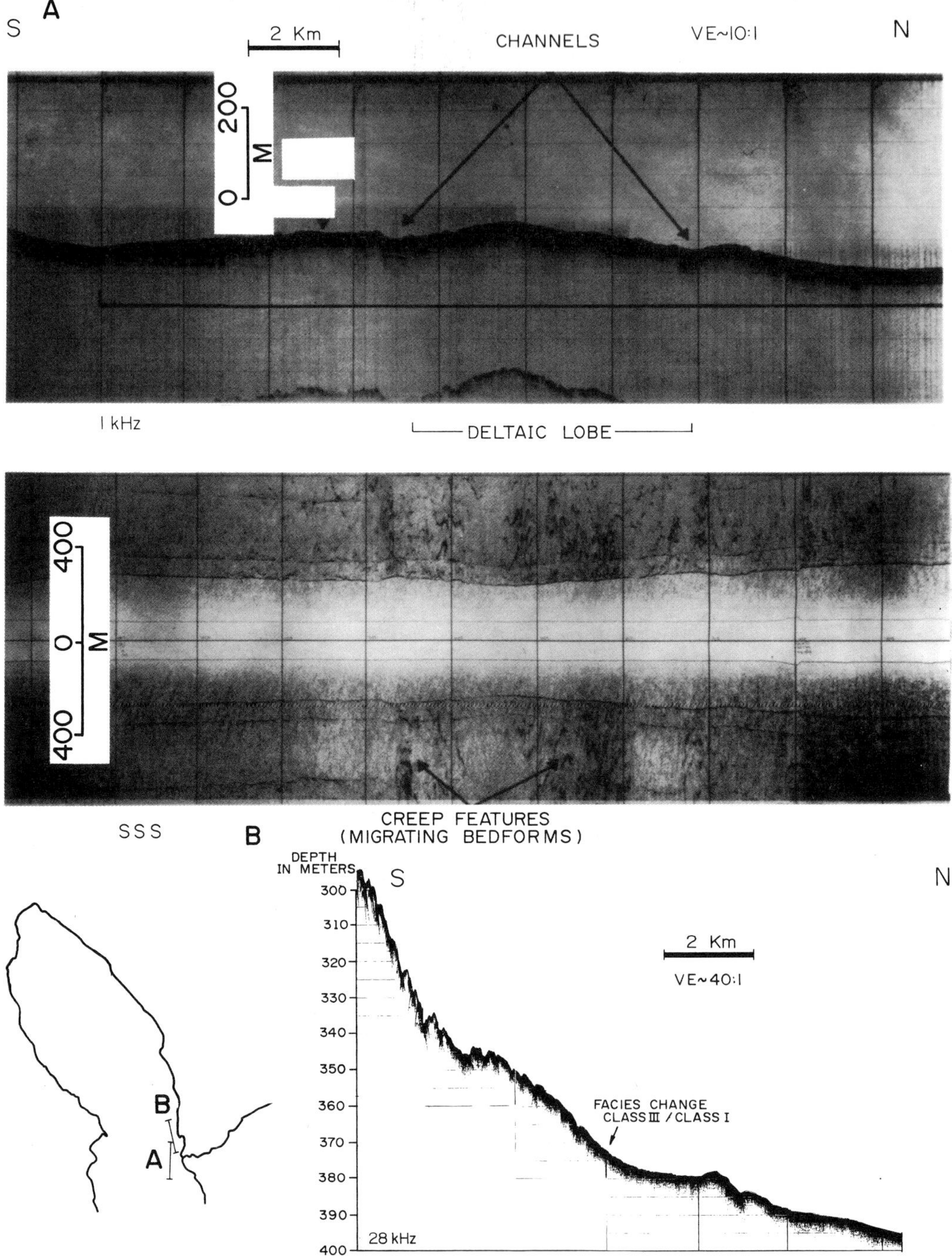

Fig. 8. (A) 1 kHz and side-scan sonar and (B) 28 kHz profiles from the Ruhuhu delta showing debris flows in the upper fan merging into a zone of corrugated lake-floor morphology interpreted as sediment waves resulting from creep processes.

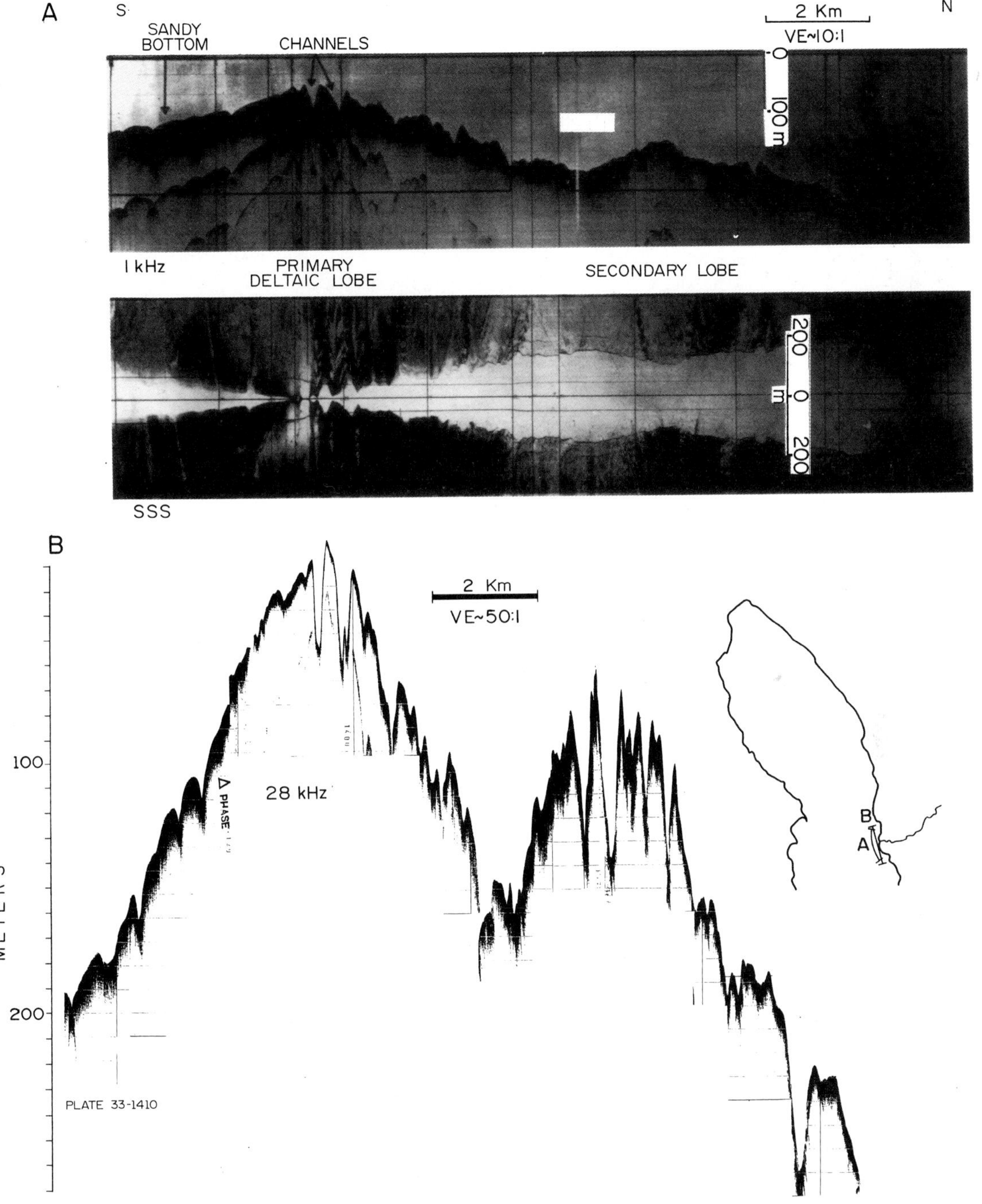

Fig. 9. (A) Transverse 1 kHz and side-scan sonar and (B) 28 kHz profiles across the Ruhuhu delta and upper fan. Deeply incised channels with well-developed levees are evident.

                                    D.L. Scott et al.

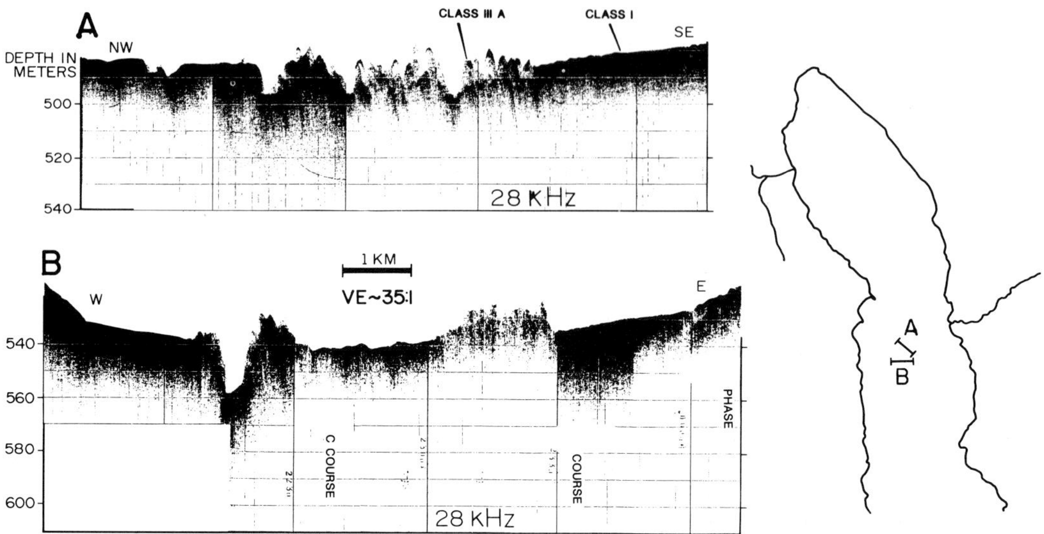

Fig. 10. 28 kHz profiles record turbidity channels and levee deposits in over 500 m of water in the Usisya–Mbamba province, documenting a large sublacustrine fan system that extends from the Ruhuhu delta. Both profiles shot coincident with MFS data, course changes less than 5° to maintain straight-line geometry.

channel and levee complex that extends into the mid-lake position (Fig. 6). This acoustic response covers approximately 900 km². The increased width of this acoustic response to the north (Fig. 4) probably records the contribution of terrigenous input from the series of axial drainage systems. Profiles across the province show that some of these channels are guided by active faults (Fig. 6). Comparison with MFS data from the same position suggests that these are long-lived structures and that this channel–levee complex may have volumetric significance within the synrift sediment package.

On the shoaling west side of this province, the North Rukuru River drains an area of about 3000 km² and forms a deltaic lobe with an area of about 200 km² (Fig. 13). Five cores taken from this delta–fan system confirm that the upper fan, spanning 20 km laterally, is composed primarily of inter-bedded sands and mud in slumps. This facies extends 5 km into the basin to a depth of 100 m. The blanket sands give way to rotational slumps that extend to a depth of 250 m before merging with hemipelagic deposits that cover the mid-delta. Contrasted with the staircase reflection geometry observed in the Ruhuhu delta data (Fig. 7), Fig. 13 demonstrates that slumps here are relatively intact blocks that have been displaced over concave-upward surfaces.

The slumps thin downslope. This thinning may be attributed to the considerable distance of sediment transfer downslope. The morphological differences between the slumps of each delta are attributed to gradient. The average slope within the North Rukuru deltaic lobe is less than 1%, whereas on the Ruhuhu delta, 7–15% is common. Figure 13 also exhibits a well-developed wave-cut scarp. These features are commonly found at the transition from multiple reflection to distinct, prolonged acoustic responses, and document one or more recent low stands (Scholz & Rosendahl, 1988; Johnson & Davis, 1989).

Sedimentation and tectonics

All three provinces show a rim of distinct, prolonged acoustic response with no sub-bottom reflections (class I) in the nearshore, shallow gradient environment and a prevalence of multiple reflection (class II) response in the mid-lake positions. However, the distribution of acoustic response type and local reflection morphology varies systematically according to underlying structure. Class I response is absent from the nearshore environments of the Usisya–Mbamba and Livingstone provinces that are coincident with underlying border faults. In the Metangula

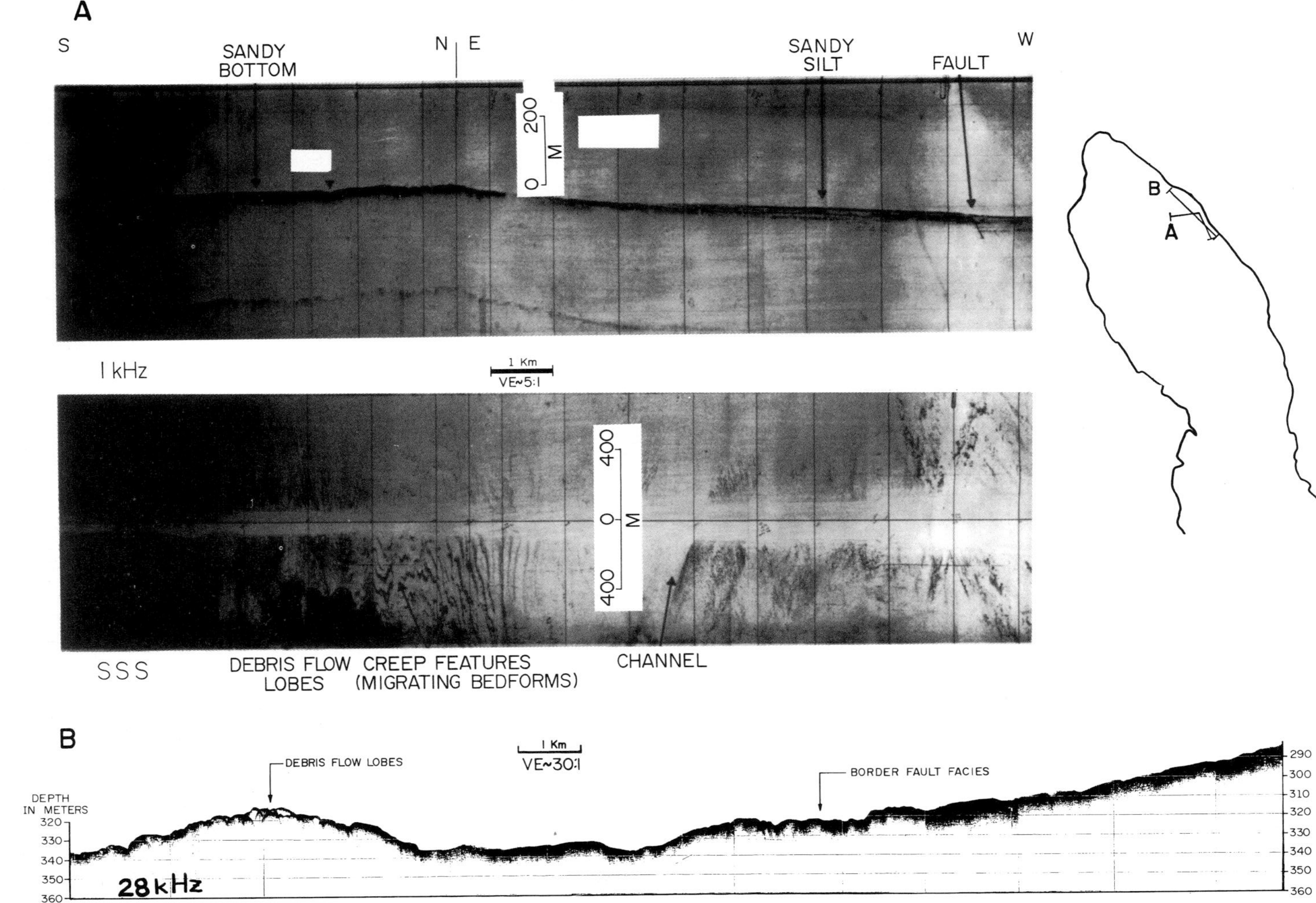

Fig. 11. Both the 1 kHz and side-scan sonar (A) data and 28 kHz (B) data document significant debris flow and creep features that create a large coarse-grained body of terrigenous sediment along the Livingstone border fault system.

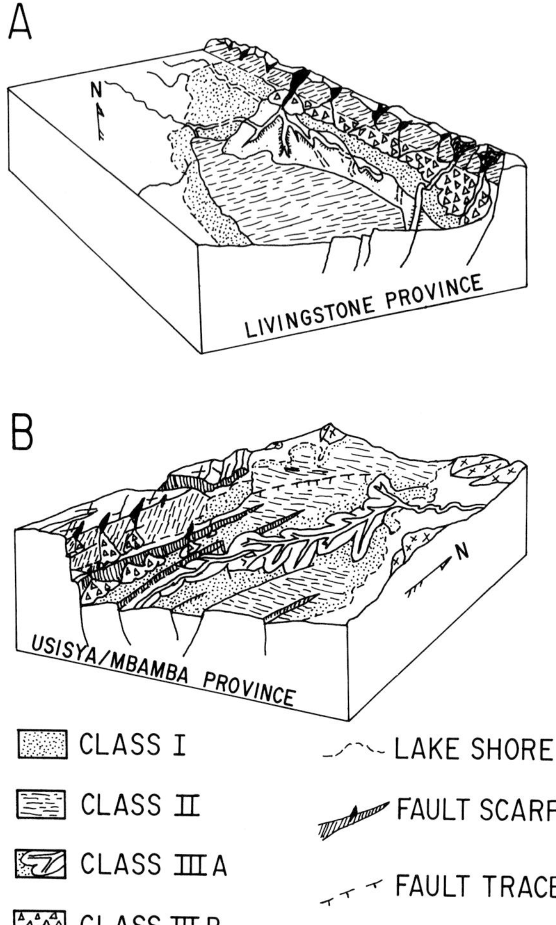

Fig. 12. Schematic representations of modern depositional processes and sedimentary structures evident in Lake Malawi. Sediment transport in both the (A) Livingstone and (B) Usisya—Mbamba Basins are dominated by traction, mass wasting, and gravity flow and creep processes.

province, the class I rim is broader on the western shore, but is none the less well developed on the eastern border fault side. This is attributable to the smaller throw on the Metangula Fault and its location well lakeward of the shore. The slightly greater lakeward extent of class I returns on the eastern shore from the dog-leg south to about 12°40′S is attributed to an increase in detrital material entering the lake from the Maniamba Trough. The centre of this province is characterized by uniform, nearly transparent lake-floor echoes with continuous sub-bottom reflections (class II). Local areas containing bedforms that are inferred to result from bottom currents (Fig. 5B), are the only exception to this nearshore—mid-lake acoustic facies distribution south of the dog-leg. North of the dog-leg, however, acoustic responses change from this simple nearshore—mid-lake distribution to an intricate collage of variation.

The acoustic character distribution in the Usisya—Mbamba province shows a clear correlation between the distribution of class IIIB hyperbolic response and the underlying border fault (compare Figs 2B, 4). Class II reflections in this province tend to be continuous over shorter distances and the water bottom reflection is increasingly more distinct to the north. Unlike the Class II response in the Metangula province, where sub-bottom reflections are distinct, parallel, and continuous for tens of kilometres, sub-bottom reflections are often weak and discontinuous. This reflection geometry is interpreted to form primarily by slumping, which often can be related directly to underlying structural activity. Local, discontinuous and sometimes disappearing seismic reflections have been attributed previously to gaseous sediments (Johnson *et al.*, 1987), or to moderate amounts of silt and sand (Damuth, 1980). In this particular case, the discontinuities may be due purely to gravity induced sediment transport attributed to a higher degree of tectonic activity.

The Livingstone province is also highly asymmetric, but down-to-the-east. The same correlation of class IIIB hyperbola and underlying border fault is evident. Mid-lake class II responses in this province are restricted to the shallower, shoaling western side (Figs 4, 6). Class II reflection character here tends to be very distinct and continuous with the exception that reflections are often truncated by faulting (Figs 6, 13). The deeper eastern side exhibits a complex acoustic response, which includes well-developed channels and levees that commonly follow the surface expression of faults (Fig. 6). The above discussion and comparison of Figs 2 and 4 suggests that there is a strong correlation between acoustic response distribution and reflection morphologies and structural activity in the northern provinces.

CONCLUSIONS

Lake Malawi is composed of three main depositional provinces: the Livingstone, the Usisya—Mbamba and the Metangula provinces. Each province is dis-

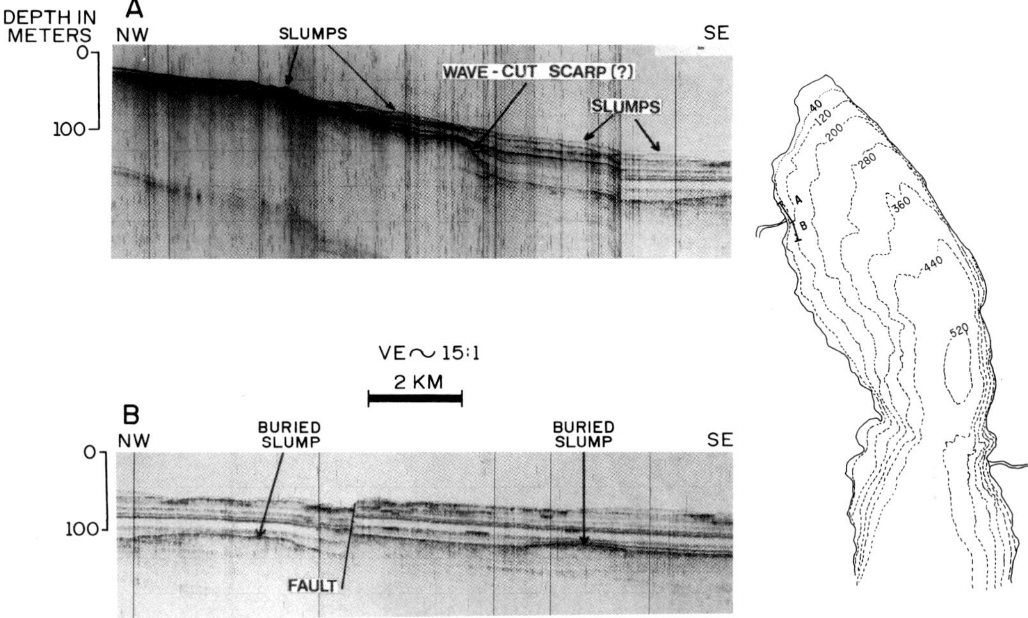

Fig. 13. The North Rukuru delta, with a gentler gradient and smaller sediment load, exhibits a different style and geometry of slumping than the Ruhuhu delta. Compare with Figs 7–9. Evidence of wave-cut scarps confirms the importance of fluctuations in lake level through time.

tinguished by unique sediment deposition mechanisms and resultant seismic expression. All provinces share in common the prevalence of class I opaque, nearshore sands in shallow areas of gentle gradient, merging downslope into zones of class II multiple reflector response. The transition zone frequently includes wave-cut terraces and buried (subaerial?) channels with well-developed levees, scoured channels or downlapping reflector geometries. These data confirm reports of major fluctuations in lake level (Scholz & Rosendahl, 1988; Johnson & Davis, 1989). They also lend support to the significance of lake-level fluctuations to the changing distribution of coarser grained facies in the synrift environment (Scholz *et al.*, 1989).

The primary depositional process in the Metangula province is that of undisturbed biogenic rain. This contrasts sharply with the dynamic and complex depositional environment in the two northern provinces. Figure 12(A, B) schematically depicts our understanding of deposition and sediment transport mechanisms operative in the Livingstone and Usisya–Mbamba provinces, respectively. In the latter province a substantial delta merges into a large sublacustrine fan system that provides sediment transport to the deepest parts of the lake, where terrestrial sediments and pelagic oozes intermix. Channels with well-developed levees, slumping, and corrugated bedforms, common to marine deltas, are observed albeit at a smaller scale. The Livingstone province is dominated by a border fault, slope apron system, which feeds detrital material to an extensive sublacustrine channel–levee complex.

Correlation of cores and acoustic response in three different types of high resolution data from Lake Malawi has constrained sediment distribution patterns. Surface morphology and reflection geometry can be used to infer depositional processes in this setting. That is, *class I* distinct, prolonged echoes correspond to a coarser, higher sand/silt ratio sediment. In the lacustrine setting, the response results from wave or current processes (nearshore) or overbank and distal fan sheet deposition (mid-lake). Multiple reflection *class II* responses generally correlate to a predominantly pelagic depositional environment. The transparency of the lake floor reflection within this class correlates to the ratio of detrital to biogenic components of the sediments. Continuity of sub-bottom reflections is largely dependent on tectonic quiescence. Areas of tectonic disturbance are expressed by irregular or hyperbolic *class III* and discontinuous *class II* responses. Variations within class III generally indicate process differences rather than sediment type. Deposition and sediment transport mechanisms recognized within this acoustic facies include turbidites, debris flows, channel−levee complexes, border-fault slope aprons and sediment waves (creep). Less commonly, class III represents bottom-current induced bed-forms or irregular erosional surfaces.

The distribution of the class III echoes correlates lake-bottom structural activity with gravity flow deposition commonly flanked by traction sediment transport. Two major sources of terrigenous sediment have been identified and described: fluvially dominated deltas and border fault, slope apron fan systems, both of which merge into turbidity channel−levee systems that transport these sediments to the deepest part of the basin. Large volumes of terrigenous sediment appear to be available only in tectonically active provinces. The spatial correlation of tectonic activity with large drainage systems may be somewhat fortuitous in Lake Malawi, where the Cenozoic rift intersects pre-rift (Ruhuhu Trough) structure.

Differences between sediment distribution and sedimentary structures in each province are largely attributable to receiving basin morphology, which is in turn primarily dependent on underlying structure and concurrent tectonic activity. In provinces underlain by well-developed half-graben, the nearshore sands are asymmetrically distributed on the shoaling side. On the opposing side and well into the deepest parts of the province, subaqueous mass-wasting and gravity induced creep processes account for the bulk of sediment transport. All of the depositional features discussed can be identified by unique acoustic signatures, which should be applicable to a wide variety of subaqueous settings.

ACKNOWLEDGEMENTS

Special thanks to Tray Marchant who got it all started. We are grateful to all the PROBE staff and students for their assistance in the data acquisition, archive, and presentation. Contributions to the basic framework and many helpful discussions are greatly appreciated from T. Davis, J. Flannery, K. Kaczmarick, C. Scholz, B. Schwede, T. Specht, W. Wheeler, & J. Versfelt. Reviews by J.-J. Tiercelin and B. Alonso greatly improved the manuscript. The governments of Malawi, Mozambique, and Tanzania are thanked for the opportunity to conduct the research and their tremendous cooperation during data acquisition. We thank the Sponsors of Project PROBE for financial assistance. Jim McGill and Chris Dills are thanked especially for all the support during the acquisition of the data and long after. Support for D. Scott was partially provided by a NSF Fellowship. T. Johnson was funded in part by NSF grant ATM-8816615.

REFERENCES

BLOOMFIELD, K., MASON, T.P.R. & HABGOOD, F. (1966) *Geological Map of Malawi*. Malawi Geological Survey, . Zomba, Malawi.

BURGESS, C.F. (1985) *The structural and stratigraphic evolution of Lake Tanganyika: a case study of continental rifting.* Unpublished Masters Thesis, Duke University, Durham, North Carolina, 46 pp.

BURGESS, C.F., ROSENDAHL, B.R., SANDER, S., BURGESS, C.A., LAMBIASE, J., DERKSEN, S. & MEADER, N. (1988) The structural and stratigraphic evolution of Lake Tanganyika: a case study of continental rifting. In: *Triassic−Jurassic Rifting* (Ed. Manspeizer, W.), pp. 859−882. Elsevier, New York.

DAMUTH, J.E. (1975) Echo character of the Western Equatorial Atlantic floor and its relationship to the dispersal and distribution of terrigenous sediments. *Mar. Geol.* **18**, 17−45.

DAMUTH, J.E. (1979) Migrating sediment waves created by turbidity currents in the northern China Basin. *Geology* **7**, 520−523.

DAMUTH, J.E. (1980) Use of high frequency (3.5−12 kHz) echograms in the study of near-bottom sedimentation processes in the deep-sea: a review. *Mar. Geol.* **38**, 51−75.

DAMUTH, J.E. & HAYES, D.E. (1977) Echo character of the east Brazilian continental margin and its relationship to sedimentary processes. *Mar. Geol.* **24**, 73−95.

ECCLES, D. (1974) An outline of the physical limnology of Lake Malawi. *Limnol. Oceanogr.* **19**, 730−742.

FLANNERY, J.W. (1988) *Seismic stratigraphy of Lake Malawi revealed by multifold seismic profiling.* Unpublished Masters Thesis, Duke University, Durham, North Carolina, 117 pp.

FLOOD, R.D. (1980) Deep-sea sedimentary morphology: modelling and interpretation of echo-sounding profiles. *Mar. Geol.* **38**, 77−92.

FLOOD, R.D. (1981) Distribution, morphology, and origin of sedimentary furrows in cohesive sediments, Southampton Water. *Sedimentology* **28**, 511−529.

FLOOD, R.D. & HOLLISTER, C.D. (1980) Submersible studies of deep-sea furrows and transverse ripples in cohesive sediments. *Mar. Geol.* **36**, M1−M9.

FROSTICK, L.E., RENAUT, R.W., REID, I. & TIERCELIN, J.J. (Eds) (1986) Sedimentation in the African Rifts. *Geol. Soc. London Spec. Pub.* **25**, 382 pp.

HARKIN, D.A. (1960) The Rungwe Volcanics at the northern end of Lake Nyasa. *Geol. Surv. Tanganyika Mem.* **II**, 172 pp.

JOHNSON, T.C. & DAVIS, T.W. (1989) High resolution seismic profiles from Lake Malawi, Africa. In: *African Rifting* (Eds Rosendahl, B.R., Rogers, J. & Rach, N.), J. Afr. Earth Sci. **8**, 415–427.

JOHNSON, T.C., HALFMAN, J.D., ROSENDAHL, B.R. & LISTER, G.S. (1987) Climatic and tectonic effects on sedimentation in a rift valley lake: evidence from Lake Turkana, Kenya. *Geol. Soc. Am. Bull.* **98**, 439–447.

JOHNSON, T.C., DAVIS, T.W., HALFMAN, B.M. & VAUGHAN, N.D. (1988) *Sediment core descriptions: Malawi 86.* Unpublished Technical Report, Duke University Marine Laboratory, Beaufort, North Carolina.

LORBER, P.M. (1984) *The Kigoma Basin of Lake Tanganyika: Acoustic stratigraphy and structure of an active continental rift.* Unpublished Masters Thesis, Duke University, Durham, North Carolina, 76 pp.

NG'ANG'A, P. (1988) *Sedimentation off deltas and border faults in Northern Lake Malawi: evidence from high-resolution acoustic remote sensing and gravity cores.* Unpublished Masters Thesis, Duke University, Durham, North Carolina, 93 pp.

ROSENDAHL, B.R. (1987) Architecture of continental rifts with special reference to East Africa. *Ann. Rev. Earth Planet. Sci.* **15**, 445–503.

SCHOLZ, C.A. & ROSENDAHL, B.R. (1988) Low lake stands in Lakes Malawi and Tanganyika, East Africa, delineated with multifold seismic reflection data. *Science* **240**, 1645–1648.

SCHOLZ, C.A., ROSENDAHL, B.R., VERSFELT, J.W., KACZMARICK, K.J. & WOODS, L.D. (1989) Seismic Atlas of Lake Malawi (Nyasa), East Africa. *Project PROBE Geophysical Atlas Series*, vol. 2. Duke University, Durham, North Carolina, 116 pp.

SCHOLZ, C.A., ROSENDAHL, B.R. & SCOTT, D.L. (1990) Development of sand-prone facies in lacustrine rift basins: examples from East Africa. *Geology* **18**, 140–144.

SCOTT, D.L. (1988) *Modern processes in a continental rift lake: an interpretation of 28 kHz seismic profiles from Lake Malawi, East Africa.* Unpublished Masters Thesis, Duke University, Durham, North Carolina, 82 pp.

SPECHT, T.D. (1987) *Architecture of the Malawi Rift, East Africa.* Unpublished Masters Thesis, Duke University, Durham, North Carolina, 56 pp.

SWAIN, C.J., AFTHAB KHAN, M.A., WILTON, T.J., MAGGUIRE, P.K.H. & GRIFFITHS, D.H. (1981) Seismic and gravity surveys in the lake Baringo, Tugen Hills area, Kenya Rift Valley. *J. Geol. Soc. London* **138**, 93–102.

TIERCELIN, J.-J., VINCENS, A. (coordinators), BARTON, C.E., CARBONEL, P., CASANOVA, J., DELIBRIAS, G., GASSE, F., GROSDIDIER, E., HERBIN, J.P., HUC, A.Y., JARDINE, S., LE FOURNIER, J., MELIERES, F., OWEN, R.B., PAGE, P., PALACIOS, C., PAQUET, H., PENIGUEL, G., PEYPOUQUET, J.P., RAYNAUD, J.F., RENAUT, R.W., DE RENEVILLE, P., RICHERT, J.P., RIFF, R., ROBERT, P., SEYVE, C., VANDENBROUCKE, M. & VIDAL, G. (1987) *The Baringo–Bogoria half-graben, Gregory Rift, Kenya. 30 000 years of hydrological and sedimentary history.* Bull. Cent. Rech. Explor. Prod. Elf-Aquitaine **11**(2), 249–540.

VERSFELT, J.W. (1988) *Basement controls in the Lake Malawi Rift Zone, East Africa.* Unpublished Masters Thesis, Duke University, Durham, North Carolina, 120 pp.

Spec. Publs Int. Ass. Sediment. (1991) **13**, 147–173

Late Pleistocene and Recent detrital sedimentation in the deep parts of northern Lake Tanganyika (East African rift)

F. BALTZER

URA 723 CNRS, Laboratoire de Pétrologie Sédimentaire et Paléontologie, Université Paris Sud, Bâtiment 504, 91405 Orsay, France

ABSTRACT

This study of the Recent sediments of northern Lake Tanganyika is based on high-resolution seismic profiling and 43 Kullenberg cores, made by Elf-Aquitaine (project Georift). In this part of the sub-basin, clastic sediments delivered via the N–S oriented Rusizi River and from lateral sources, are interbedded with laminated diatomaceous muds. Three episodes of detrital sedimentation have occurred since the late Pleistocene. Grain sizes, which range from clay to silt in most parts of the area, are coarser (sands, gravels, and mud balls) along a deep axis running N–S, parallel and close to the western shore of the lake. The predominant clay mineral, kaolinite, constitutes 50% of this mineral fraction.

The Lower Detrital Formation, documented in two cores situated on sublacustrine highs, includes the coarsest fractions, associated with smectite, abundant quartz, and feldspar. It is interpreted as a dry climatic episode older than 35 000 yr BP. A gradual evolution from this low-stand formation to the present high-stand situation, is shown by an increase in the proportion of diatomaceous muds, i.e. a tendency towards pelagic conditions. This general trend is interrupted by two detrital episodes.

An intermediate episode (Intermediate Detrital Formation, c. 5000 yr BP), composed of light brown, beige clays and silts associated with coarser layers, indicates that deposition occurred in several distinct sequences relating to turbiditic events. The number of events decreases southward in the North Basin and this formation is not found in the South Basin, which suggests a southward direction of flow. Clay minerals (smectite) and coarse fraction minerals (quartz, plagioclase, orthoclase, siderite), coarse fraction mean size, and very high linear sedimentation rates sharply deviate from the general trend of lake sediments during this period. These data all suggest a Rusizi River–Lake Kivu origin, in agreement with the pollen data and the distribution of carbonates. Data confirm that Lake Tanganyika had already reached a high stand and suggest that the mineralogy of these deposits resulted from the influx of oxidized alluvium eroded from the Rusizi–Kivu basin, and sedimented into the strongly reducing waters of the hypolimnion.

The topmost, modern sedimentary unit (Upper Detrital Formation) covers most of the lake bottom and includes dark green muds, with a wide range of grain-sizes. Fine particles are ubiquitous, whereas the coarse fraction increases within the deep, western, lateral depression, confirming the importance of this depression as a main axis of sublacustrine transport. This formation, interpreted in terms of a drier climatic episode, begins with laminated diatomaceous muds that were largely replaced by detrital deposits by 2000 yr BP, but laminated diatomaceous muds have become more and more frequent in the latest deposits. Thus, the three major sedimentary units express changes in climate, provenance, and lake level.

INTRODUCTION

Recent sediments in the deep lakes of the East African rift valley are formed predominantly by a succession of biogenic oozes and episodic detrital influxes (Rosendahl *et al.*, 1986). Owing to the division of each lake into numerous sub-basins, detrital sedimentation generally has a local origin, lateral input from drainage basins of limited extent (Yuretich, 1979,1986; Mondeguer, 1984; Cohen

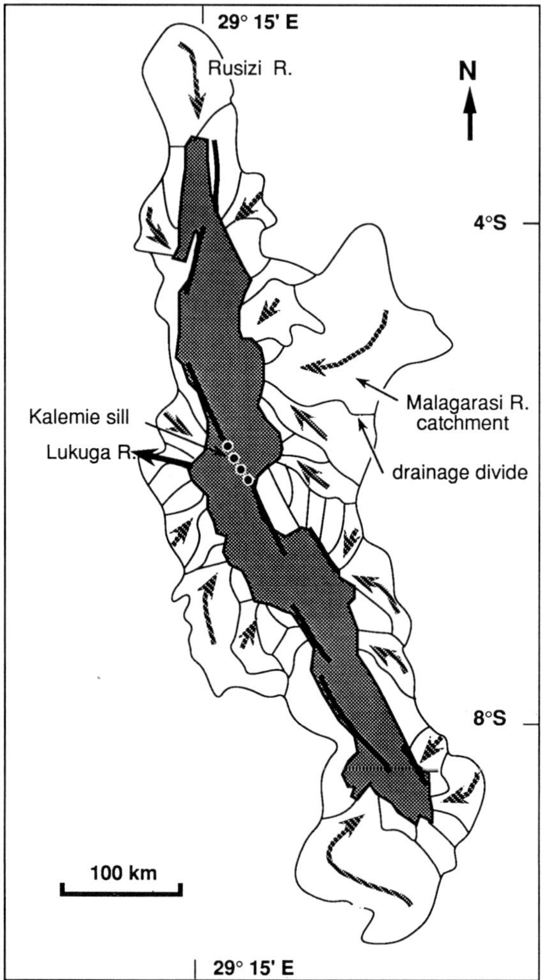

Fig. 1. Map of the catchment area of Lake Tanganyika showing relationships between the main faults, the main river basins feeding the lake, and the main directions of river drainage (modified from Frostick & Reid, 1987).

et al., 1986; Mondeguer *et al.*, 1986; Cohen, 1988). This contrasts with non-rift, elongate lacustrine basins where commonly axial transport is predominant (Sturm & Matter, 1978; Kelts & Hsü, 1980). Lake Tanganyika has a length of 673 km, a maximum width of 80 km, and a surface area of 33 000 km^2 (Serruya & Pollinger, 1983). It is the second deepest lake in the world (maximum depth, 1470 m; average depth 700 m). Its great depth and the steep slope gradients probably are the main reasons why sub-lacustrine density transport is essentially longitudinal over about 170 km in the northern sub-basin. Recent

work in Lake Malawi, based on a detailed geophysical approach (Scott *et al.*, this volume) leads to similar conclusions.

This study concerns the sedimentology and geometry of Recent detrital sediments (silty clays to gravels and mud balls) in the deep parts (mostly below 300 m) of the northern half of Lake Tanganyika, i.e. north of the Kalemie sill (Fig. 1). It is based on the study of 43 piston cores up to 6.1 m long (Fig. 2). By extrapolation of the ages obtained from cores collected on top of sills and in perched basins, the time interval recorded in these cores is considered to range from present to at least 35 000 yr BP (base of log SD 10).

These sediments are treated in terms of depositional processes relating to climatic and hydrological events. Results confirm the significance of alluvium from short lateral rivers in the detrital filling of the lake. They also indicate, however, that long distance longitudinal transport is an efficient process, carrying coarse fractions down to the deepest part of the lake, even during periods when water level was high.

The main instruments employed on the lake included an Edo Western high-resolution seismic profiler and a Kullenberg piston corer. Before opening, the cores were measured for magnetic susceptibility and gamma-ray density at the laboratory of Elf Aquitaine in Pau, France.

MORPHOTECTONIC, HYDROLOGICAL, AND CLIMATIC SETTING OF NORTHERN LAKE TANGANYIKA

Morphotectonic setting

Structural aspects

The deep rift morphology of Lake Tanganyika seems only to be 16−24 Ma old. It has allowed the deposition of 5−6 km of sediment (Rosendahl *et al.*, 1986; Rosendahl, 1987; Burgess *et al.*, 1988), superimposed on a previous morphology comprising a mosaic of small sub-basins defined by the relative positions of tilted blocks, located in an elongate, block-faulted subsiding zone (le Fournier *et al.*, 1985; Rosendahl, 1987). This earlier morphology is responsible for the existence of the secondary sub-basins typical of Lake Tanganyika (Fig. 3) and for its complex isobaths. In general, depths gradually in-

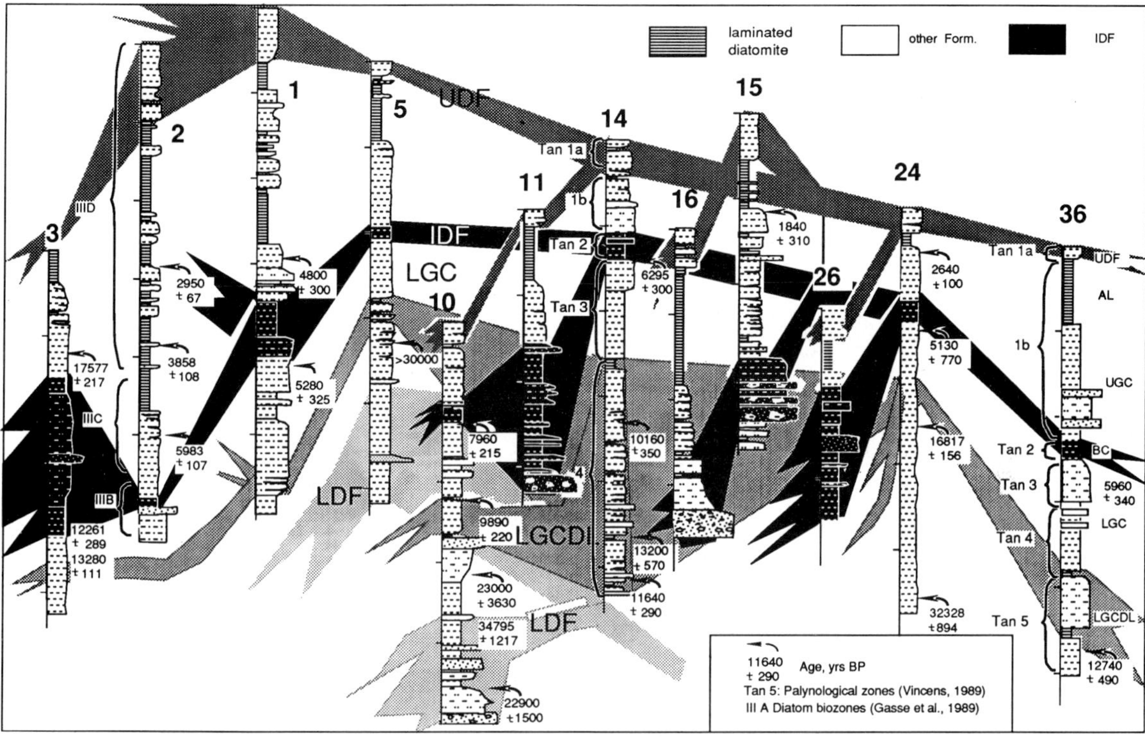

Fig. 2. Stratigraphic correlations and ages based on selected cores from the northern Tanganyika Basin.

crease from the delta of the Rusizi River in the north, to the deep sub-basin (1310 m) preceeding the Kalemie sill in the south (Fig. 3).

These fault-controlled sub-basins are asymmetrical, a steep flank contrasting with a more gently sloping side (Fig. 3). The former corresponds to a major fault scarp in the rift system, while the opposing side is the consequence of many smaller faults. The orientation of this asymmetry abruptly changes from one sub-basin to the next, resulting in a succession of steeply and gently sloping sections along the shores of the lake (le Fournier *et al.*, 1985; Rosendahl *et al.*, 1986). On the lake bottom this leads to a situation where the deepest parts are not in the middle of the lake but are located close to the steepest side (Fig. 3). This asymmetry influences different types of drainage and alluvial development on the margins of the rift.

The limits between adjacent sub-basins can be marked by bathymetric sills, which follow directions oblique to the axis of the lake. This direction is close to N 10° between the Bujumbura and the Rumonge sub-basins, and N 30° between the Rumonge and

Nyanza sub-basins. Some sills, although insignificant in terms of elevation, may place face to face the steep flanks of two asymmetrical sub-basins, forming a large, deep, apparently symmetrical double sub-basin (Reynolds & Rosendahl, 1984; le Fournier *et al.*, 1986; Rosendahl *et al.*, 1986).

Sub-basins in the area under study

The northernmost Bujumbura sub-basin rarely attains a depth of 300 m; the delta of the Rusizi River is its most prominent sedimentological feature. Only in the south, adjacent to the Rumonge sub-basin, do depths exceed 300 m.

The Rumonge sub-basin is a good example of asymmetry. The bottom of this sub-basin is a continuous NE–SW slope from the Burundi (eastern) coast to the deepest axis at the toe of the escarpment forming the Zairian (western) coast (Ubwari Peninsula). The transversal slope is moderate on the eastern side of the lake and much steeper in the vicinity of the deep axis (Figs 4, 5, 6). The axial depth rapidly increases southwards, resulting in a

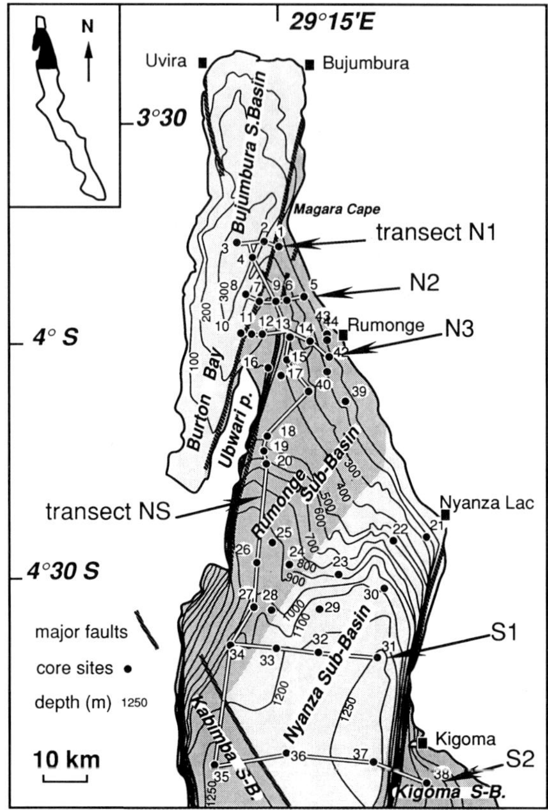

Fig. 3. Bathymetric map of northern Lake Tanganyika with distribution of structural sub-basins and location of cores.

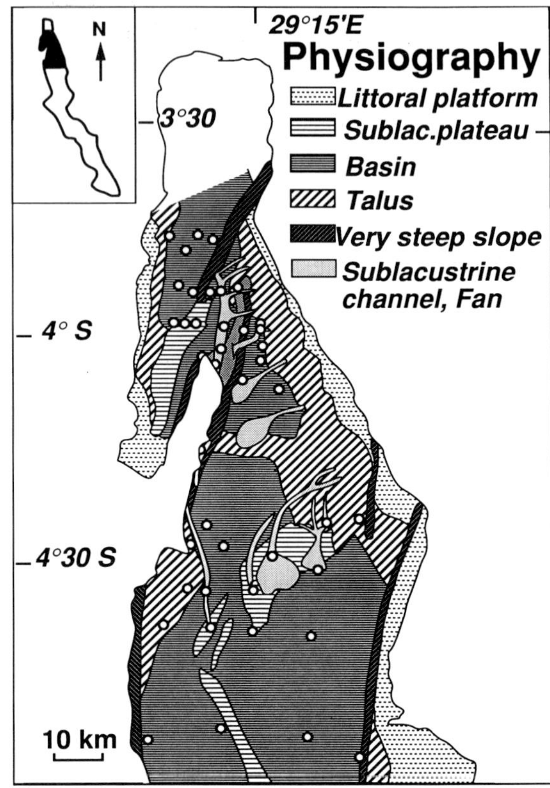

Fig. 4. Physiographic map of the northern basin of Lake Tanganyika.

longitudinal asymmetry.

Farther south, the twin Nyanza Lac and Kabimba sub-basins are the deepest in the northern half of the lake. From a morphological point of view they appear to form a single, apparently symmetrical sub-basin, with a very wide, flat bottom, and two steep, marginal slopes. Structurally, two distinct sub-basins are present, with their common limit lying along a very low (a few tens of metres), NW—SE oriented sill, i.e. oblique with respect to the rift axis.

The Kabimba sub-basin is interrupted on its southern margin by the Kalemie sill — a result of major transverse faulting (Chorovicz *et al.*, 1983; le Fournier *et al.*, 1986; Tiercelin *et al.*, 1988) — which obliquely (NW—SE) divides Lake Tanganyika into two major basins.

Drainage patterns in response to the rift morphology, and their effects on sedimentation

Both sides of the western rift branch, especially in the northern Tanganyika area, are rimmed by mountain ranges, expressing regional doming, with summits up to 2670 m on the east side and nearly 3000 m on the west. These culminations are located some 20 km from the lake. Drainage is typically a response to this morphology. This structural configuration results in two very different types of alluvium, the first carried by an extensive longitudinal system, draining the Rusizi Trough, the other transported by many short transverse streams (Fig. 1).

Longitudinal drainage

The main tributary to Lake Tanganyika, the Rusizi River, is a longitudinal stream, recently established as the outlet of Lake Kivu (Beauchamp, 1939), with

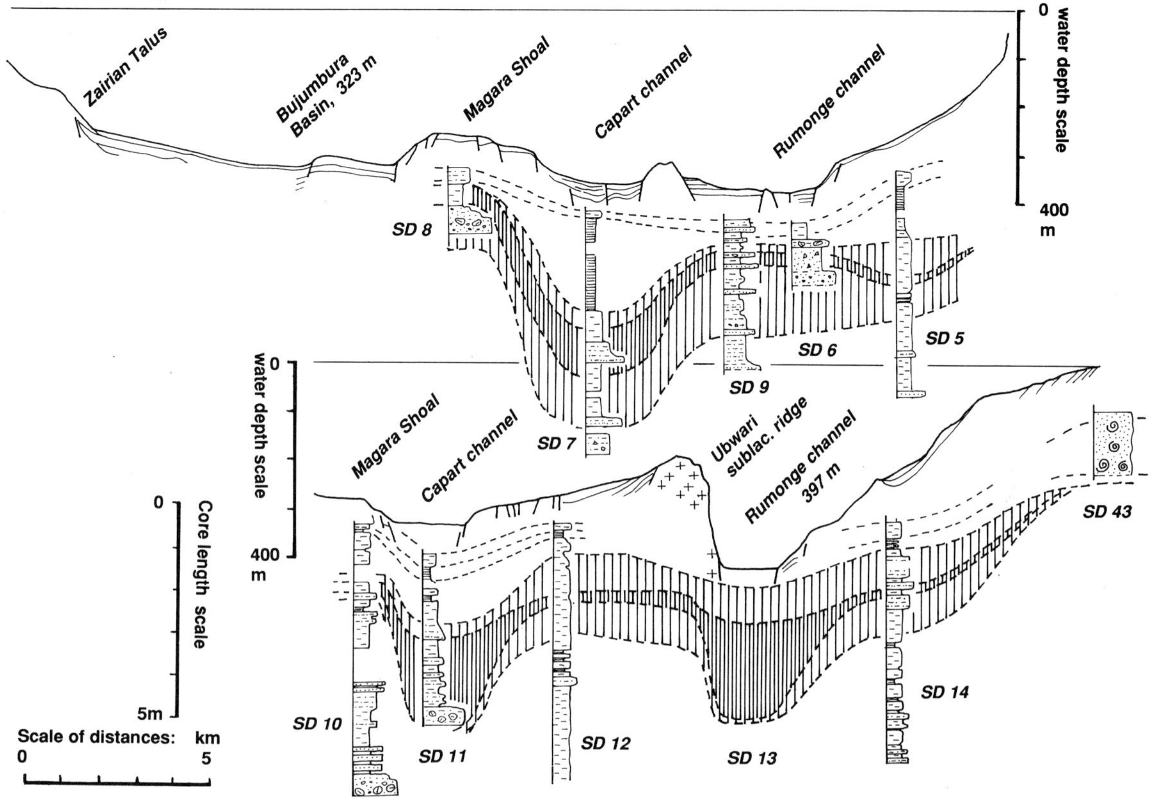

Fig. 5. Simplified structural and sedimentological transects, north 1 and north 2 (see Fig. 3 for locations).

intermittent activity during the last 5000 yr BP (Degens *et al.*, 1971; Degens & Kulbicki, 1973). It flows N–S and discharges into the north end of Lake Tanganyika. South of the Kivu, the Rusizi River crosses an important volcanic system (Kanika *et al.*, 1981), which developed in three cycles (tholeiitic, sodic alkaline, and transitional from alkaline to tholeiitic basalts) in phase with the main episodes of the rifting (Kampunzu *et al.*, 1983).

Lateral drainage

The type of alluvial discharge relating to lateral drainage depends also on the structural framework. The source rocks submitted to weathering and erosion in the lateral drainage basins of the northern part of Lake Tanganyika are mainly granites and metamorphic Precambrian basement rocks. Owing to relief and climatic conditions prevailing in this area, weathered profiles are comparatively thin and erosion of soils produces abundant quartz, kaolinite,

illite, and iron hydroxides, with feldspars and the occasional illite–smectite mixed-layer mineral. The local changes in clay mineralogy obviously depend on the nature of source rocks, and on their location with respect to relief, i.e. to tilted blocks.

In those parts of the rift where slopes are steep, catchment areas are very narrow and, except for a few gullies, no drainage organization is apparent. In the areas where slopes are more gentle, however, catchments are significantly wider and streams are larger, although most do not exceed 20 km in length. Only a few prominent rivers (Murembwe and principally Malagarasi) have significantly longer courses, these extending their catchment outside the rift, via fault-oriented valleys. These rivers develop fan-deltas. Since the present level of Lake Tanganyika was defined by the opening of the outlet of the Lukuga River, in 1878 (Devroey, 1949), these deltas must be very young, suggesting a fairly high rate of sedimentation. Remnants of older fan-deltas are also present at higher levels along the coast.

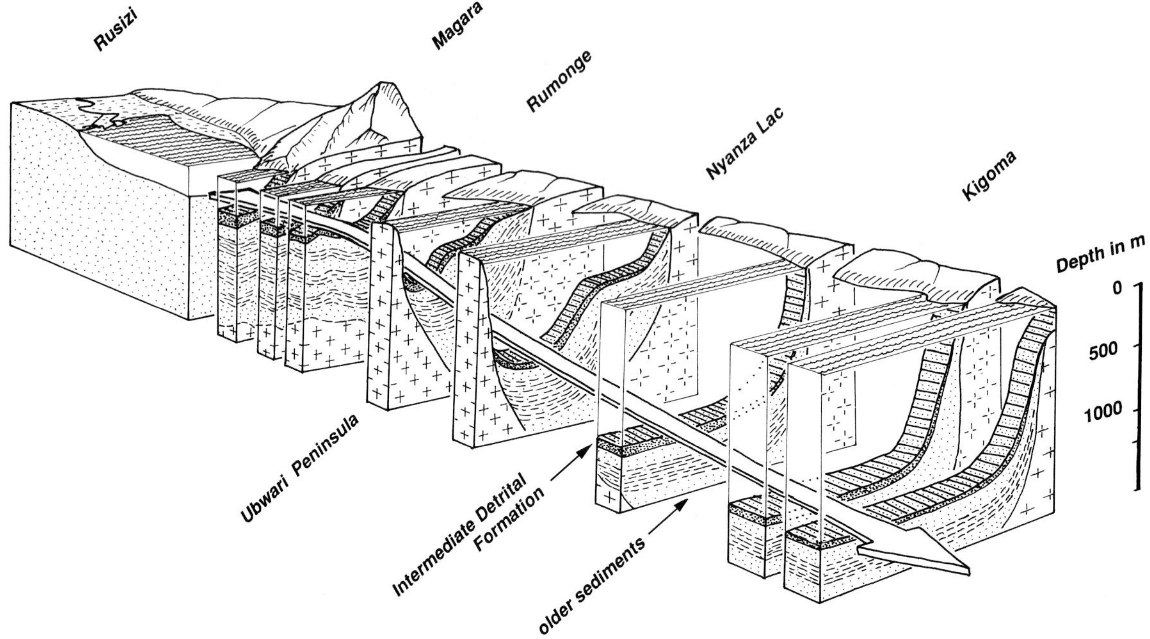

Fig. 6. Block diagram showing the distribution of the Intermediate Detrital Formation.

Hydrology and climatic conditions in Lake Tanganyika

Present conditions

With the present water level at an altitude of 775 m, Lake Tanganyika extends from latitude 30°30′ to 9°S. These latitudes and altitude result in a 'Guinean' tropical climate, with eight to nine 'tropical' and three to four dry months, 'arid' according to Peguy's classification (Peguy, 1961). Such a climate should favour a dense forest, but this is strongly limited by agriculture, and erosion rates are probably increasing.

Lake Tanganyika is a meromictic lake, with a strong, year-round stable thermocline centred at about 70 m and waters deeper than about 400 m displaying a uniform temperature of 23.3°C ± 0.05 (Degens *et al.*, 1971; Craig, 1974). Surface waters are saturated with dissolved oxygen. The oxygen content gradually diminishes with depth and is close to 0% at a depth of 100 m, while H_2S becomes abundant in the deeper lake (Degens *et al.*, 1971).

Malagarasi and Rusizi Rivers are responsible for most of the discharge into Lake Tanganyika, the latter contributing waters that account for the predominant carbonate chemistry of the waters of the lake (Degens *et al.*, 1971). Composition of Rusizi waters directly reflects the very peculiar composition of Lake Kivu waters, extremely rich in CO_2 of both biological and hydrothermal origins (Degens *et al.*, 1971; Degens & Kulbicki, 1973, Deuser *et al.*, 1973). Present Rusizi waters average 13.1 mmol/kg of CO_2 near Bukavu, while, for the Lake Kivu waters, Degens *et al.* (1973) report 12.5 at the surface and a maximum of 133.0 at a depth of 350 m. Tanganyika waters are influenced by the same source, with 5.64 mmol/kg of CO_2 at the surface and 6.38 at the nearly maximal 1460 m depth, although hydrothermal aragonite and sulphides are also reported from Tanganyika (Tiercelin *et al.*, 1989b). Waters from other tributaries to Lake Tanganyika, typified by the largest of them, the Malagarasi, contained only 1.55 meq/l (Beauchamp, 1939).

Because of the high evaporation rate, the combined discharges of the tributaries must considerably exceed the outflow of the Lukuga River; an interruption of the Rusizi input would result in a lowering of the water level below the outlet.

Quaternary conditions

For the period before 13 000 yr BP, when the Rusizi did not flow, Hecky & Degens (1973), quoted by

Stoffers & Hecky (1978), suggest a lowering of the lake level by 600 m, in rough coincidence with the −550 m sonic reflector reported by Capart (1949). However, no signs of corresponding erosion or soil formation were found in the cores by these researchers. In cores taken during the Georift project, covering a similar or larger span of time, we did not find evidence of this very low stand of the lake, suggesting an age older than *c.* 35 000 yr BP for this low stand, as suspected by Scholz & Rosendahl (1988).

In a later contribution, based essentially on one core from the southern end of Lake Tanganyika, Haberyan & Hecky (1987) infer, from palaeoecological and geochemical grounds, a low stand of −240 to −340 m during the deposition of their A2 diatom subzone from core T2 (Livingstone, 1965), from 15 000 to 13 500 yr BP. The existence of this low level is confirmed by a multidisciplinary study of the Georift MPU XII core, collected in the vicinity of core T2 (Tiercelin *et al.*, 1988). These authors also show that, during the period 25 000–22 000, a climate drier than today prevailed on land, while the lake water level remained comparatively high. Gasse *et al.* (1989), document a high water level stage c. 39 000 yr BP (core MPUIII, Tiercelin *et al.*, 1989a;) and demonstrate that the last dry episode attained its maximum c. 18 000 yr BP.

Drier climatic conditions continued to prevail up to 13 000 yr BP, and were followed by a rapid return to humid conditions. Since then a high level with minor fluctuations has been a permanent feature of the lake. Pollen analysis documents a similar climatic history (Vincens, 1989).

RECENT SEDIMENTATION IN THE NORTHERN TANGANYIKA BASIN

Cores

Kullenberg coring sites were selected according to the results of a 3.5 kHz seismic profiler. The profiler was also used to locate stations precisely. Three main groups of sites were defined as a function of physiographic conditions and correlative seismic facies.

A total of 22 cores were collected in the northern part of the area (SD 1–17, and 39–44; (Figs 3, 4, 5) — five in the Bujumbura sub-basin (SD 2–4, 8, and 10; Figs 3, 5) and 17 in the northern part of the Rumonge sub-basin. A set of three cores was collected in the deep channel of the west-central part of the Rumonge sub-basin. The remaining 18 cores (SD 21–38, Figs 3, 4, 5) are located in the Nyanza sub-basin and the southern part of the Rumonge sub-basin.

Stratigraphy

The sequences visible in nearly all cores comprise an alternation of laminated diatomites and detrital sediments. Because of the relatively important fluvial influx in the northern Tanganyika Basin, detrital sedimentation plays a significant part in the sedimentary record. This study demonstrates three detrital episodes during the last 35 000 yr BP (Fig. 2): (i) a grey coloured, Lower Detrital Formation (LDF), c. 35 000 yr BP; (ii) an Intermediate Detrital Formation (IDF), in which the Beige Clays — c. 5000 yr BP — are the central part of a triplet that includes the Lower and Upper Grey Clays; and, finally, (iii) a modern Upper Detrital Formation (UDF), consisting of dark-green muds and sands.

Laminated diatomites are concentrated in two formations: (i) in the Lower Grey Clays with diatomaceous laminae (LGCDL) and (ii) in the main body of diatomaceous algal deposits (DL), interbedded between the Upper Grey Clays (UGC) and the UDF. The thickness of algal deposits varies with local conditions: LGCDL deposits are 20 cm thick in SD 24, and 2 m thick in SD 14, and DL deposits change from 5 cm in SD 10 to 3.6 m in SD 2. This is indicated clearly by an isolith distribution map of the laminated diatomaceous muds (Fig. 7), based on measurements of thicknesses in open cores. This map also illustrates the exclusion of the laminated diatomaceous muds from areas with detrital formations, especially at the mouths of rivers, where alluvium is thicker and diatomaceous deposits are thinner.

The Lower Detrital Formation (LDF)

This formation occurs only at the base of cores showing the lowest sedimentation rates (0.1 mm/yr) based on ^{14}C dates, especially in cores SD 5 and SD 10 (Fig. 2). This formation displays changes from gravels and sands containing mud balls at the bottom of core SD 10, to fine sands and clays at the bottom of core SD 5, and pure clays in core SD 24. The relative topographic positions of these textural characteristics are logical in that SD 10 (where LDF has coarse sediments) is situated in a deeper and more central position in the lake, while SD 5, in

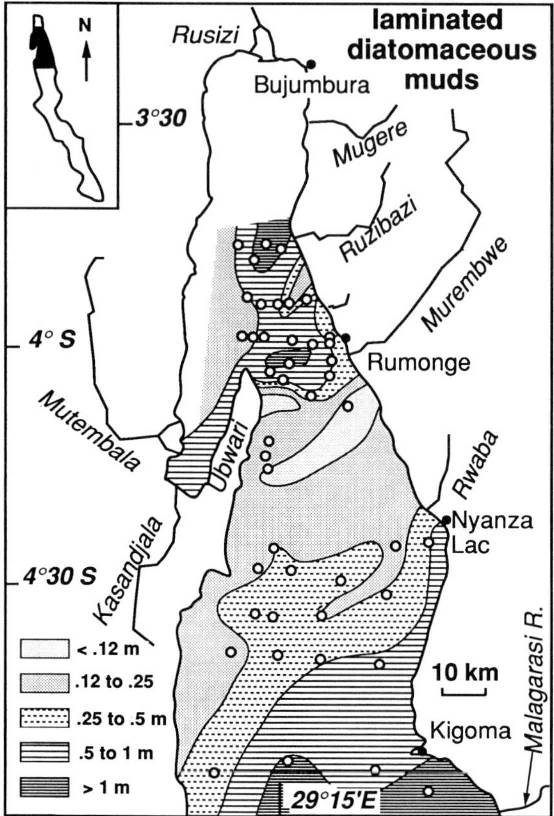

Fig. 7. Isopach map of the laminated diatomaceous muds.

with diatomaceous laminations subunit (LGCDL), present in cores SD 1, 5, 14, 24 and 36, is essentially an alternation of dark silty and clayey muds with interbedded, diatomaceous laminae, and expresses the beginning of transition from pelagic to detrital sedimentation in the IDF.

The middle, beige subunit includes pure clays, sands, gravels, and mud balls, and the layers having a significant amount of clay often display a characteristic beige colour, but light-grey and blackish clays are also found. The maximal thickness is 2 m and possibly more. Pure clays are predominant in the cores where the layer of beige clays is 1 or 2 dm thick, but strong erosional effects are indicated by mud balls and gravels in the cores where this subunit is several metres thick. The deposits of the Beige Clays vary from single — the Intermediate Detrital Formation forming one massive layer (SD 5, Fig. 8) — to multiple; the formation comprising up to eight separate layers (SD 11, Fig. 9). The contacts within the BC change in quality. From the sharpest to the most gradational they include erosional features, changes in bedforms, sharp changes of colour and, finally, progressive changes of colour.

The layers found in the Beige Clays correspond with peaks of the magnetic susceptibility logs (up to 1400 MEU at 388 cm in core SD 1 (Fig. 10)) and of the gamma-ray density log ($d = 1.9$, Fig. 10). These magnetic peaks are associated with the presence of up to 10% siderite (X-ray diffraction) in the silt fraction. Multiple layers of the beige clays are clearly visible on the magnetic susceptibility logs (SD 11, Fig. 9).

In terms of magnetic susceptibility and density logs, the grey muds above and below the Beige Clays, LGC and UGC, are transitional subunits from the BC to adjacent formations within the Intermediate Detrital Formation. For instance in SD 5 core (Fig. 8), the Lower and Upper Grey Clays are coincident with a bulge on the magnetic susceptibility log. This bulge changes sharply to a peak at the level of the Beige Clays. This situation is found in all cores having a well-developed single peak. The cores with multiple layers show no bulge at all. The dark muds therefore are interpreted as being parts of the Intermediate Detrital Formation that were deposited under similar conditions, but in smaller quantities, which permitted their reduction under the influence of the meromictic hypolimnion. The diatom assemblages, with the predominant species. *Melosira granulata* in both the dark and beige clays (F. Gasse, pers. comm., 1989) is in agreement with

which sediments are finer, was collected on the talus, closer to the shore; SD 24 is in a deep distal situation. The LDF is usually grey to blackish in colour, except for a few light-grey sand layers.

The LDF in SD 10 includes two beds, 5 and 15 cm thick, of clean, fine sand (0.2 to 0.3 mm in diameter), contrasting with the turbiditic layers by better sorting (So = $1.25-1.14$). They are unique in the cores studied, being very similar to modern beach sands.

The upper part of the LDF, found only in cores SD 5, 10, and 24, is formed essentially of dark-grey to black, undifferentiated, silty muds, referred to as Lower Black Clays (LBC).

The Intermediate Detrital Formation (IDF)

In the Intermediate Detrital Formation, the two darker units — Lower and Upper Grey Clays — form a gradual transition to and from the middle unit: the Beige Clays (BC). The Lower Grey Clays

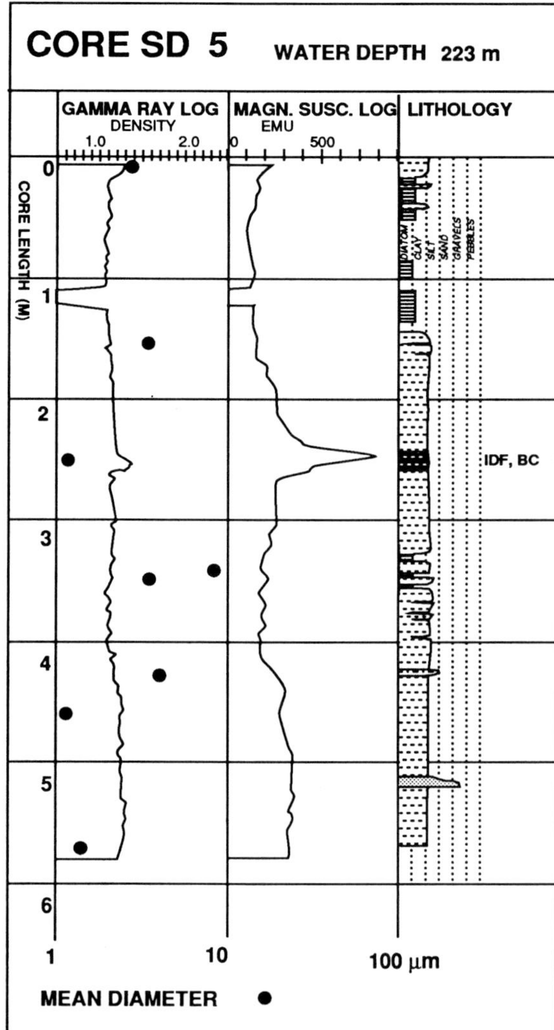

Fig. 8. Core SD 5, showing lithology, grain size, gamma ray, and magnetic susceptibility logs.

the overall similarity of sedimentary conditions, especially since this species is resistant to muddy waters.

Distribution. The IDF is found only in the northern Tanganyika Basin and apparently has no detrital equivalent south of the Kalemie sill (Degens *et al.*, 1971; Tiercelin *et al.*, 1988; Mondeguer *et al.*, 1989). The major part occurs in the Bujumbura sub-basin, north of the line joining Cape Magara to the Ubwari Peninsula (Fig. 11). Further south it is reduced in thickness within the series of sub-basins from

Bujumbura (2 m in SD 3) to Rumonge (2 m in SD 26) and Nyanza Lac (0.5 m in SD 36). The coarsest deposits fill the deeper channels, such as the Capart Channel (northern end), or the deep longitudinal axis west of the Rumonge sub-basin at the foot of the west coast. In this environment the energy conditions were high enough for the transport of gravels and mud balls as far as SD 20. Fine sands and silts are found as far as SD 26. Along any E−W section, the thickest deposits of the Intermediate Detrital Formation are not exactly in the deep channel at the foot of the Ubwari Peninsula but east of it. This distribution is typical of a sublacustrine channel bordered with levees.

Moreover, laterally from the deep axis, the sediment gradually becomes thinner and finer. This is clearly indicated by the magnetic susceptibility profiles along transect N III (location map on Fig. 3). A very fine facies of this formation, consisting of pure beige clay, forms a decimetric layer on top of isolated sublacustrine highs (SD 10 on the Ubwari shoal) and on the bottom of small scale, isolated perched basins (SD 5).

Origin. A N−S axial transportation is suggested both by the isopachs of the IDF deposits (Fig. 11) and by the number of individual layers in this formation (Fig. 12), each interpreted as sedimentary events. The isopachs of the Intermediate Detrital Formation reveal that the deposits are thickest both in the north, in the direction of the inlet of the Rusizi River, and to a lesser degree, in the south, on the abyssal plains of the Nyanza Lac and Kabimba sub-basins. However, in cores from the north, the formation is massive and includes very coarse mud balls in the channels, while in the south it is interbedded and somewhat diluted with sediment from other, organic and detrital sources (SD 35), and with a grain size not exceeding fine sand even in channels (SD 33).

The series of events responsible for the deposition of the Intermediate Detrital Formation are visible in cores in the form of an ordered distribution of grain sizes, with sharply defined maxima, and, on both the magnetic susceptibility and gamma-ray density logs (Fig. 10), in the form of clearly defined maxima. The number of events that were strong enough to be recorded varies from one site to another. The geographical distribution of the recorded events (Fig. 12) clearly demonstrates a greater frequency in the north, and a correspondingly smaller number in the south. It shows, possibly better than the isopach

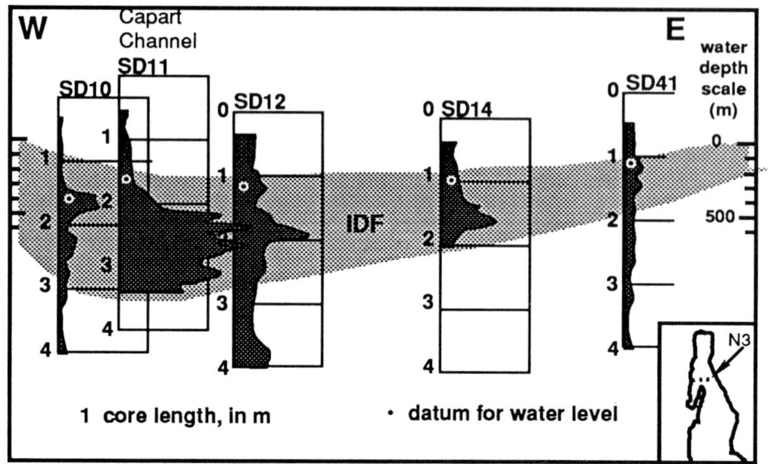

Fig. 9. Intermediate Detrital Formation: note decrease in thickness from west to east in transect N3.

map, the part played by the deeper, longitudinal channel as the main transport axis for this formation: on this map, the isograds of the maximal number of events match the contours of the deep channel. As a whole, isopachs and number of events clearly suggest that, at least during temporary, probably catastrophic, events, the longitudinal axis of northern Lake Tanganyika played a major part in the distribution of sediment, down to the deepest, central part of the sub-basins. The distribution of thickness and number of individual events in the IDF and the lack of any southern equivalent to this detrital formation are suggestive of a northern origin.

The Upper Detrital Formation (UDF)

This formation consists of three main facies. It begins usually with laminated diatomaceous muds overlain by dark-green muds. The dark-green muds transitionally change to sandy deposits in sublacustrine channels and in littoral deposits.

The laminated diatomaceous muds (DL). In areas remote from alluvial sources, especially during periods when alluvial supply was reduced, Late Holocene sedimentation in Lake Tanganyika was dominated by pelagic, rhythmic, diatomaceous mud, with green and cream couplets forming laminae of millimetric thickness, preserved from bioturbation by the anoxic environment of the hypolimnion. The green layers are a mixture of detrital, pelagic clays and diatom frustules, while the cream layers are mostly diatoms. The transition from diatomaceous couplets to detrital deposits is often gradual and

involves a reduction of the number of diatom laminae and an initial thickening of clay layers, followed by a change to silty or sandy muds.

The isoliths of the laminated diatomaceous muds also reflect the positions of the rivers mouths. The maximum cumulated thickness of the laminated diatom beds (Fig. 7) coincides with the present location of the mouths of the main rivers (Ruzibazi, Murembwe, and Malagarasi). This confirms that the last episode of detrital sedimentation (sandy or finer) is coming to an end and that the main effect of rivers is the stimulation of diatom growth. The algae could benefit either from a direct influx of nutrients with the river waters or, more probably, from nutrients otherwise trapped in the upper part of the hypolimnion and made available in the waters of the surface by turbulence provoked by the discharge of the river in the lake. Moreover, the accumulation of diatoms is favoured by the depressions formed by canyons at times when these are weakly active.

Dark-green muds, UDF proper. The upper part of cores from the sub-basins of Rumonge, Nyanza Lac, and Kabimba, and from the south of Bujumbura sub-basin is usually a dark, massive mud (Fig. 2), which overlies either laminated diatomaceous mud or the Upper Grey Clays of the 'Intermediate Detrital Formation'. This UDF has an important detrital fraction in which clay-sized particles are predominant, but where silty and even sandy quartzose beds are not rare and become predominant in nearshore sites. When muddy, this formation frequently contains scattered diatomaceous laminae and thin, white aragonitic layers, which are similar

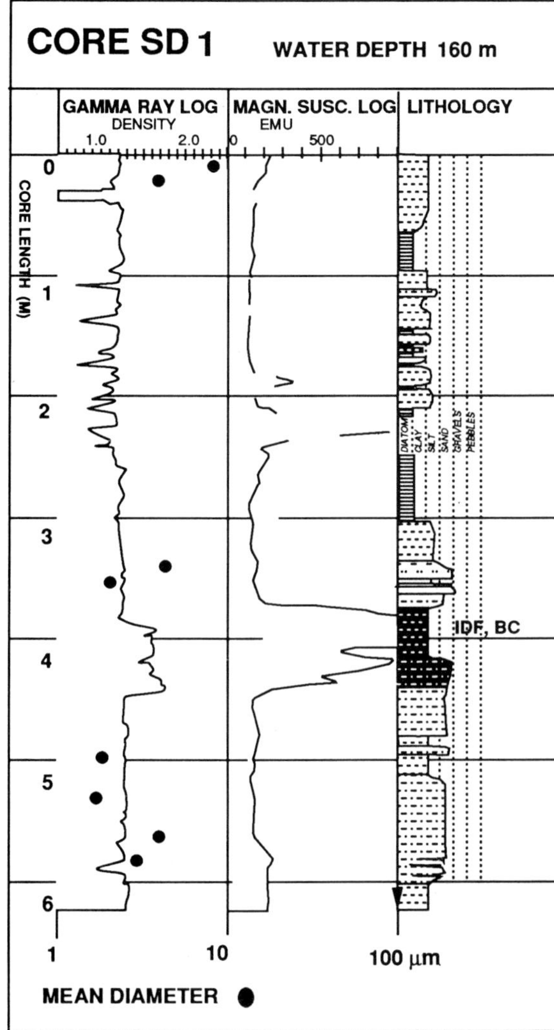

Fig. 10. Core SD 1, showing lithology, grain size, gamma ray, and magnetic susceptibility logs.

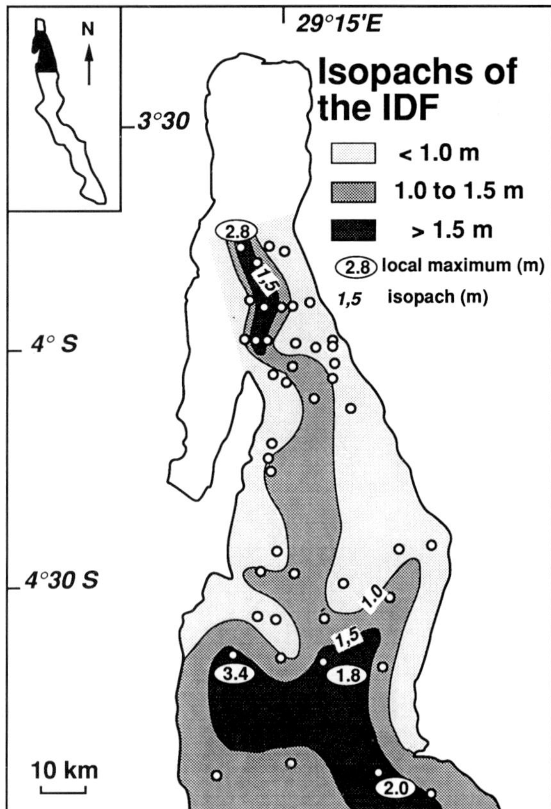

Fig. 11. Isopach map of the Intermediate Detrital Formation.

to the layers described by Degens *et al.* (1971) and Stoffers & Hecky (1978), in cores collected south of Kigoma.

At the time of collection, this sediment was a dark green, very soft mud, usually homogeneous but locally with a clotted aspect and always very different from the underlying laminated diatomaceous muds. When the cores were opened in the laboratory several months after collection, however, the dark green colour had changed to dark brown, occasionally greenish. The original greenish colour was ob-

served also for the sands and gravels that occur locally in this formation (core SD 4). In those cores displaying a gradual transition from the underlying laminated diatomaceous muds to the overlying detrital formation, the upper parts of cores show an increasing proportion of thin mud layers interbedded in the diatom laminae, the sediment evolving progressively into a muddy or coarser detritus. A contrasting, fining upward evolution, is visible at the upper part of this detrital formation, deposited in very recent times, in the form of a gradual change from sand to silt, silt to clay, and even to clay with diatom laminae.

The return to conditions favouring the deposition of diatom laminae suggests a significant reduction in the rate of detrital input. Therefore, while the UDF continues growing today, its growth is less rapid than in the past, suggesting a recent change from

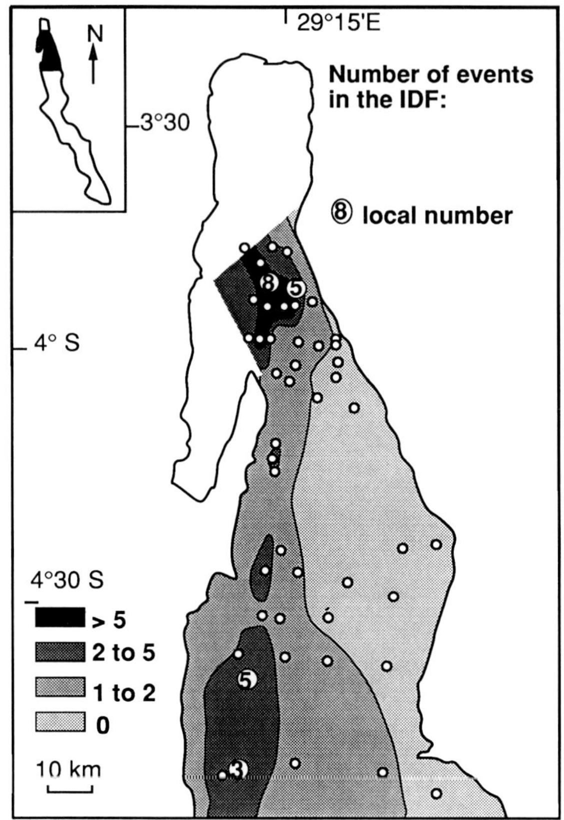

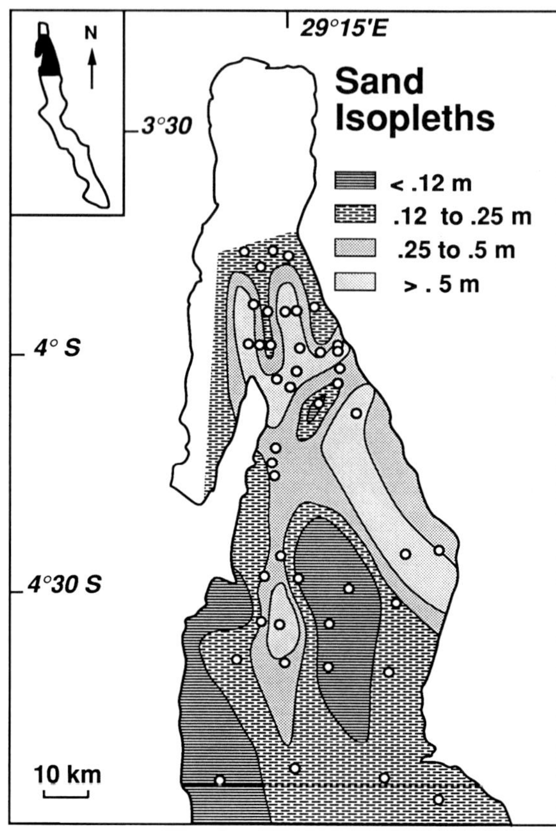

Fig. 12. Geographical distribution of the number of deposition events in the Intermediate Detrital Formation.

Fig. 13. Isopach map of the sandy deposits of the northern Tanganyika Basin.

arid to humid conditions. This conclusion is consistent with that of Haberyan & Hecky (1987), based on diatom communities, indicating a dryer climate from 3500 to 1400 yr BP, relative to present-day conditions.

Sands. Sandy facies constitute the coarsest sediments of the Upper Detrital Formation, and represent littoral platform deposits and specific high-energy sublacustrine sedimentary environments, such as channel fills and deep fans (Figs 4, 5, 13). Much of this sediment occurs adjacent to the eastern shore of the northern Tanganyika Basin and therefore is correlated with the alluvium from this shore. This is confirmed by the distribution of the cumulated thickness of the quartz sand layers based on visual inspection (Fig. 13). On this map, the sandy bodies are considered as a single, massive body, although each of them has a limited lateral extension and may be discontinuous from one core to another.

Sequential changes in cores from different physiographic conditions

Cores from the southern Bujumbura and the northern Rumonge sub-basins

Considerable sequential changes reflect local physiographic conditions. Cores from medium slope taluses, sublacustrine plateaux, and basins are typified by core SD 1 (Figs 2, 10), collected on the protected talus between the Burundi coast and the southward continuation of Cape Magara. The basic sequence displayed by SD 1 also occurs, but in a condensed form, in areas remote from detrital influx (sills, shores distant from major rivers: SD 5 and 10). Other observed changes involve higher sedimentation rates of algal (SD 2) or mineral (e.g. SD 3, 11, 15 or 26) origins. The cores expressing maximal mineral influx occur near basin centres (SD 3 and 4) and in channels (SD 13, 15, 17, and 22).

Cores collected in sublacustrine channels display very heterogeneous coarse material — sands rich in quartz, mica, quartz gravels, and some mud balls — in their lower section while their upper section generally consists of laminated diatomites and clays, the series almost always ending with dark green muds of the UDF (Core SD 11, Fig. 5). In contrast with the widespread distribution of fine mud that generally overlies laminated diatomite in the upper layers of channel fillings, the channel draining the toe of the escarpment of Nyanza Lac fault (core SD 22) lacks such upper deposits; it displays a fine sand layer on top of coarse immature sands, gravels, and mud balls. Nevertheless, both sequences tend to fine upwards.

Cores collected on littoral platforms (SD 43, Fig. 5) are formed mainly of pelecypod and gastropod debris associated with a lithic fraction of quartzose sands and gravels.

Cores from deep basins

In the deep parts of the lake, cores still display major differences in thickness and texture, again depending on location. Core SD 26 was collected at a depth of 970 m, near the southwest, channel-shaped, deep end of the Rumonge sub-basin. It is located on a deep fan, below the steepest slope present along this basin (Fig. 3). The basal layer, the major component of this core, is Beige Clay, but with a much finer texture than in SD 11. Most clays form fining upwards sequences in association with very fine sand and silts. Magnetic susceptibility figures are the highest in this set of cores (up to 2200 EMU), while density of the sediment rises up to 1.7. A succession of grey clays of the UGC, laminated diatomites, and dark green mud overlie the basal BC sequence.

The deepest and most distal (170 km from Rusizi) coring sites are located at the latitude of Kigoma. Core SD 36, in the northern part of the Kabimba sub-basin, comes from a relative high (1285 m), forming a subtle relief between two maximal depths (1285 m in SD 35 and 1270 m in SD 37). It displays a sequence beginning with the black clays of the LDF and is easily correlated with the sequences of cores located on highs situated farther north, either in the Rumonge (SD 24, 915 m deep) or in the Bujumbura sub-basins (SD 10, 275 m deep).

Distribution of textures in the detrital formations

The Upper and Intermediate Detrital Formations are especially interesting in terms of sedimentary processes, by comparison with other physiographic units (Fig. 4). The Upper Detrital Formation, because it is associated with all lacustrine environments, from the shore to the deepest channels and basins, expresses present transportation processes in channels. The Intermediate Detrital Formation, because of its thickness, provides a better record of lateral, interchannel deposits.

Textural distributions in shore (deltas, beaches) and talus deposits

At the mouths of the gullies displayed by the steeply sloping sections of the coast, sediments are limited to a talus breccia, including metric-sized blocks, and to some very small alluvial fans composed of extremely heterogeneous debris flows. Owing to the narrowness of their catchment, these areas contribute little to the bulk of lacustrine sedimentation (Figs 1, 3) and the straight shoreline is hardly modified by sedimentation.

In contrast, many small fan-deltas modify the general outline of the gently sloping parts of the coast. There, streams are building coarse fan-deltas in which debris flow deposits reflect catastrophic events favoured by relief and climatic conditions. Under ordinary stream-flow conditions, those rivers carry mainly sands and silts eroded from weathered granitic rocks. A very efficient northwards longshore drift, evidenced by long beaches, sand spits and hooks, reworks these deltas and forms lacustrine lagoons. This affects sediment textures, with these deltas displaying river clayey to silty sands and gravels (So = 1.68; Fig. 14), while the associated, modern beaches display very well-sorted sands (So = 1.15; Fig. 14). The abundance of mollusc shells results in a poorer sorting in platform sands (SD 43, upper sand layer, So = 1.54; Fig. 14). The 'lagoons' sheltered by sand spits receive very fine clays, which rapidly evolve into black, smectite-rich soils.

Talus deposits consist mostly of muddy sediments with occasional sandy layers, closely resembling very fine beach sands, suggesting their deposition to be an effect of storms (SD 39, upper sand layer, So = 1.22; Fig. 14).

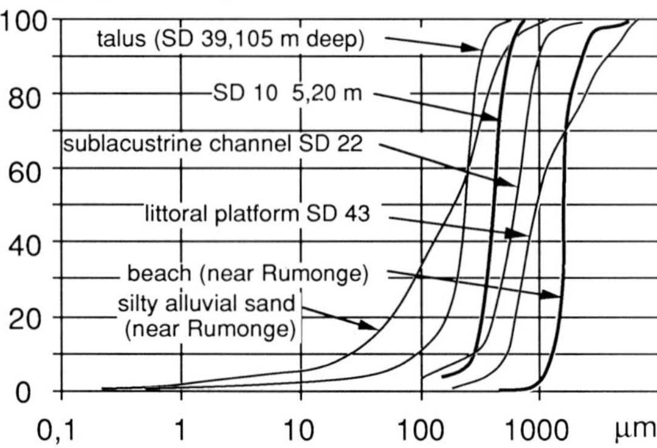

Fig. 14. Examples of grain-size distributions in sands from the northern Tanganyika Basin.

Textural variations in deep channels

In deep channels, deposits are comparatively thinner and coarser, most probably as a consequence of the acceleration of the transporting current, and winnowing of the finer fraction of the suspensions and bed loads (SD 18–20, Fig. 11). Although most channels display modern mud or diatomite filling, the bottoms of some deep, canyon-shaped channels, are presently covered with sands. The latter are devoid of detrital mud or diatomaceous mud, and their sand cover suggests that the transportation of such sediment in the form of density currents is still active. Their sands are well sorted, but less so than beach sands (SD 22, So = 1.33; Fig. 14). Most channels follow a transverse course down the eastern slope of the lake bottom, ending in small, deep, sublacustrine fans (Fig. 13).

Grain size distributions of sediments immediately respond to changes in the longitudinal slope and to changes in the width of the main, N–S channel, along the Ubwari Peninsula. The response of grain size to changes in the mean slope is shown by an increase from mud to sand at sites SD 18–20, where the average slope is the steepest (0.03) among the values observed along the longitudinal axis (averaging 0.008 from site SD 14 to SD 35). Where the channel narrows, a local increase in grain size is observed. For instance, in the north of the Rumonge sub-basin, the dark green, Upper Detrital Formation is essentially a muddy sediment at site SD 2, while it becomes sandy at site SD 9, and probably gravelly at site SD 13 (according to our interpretation of seismic profiling and confirmed by several unsuccessful

coring attempts). The coarser grain sizes are possibly the result of sorting under the influence of accelerated sublacustrine turbidity currents. The velocity of these currents is increased in narrower reaches by Venturi effect and also where a steeper slope enhances the current-generating gravity component.

Textural distribution in deep, lateral, non-channel deposits

Laterally from the deep longitudinal axis, as water depth diminishes, the Intermediate Detrital Formation exhibits a gradual decrease in thickness (Fig. 6), number of events (Fig. 12), and mean grain size. In the northern part of the area studied (SD 1–3), this decrease from west to east (Fig. 9) occurs in cores from comparatively shallow depths (135 m in SD 41). This formation also is present, mostly as very fine clay, in minor depressions (perched basins), inactive channels branching on the main sublacustrine valley, and on the tops of sublacustrine highs.

Both the Intermediate and the Upper Detrital Formation occur even on the culmination of isolated highs (site SD 10), or in isolated, perched basins (SD 5, Fig. 5). This indicates that transportation occurred not only in deep waters, close to the bottom, but also much higher in the water column. These formations also exist at comparatively shallow sites (130 m deep) and thus the corresponding sediments were necessarily transported during a high stand of lake level. They may have been transported either in turbidity currents or in originally hyperpycnal turbid waters, subsequently floating as 'interflows' (Sturm & Matter, 1978; Pharo & Carmack, 1979) or as a detached nepheloid layer at an intermediate depth in the lake, as suggested for certain Ebro River deposits in the Mediterranean (Nelson & Maldonado, 1988).

Distribution of average textures as a function of time

An average of mean sizes has been calculated for every sedimentological unit and plotted as a function of average time (Fig. 15). Sediments from the Lower Detrital Formation (37 000 yr BP) average 40 µm in diameter, while the Lower Black (18 000 yr BP) and Lower Grey Clays with diatomaceous laminae (7800 yr BP) average 5–8 µm. The deposition of the IDF is a high-energy episode during which the average mean size increased very rapidly to 55 µm (Lower Grey Clay turbidites) and immediately after, fell

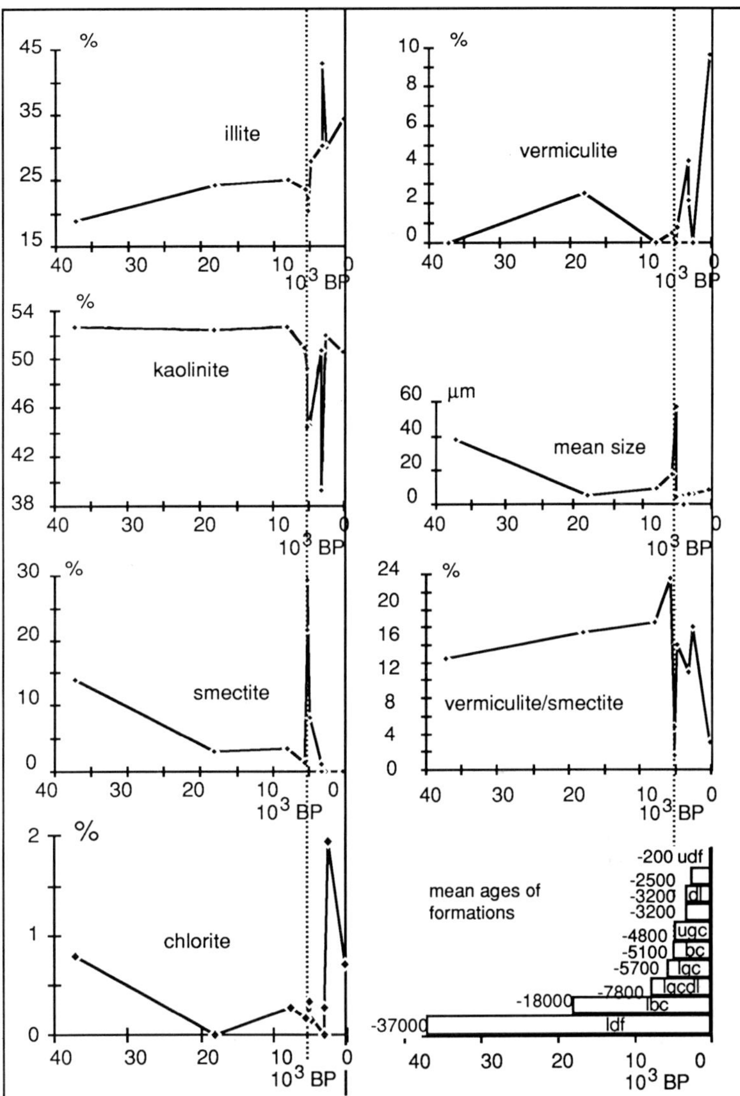

Fig. 15. Clay mineralogy (average by formations) as a function of time in selected cores from the northern Tanganyika Basin.

again to 5 µm (Fig. 15) during the deposition of the Beige Clay (5100 yr BP). Mean sizes of the upper formations, Upper Grey Clays and diatomaceous muds display a regular, moderate increase up to 10 µm (Upper Detrital Formation).

Mineralogy and geochemistry

Non-clay minerals

The non-clay, crystalline fraction typically includes quartz, feldspars (plagioclase and alkaline feldspars), with some siderite and pyrite, restricted to particular environmental conditions. The large proportion of diatom frustules and/or clay particles in sediments of Lake Tanganyika result in a dilution of this non-clay fraction. Dilution by diatom frustules is lower on sills, as also noted by Degens *et al.* (1971), resulting in a correlative increase in the crystalline fraction.

Quartz is the predominant mineral in the sediments of coarse formations, constituting 80–100% in the mineral, non-clay fraction, resulting in an average 30% of total sediment, including clays and

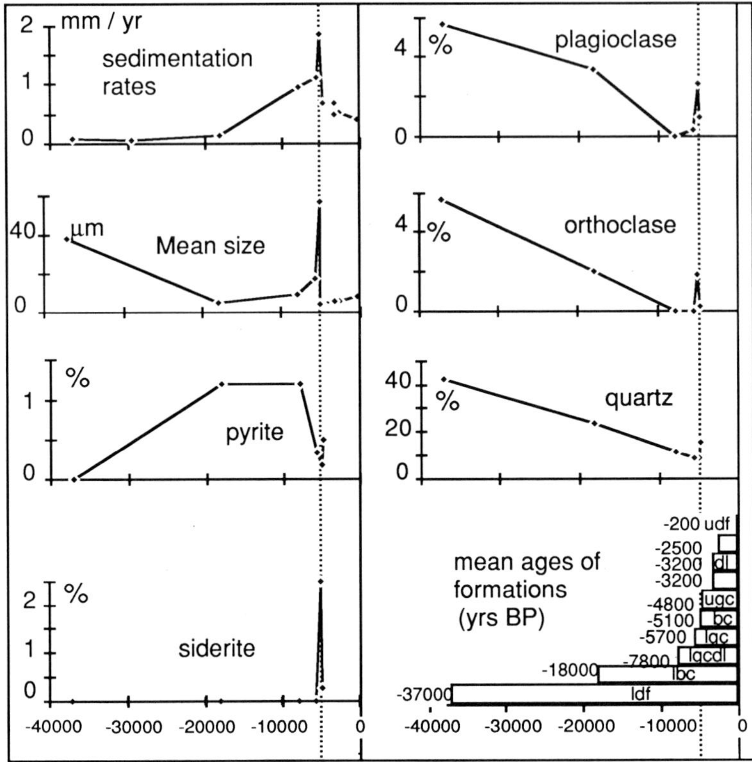

Fig. 16. Non-clay mineral fraction (average by formations) of sediments as a function of time in cores SD 1, SD 10 and SD 36.

diatom frustules (Fig. 16). Quartz, plagioclase, and orthoclase have nearly parallel distributions and the average proportions of these detrital minerals steadily decrease with time. Between 35 000 and 8000 yr BP, quartz regressed from 45 to 10%, plagioclase and orthoclase each regressed from 6 to 0%. A sharply contrasting situation is displayed by a period centred on 5000 yr BP, with the deposition of the IDF layers, and especially the Beige Clay, in which quartz 'instantly' increases from 9 to 17%, and orthoclase and plagioclase from 0 to 1.8 and 2.3, respectively (Fig. 16).

Siderite and mangano-siderite are an important component (up to 11% of total sediment) in the BC, and siderite is present in smaller amounts in the other members of the IDF, but was not found either in older or younger formations. This results in a very local distribution (Fig. 16), in exact correspondence with the quartz, feldspars and grain size maxima found at 5000 yr BP.

Pyrite is a constant component, with the exception of the LDF where the average is very close to 0%. It displays distribution trends that are opposite to mean size trends, with a maximum associated with the deposition of both the Lower Black and the Lower Grey Clays with diatomaceous laminae (up to 3%). In contrast with siderite, pyrite experienced a sharp minimum during the sedimentation of the IDF.

Scattered, individual millimetric layers of amygdaloid aragonite are conspicuous in the UDF and/or the UDFDL of most cores.

Clay minerals

Results are presented as percentages of X-rays diffracted by the 001 layer of the minerals, without other calibration (Fig. 15). Kaolinite is the major component of all cores, with average percentages of about 52%. These proportions change little from one formation to another, except a 5% average diminution during the last 10 000 years and two anomalous minima, obviously departing from the general trend. Kaolinite is present in similar proportions (50%) in the soils of the drainage basins and in the sediments from the deltas of the Burundi shore. This mineral remained constant during the stratigraphic period covered by this study, probably because it forms mostly in the high, well drained, distal areas of the basins.

Illite, regardless of the anomalies, gradually increases with time from 15%, 35 000 years ago, to 34% presently (Fig. 15). The distributions of vermiculite and vermiculite−smectite are somewhat complementary, suggesting diagenetic changes from one mineral structure to the other. Combined, the distributions of these minerals display a trend similar to that of illite. Illite and chlorite (the latter found in detrital sediments, not exceeding 5%) are typical of little or no chemical alteration, while vermiculite characterizes low-grade alteration of micas (Chamley, 1989). Illite, after kaolinite, is the second most frequent clay mineral of the drainage basins, except for lowlands where it is diagenetically replaced by smectite.

The frequency of smectite as a function of time has a general, decreasing trend, from 15%, 35 000 years ago, to nearly 0%, 3000 years ago onwards.

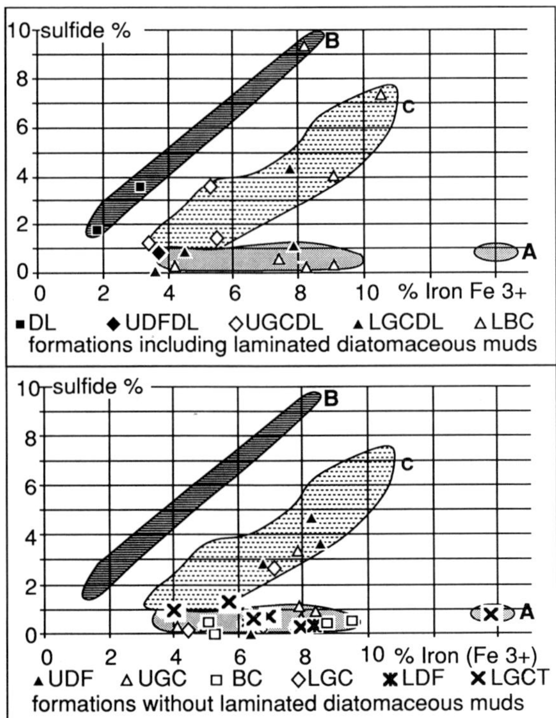

Fig. 17. Sulphides versus iron (expressed as Fe^{3+}) in sediments from various late Pleistocene to Holocene detrital and pelagic formations of the northern Tanganyika Basin.

Notwithstanding distribution anomalies, this decrease is symmetrical with the increase trend noted for illite. This is not likely to be the effect of a post-depositional transformation of detrital clay minerals because this distribution holds true for all cores, collected in environments ranging from shallow to very deep. Unexpectedly, smectite in fact is associated with high-energy detrital sedimentation: the overall frequency distribution of smectite as a function of time resembles the distribution of mean sizes (Fig. 15) and the periods when smectite was the more abundant correspond to periods when the sediment was coarsest. This view that smectite was associated with high energy, coarse sedimentation is also supported by the predominance of smectite in the mud balls found in the turbidites both of the LDF and the IDF. In modern environments of the Burundi shores of Lake Tanganyika, smectite (30% on average, and up to 50%) is found in the fine clays of the marshes developed behind the sand spits associated with fan-deltas. Such marshes are a typical location for the formation of smectite (Paquet, 1970).

The decrease of smectite as a function of time was interrupted by the deposition of the IDF, a contrasting episode, less than 2000 years long, during which the average smectite frequency passed from 0 to 30%, and following which smectite regressed to 0%. This smectite maximum accounts for the corresponding minima previously noted in the distributions of other clay minerals at the same epoch (Fig. 15).

Geochemical analyses

The chemical analyses of major elements were performed on the fraction of sediments finer than 50 μm, using a Philips scanning electron microscope equipped with an X-ray energy scanner. Samples were crushed to a standard 0.1 mm diameter controlled by sieving, and subsequently moulded into pellets 3 mm in diameter.

The most abundant element in the sediments studied is silicium. Maximal values occur, not surprisingly, in the layers with abundant diatom debris (LGCDL, DL). The admixture of biological silica in the detrital formations results in a lowering of the percentages of all other components during the deposition of the laminated diatomaceous muds, 2550 yr BP (Fig. 17). All cations display stable compositions during most of the period studied, the only long-term modification being a relative diminution owing to the influx of silica associated with the sedimentation of diatom-rich layers. The period under the influence of silica sedimentation was interrupted when the occurrence of cations, such as Al, Mg, Ti, and Mn, abruptly reached maximum values, higher than values attained prior to silica sedimentation, in exact correspondence with the deposition of the IDF. Especially abundant were Ti, Al, and Mn (Fig. 17), in remarkable concordance with data from Lake Kivu (Degens & Kulbicki, 1973). Elements Fe, Ca, and Sr, although a relative maximum is also observed at this time, do not exceed pre-silica figures.

Sulphur in the form of sulphides is present in the sediment (expressed as SO_3 in the logs). It is more abundant in the Upper Detrital Formation and in the diatomaceous sediments and much less so in the IDF (Figs 16, 17). Iron is present in significant amounts throughout the column. It is more abundant in the detrital formations, especially in the Beige Clays ($5-9\%$, expressed as Fe_2O_3 although part of it was in the form of Fe^{2+} in the sediments), and much less in the diatom-rich layers ($2-3\%$). Iron occurs in different forms along the cores. In the UDF it is associated with sulphides, a point consistent with the occurrence of pyrite, already evidenced by X ray diffraction on the coarse fraction. In the Beige Clays it is associated with carbonate ions in the form of siderite and with silicate to form clay minerals (smectite).

The various associations of iron and sulphur are expressed on a graph (Fig. 17) displaying the distribution of sulphur as a function of iron for selected sediments of each formation. Three different distributions are visible.

1 A distribution (A, Fig. 17) in which sulphur is usually less than 1% and varies independently from iron; this distribution characterizes sediments from typically detrital, turbiditic formations, i.e. part of the Upper Detrital Formation, most of the Intermediate Detrital Formation (all samples from Beige Clays and most of the Upper and Lower Grey Clays belong to it), and most of the Lower Detrital Formation.

2 A distribution (B, Fig. 17) where sulphide is a direct function of iron, with a 0% minimum for both elements. This distribution is typical of the diatomaceous laminated muds, that is of pelagic deposits, of the DL and LBC formations.

3 A distribution (C, Fig. 17) in which sulphur is a direct function of iron, but only above 4% iron, indicating that part of the latter is linked to other components, such as carbonate in the form of siderite or clays in the form of smectite. This distribution is typical of transitional deposits, from detrital to pelagic formations, for which the presence of interbedded, laminated diatomaceous muds is indicated by the DL suffix: Upper Detrital Formation DL, Upper (and Lower) Grey Clays DL; Lower Black Clays. Distribution C also affects certain samples in which no laminated diatomaceous muds were visible, but which correspond to essentially transitional deposits (LGC and UGC, for instance).

^{14}C chronology

Previous work

Previous work demonstrated that Holocene sedimentation rates changed from one period to another in the same core, and from cores in deep basins to cores on top of sills. Sedimentation rates range from 0.3 to 1.4 mm/yr for core MPU 12 from the Mpulungu sub-basin of southern Lake Tanganyika (Tiercelin *et al.*, 1988). In the deep parts of the central lake, sedimentation rates attain 0.3 to 0.5 mm/yr, and 10 times less on the sills (Degens *et al.*, 1971). A very high average sedimentation rate, 1.98 mm/yr is reported from Burton Bay, in the northern Tanganyika Basin (Mondeguer *et al.*, 1986).

This study

The ^{14}C ages were measured by associates of J.Ch. Fontes in Orsay (Laboratoire d'Hydrologie et de Géochimie isotopique), on the total organic matter of selected samples. Most samples were submitted to the classic ^{14}C methods. A length of about 1 dm of the half core was necessary to obtain a sufficient amount of carbon for a reasonable degree of precision. In spite of this, samples older than 10 000 years gave very large uncertainties. Older samples (LDF), and samples having very little organic matter (laminated diatomites and beige clays), were analysed by Accelerator Mass Spectrometry (AMS) unit in Orsay/Gif.

Age of the Lower Detrital Formation

According to dating of cores SD 5 and SD 10 (north of the Rumonge sub-basin), this formation is consistently older than 30 000 yr BP: > 30 000 yr BP dated by classic ^{14}C above the upper limit of the LDF in

SD 5; c. 35 000 yr BP within the LDF, dated by AMS in SD 10. Measurements on SD 24 (shoal between the sub-basins of Rumonge and Nyanza) give 32 300 yr BP within the LDF and 16 800 yr BP near the top. However, the core is a massive, undifferentiated clay in which the upper limit is based on magnetic susceptibility only.

Age of the Intermediate Detrital Formation

Ages obtained from organic matter in the pelagic layers immediately above and below the Intermediate Detrital Formation indicate that the deposition of this formation took place in a comparatively short time, i.e. between 5280 ± 325 and 4800 ± 300 yr BP if we consider samples from the core SD 1. This core was exceptionally favourable because it contained bulk organic matter in sufficient quantity near the limits of the IDF. The dates obtained on SD 1 are consistent with results from the other cores but more precise. Other cores give a lower age limit of 5130 ± 770 yr BP (SD 24), 5960 ± 340 yr BP (SD 36), 6295 ± 300 yr BP (SD 14), and 7960 ± 215 yr BP (SD 10). Especially in SD 10, a very low sedimentation rate maximizes the error owing to the use of long samples from the over and underlying pelagic beds. An age of 3858 ± 108 yr BP was obtained on the laminated diatomites at 1000 mm above their limit with the IDF. The sedimentation rate in these diatomites (1 mm/year) leads to an age of 4850 yr BP for this limit. These results suggest a very short duration, about 500 years, for deposition of the IDF.

The IDF contains relatively little organic matter and ^{14}C measurements required use of the accelerator technique. The ages of the IDF sediments are systematically older than the ages indicated by the associated pelagic layers. In the core SD 3, the bulk organic matter of three samples from the IDF gave: $13\,280 \pm 111$ yr BP and $12\,261 \pm 289$ yr BP at a depth of 309 cm; $17\,577 \pm 217$ yr BP at a depth of 122 cm.
1 The discrepancy of 1000 years between the two samples from 309 cm suggests the effect of particulate, detrital organic matter on the age of relatively small samples.
2 The 5000-year older age on the sample from 122 cm is consistent considering that the IDF resulted from the reworking of older sediments.

Assuming a 500-year period for deposition of the IDF, this results in a sedimentation rate of 2.45 mm/yr in the SD 1 core, where the IDF is 1.17 m thick, and up to 6.25 mm/yr in the deep axis, where it is 2 m thick (SD 3).

Age of the Upper Detrital Formation

This formation is modern, although we have already noted a tendency to a reduction of the corresponding detrital influx in present times, with sediments tending to be finer and to contain more scattered, laminated diatomites at the very top of the cores. Ages are based on sedimentation rates in the laminated diatomites. Since there is a gradation from pure laminated diatomite to pure detrital sediments, and since diatoms are very numerous in the UDF itself, the exact time at which the detrital sediments became predominant differs from site to site. When the upper sediment is a pure, massive, dark mud the transition from laminated diatomite to detrital sediments occurs at c. 2000 yr BP. When we consider scattered diatomite layers in dark mud, the upper formation commonly dates back to 2500 yr BP, and occasionally even to 5000 yr BP.

Millimetric aragonite layers (a lower one and a closely associated pair) occur systematically in the massive, green, muddy facies of the UDF but not in the underlying, laminated diatomites, nor in the top mud containing scattered algal laminae. Thus, the aragonite layers are closely associated with the muddy facies of the UDF in the northern Tanganyika Basin. Moreover, they are easily correlated with the three aragonite layers mentioned farther south, both in the northern and the southern basins (Stoffers & Hecky (1978) on cores from Degens *et al.* (1971)), which also are grouped into a pair and one separate, deeper layer. The aragonite layers are contained in Unit I (Degens *et al.*, 1971), a metric unit formed of laminated diatomites, which therefore is a lateral equivalent to the UDF.

Average sedimentation rates as a function of time

A diagram of the average, linear sedimentation rates, plotted as a function of time (Fig. 16) combines an evolution including compaction effects and a temporary, superimposed increase on sedimentation rates. On this diagram the curve joining points corresponding to 30 000 yr BP, 18 000 yr BP, and present, indicates sedimentation rates changing from 0.4 mm/yr to 0.1 mm/yr. The ascending trend of this curve is an expression of compaction, more effective in the deeper parts of the sedimentary record. Gradually increasing sedimentation rates, associated with increasing average grain sizes, occurred during the period from 10 000 to 4000 yr BP. A sharp exaggeration of both curves (Fig. 16) results in distinctive peaks close to 5000 yr BP (sedimentation rate: 1.8 mm/yr), immediately followed by a return to average conditions.

DISCUSSION

General climatic evolution suggested by the sedimentary record

In general, the sedimentary record available for this study demonstrates that the percentages and sizes of detrital particles in the total sediments decrease with time while the proportion of diatom frustules increases correspondingly. This change from detrital to pelagic conditions indicates a rising lake level and therefore suggests a climatic change from dry to more humid. This interpretation is supported by mean grain size (Fig. 16), the average value of which changed from silt to clay during the same period, and is supported also by the gradual decrease of the proportion of detrital material versus biogenic deposits. It is indicated for instance by the divergent paths of the curve for Al, reflecting the detrital fraction only, and of the curve for Si, representing both detrital and biological contributions (Fig. 19). The gradual decrease in the proportion of feldspars versus quartz attests to an increased weathering of the soil profiles and is thus another indication of more humid climatic conditions.

The general decrease of smectite with time is further evidence that can be interpreted in terms of a rising, initially low, lake level. We have already noted that smectite is forming at the present high level of the lake, in marshy lowlands associated with fan-deltas. In periods of low stands, such fan-deltas become high terraces, which are subject first to slumping (favoured by the outflow of ground-waters during the retreat of the lake) and subsequently to erosion. This destructional evolution of deltas at the beginning of regressive episodes explains the coincidence of smectite with high-energy detrital deposits. Moreover, the coincidence with lower levels of the lake is in perfect agreement with the arid climatic conditions that prevailed at the time of the Lower Detrital Formation.

The problem of the occurrence of slightly more kaolinite in sediments deposited during this dry period must be considered in terms of erosion of soil profiles in high, well-drained parts of the basins. The soil horizons eroded at the beginning of such dry periods concern the upper part of profiles that evolved under the previous, more humid conditions. Under a tropical climate, these conditions are known to favour monosiallitization, i.e. formation of kaolinite, especially on cratonic rocks (Pedro, 1968). With time, erosion, favoured by the rift morphology, attains deeper, less mature horizons in which micas of the original rock are still in the form of 2/1 clay minerals.

Evidence of emersion in the Lower Detrital Formation

In the cores we failed to note any conclusive evidence of emersion, such as desiccation cracks, roots, oxidation, etc., typical of shallower tropical lakes (Talbot *et al.*, 1984). Since we noticed that beach processes are a common feature along gently sloping sections of the shores, the question arises as to the possible presence of ancient beach deposits in deeper sediments of the northern Lake Tanganyika. At the lower part of one core, displaying ages over 30 000 yr BP (SD 10, Fig. 2), we did observe thin layers of a clean, very well sorted (So = 1.14), fine sand that is very similar to modern beach or beach-related sands (Fig. 14). If these sands are correctly interpreted as being beach deposits, we have an indication that, at this period, the shoreline of the lake was at least 275 m lower than present. Obviously, further investigations are needed to confirm this point, although this level is in good agreement with the level expected for the last low stand of Lake Tanganyika.

These sands would be the expression of the transgression subsequent to a low stand of the lake and suggest that the LDF was deposited under dryer conditions. The most recent, generally admitted, lowering of lake level lasted from 15 000 to 13 500 yr BP and reached −240 to −340 m (Haberyan & Hecky, 1987; confirmed by Tiercelin *et al.*, 1988). Both SD 10 and SD 5 sites are shallow (−275 and −225 m versus present lake level, respectively) and thus are likely to have been subject to emersion.

Sedimentary dynamics in Lake Tanganyika under the high-level conditions of the Intermediate and Upper Detrital Formations

The sedimentary records of the three detrital formations include very coarse channel deposits, which, in the case of both the IDF and the UDF, are continued laterally by finer deposits, including fine clays in relatively high bathymetric situations. The bathymetric contrast between the locations of these two simultaneous types of sediments, indicates:

1 That water level in the lake did not differ significantly from the modern level at the time of deposition of the IDF and the UDF, being higher than -135 m. Therefore, deposition was entirely lacustrine and certainly not the result of river floods during a low stage of the lake, because the highest deposits are hundreds of metres above the possible height for any fluvial flood.

2 That turbidity currents probably played an important part, being the only agent able to transport coarse debris in channels and, at the same time, to disperse important quantities of clay-size particles at intermediate and high levels in the water column, although for the latter, the hypothesis of turbid, river-water floating at intermediate depth cannot be discarded.

Owing to the elongate form and great depth of Lake Tanganyika, a record of the geometry of extensive longitudinal turbiditic events is available and this should favour an improved understanding of *in situ* sedimentary processes. Along any transverse section in the basins studied, the number of sedimentary events is highest in the vicinity of the deepest axis and decreases away from it, correlative to a reduction in depth (Fig. 18). Therefore, the number of events can be graphed as a function of depth and this gives both an indication of the depth of water where events were more frequent as well as an indication of the maximal elevation attained by one event. It can be assumed that this maximal elevation, which is observed in a lateral situation, represents the minimum height attained by the turbidity current above the deep axis. This is the rationale on which Fig. 18 is based, giving isopleths of the number of events as a function of depth along the axis of Lake Tanganyika.

As shown by Fig. 18, the maximum number of events usually is found a little higher than the deepest channels. In sediments of these channels, the number of recorded events may be zero where the longitudinal slope is steep, because of subsequent erosion and reworking of the Beige Clay in the channels. On any profile the number of events decreases upwards but this decrease, while moderately rapid in the north of the northern Tanganyika Basin, is very rapid in the south. Considering a level of 300 m above the lowest level where the Intermediate Detrital Formation is present, the number of recorded events is found to decrease from north to south: three in profile 1, two in profile 3, one in profile 8. Moreover, at least one event is recorded at about 100 m depth in most of the northern part of the lake.

This geometric expression of the sedimentary record in the beige clays of the northern basin of Lake Tanganyika is consistent with transport by large turbidity currents flowing in a N$-$S direction.

Turbidity currents, as an agent of detrital sedimentation

We have considerable evidence that the IDF was prominent in the late Pleistocene and Holocene history of Lake Tanganyika, not only as a unique high-energy episode, but also as a mineralogically and geochemically distinct episode. We noted a marked increase in feldspar percentages (cumulated averages: 4% of total sediment for plagioclase and orthoclase, instead of 0% in the underlying layers). Quartz, in spite of a visible increase, accounts for

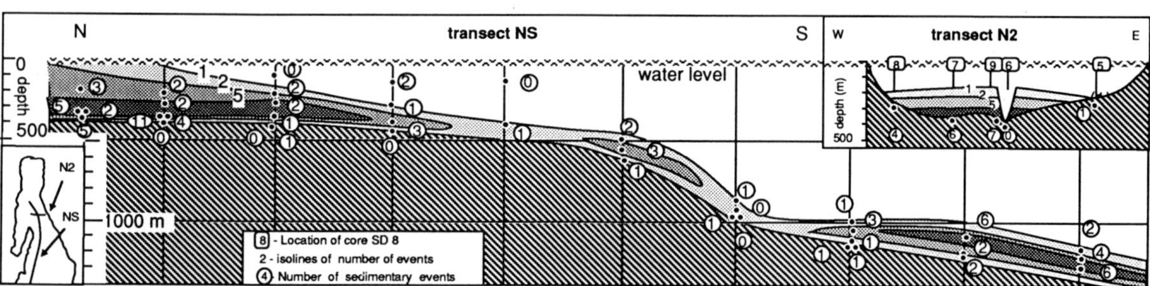

Fig. 18. Distribution of the number of sedimentary events in the Intermediate Detrital Formation as a function of depth: along the N$-$S axis of Lake Tanganyika (projection of transverse transects on the longitudinal transect); (inset) along an E$-$W transect (transect N2, see Fig. 3 for locations).

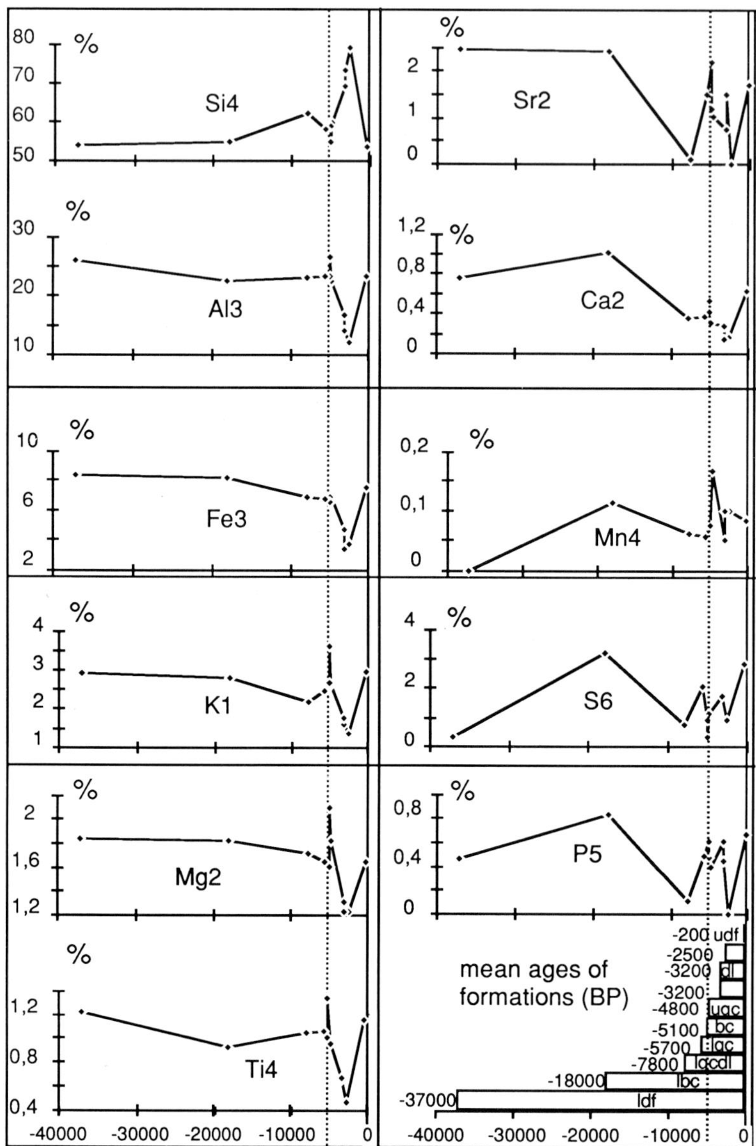

Fig. 19. Geochemical analyses (average by formations) as a function of time on selected cores from the northern Tanganyika Basin.

only 17% of total sediment, a surprising figure under a tropical climate, compared with the total 4% of feldspars. This corresponds to the dilution of the local, quartz-rich sands by an allochthonous, massive influx of minerals whose chemical composition is indicative of a basic origin rather than granitic (richness in feldspars, especially plagioclase, paucity in quartz). This sedimentary episode was so important that the average sedimentation rate reached 1.8 mm/yr while local values of 2.45 mm/yr were recorded in the SD 1 core and up to 6.25 mm/yr in the deep axis (SD 3).

Considering the volume and homogeneity of this short lived but important sedimentary event, the problem of the source of material arises. It should be a low-land source at modern lake level, with considerable mud and silt deposits to account for the volume deposited in such a short time, over 170 km of the length of the lake. The only source in terms of available volume of loose sediment is perhaps down-cutting in the Rusizi Valley. The basic composition is in agreement with the volcanic episodes reported at the same period in the Kivu area (Degens & Kulbicki, 1973) and moreover, Al, Mg, Mn, Ti, and Fe (Fig. 19) display maxima in phase with maxima in Lake Kivu (Degens & Kulbicki, 1973). This episode is synchronous with the overflow of Lake Kivu, but it predates slightly the evidence of this overflow documented by the evolution of carbonates in southern basin cores (Stoffers & Hecky, 1978), a point clearly in agreement with a N−S evolution. This agrees also with the interpretation of Vincens (1989), which shows that pollen of the Beige Clay was different from pollen of adjacent formations and came probably from the Rusizi. Waters from Lake Kivu rushed via the Rusizi Valley into Lake Tanganyika, inducing a major, temporary interruption in the general, calm trend of lacustrine sedimentation, which at that time tended to be starved in detrital influx and dominated by laminated diatomaceous muds. This catastrophic event provoked density flows, including major turbidity currents, in the lake.

We have already noted that smectite was the predominant clay mineral of the IDF. The origin of this smectite must logically be a lowland source during this episode, as it was during the earlier evolution of the lake. This time, however, the average smectite proportion is higher, completely departing from the general trend, and therefore the lowland had to cover a very large area. This again suggests a Rusizi Valley origin.

Turbidity currents, as an agent of chemical sedimentation

The oxidizing properties of waters forming turbidity currents contribute to the oxygen content of the lake-bottom environments. In a rift lake from the Rubielos de Mora Basin (Spain), Anadón *et al.* (1988) have reported burrowed layers within laminated series. They assumed that turbiditic events transported the oxygen needed to support the life responsible for this bioturbation. In this study we did not find any indication that such temporary oxic water influx had consequences on sublacustrine life, although we did find several indications suggestive of the influence of such waters on the oxidation reduction potential of the recently deposited sediments.

Multiple, either abrupt or gradational, changes in colours were mentioned in the description of the IDF. They are a first indication of the occurrence of strong contrasts in oxidation reduction potential between the incoming, sediment-load density currents and the deep hypolimnic waters. Under such conditions, the oxygenated state, not only of the incoming water but also, mainly, of the sedimentary load (iron oxides and hydroxides), will produce a relative rise in Eh, with gradational effects, depending on the influx, sufficient or not sufficient to counteract the reducing influence of the hypolimnic environment. In the case of a small influx, the sedimentary load is reduced, sulphides such as pyrite or greigite are formed and the sediment acquires greenish and blackish colours.

The mutual exclusion of pyrite and siderite noted earlier (Fig. 16) provides interesting support for this view. Pyrite is typically associated with layers in which sulphur is a direct function of iron (distributions B and C, Fig. 17), such as the UDF and LBC. In the IDF it occurs in a limited amount only in the LGC and the UGC. It is totally absent from the Beige Clays in the individual layers where siderite is at a maximum. Siderite is typical of layers in which sulphur is in minor amounts and distributed independently from iron (distribution A, Fig. 17). Siderite, although not uncommon in ancient non-marine sediments, is not observed precipitating today (Kelts & Hsü, 1978). It has been reported for late Quaternary coarse deposits from 'weakly reducing' environments of Lake Kivu (Stoffers, 1975; Stoffers & Hecky, 1978). Therefore there is a possibility that the siderite found in Lake Tanganyika is reworked from sediments of Lake Kivu. The mutual exclusion of pyrite and siderite, however, rather suggests that changes in environmental conditions resulted in changes of local authigenesis. This matter obviously requires discussion in terms of oxidation-reduction potential, in view of the strongly reducing conditions that prevail in the hypolimnion of a tropical meromictic lake such as Tanganyika.

A standard Eh−pH diagram established for 1 atm pressure (instead of 10 to 120 atm), $\Sigma\ CO_2 = 10°$ and $\Sigma\ S = 10^{-6}$ (Garrels & Christ, 1965) illustrates that (i) if siderite is to have an important field of stability, dissolved carbonate must be very high and reduced sulphur extremely low, (ii) a considerable field of pyrite remains, (iii) under these conditions, siderite may be a criterion of very strong reducing conditions or of moderate reducing conditions.

Pyrite is clearly related to layers deposited during periods of moderate detrital supply, and, in the case of the muddy facies of the UDF, during periods favouring the deposition of thin, individual laminae of aragonite. In these periods, although carbonate ions were present, they resulted in the deposition of calcium carbonate in the form of aragonite only, while sulphides combined with iron to form pyrite and certainly other, less stable, sulphides, not identified by standard X-ray procedures. The Eh−pH diagram suggests strong reducing conditions.

Siderite is related to the IDF, i.e. in correspondence with turbiditic events, and, therefore, reducing conditions are not likely to have prevailed during the deposition of this unit. The sedimentary influx was transported in the form of relatively oxygenated density currents, with a plume of fine oxide-rich alluvium, large enough to counteract the strongly reducing environment that prevailed in the bulk of the hypolimnion, resulting finally in the moderately reducing environment favouring the precipitation of siderite. In terms of the Eh−pH diagram, it suggests that this siderite is a criterion of a moderately reducing environment (Garrels & Christ, 1965). This is consistent also with the limited, total organic carbon — significantly smaller than that of both the other members of the IDF and of the other detrital formations (Huc *et al.*, 1989).

Another important conclusion of the Eh−pH diagram (Garrels & Christ, 1965) is the necessity for a very high carbonate-ion concentration to form siderite. Rusizi River and Lake Kivu provide an adequate source — in periodicity and abundance — for the carbonate ions necessary to form siderite. Iron is available from soils. We noted earlier that, today, the carbonate-ion concentration of waters from these sources is twice that of Lake Tanganyika waters. The authigenesis of siderite might explain the special abundance of this mineral in very fine layers of the Beige Clay. The fine oxide and hydroxide particles included in the plumes of turbidity currents, mixed with the reducing environment of the hypolimnion, could result in the appropriate Eh conditions. This hypothesis has important bearings on the lake level 5000 yr BP. If the thin layer of Beige Clay found in SD 40, 135 m deep was really chemically modified by reducing, hypolimnic conditions, it implies that at this time the lake surface was at least 100 m above the deposition site, i.e. very close to its present level. This relatively stable high level is a widely accepted concept (Haberyan & Hecky, 1987; Gasse *et al.*, 1989; Casanova & Thouin, 1990).

The last detrital episode, the Upper Detrital Formation

The UDF differs from the IDF by the occurrence of pyrite and aragonite instead of siderite, probably because of the moderate detrital influx produced by the UDF, in contrast with the catastrophic events mentioned for the IDF. The moderate sedimentation rate provided time for the renewal of sulphide ions at the bottom of the lake, in the vicinity of the iron hydroxides eroded from soil profiles. The UDF is clearly a temporary interruption in the diatomaceous sedimentation resulting from a dry climatic episode. The upper layers of the UDF often include diatomaceous laminae, suggesting that more humid conditions tend to prevail presently. This agrees with results from the north basin based on diatom and stromatolite communities (Haberyan & Hecky, 1987; Casanova & Thouin, 1990). Aragonitic layers are present in the northern basin, both in the UDF proper and in the laminated diatomites. In the northern part of the southern basin, high magnesium calcite is found in place of aragonite, and no carbonates occur in the extreme south of Lake Tanganyika (Stoffers & Hecky, 1978). This horizontal distribution of carbonates and the time correlations with events in the Lake Kivu region were interpreted by these authors in terms of precipitation from water as a consequence of the last activity episode of the Rusizi River. This leads to the conclusion that the UDF is also largely a result of sedimentation from this river.

CONCLUSION

Studies of Lake Tanganyika sediments aimed at establishing a continuous record of climatic conditions during the Quaternary, recorded by successions of climate-sensitive floras and geochemical events. The three recent detrital episodes discussed also suggest contrasting climatic influences, which are correlated with episodes recorded at the southern end of the lake. They coincide with gradually increasing lake levels.

The Lower Detrital Formation can be correlated with dry climatic conditions and possibly a comparatively low stand of the lake, suggested by possible beach deposits in sediments presently found at a depth of 275 m and by a relative abundance of smectite. Predominant smectite is reported also in the central part of the lake in the deep part of

Kullenberg cores (units 3 and 4) by Stoffers & Hecky (1978), but not in the south (Tiercelin *et al.*, 1988). The climatic interpretation is in agreement with results from the southern basin of Lake Tanganyika, although smectite is not reported from there (Tiercelin *et al.*, 1988; Gasse *et al.*, 1989).

The Intermediate Detrital Formation is a distinctive detrital episode, which lasted about 500 years. During this episode, all mineralogical, chemical, and sedimentological parameters of northern Lake Tanganyika sediments sharply deviated from their general trends, as a result of an overflow of Lake Kivu. At this time (4800–5300 yr BP), lake level was probably similar to modern level, and certainly not lower than −135 m. This level was high enough to allow the development of a thermocline and of subsequent stagnation in the southern areas of the lake, in sites that today are moderately deep (440 m versus present level), as indicated by the preservation of a continuous column of diatom laminations (Haberyan & Hecky, 1987).

The Upper Detrital Formation corresponds to a very recent dryer climatic episode, followed by a return to more humid conditions.

The distribution of detrital sediment on the bottom of Lake Tanganyika is controlled largely by morphological factors. Sediments from the lateral drainage systems are controlled entirely by rivers and channel courses inherited from the original block faulting. The most important of these occur in the deepest part of the lake, thanks to gravity flows, including mass flow, turbidity currents and probably grain flow. They reach the deep longitudinal axis, which has played an important role in the recent sedimentary record. Starting from the Rusizi Valley, density currents, including high concentration turbidity currents, reached the deepest parts of the northern half of Lake Tanganyika, 1250 m below present sea level and at least 170 km downstream. The coarsest fractions were restricted to the deep axis but the finer ones were distributed over the entire bottom of the lake. With time, a series of such sedimentary events could result in the widespread distribution of similar units whose principal components have an economic potential; extensive impermeable clays could seal both organic oil-source material and lenticular reservoir sand bodies.

ACKNOWLEDGEMENTS

This study is a part of the 'Georift' project launched by Elf Aquitaine. The author is grateful to SNEA(P) for allowing publication. This research is based on field work with J. le Fournier and D. Masson. Selected samples were analysed for clay and coarse minerals by X-ray diffraction (D. Masson in Orsay University and J.P. Séverac in Boussens at the Elf Aquitaine laboratory). Chemical analysis of major elements was performed on a Philips scanning electron microscope equipped with an X-ray energy analysis scanner (P. Tremblay in Orsay) and trace elements and rare earths by neutronic activation (Ch. Bonnot-Courtois and J.L. Joron, in Saclay). Organic material was analysed by A.Y. Huc and M. Van den Broucke (IFP, Rueil-Malmaison) using the Rockeval system. Radiocarbon dates (standard and Tandetron particle accelerator, when necessary) were performed on the bulk organic material of the muds by J.C. Fontes, in University Paris-Sud-Orsay. The data contributed by these colleagues are gratefully acknowledged. The author is greatly indebted to P. Anadón, B.H. Purser, J.C. Plaziat and J.-J. Tiercelin for helpful suggestions that have improved this manuscript.

REFERENCES

ANADÓN, P., CABRERA, LL. & JULIÁ, R. (1988) Anoxic–oxic cyclical lacustrine sedimentation in the Miocene Rubielos de Mora Basin. In: *Lacustrine Petroleum Source Rocks* (Eds Fleet, A.J., Kelts, K. & Talbot, M.R.), Geol. Soc. London Spec. Publ. **40**, 353–367.

BEAUCHAMP, R.S.A. (1939) Hydrology of Lake Tanganyika. *Int. Rev. Hydrobiol.* **39**, 316–353.

BURGESS, C.F., ROSENDAHL, B.R., SANDER, C.A., LAMBIASE, J., DERKSEN, S. & MEADER, N. (1988) The structural and stratigraphic evolution of Lake Tanganyika: a case study of continental rifting. In: *Triassic–Jurassic Rifting* (Ed. Manspeizer, W.), Dev. Geotectonics **22**, 859–881.

CAPART, A. (1949) Sondages et carte bathymétrique. *Exploration Hydrobiologique du Lac Tanganyika*, vol. II, fasc. II, pp. 1–16, Institut Royal de Sciences Naturelles de Belgique, Brussels.

CASANOVA, J. & THOUIN, C. (1990) Biosédimentologie des stromatolites holocènes du Lac Tanganyika (Burundi). Implications hydrologiques. *Bull. Soc. Geol. Fr.* **(8)6**, 647–655.

CHAMLEY, H. (1989) *Clay Sedimentology*. Springer Verlag, Berlin, 623 pp.

CHOROVICZ, J., LE FOURNIER, J., LE MUT, C., RICHERT, J.P., SPY-ANDERSON, F.L. & TIERCELIN, J.-J. (1983) Observation par télédétection et au sol de mouvements décrochants NW–SE dextres dans le secteur transformant Tanganyika–Rukwa–Malawi du rift est-africain. *C. R. Acad. Sci. Paris (II)*, **296**, 997–1002.

COHEN, A.S. (1988) Facies relationships and sedimentation in large rift lakes and implications for hydrocarbon ex-

F. Baltzer

ploration: examples from lakes Turkana and Tanganyika. *Palaeogeogr. Palaeoclimatol. Palaeoecol.* **70**, 65−80.

COHEN, A.S., FERGUSON, D.S. GRAM, P.H., HUBLER, S.L. & SIMS, K.W. (1986) The distribution of coarse grained sediments in modern Lake Turkana, Kenya: implications for clastic sedimentation models of rift lakes. In: *Sedimentation in the African Rifts* (Eds Frostick, L.E., Renaut, R.W., Reid, I. & Tiercelin, J.-J.), Geol. Soc. London Spec. Publ. **25**, 127−139.

CRAIG, H. (1974) Lake Tanganyika geochemical and hydrographic study: 1973 expedition. *Scripps Inst. Oceanogr. Ref. Ser.* **75**, 5.

DEGENS, E.T. & KULBICKI, G. (1973) Hydrothermal origin of metals in some East African rift lakes. *Miner. Deposita* **8**, 388−404.

DEGENS, E.T., VON HERZEN, R.P. & WONG, H.K. (1971) Lake Tanganyika: water chemistry, sediments, geological structure. *Naturwissenschaften* **58**, 229−241.

DEGENS, E.T., VON HERZEN, R.P., WONG, H.K., DEUSER, W.G. & JANNASCH, H.W. (1973) Lake Kivu: structure, chemistry and biology of an East African rift lake. *Geol. Rundsch.* **62**, 245−277.

DEUSER, W., DEGENS, E.T., HARVEY, G.R. & RUBIN, M. (1973) Methane in Lake Kivu: new data bearing on its origin. *Science* **181**, 51−54.

DEVROEY, E. (1949) A propos de la stabilisation des eaux du Lac Tanganyika. *Inst. R. Coll. Belge, Sci. Tech., Mem.* **5-3**, 1−135.

FROSTICK, L.E. & REID, I. (1987) Tectonic control of desert sediments in rift basins: ancient and modern. In: *Desert Sediments: Ancient and Modern* (Eds Frostick, L.E. & Reid, I.). Geol. Soc. London Spec. Publ. **35**, 53−68.

GARRELS, R.M. & CHRIST, C.L. (1965) *Solutions, Minerals and Equilibria*. Harper & Row, New York, 450 pp.

GASSE, F., LEDEE, V., MASSAULT, M. & FONTES, J.Ch. (1989) Water level fluctuations of Lake Tanganyika in phase with oceanic changes during the last glaciation and deglaciation. *Nature* **342**, 57−59.

HABERYAN, K.A. & HECKY, R.E., (1987) The Late Pleistocene and Holocene stratigraphy and paleolimnology of Lakes Kivu and Tanganyika. *Palaeogeogr. Palaeoclimatol. Palaeoecol.* **61**, 169−197.

HECKY, R.E. & DEGENS, E.T. (1973) Late Pleistocene−Holocene chemical stratigraphy and paleolimnology of the Rift Valley lakes of Central Africa. Unpublished report, *Woods Hole Oceanogr. Inst. Tech. Rep.*, WHOI 73−28.

HUC, A.Y., LE FOURNIER, J., VANDENBROUCKE, M., BESSEREAU, G., BERNON, M., DA SILVA, M. & FABRE, M. (1989) *Northern Lake Tanganyika: a Conceptual Model of a Rift Lake*. Rep. Institut Français du Pétrole 37253, Rueil Malmaison, 32 pp.

KAMPUNZU, A.B., VELLUTINI, P.J., CARON, J.P., LUBALA, R.T., KANIKA, M. & RUMVEGERI, B.T. (1983). Le volcanisme et l'évolution structurale du Sud Kivu (Zaire). In: *Rifts et Fossés anciens* (Eds Popoff, M. & Tiercelin, J.-J.), Bull. Cent. Rech. Explor. Prod. Elf-Aquitaine **7**, 257−271.

KANIKA, M., KAMPUNZU, A.L., CARON, J.P. & VELLUTINI, P.J. (1981). Données nouvelles sur le volcanisme de la Haute Ruzizi (Sud Kivu, Zaïre). *C. R. Acad. Sci. Paris (II)* **292**, 1277−1282.

KELTS, K. & HSÜ, K.J. (1978) Freshwater carbonate sedimentation. In: *Lakes: Chemistry, Geology, Physics* (Ed. Lerman, A.), pp. 295−323. Springer Verlag, Berlin.

KELTS, K. & HSÜ, K.J. (1980) Resedimented facies of 1875 Horgen slumps in Lake Zurich and a process model of longitudinal transport of turbidity currents. *Eclogae Geol. Helv.* **73**, 271−281.

LE FOURNIER, J., CHOROWICZ, J., THOUIN, C., BALTZER, F., CHENET, P.Y., HENRIET, J.P., MASSON, D., MONDEGUER, A., ROSENDAHL, B., SPY-ANDERSON, F.L. & TIERCELIN, J.-J. (1985) Le bassin du Lac Tanganyika: évolution tectonique et sédimentaire. *C. R. Acad. Sci. Paris (II)* **301**, 1053−1058.

LE FOURNIER, J., KERYSAOUEN, G., MASSON, D. & MONDEGUER, A. (1986) *Georift, Evolution structurale et sédimentaire du Rift Est Africain*. Rapport interne Elf-Aquitaine, DRAG Boussens, 93 pp.

LIVINGSTONE, D.A. (1965) Sedimentation and the history of water-level change in Lake Tanganyika, *Limnol. Oceanogr.* **10**, 607−610.

MONDEGUER, A. (1984) *Un bassin sédimentaire en contexte extensif et décrochant dans le rift est africain: la Baie de Burton (fossé Nord du Lac Tanganyika), approche sédimentologique et structurale*. Mémoire de Diplome d'Etudes Approfondies, 95 pp.

MONDEIGUER, A., TIERCELIN, J.-J., HOFFERT, M., LARQUE, P., LE FOURNIER, J. & TUCHOLKA, P. (1986) Sédimentation actuelle et Récente dans un Petit Bassin en Contexte extensif et décrochant, la Baie de Burton, Fossé Nord Tanganyika, Rift Est Africain. *Bull. Cent. Rech. Exp. Prod. Elf Aquitaine* **10**, 229−247.

MONDEGUER, A., RAVENNE, C., MASSE, P. & TIERCELIN, J.-J. (1989) Sedimentary basins in an extension and strike-slip background: the 'South Tanganyika troughs complex', East African Rift. *Bull. Soc. Géol. Fr.* **(8)5**, 501−522.

NELSON, C.H. & MALDONADO, A. (1988) Factors controlling depositional patterns of Ebro turbidite systems, Mediterranean Sea. *Bull., Am. Assoc. Petrol. Geol.* **72**, 698−716.

PAQUET, H. (1970) Evolution géochimique des minéraux argileux dans les altérations et les sols des climats méditerranéens et tropicaux à saisons alternées. *Mém. Serv. Carte Geol. Alsace Lorraine* **30**, 1−212.

PEDRO, G. (1968) Distribution des principaux types d'altérations chimiques à la surface du globe. *Rev. Geogr. Phys. Geol. Dyn.* **10**, 457−470.

PEGUY, Ch.P. (1961) *Précis de Climatologie*. Masson et Cie, Paris, 348 pp.

PHARO, C.H. & CARMACK, E.C. (1979) Sedimentation processes in a short residence-time lake, Kamloops Lake, British Columbia. *Sedimentology* **26**, 523−541.

REYNOLDS, D.J. & ROSENDAHL, B.R. (1984) Tectonic expressions of continental rifting. *Trans. Am. Geophys. Union* **65**, 1116.

ROSENDAHL, B.R., REYNOLDS, D.J., LORBER, P.M., BURGESS, C.F., McGILL, J., SCOTT, D., LAMBIASE, J.J. & DERKSEN, S.J. (1986) Structural expressions of rifting. Lessons from Lake Tanganyika, Africa. In: *Sedimentation in the African Rifts* (Eds Frostick, L.E., Renaut, R.W., Reid, I. & Tiercelin, J.-J.), Geol. Soc. London Spec. Publ. **25**, 127−139.

ROSENDAHL, B.R. (1987) Architecture of continental rifts

with special reference to East Africa. *Ann. Rev. Earth Planet. Sci.* **15**, 445–503.

SCHOLZ, C.A. & ROSENDAHL, B.R. (1988) Low lake stands in Lakes Malawi and Tanganyika, East Africa, delineated with multifold seismic data. *Science* **240**, 1645–1648.

SERRUYA, C. & POLLINGER, U. (1983) *Lakes of the Warm Belt.* Cambridge University Press, Cambridge, 570 pp.

STOFFERS, P. (1975) *Sedimentologische, geochemische und paläoklimatische Untersuchungen an ostafrikanischen Riftseen.* Habilitationsschrift Universität Heidelberg, 117 pp.

STOFFERS, P. & HECKY, R.E. (1978) Late Pleistocene–Holocene evolution of the Kivu–Tanganyika Basin. In: *Modern and Ancient Lake Sediments* (Eds Matter, A. & Tucker, M.E.), Spec. Publ. Int. Assoc. Sediment. **2**, 43–55.

STURM, M. & MATTER, A. (1978) Turbidites and varves in Lake Brienz (Switzerland): deposition of clastic detritus by density currents. In: *Modern and Ancient Lake Sediments* (Eds Matter, A. & Tucker, M.E.), Spec. Publ. Int. Assoc. Sediment. **2**, 147–168.

TALBOT, M.R., LIVINGSTONE, D.A., PALMER, P.G., MALEY, J., MELACK, J.M., DELIBRIAS, G. & GULLIKSEN, S. (1984) Preliminary results from sediment cores from Lake Bosumtwi, Ghana. In: *Palaeoecology of Africa and the Surrounding Islands* (Eds Coetzee, J.A. & van Zinderen Bakker, E.M.), pp. 173–192. Balkema, Rotterdam.

TIERCELIN, J.-J., MONDEGUER, A., GASSE, F., HILLAIRE-MARCEL, CL., HOFFERT, M., LARQUE, PH., LEDEE, V., MARESTANG, P., RAVENNE, C., RAYNAUD, J.F., THOUVENY, N., VINCENS, A. & WILLIAMSON, D. (1988) 25 000 ans d'histoire hydrologique et sédimentaire du lac Tanganyika, Rift Est Africain. *C. R. Acad. Sci. Paris (II)* **307**, 1375–1382.

TIERCELIN, J.-J., SCHOLZ, C.A., MONDEGUER, A., ROSENDAHL, B.R. & RAVENNE, C. (1989a) Discontinuités sismiques et sédimentaires dans la série de remplissage du fossé du Tanganyika, Rift Est-africain. *C. R. Acad. Sci. Paris (II)* **309**, 1599–1606.

TIERCELIN, J.-J., THOUIN, C., KALALA, T. & MONDEGUER, A. (1989b) Discovery of sublacustrine hydrothermal activity and associated massive sulfides and hydrocarbons in the north Tanganyika trough, East African Rift. *Geology* **17**, 1053–1056.

VINCENS, A. (1989) Paleoenvironnements du Bassin Nord Tanganyika (Zaire, Burundi, Tanzanie) au cours des 13 derniers mille ans: apport de la Palynologie. *Rev. Palaeobot. Palynol.* **61**, 69–88.

YURETICH, R.F. (1979) Modern sediments and sedimentary processes in Lake Rudolf (Lake Turkana) eastern Rift Valley, Kenya. *Sedimentology* **26**, 313–331.

YURETICH, R.F. (1986) Controls on the composition of modern sediments, Lake Turkana, Kenya. In: *Sedimentation in the African Rifts* (Eds Frostick, L.E., Renaut, R.W., Reid, I. & Tiercelin, J.-J.), Geol. Soc. London Spec. Publ. **25**, 141–152.

Spec. Publs Int. Ass. Sediment. (1991) **13**, 175−195

Shore-zone sedimentation and facies in a closed rift lake: the Holocene beach deposits of Lake Bogoria, Kenya

R.W. RENAUT* *and* R.B. OWEN[†]

**Department of Geological Sciences, University of Saskatchewan, Saskatoon, Saskatchewan, S7N OWO, Canada*
[†]Department of Geography and Earth Science, Chancellor College, University of Malawi, P.O. Box 280, Zomba, Malawi

ABSTRACT

Small beach bars, spits, and barriers, composed predominantly of coarse sand, granules, and pebbles, are present along much of the modern shoreline of Lake Bogoria, a perennial saline, alkaline lake in the Kenya Rift Valley. Study of their distribution and composition indicates that most of the sediments are derived from peripheral fan-deltas, including material brought down in flood, and that derived by erosion of older exposed fan-delta sediments. Much of the sediment is redistributed by longshore currents, induced by winds funnelled along the axis of the lake.

Surrounding the lake, a series of regressive littoral terraces, composed of angular gravels, record shoreline sedimentation associated with former higher lake levels during the Holocene. During terrace formation, many fan-delta platforms were drowned and shoaling effects reduced, thereby increasing wave energy around much of the shoreline. The terraces record a complex history of Holocene lake-level fluctuations.

Although small saline lakes are unfavourable locations for beach development, a combination of factors, including the elongate form, significant fluvial inflow, and steep margins to supply coarse debris to the shore zone, permit a narrow zone of littoral clastic accumulation.

In terms of resources, shoreline clastic sediments in rifts can host metalliferous placer deposits, and can be sources of lime and aggregate. Their potential as hydrocarbon reservoirs increases with lake size and the quartz content of the catchment rocks. Littoral sands in volcanic-rich catchments are unfavourable owing to their high reactivity during diagenesis.

INTRODUCTION

Although beach sediments are usually only minor facies of rift basins, their recognition in the geological record can contribute important palaeoenvironmental information. Beach deposits provide conclusive evidence for the location of palaeoshorelines. In conjunction with other data, they can help to determine whether rift lakes have been hydrologically open or closed systems, and sometimes may be used to infer former lake water depths. Economically, littoral clastic sediments of large rift lakes are a potential reservoir for hydrocarbons. In large rift lakes and during expanded phases of smaller, closed rift lakes, they also may host placer deposits of heavy minerals.

Littoral clastic sediments surrounding modern African rift lakes have long been used to infer late Quaternary palaeoshorelines (e.g. Gregory, 1921; Leakey, 1931; Nilsson, 1931; Butzer *et al.*, 1972; Grove *et al.*, 1975; Kamau, 1977; Owen *et al.*, 1982), but the beach sediments themselves generally have received little attention. Studies of littoral sediments have been undertaken in a few large rift lakes, notably the Dead Sea (e.g. Bowman, 1971) and Lake Turkana (e.g. Butzer, 1980; Owen, 1981; Reid & Frostick, 1985). However, those of the smaller, closed saline lakes that are common in the dry Kenya and Ethiopian rifts have been neglected. This paper describes the Holocene littoral clastic sediments of Lake Bogoria, a small (34 km^2) saline, alkaline lake in the central Kenya Rift.

In the large African rift lakes (e.g. Lakes Tanganyika, Malawi, and Turkana; Fig. 1A), the relatively high energy levels in the littoral zone can lead to beach development wherever the sediment supply is adequate and the slope of the shore zone is not too steep. In contrast, the smaller closed lakes, many of which are strongly saline, commonly lack beaches and are bordered by rooted littoral marsh or saline mudflats. Unless coarse detritus is introduced by ephemeral streams, sheet-floods, or debris flows, little sediment is available for beach development. Low wave energy restricts littoral erosion and the sorting of coarse particles. Furthermore, shorelines are inherently unstable, advancing and retreating in response to lake level fluctuations. Consequently, littoral clastic sediments have rarely been described from modern and ancient saline lakes, including those in rifts (Hardie *et al.*, 1978; Allen & Collinson, 1986).

Lake Bogoria is a shallow, closed, perennial saline lake with well-developed beach deposits. There are two main clastic littoral facies. Surrounding the shoreline are flights of lake-shore terraces composed of *coarse angular gravels and angular ortho-*

conglomerates that record higher lake levels of late Pleistocene and Holocene age. Modern beaches and abandoned beach ridges on fan-deltas are discontinuous, narrow, linear accumulations mostly of poorly sorted *coarse sands, granules and pebbles, and pebbly sandstones.*

The aims of this paper are (i) to describe the morphological and sedimentological characteristics of the modern beaches and older littoral terraces, (ii) to address problems of their preservation and recognition in the geological record as distinct facies, and (iii) to comment upon the resource potential of beaches in rift lakes.

GEOLOGICAL AND ENVIRONMENTAL SETTING

Lake Bogoria lies within an asymmetric half-graben in the Kenya Rift Valley, 20 km north of the Equator (Fig. 1B). It is confined to the east and south by the 700 m high, N–S to NE–SW trending, Bogoria–Emsos escarpments, and on the more subdued rift floor to the west by N–S trending grid-faulted lavas (Fig. 2). The catchment rocks consist predominantly of basalts, trachytes, trachyphonolites, and phono-

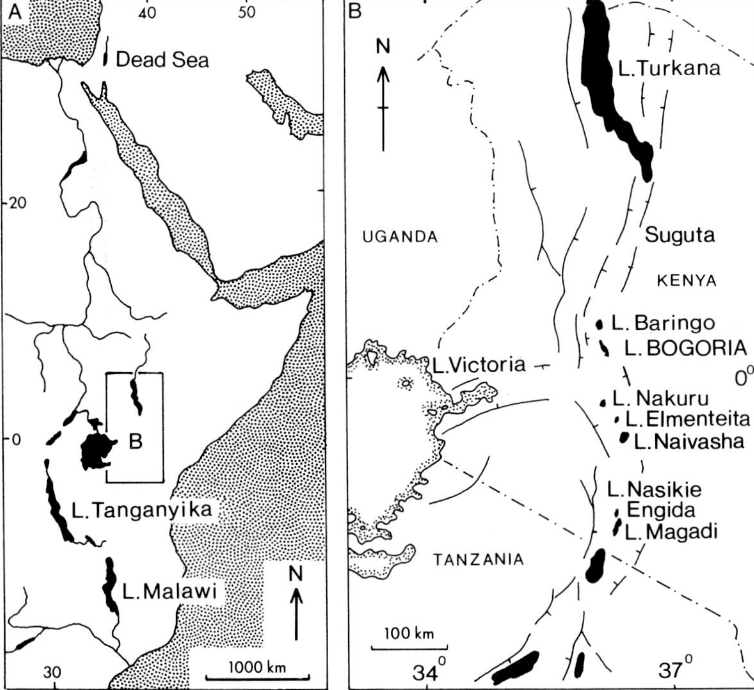

Fig. 1. (A) Location of the large rift lakes mentioned in the text. (B) The Kenya Rift Valley, showing the location of Lake Bogoria.

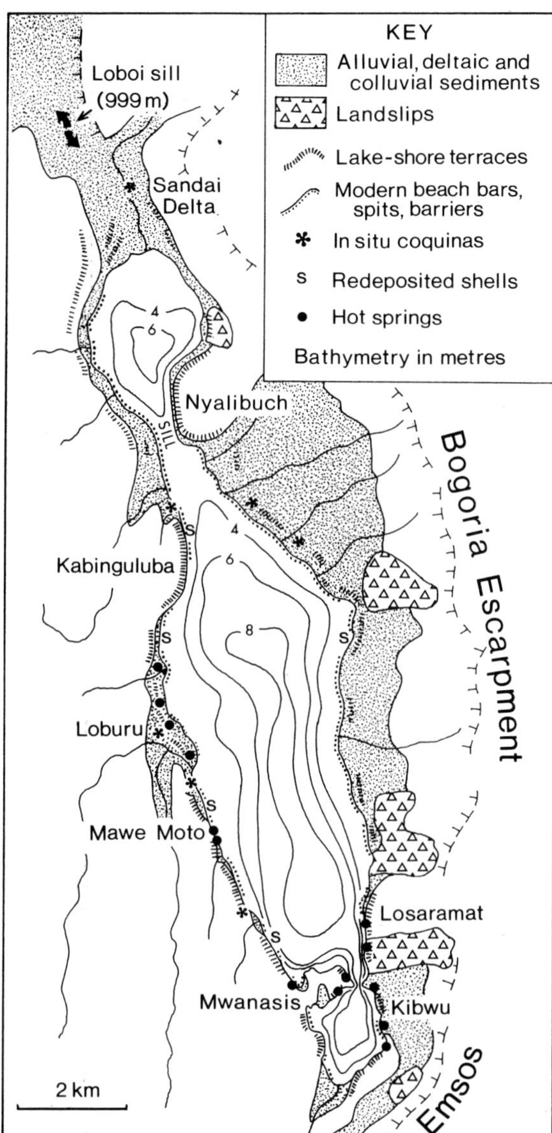

Fig. 2. Lake Bogoria, showing the distribution of the littoral sediments.

lites of Miocene to Pleistocene age (McCall, 1967; Griffiths, 1977; Williams & Chapman, 1986).

The elongate lake has three sub-basins, separated by two shallow, tectonically controlled sills (Nyalibuch and Mwanasis, Fig. 2). The maximum and mean depths in 1978 were 11.5 m (in the central sub-basin) and 5.5 m, respectively. The maximum length is 17.6 km; the width ranges from 0.5 to 3.6 km. The lake waters have no surface outlet and there is no evidence for significant subsurface drainage. Lake level lies at c. 990 m (August 1977) and fluctuates seasonally about 50–100 cm, but periodically undergoes short-term changes of several metres (Tiercelin *et al.*, 1980). Lake level has fluctuated between c. 988 and c. 996 m during this century (Young & Renaut, 1979; Renaut *et al.*, 1987). The lowest point along the watershed lies at 999 m near Loboi Village ('Loboi Sill' in Fig. 2). Whenever the lake reaches this level its waters will overflow northward into the Lake Baringo basin.

The lake is fed by direct precipitation, ephemeral streams, and about 200 hot springs, most of which are located along the shoreline. Sublacustrine hydrothermal discharges are also present (Renaut *et al.*, 1987; Renaut & Owen, 1988). Mean annual precipitation is 600–900 mm/yr, which compares with an evaporation rate of c. 2600 mm/yr (East African Meteorological Department, 1975). The seasonal Sandai-Waseges River, which enters the lake from the north (Fig. 2), is the principal affluent stream. The lake waters are meromictic, saline (up to 90 g/l total dissolved solids) and alkaline (pH range: 9.2–10.3), with Na^+, CO_3^{2-}, HCO_3^-, and Cl^- the principal ions.

Sedimentation within the lake consists of two main zones: a peripheral zone of bedded sands, silts, and clays derived from the Sandai delta and marginal fan-deltas, and an inner zone where alternating sapropelic muds and evaporites (principally trona and nahcolite) have formed during the Holocene (Tiercelin *et al.*, 1981,1982; Renaut *et al.*, 1986; Tiercelin & Vincens, 1987). Details of the modern depositional environments are given in Tiercelin *et al.* (1980), Tiercelin (1981), Renaut (1982), and Tiercelin & Renaut (1987).

MODERN BEACH SEDIMENTS

Littoral environments of Lake Bogoria

Donovan (1975) recognized two types of shoreline in lake basins. A 'coincident margin' occurs where the lake perimeter coincides with the basin margin; a 'non-coincident margin' is present where the lake is fringed by alluvial sediments. Both types of shoreline are present at Lake Bogoria.

Coincident margins dominate the western margin of the lake, where linear headlands, formed by lakeward-dipping fault-blocks of lava, extend to the lake shore with only a narrow (0–15 m), intervening, thin veneer of sediments. Similar coincident margins are found along sectors of the southern and north-eastern shores where, in places, they are formed by major landslips (Fig. 2).

Non-coincident margins dominate the eastern shoreline. Below the Bogoria Escarpment, the shore zone is a broad dissected plain up to 1 km wide, representing the subaerial toe-zone of a group of coalescent fan-deltas (Fig. 2). Several small fan-deltas (< 0.5 km^2), fed by ephemeral streams, also punctuate the western shoreline. The larger Sandai delta forms the northern lake shore. These gently sloping ($< 4°$) delta plains are composed of inter-bedded gravels, sands, and silts, most of which were deposited subaqueously during higher Holocene lake levels (Renaut, 1982; in preparation). Patches of rooted littoral marsh (predominantly *Cyperus laevigatus*) are present locally along the lake shore of the fan-deltas.

Morphology of the modern beaches

Modern beaches, variously composed of coarse sands, granules, and small pebbles, are present along c. 50% of the shoreline. They occur in three forms — fringing beach bars (swash bars) attached to the shoreline, spits, and narrow offshore barriers.

The *beach-bars* (Figs 3, 4) are linear, asymmetric mounds up to several hundred metres in length and from < 5 cm to 4 m in width. They occur discontinuously along the margins of most fan-deltas and,

locally, at the base of interdeltaic volcanic headlands. Along delta margins they usually rest upon partially lithified, lacustrine (delta front) sandy silts or clayey silts. At the base of volcanic headlands, the substrate may be bedrock, older gravels, or weakly lithified silts and sands. Many bars along the western shore terminate and/or diminish in width, a short distance (50–400 m) south of the fan-deltas.

The bars range from narrow (< 5 cm) strandlines of granules and small pebbles through to well-developed mounds up to 60 cm high, with a narrow, steep beach face and a broad backshore berm (Fig. 4). The beachface is typically stepped in profile, with numerous swash marks, and is commonly littered with plant detritus. The berm crest normally lies lakeward of the bar centre. Berms dip landward at $2–10°$. Some bars are backed by small (1–5 m wide by 10 to > 500 m long), marshy, ephemeral alkaline lagoons. The lagoonal sediments are typically mud-cracked laminated silts and clays < 30 cm thick with abundant efflorescent salts and root marks. Small, lobate washover fans up to 2 m wide and composed of distally fining, sand, fine gravels, and granules, are present locally at deltaic sites (e.g. north Kabinguluba, Fig. 2).

The *spits* differ little from the bars in morphology, dimensions, and composition. They are most common along delta margins, in places extending across ephemeral distributary channels, thereby enclosing alkaline lagoons (Fig. 5). Along the western shore, many ephemeral stream channels have been diverted southward by spit development.

Fig. 3. Small beach bar along the northwest shoreline of the central sub-basin. The highest part of the bar (berm crest) lies < 20 cm from the swash zone. A small lagoon lies to the left of the bar. Note oblique approach of wave-front orthogonals.

Fig. 4. Beach bar (granules), 300 m south of Loburu fan-delta, mid-western shoreline. Note the multiple swash marks, the coarser gravel at the berm crest, and broad backshore berm. Coarse, unsorted gravel litters the shore behind the bar.

Small cuspate and double spits (Zenkovich, 1971) occur along the eastern shore (plate 6.6 in Tiercelin & Renaut, 1987).

Barrier bars, up to 150 m long and 1.4 m wide, occur up to 12 m from the inner shoreline on gently sloping ($<2°$) fan-delta platforms east of the lake (Fig. 6). Small attached and unattached concave barriers also occur within small bayheads between lobes at deltaic sites. Small attached bayhead beaches, generally composed of coarser particles (granules to medium pebbles) than the beach bars, are present between volcanic headlands at Losaramat, Mwanasis, and Kibwu (Fig. 2).

Significantly, siliciclastic beaches are absent along the mudflats of the Sandai delta, where plant debris is normally the only littoral accumulation. Measured cross-sections of several bars are shown in Figs 7, 8; their dimensions are summarized in Table 1.

Texture, fabric, and composition

Figure 8 shows excavated sections through typical beach bars and associated sediments. Most bars are composed of particles between medium sand and pebbles (up to c. 20 mm), with coarse sand and granules the predominant fractions (Fig. 9A). Large,

Fig. 5. Small beach bar (looped barrier) that has enclosed an alkaline lagoon. The lagoon is the distributary channel of an ephemeral delta. Note abundant plant debris and salt efflorescence in foreground. Northwest shoreline, central sub-basin.

Fig. 6. Small attached barrier enclosing very shallow lagoon (15 cm), Nanyonchengare fan-delta, southeastern shoreline, central sub-basin.

randomly distributed gravel clasts (cobbles) are commonly present in the wave-sorted granules and sands. Particle size varies normal to the shoreline, and is usually coarsest at the berm crest, making representative textural parameters difficult to obtain (Fig. 9B).

Sedimentary structures within most small modern bars are poorly developed. In cross-section, the open framework granules and pebbles of the bars and washover fans commonly appear massive (Fig. 8C). Large bars, particularly where sandy, show some internal stratification. Behind the berm crest,

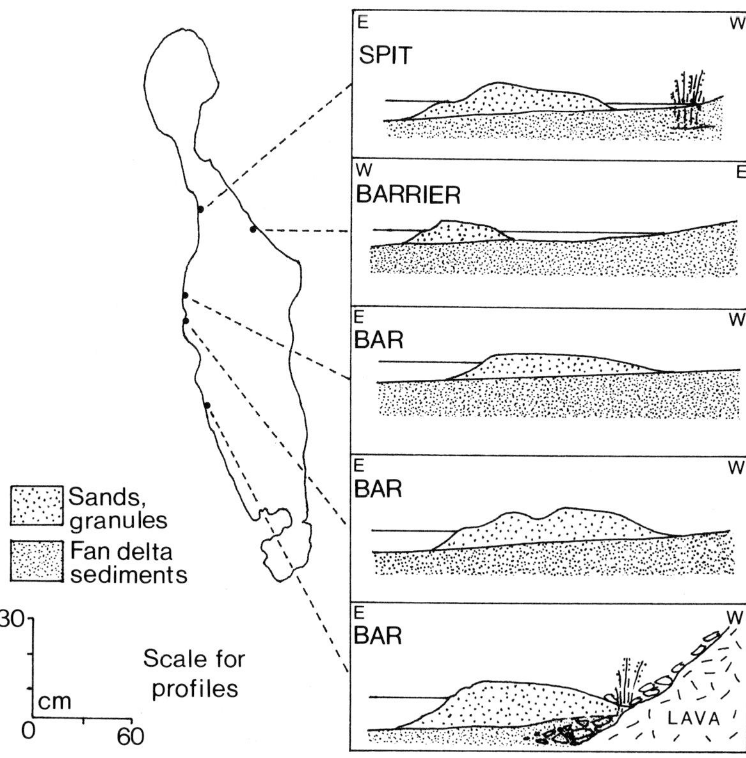

Fig. 7. Levelled profiles across beach bars around Lake Bogoria.

Table 1. Dimensional data (in cm) for modern bars, spits, and barriers around Lake Bogoria. Height refers to the maximum height of the berm crest above the substrate

		Mean	Standard deviation	Minimum	Maximum
Width	(W)[a]	36.3	27.2	5	312
	(E)	39.4	24.6	5	172
Height	(W)	13.6	3.9	1	25
	(E)	8.9	4.8	1	17
Width/height	(W)	5.2	3.7	2.1	28
ratio	(E)	4.5	2.7	2.8	27

[a] (W) Western shoreline (18 measurements); (E) eastern shoreline (17 measurements).

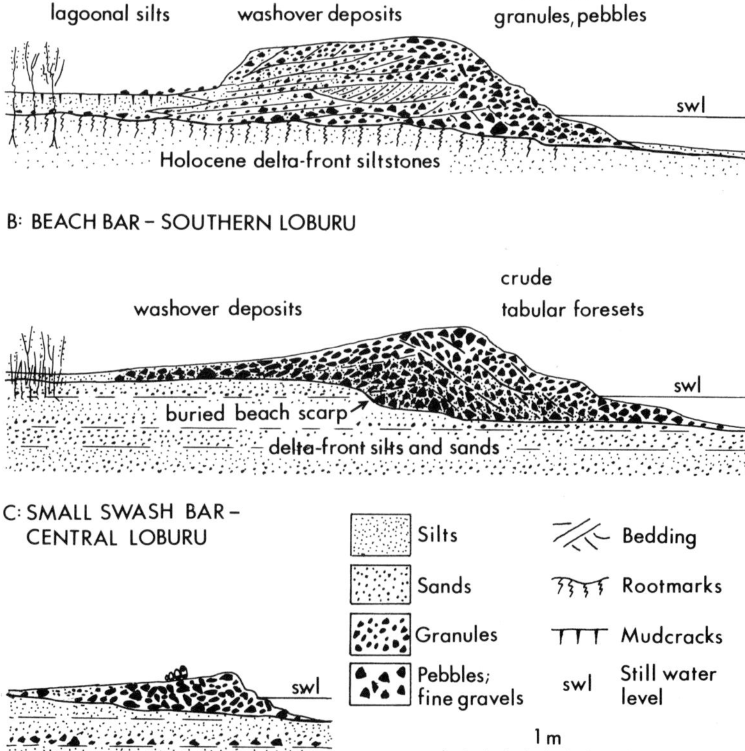

Fig. 8. Excavated cross-sections through beach bars, western shoreline of Lake Bogoria.

a weak subhorizontal stratification, or low angle (2–4°), landward-dipping laminae are revealed. Many laminae are truncated by overlying steeper sets. Small scours are common between sets. On the lakeward margin, coarse sands, granules, and pebbles are either massive or show crude, lakeward-dipping tabular foresets (c. 5–30°; Fig. 8A,B). A poorly sorted, infiltrated matrix of feldspathic silt and sand is common within the back-berm sediments.

Most clasts are volcanic rock fragments. Other particles include alkali feldspar (detrital phenocrysts), detrital limestones (travertine and stro-

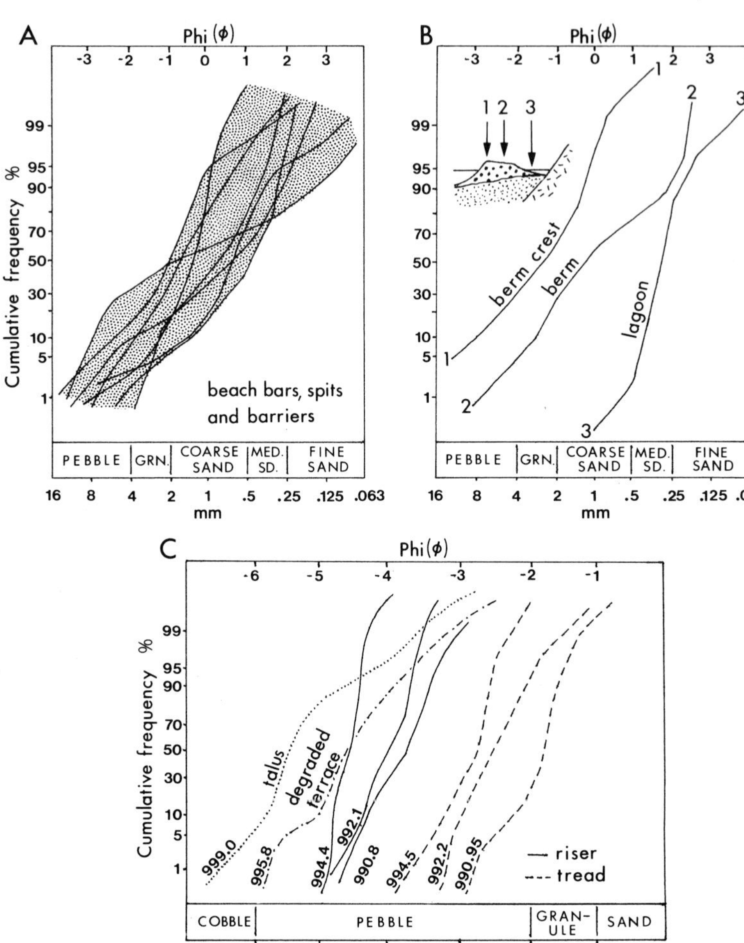

Fig. 9. Cumulative grain-size curves for beach sediments, Lake Bogoria. (A) Modern beach bars, spits, and barriers from various localities; (B) variation within a single bar at Kabinguluba; (C) lake-shore terrace gravels, Kabinguluba. Levelled height of bar (in metres above sea-level) is given.

matolitic), bone debris, reworked gastropod shells, rhizoliths (calcite and fluorite), and obsidian and chert artefacts of late Neolithic affinity. Quartz is a minor component because most lavas in the catchment are quartz-poor. Particles range from subangular to subrounded. Field observations suggest that degree of rounding depends mostly on lithology, particle size, and derivation. For the granule and gravel fractions, phonolite and trachyte clasts are generally more angular than basalt. Rounded grains are most common on beaches at, or near, deltaic sites, suggesting that many clasts have been previously abraded by fluvial processes.

FORMATION OF THE MODERN BEACHES

Modern wave and current activity at Lake Bogoria

Water circulation within Lake Bogoria is poorly known. The dominant winds come from the east and northeast, but are funnelled along the basin axis. They are typically north to south by day, commonly reversing in the early evening. By day, wave fronts are strongly refracted and approach the eastern and western shores obliquely from the north (Fig. 3). Orthogonals are usually parallel at Mwanasis, along the line of maximum fetch (Fig. 2). Considerable refraction probably occurs where waves impinge and shoal upon the shallow (< 2 m) Nyalibuch sill. During convectional storms, an E−W wave motion was often observed. During storms wave heights can reach up to 50–60 cm (J.-J. Tiercelin, pers. comm.), but the waves are of low energy. Wave energy is concentrated along the base of the volcanic headlands but is dissipated by shoaling on the more gently sloping deltas.

Currents and circulation within the lake may be generated by the inertial flow of the Sandai-Waseges River and wind shear. Swells of short duration, accompanying late afternoon wind reversals or a velocity reduction, may be evidence of small seiches, but this has not been confirmed. Aerial photographic evidence, including suspended sediment plumes and organic foam lines, suggests that an anticlockwise circulation may exist in the central and northern sub-basins. Whether this is constant, or changes daily or seasonally is unknown. There is sedimentary evidence for periodic density currents within the lake (Tiercelin *et al.*, 1981,1982) and possible evidence for upwelling along the northern Mwanasis Peninsula (see below).

Beach construction and sediment provenance

The sand and granule beaches are the product of normal swash constructive processes. Waves breaking in the nearshore zone carry fine sediment in suspension while coarse particles roll up the beach face. Gravity, friction, and percolation into the beach face cause the water to stop and a narrow line of sediment (swash mark) is deposited (Komar, 1976; Kirk, 1980; Davies, 1985). Deceleration normally produces decreasing grain size from the base of the beachface to the top, but the largest clasts, moved

during storms, are more common near the berm crest. The internal stratification seen on large beach bars (Fig. 8) may be explained by sediment accretion on the shoreface and periodic washover during periods of slightly elevated lake level or storms (Fig. 10).

The morphology and distribution of the beaches, together with compositional evidence, indicate significant longshore drift of sediment along the axis of the basin. Given the diurnal reversals in wind direction, drift might operate in either direction. However, along the western shore, most evidence supports a predominant north-to-south sediment drift. Many fan-delta distributary channels are diverted southward by spits and barriers. Clasts of limestone precipitated at the Loburu and Mawe Moto hot springs are more common in modern beaches to the south of each hydrothermal site (Fig. 2) than to the north. Molluscs are being eroded and transported southward from Holocene delta sediments (see below).

Evidence from the eastern shore is inconclusive. Cuspate double spits can result from convergence of opposing currents (Zenkovich, 1971), but also can be produced by inundation of attached bars along an irregular protruding shoreline. Given the frequent minor oscillations of lake level in closed lakes, spits and barriers produced by a slight rise in lake level must be very common.

The occurrence of better rounded clasts in beaches at and near fan-deltaic sites, together with the decrease in beach width south of fan-deltas, suggest that most of the modern beach sediment originates at deltaic sites and is transported southwards. The

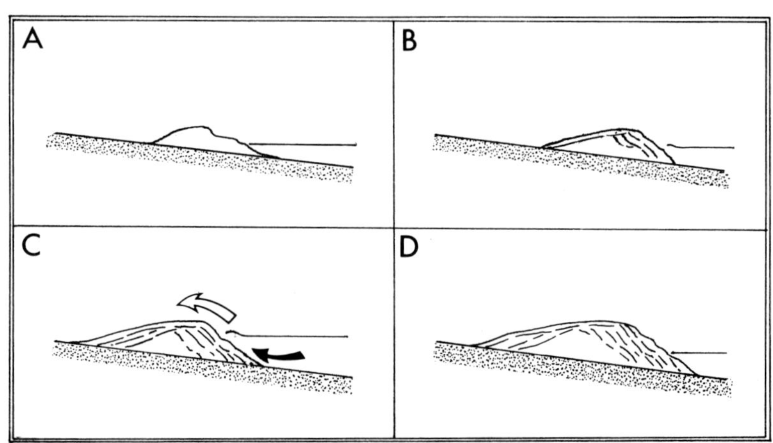

Fig. 10. Simplified model for construction of small beach bar stratification. (A) Accumulation of coarse sediment in swash zone; (B) minor progradation — deposition of tabular, lakeward-dipping accretionary foresets in swash zone; (C) periodic increased wave energy or slight rise in level leads to erosion, spillover, and deposition of landward-dipping laminae; (D) return to original level and foreset accretion. This model assumes adequate sediment supply by longshore drift and lake level more or less at equilibrium.

Fig. 11. Dispersed, lava pebbles on the foreshore of southern Loburu fan-delta. Many pebbles are subrounded and have been eroded from older ephemeral stream deposits. Large block at bottom left is c. 20 cm long.

beach sediments include granules and fine gravels washed into the lake by ephemeral streams in flood, unchannelled flow, and debris flows, together with coarse debris eroded from older, incised, deltaic sediments.

The shorelines of the Sandai delta and northern Mwanasis Peninsula (Fig. 2) are exposed to the greatest fetch, yet neither has significant beaches. Wave energy is dissipated on the gently sloping Sandai delta-front, but more significantly, neither site has a local fluvial source of coarse clastic sediment.

Under the present low-energy conditions, little beach debris can be eroded from the substrate of volcanic headlands. Random coarse gravels within the beaches derive from older lake-shore terraces upslope. At the southern lobe of Loburu fan-delta (Fig. 2), unsorted, but well rounded, coarse gravels mantle the distal delta-plain and shoreline, and are being eroded from old distributary channel bar deposits. They are widely dispersed, demonstrating the inability of modern swash processes to sort coarse debris in the littoral zone (Fig. 11).

Rippled sands

Small, isolated pockets of rippled medium and coarse sands from 2 to 15 cm thick are found in the near-shore (shoreface) zone on many fan-deltas in water from 5 to 60 cm deep (e.g. south Loburu). Most common are asymmetric, transverse sinuous, and transverse catenary ripples (Allen, 1968) parallel to the shoreline. Silts and organic debris commonly occupy ripple troughs.

It is possible that some littoral sediment is derived from shallow offshore sediments. Coring has shown that submerged coarse sediments (e.g. turbidite sands, paraconglomerates) do exist locally offshore from fan-deltas (Tiercelin *et al.*, 1987). However, to bring this coarse sediment into the littoral zone would probably require a fall in lake level, which in turn would further decrease littoral energy (see below). Thus, offshore sediments are unlikely to be a significant source of beach deposits today.

LAKE-SHORE TERRACES AND ABANDONED SHORELINES

Flights of abandoned, lake-shore beach terraces (Fig. 12), composed predominantly of gravels, surround c. 45% of the shoreline, up to c. 9 m above lake level. They are best developed on interdeltaic volcanic headlands on substrates that slope from 5–20°. Equivalent high-level shorelines on fan-deltas are represented by fine gravel beach-terraces and small wave-cut beach-scarps.

The terraces form by wave erosion, sorting of gravels on the foreshore, and removal of fines (Fig. 6 in Bowman, 1971). When these processes occur under a fluctuating, but falling, lake level with intermediate still stands, a series of regressive beach terraces can form (Bowman, 1971; Donovan & Archer, 1975). Each terrace comprises a gently

Fig. 12. Lake-shore terraces, northern Kabinguluba. Most of the terraces, which extend upward almost to the tree line, are partially grassed over. A small granule beach occurs along the modern shoreline. Note the abundant plant debris along the foreshore.

sloping platform or 'tread' and a steeper 'riser'. The lower limit of the tread represents the breaker zone, while that of the riser represents the normal upper limit of swash. Many treads are now partially grassed over (Fig. 12).

Measured tread widths range from 5 to 170 cm, with risers from 8 to 85 cm. Levelled treads at five sites along the western shore have a mean slope of 7°, ranging from 3.2 to 11°. Risers range from 10.2 to 23.6°, with a mean slope of 14.8°. The steepest risers occur on the highest terraces. The thickness of the gravels was often difficult to determine, but is usually < 1.5 m.

Particle size, sorting, and angularity of clasts are generally higher than for the modern beaches (Fig. 13). Granulometric data for a well-preserved terrace sequence are shown in Fig. 9C. The highest terraces have the coarsest clasts. Texturally, the lowest terraces differ little from the modern beaches. Gravels in risers are coarser than those forming the adjacent treads by a ratio of about 1.6:1. Most clasts are angular to subangular. Clast shape is variable, but blades and discs are predominant where the substrate is phonolite or trachyphonolite.

Unlike the modern beaches, most coarse gravels are monomictic and are derived from the underlying

Fig. 13. Lake-shore terrace gravels from a terrace at c. 994 m at Kabinguluba. Note high angularity of clasts (mostly trachyphonolite). These gravels lack stromatolitic coatings and may, therefore, be of late Holocene age. Scale bar: 10 cm.

substrate. They appear to have experienced little lateral transportation. Where exposed, the volcanic substrates are commonly weathered along joints and planes of fissility. Coarse rounded pebbles occur on terraces near deltas, indicating littoral reworking of fluvially derived materials. The greatest compositional diversity occurs in terrace gravels to the south of the small fan-deltas. Near the Loburu and Mawe Moto hot springs (Fig. 2), hydrothermal carbonate clasts are present, mostly on the lower terraces (< 994 m) south of each site, but a few clasts are also found to the north. Other clasts include abraded subfossil mammal bones, carbonate rhizoliths, silicified wood, obsidian artefacts, pottery sherds, molluscs, and stromatolitic limestone.

Most gravels are imbricated, with clasts dipping shoreward. Buried clasts tend to be aligned parallel to the shoreline. Sectioning the terraces, without disturbing the gravels, proved difficult, but confirmed that an infiltrated matrix of sand and silt is normally present at 15–30 cm depth, and that imbrication continues below the surface. In section, most gravels appear massive or show crude lakeward-dipping tabular foresets (Fig. 14). Most gravels are unconsolidated, except where cemented by stromatolitic limestone crusts (see below). Angular beach gravels along the southern shore of Mwanasis (Fig. 2) are cemented by hydrothermal opaline silica (Renaut & Owen, 1988).

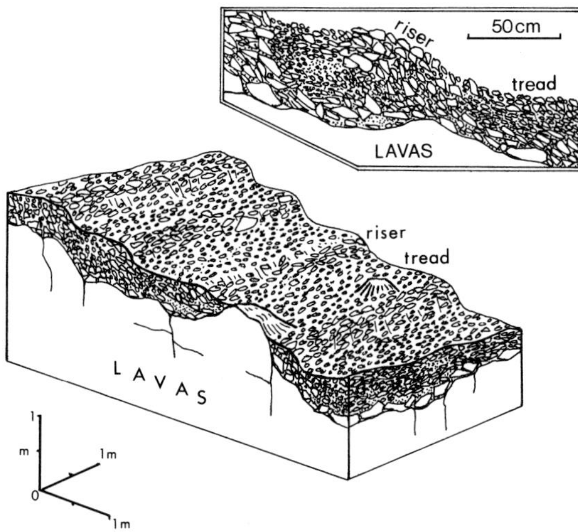

Fig. 14. Cross-section through lake-shore terraces, c. 200 m south of Mawe Moto hot springs. Inset shows detail of structure (drawn from photograph).

Abandoned shorelines on fan-deltas

Fan-deltas exhibit both depositional and erosional palaeoshoreline features. Abandoned beaches are represented by linear and curved bars of sand and fine gravel up to 50 cm thick and 1–5 m wide. Most rest upon broad terraces that represent exposed surfaces of subaqueously deposited, fan-delta sediments. Several of these surfaces are separated by small (< 1 m) wave-cut beach-scarps (Fig. 15).

The abandoned beaches occur in two situations. The thickest and best preserved deposits are found at or near the base of the beach-scarps. Thinner, chenier-like, beach ridges can be recognized across the exposed terraces of Holocene delta sediments. The sediments forming the beaches are finer than those of equivalent terraces at the same level on the volcanic headlands. Most are polymictic and composed of poorly sorted, small, subrounded to subangular pebbles and granules with a sandy matrix. They are texturally similar to, but coarser than, sediments forming the modern beaches. Where well preserved, the sandy sediments locally show planar laminae dipping landward that overlie, and are interbedded with, shoreward-dipping tabular sets (Fig. 15A,B). Units with ripple-cross laminae, lag gravel lenses, and mud-silt flasers are intercalated within the sands. Some deposits are cemented by vadose calcite, forming pebbly sandstones.

Formation of the lake-shore terraces

The lake-shore terraces clearly were formed under conditions of higher littoral energy and turbulence than experienced today. The upward increase in clast size reflects the higher wave and current energy with an expanded, fresher lake. Lake-shore terraces and wave-eroded beach-scarps were levelled at 18 sites around the lake shore using surveying equipment (Fig. 16). On volcanic headlands the lake-shore terrace gravels can be distinguished reliably up to 995 m, above which they merge into bedrock, talus, or colluvium. Correlation of the palaeoshorelines above 995 m is poor, probably reflecting the multiplicity of levels, disturbance by slope processes, and the difficulty of differentiating risers from treads.

Palaeoshorelines can be recognized up to about 998 m, which is close to the level at which the lake would overflow northward (999 m). At its maximum level, the lake would have extended northward, increasing the fetch by at least 1.5 km, and the depth to a maximum of 21 m. Most expansion would have

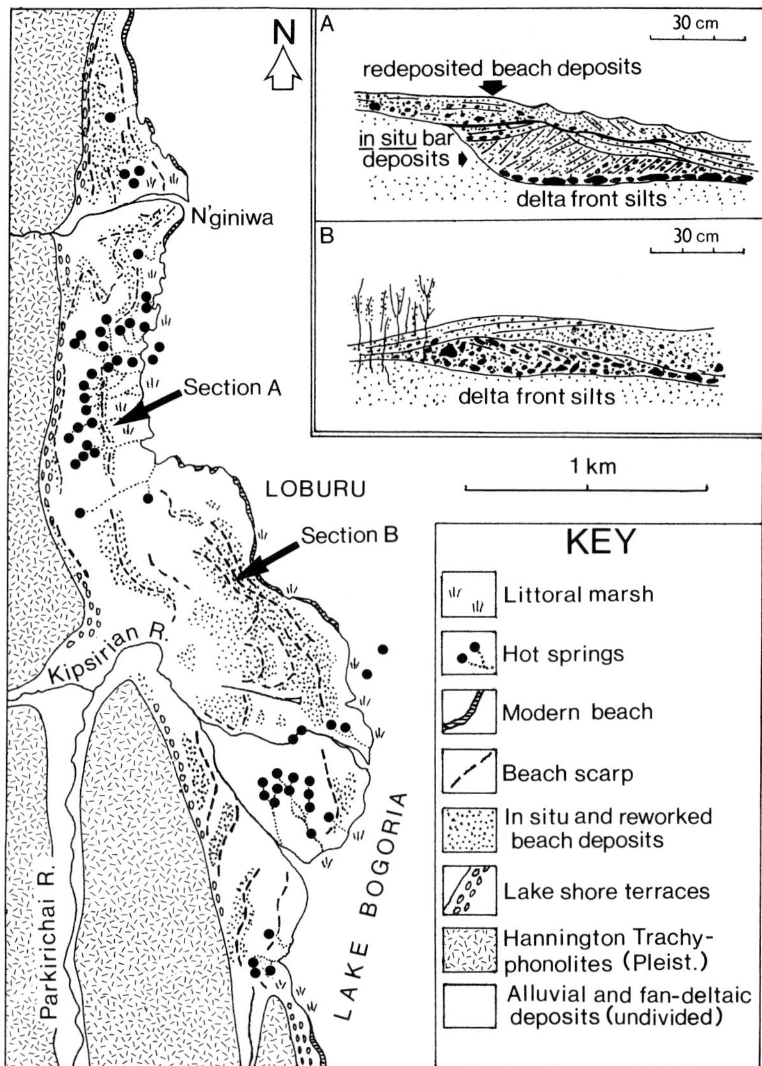

Fig. 15. Loburu fan-delta, showing distribution of beach scarps and beach sediments. Inset shows two sections of abandoned beach deposits.

been along the basin axis, rather than in width. At the maximum level, most of the modern delta plains would have been submerged. Waves could then have lapped up against and sorted gravels on the steep volcanic headlands, without much of the energy dampening effects and shoaling of an intervening sediment apron. Longshore currents may have moved fine gravels along the treads. Therefore, during the periods of higher lake level the proportion of coincident margins increased.

Age of the lake-shore terraces

Terraces can only develop when lake level achieves a temporary equilibrium. The time required to form individual terraces is unknown. The terraces and beach-scarps do not represent temporary stillstands of a single recession, but rather record many oscillations in lake level in which low and intermediate levels, at least, have been reoccupied repeatedly. During reoccupation, littoral gravels may have been reworked, sometimes totally obliterating and/or reconstructing terraces. Dating individual terraces is,

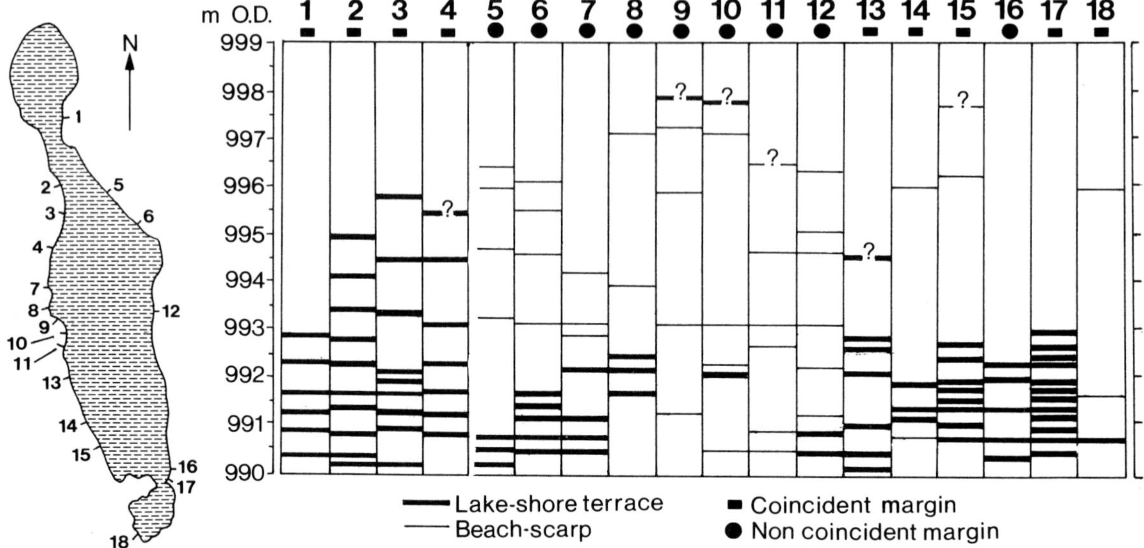

Fig. 16. Correlation chart of levelled lake-shore terraces and beach scarps from selected sites around Lake Bogoria. Levelling was undertaken using surveying instruments.

therefore, a difficult task. The contained molluscs and stromatolites, although datable by radiocarbon method, both present problems of derivation and interpretation.

Evidence from molluscs

Molluscs, principally the gastropod *Melanoides tuberculata*, which accounts for >95% of the fauna, and *Corbicula africana*, are present both on and within the interstices of beach gravels of intermediate level terraces (<994 m) at several sites around the shoreline. Some have been reworked and incorporated locally in the modern beaches. Lensoid, weakly lithified coquinas up to 15 cm thick occur *in situ* in fan-deltaic sediments at several sites around the shoreline at levels between 992 and 994 m (Fig. 2). Most shells in the littoral-terrace gravels lie within 100 m south of the *in situ* coquinas.

It seems probable that, periodically, shell-bearing sediments have been eroded, and the shells have then been transported and redeposited in younger littoral gravels. This process was observed at northern Kabinguluba (Fig. 2), where shells, radiocarbon dated at c. 10 500 yr BP (Young & Renaut, 1979), are being incorporated in the modern granule beach with only minor abrasion (Fig. 17). In an ancient analogous situation, this could lead to misinter-

pretation of the palaeosalinity of the lake at the time when the granule beaches were forming. In a similar manner, vertebrae of catfish (*Clarius* sp.), washed in from the Sandai River, are today being incorporated in modern beach gravels along the northwest shore.

Evidence from stromatolitic limestone crusts

Many beach gravels and shoreline bedrock outcrops are encrusted by grey, stromatolitic limestone crusts, usually from <1 to 5 cm thick. At northern Mwanasis, they are up to c. 30 cm thick, at a possible site of upwelling of nutrient-rich waters from the monimolimnion of the central sub-basin. The stromatolites are present around much of the shoreline from lake level up to c. 999 m. Two main generations of crust have been recognized (Tiercelin, 1981; Renaut, 1982; Casanova, 1986a). The oldest crusts are thin (c. 2 cm), occur up to 999 m, and have been dated at 5290–5200 yr BP. The second generation crusts are thicker (up to 30 cm), occur up to 995 m, and represent a period of relatively stable lake waters at 5040–4750 yr BP. The crusts contain filaments of *Phormidium* sp. Most calcite was precipitated biochemically, rather than trapped and bound (Casanova, 1986a,b; Vincens *et al.*, 1986; Casanova & Renaut, 1987).

Fig. 17. Early Holocene *Melanoides tuberculata* being incorporated into beach gravels, Kabinguluba. Shoreline at right of photograph. Scale bar: 10 cm.

The crusts demonstrate that well-developed littoral gravel terraces were in place prior to 5300 yr BP. For stromatolites to form on high-level terrace gravels implies lower littoral energy during the period of encrustation than when the gravels were being actively sorted into risers and treads. This demonstrates that when lake level is stable for a prolonged period, relatively high current and wave energy is limited to the narrow active swash zone. Submerged terrace gravels remain sufficiently stable for extensive encrustation by cyanobacterial mats.

This evidence shows that lake-shore terraces had formed up to the level of the Loboi Sill (999 m) at some time prior to 5300 yr BP. Lake level was generally high and fluctuating during the early and middle Holocene (Tiercelin *et al.*, 1981,1982; Tiercelin & Vincens, 1987). From their state of preservation, much of the terrace development may date from this period. However, initial development during earlier, late Pleistocene high lake levels cannot be ruled out. Lake level was at least 5 m higher, and possibly up to 8 m higher than today, earlier this century (Johnston, 1902; Young & Renaut, 1979; Renaut *et al.*, 1987). Thus, considerable reworking may have occurred, even during the last 50–100 years.

PRESERVATION AND RECOGNITION OF THE LITTORAL SEDIMENTS

This study has shown that there are two main facies associated with the littoral environments of Lake Bogoria — a *coarse angular gravel and angular orthoconglomerate facies*, associated with coincident margins, and a *coarse-sand–granule–pebble and pebbly sandstone facies*, associated with non-coincident, fan-deltas. The main characteristics that define each facies are summarized in Table 2.

The potential for the preservation and recognition of each group of sediment as a littoral facies in the geological record is dependent on many, often inter-related factors, including lake-level fluctuations (their magnitude, rate, and duration), sediment supply, and wave climate. Figure 18 summarizes the main consequences of a simple rise and fall in lake level on the littoral environments.

Coincident margins: coarse angular gravel and angular orthoconglomerate (sedimentary breccia) facies

The preservation potential of the lake-shore terrace gravels as an interpretable facies depends, to a great extent, on the length of subaerial exposure prior to burial. Today, the gravel terraces are being degraded by slope processes, and disrupted by tree roots and

Table 2. Summary of the sedimentary characteristics of the main littoral facies at Lake Bogoria

Feature	Modern beaches	High-level beaches: fan-deltas (non-coincident)	High-level beaches: littoral terraces (coincident)
Grain size of clasts	Coarse sand to fine gravel (pebble)	Coarse sand to coarse gravel	Granules to boulders
Dominant size fraction	Coarse sand to granules	Granules to pebbles	Coarse gravels
Matrix	Silt, fine sand	Silt to coarse sand	Silt to coarse sand
Sorting	Usually poor, locally moderate	Poor to moderate	Poor to moderate
Clast roundness	Subangular to subrounded	Subangular to subrounded	Predominantly angular
Clast shapes	Variable with lithology	Variable with lithology	Variable, but many discs, blades
Clast composition	Volcanic rock fragments K-feldspar Limestone	Volcanic rock fragments K-feldspar Limestone Mud intraclasts	Volcanic rock fragments Limestone
Geometry of sediment body	Narrow shoestring	Shoestring Irregular sheet	Shoestring Irregular sheet
Bed thickness: average	10 cm	30 cm	50–60 cm?
minimum	1 cm	5 cm	<20 cm
maximum	40 cm	100 cm	>100 cm
Basal contact	Usually non-erosive	Commonly erosive	Erosive or non-erosive
Colour	Dark grey Reddish brown	Dark grey Reddish brown	Dark grey
Bedding types	Massive Subhorizontal laminae Tabular foresets Ripple laminae Reverse grading Wavy bedding	Massive Subhorizontal laminae Tabular foresets Flaser beds Ripple laminae Wavy bedding	Massive Tabular foresets Reverse grading
Sedimentary structures	Small scours	Imbrication Small scours Root marks Rhizoliths	Imbrication
Fossils; other features	Reworked gastropods Derived fish bones Fragmented diatoms Artefacts	Gastropods Derived fish bones Fragmented diatoms Artefacts	Reworked gastropods Stromatolitic crusts Diatoms in matrix? Artefacts

trampling by large mammals. Runoff infiltrates the gravels without disturbing them significantly and is slowly contributing, or adding to an existing, silty sand matrix. The oldest, highest terraces (>997 m) are partially degraded and merge into colluvium. The gravel clasts, notably in the lowest terraces, are undergoing salt weathering. Efflorescent trona ($NaHCO_3.Na_2CO_3.2H_2O$) and thermonatrite ($Na_2CO_3.H_2O$) crystals are common within cracks in the clasts. Thus, prolonged exposure, unless under hyperarid conditions, would probably result in gradual mass wasting of the gravels, obliterating the terrace morphology and much of the depositional fabric.

Rapid rise in lake level, followed by a period of equilibrium, could lead to preservation of the terraces, as demonstrated by the encrusting stromatolites. Once littoral gravels lie below normal wave-base, usually only a short distance offshore (Sly, 1978, p. 73), the potential for preservation increases, although they may still be subjected to longshore currents or slumping associated with earthquakes or increased pore fluid pressures. Fault-induced subsidence and burial by transgressive finer

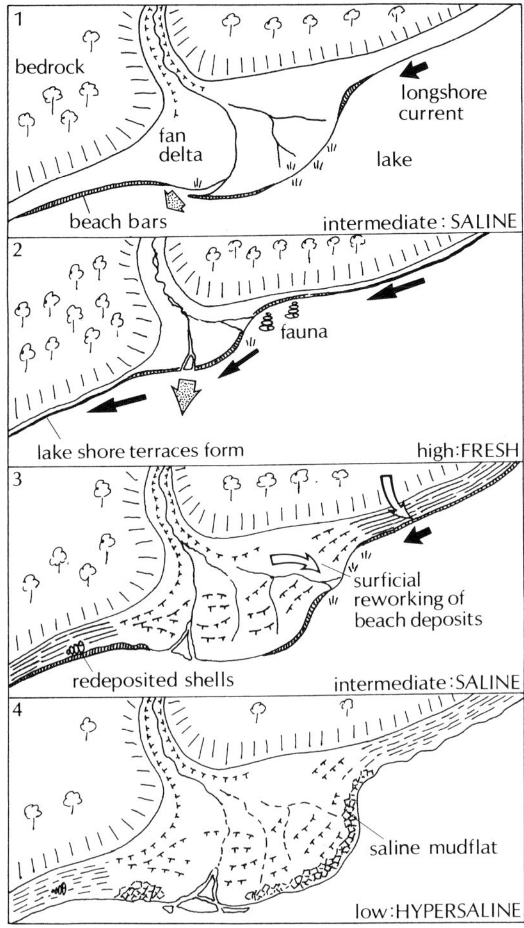

Fig. 18. Effects of lake-level change on littoral sedimentation — a simplified model, assuming climatic control of lake fluctuation. (1) Initial condition, similar to modern Lake Bogoria, with saline waters and intermediate lake level. (2) Rise in lake level and expansion leads to increased wave and current energy and strong longshore currents. Fan-delta platform is submerged. Fresh waters lead to niches for macrofauna. Aggradation in fan-delta feeder channel. High sediment yield results in deltaic sedimentation across platform (may reduce with time, if increased vegetation stabilizes soils). Development of high-level lake-shore terraces. (3) Fall in lake level leads to reduced wave and current energy and exposure of fan-delta plain. Recessional beaches and terraces develop. Incision of valley fill and erosion of emergent fan-delta deposits provide coarse sediment, some of which is reworked into beaches. Runoff on delta-plain leads to redeposition of some abandoned beach sediments. (4) Fall in lake level associated with increased aridity. Wave energy and longshore currents largely ineffective. Lake margins transformed in places to saline mudflats. Salt weathering of littoral gravels. Periodic delivery of coarse sediment to littoral zone by flash floods and debris flows.

lacustrine sediments or even subaqueous evaporites would enhance preservation of the terrace gravels.

Unlike the littoral deposits in large rift lakes, which generally are texturally more mature, with well-preserved structures (Owen, 1981; Reid & Frostick, 1985), recognition of the gravels as beach deposits in the geological record could be difficult. With high angularity and an infiltrated matrix, the resultant angular orthoconglomerates (sedimentary breccias) would not readily denote littoral sedimentation. Where resting on a volcanic substrate, they would resemble a talus or colluvial accumulation. However, as observed by Morrison (1964) and Tucker (1978), beach gravels are commonly highly angular in lakes because of low levels of wave activity.

Where the gravels are finer, fluvially derived, or have undergone prolonged abrasion, subangular to subrounded orthoconglomerates would result. Although they might resemble alluvial gravels, they would probably be better segregated into discrete beds with lenticular bedding.

In both cases, the matrix may provide clues. The matrix associated with lacustrine transgression might be fossiliferous, suggesting a littoral origin. Angular littoral gravels at Mwanasis and Kibwu (Fig. 2) have an infiltrated diatomaceous matrix (Renaut & Owen, 1988). In contrast, the matrix of littoral gravels associated with regression might lack lacustrine fossils, unless formed syndepositionally.

Non-coincident margins: coarse-sand−granule− pebble and pebbly sandstone, fine orthoconglomerate facies

The frequent, but highly variable, lake-level fluctuations experienced in closed-basin lakes, tied both to climate and tectonics, make construction of simple vertical facies sequences in these fan-deltas very difficult. Examination of more than 20 sections in the Loburu and eastern fan-deltas has shown that littoral deposits can be preserved both within regressive and transgressive sequences (Fig. 19).

As with coincident margins, recognition of the littoral deposits depends greatly on the degree of post-depositional modification. With fall in lake level, the deposits are prone to subaerial reworking by sheetwash, erosion by shifting channels, and by bioturbation. The littoral sediments are commonly redistributed as rippled sand sheets of variable thickness (< 5−30 cm) across the fan-delta plains, leaving irregularly strewn pebble lags.

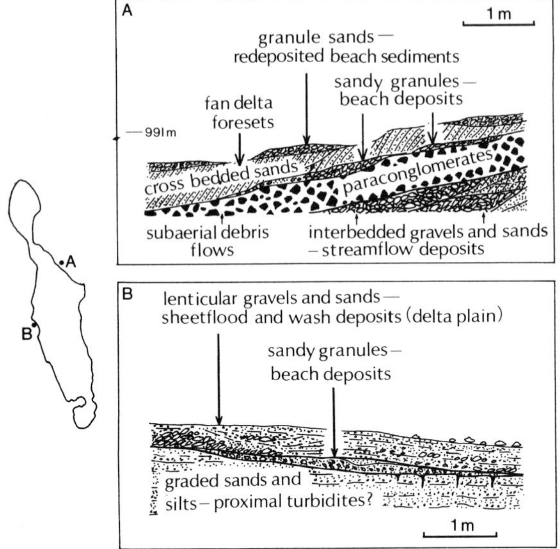

Fig. 19. Cross-sections, showing relationship of beach sediments to other fan-delta facies. In section A, beach deposits are preserved in a transgressive sequence, overlain by subaqueous deltaic sediments. In section B, they are overlain by regressive subaerial delta-plain deposits.

If part of a transgressive sequence, the beach deposits may be preserved or reworked. Rapid rise of lake level, leaving the sediments below wave-base, followed by burial beneath stream-derived fan-deltaic sediments, may lead to preservation intact (Fig. 19A). In contrast, a more gradual rise may permit multiple reworking of the littoral sediments. If in an area with low amounts of available sand and fine gravel, chenier-like ridges or barrier bars may advance across the platform in step with lake-level rise. Thus former beaches may be reconstructed at a new level or, perhaps, be transformed into rippled shoreface sands. Similar behaviour has been interpreted for transgressive marine beach sequences (e.g. Kraft, 1971; Sanders & Kumar, 1975).

On fan-deltaic sites, high-level beach sands and gravels resemble poorly sorted, ephemeral stream bar deposits, particularly where the gravels originated as fluvially rounded pebbles. However, the wave-worked gravels are probably better segregated into discrete beds and the bedding less lenticular than alluvial gravels (cf. Clifton, 1973; Bourgeois & Leithold, 1984). Lying in a site of active sedimentation, they are more likely to be buried rapidly and preserved than those along coincident margins.

Beach deposits have been recognized within ancient alluvial fan and fan-delta deposits by several workers (Tucker, 1977; Nemec *et al.*, 1984; LeTourneau & Smoot, 1985; Smoot *et al.*, 1985). Similar to those at Bogoria, they are clast-supported conglomerates, with a well-sorted infiltrated matrix, and are commonly associated with sandy deposits exhibiting parallel bedding and oscillation ripples.

BEACH DEPOSITS IN OTHER KENYA RIFT LAKES: COMPARISONS AND RESOURCE POTENTIAL

With the exception of Lake Turkana, the other Kenya Rift Valley lakes (Fig. 1B) have very poorly developed beaches. Lakes Nakuru and Elmenteita are bordered by saline mudflats. They are shallower than Bogoria, more rectangular in shape, with unconfined margins. Lake Magadi, although elongate and confined by lava escarpments, is a hot-spring fed, ephemeral salt pan with very low clastic input (Eugster, 1980). Lake Nasikie Engida (Little Magadi), which like Bogoria is hot-spring fed and perennial, has narrow discontinuous beaches. Lakes Baringo and Naivasha are subrectangular in shape. Although closed, their waters are dilute. Their shorelines have extensive littoral marsh and swamp, which trap sediment but limit beach development. Local sand−granule−pebble strandlines are present at both lakes. Lake Turkana, which is 7500 km^2 in area and 125 m deep, and only subsaline (2.5 g/l TDS), has extensive large sandy beaches (Butzer, 1980; Owen, 1981; Reid & Frostick, 1985).

All the saline, Kenya Rift Valley lakes had significantly fresher precursors and had higher levels during the early and middle Holocene. High-level strandlines, both depositional and erosional, are present around each of the lakes and are products of sediment sorting by waves in the deeper waters of expanded lakes (Butzer *et al.*, 1972; Kamau, 1977; Eugster, 1980; Owen *et al.*, 1982; Casanova, 1986a; Owen & Renaut, 1986; Renaut & Owen, 1987).

The resource potential of the beach deposits in the Kenya Rift Valley lakes appears to be largely a function of scale, increasing with size of lake and littoral wave energy. There are at least four potential resources associated with the beach sediments.

Reid & Frostick (1985) have shown the importance of wave processes in concentrating heavy minerals in large sandy swash bars at northeastern Lake Turkana, where concentrations of metalliferous minerals locally reach 90%. At Bogoria, mafic mineral separates normally account for only 5–15% of the sand fraction.

Coquinas, composed of wave-sorted mollusc shells are common locally in Holocene littoral deposits at Lake Bogoria and Lake Baringo (Renaut & Owen, 1987). They are abundant, although less concentrated, in the unconsolidated Holocene beach sands of Lake Turkana (Owen, 1981). In places, they account for >90% by volume of the sediment and are a potential source of lime. The Holocene gravels of Lake Bogoria, Lake Naivasha, and the sands of Lake Turkana are possible sources of aggregate for construction.

In large ancient rift lakes, the littoral sands, such as those that surround Lake Turkana, constitute a potential reservoir facies for hydrocarbons sourced within the adjacent lake. At Lake Turkana, littoral sands several metres thick extend offshore with ribbon and sheet-like geometries, and are overlain and underlain by lacustrine muds.

The effectiveness of littoral sands as a reservoir is probably strongly dependent on their composition. The sands of NE Lake Turkana are mostly litharenites, arkosic arenites, and subarkoses (Owen & Renaut, 1986). Although well sorted with high primary porosities, compaction and diagenetic alteration of feldspars and mafic minerals to clay minerals and iron oxides are likely to result in early loss of porosity. At Bogoria, where quartz is a minor component of most sands, the volcanic rock fragments, which constitute most sands and granules, break down physically to silts and fine sands composed of feldspars. These in turn may undergo early diagenetic alteration to clay minerals and zeolites (Renaut, 1982; Renaut *et al.*, 1986).

Therefore, coarse littoral sediments in rifts are most likely to form a reservoir facies where the catchment contains a high percentage of quartz-bearing rocks, such as Precambrian metamorphic shields, as for example at Lakes Malawi and Tanganyika. Where volcanic rocks constitute a major proportion of the catchment, as at Lake Bogoria and parts of Lake Turkana, the diagenetic alteration of feldspars and volcanic rock fragments in sands will probably lead to early loss of potential reservoir porosity.

DISCUSSION AND CONCLUSIONS

Although beaches are common, but minor, features of the large, deep Afro-Arabian rift lakes, such as the Dead Sea, and Lakes Turkana, Tanganyika and Malawi, this study has shown that they can also form in small, shallow rift lakes, both fresh and saline. The elongate form of many rift lakes plays a significant role. Where the rift has steep margins, winds can be funnelled along the axis, generating waves. Inflowing streams, which commonly enter lakes from their extreme ends (e.g. Lakes Bogoria, Turkana, and Malawi), may amplify circulation patterns (Sly, 1978). The linear form also provides the maximum possible fetch for a given surface area. Together these can result in the development of effective longshore currents.

Two main shoreline facies have been recognized at Lake Bogoria: poorly sorted coarse sands, granules and pebbles, and pebbly sandstones, associated with fan-deltas, and coarse angular gravels and angular orthoconglomerates (sedimentary breccias) that formed along coincident margins during periods of high lake level.

Provided that sediment of appropriate grade is available in the littoral zone at any given time and lake level, beaches may develop. If littoral sediments are too coarse for available littoral energy, unsorted lag deposits litter the shore zone. If wave energy exceeds that required to produce beaches at a given level, erosion may take place, producing wave-cut beach-scarps on fan-deltas. At Lake Bogoria, most modern beach sediment is provided directly, or indirectly (by erosion of older exposed sediments), by peripheral fan-deltas.

Lake level fluctuations, responding to climatic or tectonic controls, can increase or decrease littoral energy. Rising lake level can increase fetch and transform the configuration of the shoreline. Non-coincident margins may become coincident until delta progradation and aggradation respond to the new level. During the 'catch-up' phase, frictional shoaling effects on submerged fan-deltas are reduced. Increased wave energy, turbulence, and longshore currents along the shoreline then permit erosion and sorting of coarser sediments, but at Bogoria this energy has been insufficient to produce much lateral movement of gravels. An overall upward-coarsening sequence of gravel terraces results.

Small closed rift lakes, such as Bogoria, generally do not appear to develop broad progradational shoreline sequences because of the frequent fluctuations in level. Where they do occur implies either prolonged equilibrium at a particular lake level (which in turn implies a climatic balance between inflow and evaporation), or alternatively that the lake has reached its overflow level and is behaving as an open, rather than a closed, system.

ACKNOWLEDGEMENTS

This research was supported by grants from the Natural Sciences and Engineering Research Council (Canada), the Natural Environment Research Council (UK), and the Royal Society (UK). We thank The Government of Kenya for permission to undertake field work in the Lake Bogoria National Reserve. We are very grateful to Drs Cristino Dabrio and Jean-Jacques Tiercelin for their very helpful reviews of an early version of the manuscript.

REFERENCES

ALLEN, J.R. (1968) *Current Ripples: their Relation to Patterns of Water and Sediment Motion.* North Holland, Amsterdam, 433 pp.

ALLEN, P.A. & COLLINSON, J.D. (1986) Lakes. In: *Sedimentary Environments and Facies*, 2nd edn. (Ed. Reading, H.G.), pp. 63–94. Blackwell, Oxford.

BOURGEOIS, J. & LEITHOLD, E.L. (1984) Wave-worked conglomerates — depositional processes and criteria for recognition. In: *Sedimentology of Gravels and Conglomerates* (Eds Koster, E.H., & Steel, R.J.), Can. Soc. Petrol. Geol. Mem. **10**, 331–343.

BOWMAN, D. (1971) Geomorphology of the shore terraces of the late Pleistocene Lisan Lake (Israel). *Palaeogeogr. Palaeoclimatol. Palaeoecol.* **9**, 183–209.

BUTZER, K.W. (1980) The Holocene lake plain of north Rudolph, East Africa. *Phys. Geogr.* **1**, 42–58.

BUTZER, K.W., ISAAC, G.L., RICHARDSON, J.L. & WASHBOURN-KAMAU, C.K. (1972) Radiocarbon dating of East African lake levels. *Science* **175**, 1069–1076.

CASANOVA, J. (1986a) *Les Stromatolites continentaux: paléoecologie, paléohydrologie, paléoclimatologie. Application au Rift Gregory.* Unpublished thesis, Université Aix-Marseille II, 256 pp.

CASANOVA, J. (1986b) East African Rift stromatolites. In: *Sedimentation in the African Rifts* (Eds Frostick, L.E., Renaut, R.W., Reid, I. & Tiercelin, J.-J.), Geol. Soc. London Spec. Publ. **25**, 201–210.

CASANOVA, J. & RENAUT, R.W. (1987) Stromatolites du milieu lacustre. In: *Le demi-graben de Baringo-Bogoria, Rift Gregory, Kenya: 30000 ans d'histoire hydrologique et sédimentaire* (Eds Tiercelin, J.-J. & Vincens, A.), Bull. Cent. Rech. Exp. Prod. Elf Aquitaine **11**, 490–498.

CLIFTON, H.E. (1973) Pebble segregation and bed lenticularity in wave-worked versus alluvial gravel. *Sedim. Geol.* **20**, 173–187.

DAVIES, R.A. (Ed.) (1985) *Coastal Sedimentary Environment*, 2nd edn. Springer, New York, 716 pp.

DONOVAN, R.N. (1975) Devonian lacustrine limestones at the margin of the Orcadian Basin, Scotland. *Q. J. Geol. Soc. London* **131**, 489–510.

DONOVAN, R.N. & ARCHER, R. (1975) Some sedimentological consequences of a fall in the level of Haweswater, Cumbria. *Proc. Yorks. Geol. Soc.* **40**, 547–562.

East African Meteorological Department (1975) *Climatological statistics for East Africa, Part 1: Kenya.* East African Community, Nairobi.

EUGSTER, H.P. (1980) Lake Magadi, Kenya and its precursors. In: *Hypersaline Brines and Evaporitic Environments* (Ed. Nissenbaum, A.), pp. 195–232, Elsevier, Amsterdam, 270 pp.

GREGORY, J.W. (1921) *The Rift Valleys and Geology of East Africa.* Seeley Service, London, 479 pp.

GRIFFITHS, P.S. (1977) *The Geology of the Lake Hannington–Perkerra River area, Kenya Rift Valley.* Unpublished PhD thesis, University of London, 187 pp.

GROVE, A.T., STREET, F.A. & GOUDIE, A. (1975) Former lake levels and climatic changes in the rift valley of southern Ethiopia. *Geogr. J.* **141**, 171–202.

HARDIE, L.A., SMOOT, J.P. & EUGSTER, H.P. (1978) Saline lakes and their deposits: a sedimentological approach. In: *Modern and Ancient Lake Sediments* (Eds Matter, A. & Tucker, M.E.), Spec. Publ. Int. Assoc. Sedimentol. **2**, 7–41.

JOHNSTON, H. (1902) *The Uganda Protectorate, Volume 1.* Hutchinson, London, 470 pp.

KAMAU, C.K. (1977) The Ol Njorowa Gorge, Lake Naivasha basin, Kenya. In: *Desertic Terminal Lakes* (Ed. Greer, D.C.), pp. 297–307. Utah Water Research Laboratory, Logan, 436 pp.

KIRK, R.M. (1980) Mixed sand and gravel beaches: morphology, processes and sediments. *Progr. Phys. Geogr.* **4**, 189–210.

KOMAR, P.D. (1976) *Beach Processes and Sedimentation.* Prentice Hall, Englewood Cliffs, NJ, 429 pp.

KRAFT, J.C. (1971) Sedimentary patterns and geologic history of a Holocene marine transgression. *Geol. Soc. Am. Bull.* **82**, 2131–2158.

LEAKEY, L.S.B. (1931) East African lakes. *Geogr. J.* **77**, 497–514.

LETOURNEAU, P.M. & SMOOT, J.P. (1985) Comparison of ancient and modern lake-margin deposits from the Lower Jurassic Portland Formation, Connecticut, and Walker Lake, Nevada (abstr.). *Geol. Soc. Am. Abstr. Prog.* **17**, 13.

MCCALL, G.J.E. (1967) Geology of the Nakuru–Thomson's Falls–Lake Hannington area. *Rep. Geol. Surv. Kenya* **78**, 122 pp.

MORRISON, R.B. (1964) Lake Lahontan: geology of the southern Carson Desert, Nevada. *U.S. Geol. Surv. Prof. Pap.* **401**.

NEMEC, W., STEEL, R.J., POREBSKI, S.J. & SPINNANGR, A. (1984) Domba Conglomerate, Devonian, Norway: process and lateral variability in a mass flow-dominated, lacustrine fan-delta. In: *Sedimentology of Gravels and Conglomerates* (Eds Koster, E.H. & Steel, R.J.). Can. Soc. Petrol. Geol. Mem. **10**, 295–320.

NILSSON, E. (1931) Quaternary glaciations and pluvial lakes in British East Africa. *Geogr. Annlr.* **13**, 249–349.

OWEN, R.B. (1981) *Quaternary diatomaceous sediments and the geological evolution of lakes Turkana, Baringo and Bogoria, Kenya Rift Valley.* Unpublished PhD thesis, University of London, 465 pp.

OWEN, R.B. & RENAUT, R.W. (1986) Sedimentology, stratigraphy and palaeoenvironments of the Holocene Galana Boi Formation, northeast Lake Turkana, Kenya. In: *Sedimentation in the African Rifts* (Eds Frostick, L.E., Renaut, R.W., Reid, I. & Tiercelin, J.-J.), Geol. Soc. London Spec. Publ. **25**, 311—322.

OWEN, R.B., BARTHELME, J.W., RENAUT, R.W. & VINCENS, A. (1982) Palaeolimnology and archaeology of Holocene deposits north-east of Lake Turkana, Kenya. *Nature* **298**, 523—529.

REID, I. & FROSTICK, L.E. (1985) Beach orientation, bar morphology and the concentration of metalliferous placer deposits: a case study, Lake Turkana, N. Kenya. *J. Geol. Soc. London* **142**, 837—848.

RENAUT, R.W. (1982) *Late Quaternary geology of the Lake Bogoria fault-trough, Kenya Rift Valley.* Unpublished PhD thesis, University of London, 498 pp.

RENAUT, R.W. & OWEN, R.B. (1987) Le bassin de Baringo: Niveaux émergés. In: *Le demi-graben de Baringo-Bogoria, Rift Gregory, Kenya: 30 000 ans d'histoire hydrologique et sédimentaire* (Eds Tiercelin, J.-J. & Vincens, A.), Bull. Cent. Rech. Exp. Prod. Elf Aquitaine **11**, 508—514.

RENAUT, R.W. & OWEN, R.B. (1988) Opaline cherts associated with sublacustrine hydrothermal springs at Lake Bogoria, Kenya Rift Valley. *Geology* **16**, 699—702.

RENAUT, R.W., TIERCELIN, J.-J. & OWEN, R.B. (1986) Mineral precipitation and diagenesis in the sediments of the Lake Bogoria basin, Kenya Rift Valley. In: *Sedimentation in the African Rifts* (Eds Frostick, L.E., Renaut, R.W., Reid, I. & Tiercelin, J.-J), Geol. Soc. London Spec. Publ. **25**, 159—176.

RENAUT, R.W., TIERCELIN, J.-J. & PAGÉ, P. (1987) Alimentation, hydrologie. In: *Le demi-graben de Baringo-Bogoria, Rift Gregory, Kenya: 30 000 ans d'histoire hydrologique et sédimentaire* (Eds Tiercelin, J.-J. & Vincens, A.), Bull. Cent. Rech. Exp. Prod. Elf Aquitaine **11**, 284—309.

SANDERS, J.E. & KUMAR, N. (1975) Evidence of shoreface retreat and in place 'drowning' during Holocene submergence of barriers, shelf off Fire Island, New York. *Geol. Soc. Am. Bull.* **86**, 65—76.

SLY, P.G. (1978) Sedimentary processes in lakes. In: *Lakes: Chemistry, Geology, Physics* (Ed. Lerman, A.), pp. 65—89. Springer, New York, 366 pp.

SMOOT, J.P., LETOURNEAU, P.M., TURNER-PETERSON, C.M. & OLSEN, P.E. (1985) Sandstone and conglomerate shoreline deposits in Triassic—Jurassic Newark and Hartford basins of Newark Supergroup (abstr.). *Bull., Am. Assoc. Petrol. Geol.* **69**, 1448.

TIERCELIN, J.-J. (1981) *Rifts continentaux, tectonique, climats, sédiments. Exemples: la sédimentation dans le Nord du Rift Gregory (Kenya) et dans le Rift de l'Afar (Ethiopie) depuis le Miocène.* Unpublished thesis, Université Aix-Marseille II, 260 pp.

TIERCELIN, J.-J. & RENAUT, R.W. (1987) La sédimentation actuelle. In: *Le demi-graben de Baringo-Bogoria, Rift Gregory, Kenya: 30 000 ans d'histoire hydrologique et sédimentaire* (Eds Tiercelin, J.-J. & Vincens, A.), Bull. Cent. Rech. Exp. Prod. Elf Aquitaine **11**, 310—324.

TIERCELIN, J.-J. & VINCENS, A. (Eds) (1987) Le demi-graben de Baringo—Bogoria, Rift Gregory, Kenya: 30 000 ans d'histoire hydrologique et sédimentaire. *Bull. Cent. Rech. Exp. Prod. Elf Aquitaine* **11**, 249—540.

TIERCELIN, J.-J., LE FOURNIER, J., HERBIN, J.P. & RICHERT, J.P. (1980) Continental rifts: modern sedimentation, tectonic and volcanic control. Example from the Bogoria—Baringo graben, Gregory Rift, Kenya. In: *Geodynamic Evolution of the Afro-Arabic Rift System*, vol. 47, 143—164. Academia Nazionale dei Lincei, Rome.

TIERCELIN, J.-J., RENAUT, R.W., DELIBRIAS, G., LE FOURNIER, J. & BIEDA, S. (1981) Late Pleistocene and Holocene lake levels in the Lake Bogoria basin, northern Kenya Rift Valley. *Palaeoecol. Afr.* **13**, 105—120.

TIERCELIN, J.-J., PERINET, G., LE FOURNIER, J., BIEDA, S. & ROBERT, P. (1982) Lacs du rift est-africain, exemples de transition eaux douces-eaux salées: le lac Bogoria, rift Gregory, Kenya. *Mem. Soc. Géol. Fr.* **144**, 217—230.

TIERCELIN, J.-J., RENAUT, R.W., PAQUET, H. & MELIERES, F. (1987) Le lac Bogoria: sédimentation minerale. In: *Le demi-graben de Baringo-Bogoria, Rift Gregory, Kenya: 30 000 ans d'histoire hydrologique et sédimentaire* (Eds Tiercelin, J.-J. & Vincens, A.), Bull. Cent. Rech. Exp. Prod. Elf Aquitaine **11**, 324—327.

TUCKER, M.E. (1977) The marginal Triassic deposits of South Wales: continental facies and paleogeography. *Geol. J.* **12**, 169—188.

TUCKER, M.E. (1978) Triassic lacustrine sediments from South Wales: shore-zone clastics, evaporites and carbonates. In: *Modern and Ancient Lake Sediments* (Eds Matter, A. & Tucker, M.E.), Spec. Publ. Int. Assoc. Sedimentol. **2**, 205—224.

VINCENS, A., CASANOVA, J. & TIERCELIN, J.-J. (1986) Palaeolimnology of Lake Bogoria (Kenya) during the 4500 BP high lacustrine phase. In: *Sedimentation in the African Rifts* (Eds Frostick, L.E., Renaut, R.W., Reid, I. & Tiercelin, J.-J.), Geol. Soc. London Spec. Publ. **25**, 323—330.

WILLIAMS, L.A.J. & CHAPMAN, G.R. (1986) Relationships between major structures, salic volcanism and sedimentation in the Kenya Rift from the equator northwards to Lake Turkana. In: *Sedimentation in the African Rifts* (Eds Frostick, L.E., Renaut, R.W., Reid, I. & Tiercelin, J.-J.), Geol. Soc. London Spec. Publ. **25**, 59—74.

YOUNG, J.A.T. & RENAUT, R.W. (1979) A radiocarbon date from Lake Bogoria, Kenya Rift Valley. *Nature* **278**, 243—245.

ZENKOVICH, V.P. (1971) A theory of the development of accumulation forms in the coastal zone. In: *Introduction to Coastline Development* (Ed. Steers, J.A.), pp. 94—116. MacMillan, London, 229 pp.

Stratigraphy and Rhythmicity in
Lacustrine Deposits

Spec. Publs Int. Ass. Sediment. (1991) **13**, 199–221

Ephemeral lakes, mud pellet dunes and wind-blown sand and silt: reinterpretations of Devonian lacustrine cycles in north Scotland

D.A. ROGERS*,[1] *and* T.R. ASTIN[†]

* *Department of Earth Sciences, University of Oxford, Parks Road, Oxford OX1 3PR, UK*
[†] *Postgraduate Research Institute for Sedimentology, Reading University, Whiteknights, Reading RG6 2AB,*
UK

ABSTRACT

Facies analysis of a representative section of the classic cyclic lake deposits of the Devonian Orcadian Basin, Scotland, has produced significant palaeoenvironmental reinterpretations and an enhanced understanding of the short and long-term cyclicity of lake level and chemistry. The most abundant lake facies, 'flagstone' (thinly interstratified mud and coarse silt/very fine sand), is recognized as an intercalation mostly of lake suspension deposits and of coarser sediment blown across dried-up lake floors. Supposedly classic examples of subaqueous ('syneresis') shrinkage cracks are reinterpreted as gypsum pseudomorphs and desiccation cracks. Other evidence for evaporation, exposure, and the importance of wind transport includes halite pseudomorphs and possible wind-blown mud pellet bedforms ('clay dunes'). Thus, contrary to previous models, the Orcadian lake was ephemeral for most (c. 90%) of its history. The well-known c. 10 m cycles, recognized here as abrupt alternations of ephemeral and permanent lake facies, are the result of variations in lake level, ephemerality, and chemistry, controlled by climatic cyclicity. Though the transition between the two states of the lake was rapid, it is quite possible that the climatic variation was gradual. At least two more scales of cyclicity can be identified within the facies and attributed to weather and seasonality. This view of the basin centre environment has important implications for the understanding of lacustrine−fluvial interactions in marginal Orcadian successions and for the modelling of successions offshore and in neighbouring basins; the set of depositional processes invoked had the potential to produce reservoir-scale sand accumulations far from a basin edge.

INTRODUCTION

The Middle Devonian Orcadian Basin of north Scotland is a classic ancient lake basin. Towards the centre of the Orcadian outcrop, in Caithness and Orkney (Fig. 1), the lake deposits take the form of repetitive cycles, averaging 10 m thick (Table 1), of sandier and shalier sediments (Crampton & Carruthers, 1914; Wilson *et al.*, 1935; Fannin, 1970; Donovan, 1971,1978,1980; Foster, 1972; Donovan *et al.*, 1974; Plimmer, 1974; Astin, 1990). Donovan, Foster, Fannin, and Plimmer recognized facies deposited in deep permanent lakes, in shallow permanent lakes with varying amounts of sand input,

and in ephemeral lakes. The implication of these studies was that over a third of most Orcadian lake-dominated successions (and up to 73% of some — Table 1) was deposited from permanent standing water. The cyclic arrangement of these facies, often incomplete, has been attributed variously to an interplay of tectonic and climatic changes (Fannin, 1970; Mykura, 1976), to tectonics (Crampton & Carruthers, 1914), to autocyclicity (implicitly, Crampton & Carruthers, 1914), and to climatically controlled variations in lake level and ephemerality (Donovan, 1975,1978,1980; Hamilton & Trewin, 1988; Astin, 1990), possibly with an orbital (Milankovitch cyclicity) control.

Despite the attention paid to the cycles, the only comprehensive descriptions published of basin

---

[1] Present Address: Lomond Associates, De Quincey House, 48, West Regent Street, Glasgow G2 2RA, UK

Table 1. Cycle and facies thicknesses in the main central Orcadian Middle Devonian lake successions: compilation of data from previous accounts

| Stratigraphic unit | | Locality (references) | Cycle thickness (m) | | Permanent lake laminite beds | | | 'Subaqueous cracked' flags reattributed here to *ephemeral* lakes (percentage of unit) | Undisputed ephemeral lake and river facies (percentage of unit) |
| | | | | | | Thickness (m) | | | |
			Mean	SD	Percentage of unit	Mean	SD		
Lower Stromness Flags		Hoy and western mainland	8.1	5.6	8	0.6	—	32	67
Hoy Cycles		Orkney	X	X	5	0.5	—	39	60
Higher Upper Stromness Flags		(Fannin, 1970)	X	X	15	1.3	—	9	76
Upper Stromness Flagstone Fm		Brough Head	11.8	6.0	—	—	—	—	—
Upper Stromness Flagstone Fm		Evie	13.3	3.9	—	—	—	—	—
Rousay Flagstone Fm		West Rousay	9.8	3.7	—	—	—	—	—
Rousay Flagstone Fm		South Eday (Astin, 1990)	12.6	4.5	—	—	—	—	—
Grims Ness subdivision	Rousay	Orkney			6[a]	1.5[a]	1.7[a]	67[a]	27
Skel Wick subdivision	Flags	(Plimmer,	'Usually 5–20'		6[a]	0.8[a]	0.6[a]	31[a]	63
Weather Ness subdivision		1974)			7[a]	0.8[a]	0.7[a]	28[a]	65
Latheron Subgroup, Upper Caithness Flags		Brims to Dounreay, northern Caithness (Donovan, 1980)	6.1	1.8	4	0.4	0.1	48	48
Clyth Subgroup	Lower	Northeast	—	—	1	—	—	27	72
Lybster Subgroup	Caithness	Caithness (Donovan	—	—	2.5	—	—	35.5	62
Robbery Head Subgroup	Flags	*et al.*, 1974)	—	—	1.5	—	—	57	41.5
Latheron Subgroup	Upper	As above	—	—	16	—	—	6	78
Ham-Scarfskerry Subgroup	Caithness		—	—	12	—	—	61	27
Mey Subgroup	Flags		—	—	4	—	—	28	68
Middle ORS flags generally		(Donovan, 1978)	'8–9' ('range 3–20')						

(−) Data not available; (X) data superceded.
[a] Slightly inaccurate, since 18–20% of Plimmer's 'L2 laminite' beds have lenticular cracks.

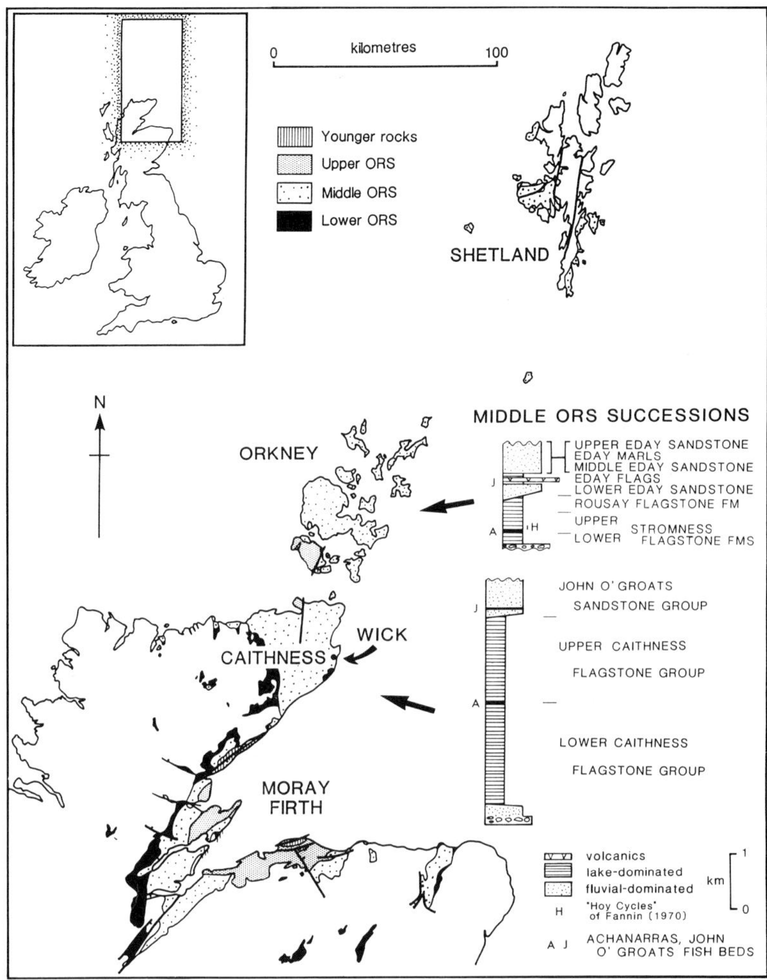

Fig. 1. The Devonian of the Orcadian Basin, northern Scotland, with stratigraphic columns for the Middle Old Red Sandstone (ORS) of Orkney, after Astin (1990), and Caithness, after Donovan *et al.* (1974). The latter is composite; maturity data indicate that the total Devonian thickness at any point was much less (S.J. Hillier, pers. comm.).

centre facies have been of permanent lake 'fish beds' (Rayner, 1963; Trewin, 1986; Duncan & Hamilton, 1988; Janaway & Parnell, 1989), which comprise 16% at most of any succession (Table 1). The present paper aims to provide a detailed description of the rest of the finely interlaminated and interbedded lake deposits ('flagstones') and their associated sand bodies. Their facies analysis leads to a reinterpretation of Orcadian palaeoenvironments, implying that the vast majority of the succession was deposited on ephemeral lake floors. This has implications for the prediction of reservoir sand body occurrence and geometry in such 'distal' playa lake successions, for the understanding of more marginal Orcadian sequences, where lake deposits intercalate with alluvium, and for the interpretation of the previously

recognized cyclic arrangements of facies and of bed types within facies.

The section studied is on the coast of Caithness at South Head of Wick (ND372493; Fig. 1), where sedimentary structures show up unusually well in the weathered faces (Fig. 2) of flagstone quarries in rocks that have been indurated in an area of high thermal maturity (S.J. Hillier, pers. comm.). There is 46 m of continuous vertical exposure (three cycles), nearly 19 m of which is sufficiently well exposed to have been logged lamina by lamina (Fig. 3). Lateral variations can be studied for between 200 and 350 m. The section occupies a relatively central position within the basin, at least with respect to present outcrops (Fig. 1). Its exact stratigraphic position is unknown, but is probably quite low in

Fig. 2. South Head quarry, Wick: part of the main interval measured in detail, coarsening upward from simple flagstones (facies 2) at about 14 m on Fig. 3, through flagstones with turbate sandy muds and muddy sands (facies 5) to the top of the upper thick sandstone body (facies 6) at 18.97 m.

the sequence, below a basin-wide marker, the Achanarras Fish Bed (Donovan *et al.*, 1974; Rogers *et al.*, 1989). It is representative of the central Orcadian sequence as a whole to the extent that the laterally persistent, fine grained, finely stratified 'flagstone' and laminite facies with their gradational contacts are typical; the volumetrically subordinate sand-dominated facies are much more variable from one section or cycle to another. Also, the section contains both complete, laminite-bearing cycles and one where a laminite is absent.

The sequence logged is composed of six facies. Five are intergradational categories of laminite and 'flagstone'. The sixth forms laterally extensive sand bodies with abrupt boundaries. All the facies occur in tabular units extending across the whole, 200–350 m wide exposure.

FACIES ANALYSIS

Facies 1: laminites

These limestones (Fig. 4A) consist of parallel inter-laminations of light-coloured micrite and dark organic matter. Clastic material (quartz, mica, feldspar, clay) of clay to coarse silt grade is present in both, comprising 20–50% of rock volumes in the samples studied. The dark-light couplets average between 0.25 and 0.5 mm in thickness and are laterally persistent. Fish skeletons and fragments are concentrated in certain laminae, as are clusters of pyrite. Staining of specimens and weathering sometimes reveal bundles, 2–10 mm thick, of dolomite- and calcite-rich laminae. Janaway & Parnell (1989) describe the geochemistry of the lower bed (the 'Trinkie laminite'). There is less than 10% dolomite through most of the bed, but up to 50% in the basal and top few cm.

Interpretation

This facies is a typical carbonate-rich variety of the 'laminites', 'rhythmites', or 'fish beds' described from many central Orcadian lake successions (Rayner, 1963; Fannin, 1970; Donovan, 1971,1980; Foster, 1972; Donovan *et al.*, 1974; Plimmer, 1974; Trewin, 1986; Duncan & Hamilton, 1988; Janaway & Parnell, 1989) and forming 1–16% of total thicknesses (Table 1). Their carbonate and terrigenous components both range from less than 10 to over 90%, while their organic carbon content can reach 5% (Marshall *et al.*, 1985; Parnell, 1985). For classic Orcadian laminites, such as those at Wick, with no evidence of wave activity, we agree with the interpretations made by previous workers, and therefore these will not be discussed in any detail. These are the subwave-base deposits of *permanent* lakes, which were for most of the time thermally or chemically stratified with anoxic bottom waters. Rayner (1963) has explained how most of the couplets of carbonate and organic laminae probably represent annual 'non-glacial varves', but some may record the occurrence of more than one algal bloom in a year owing to, for instance, mixing by occasional severe storms (Trewin, 1986). Clastic supply may have been continuous, seasonal, or episodic; laminae have been described in this facies that are attributable to storm and river-generated overflows, interflows, density currents, wind transport, and settling following suspension by storms (Fannin, 1970; Donovan,

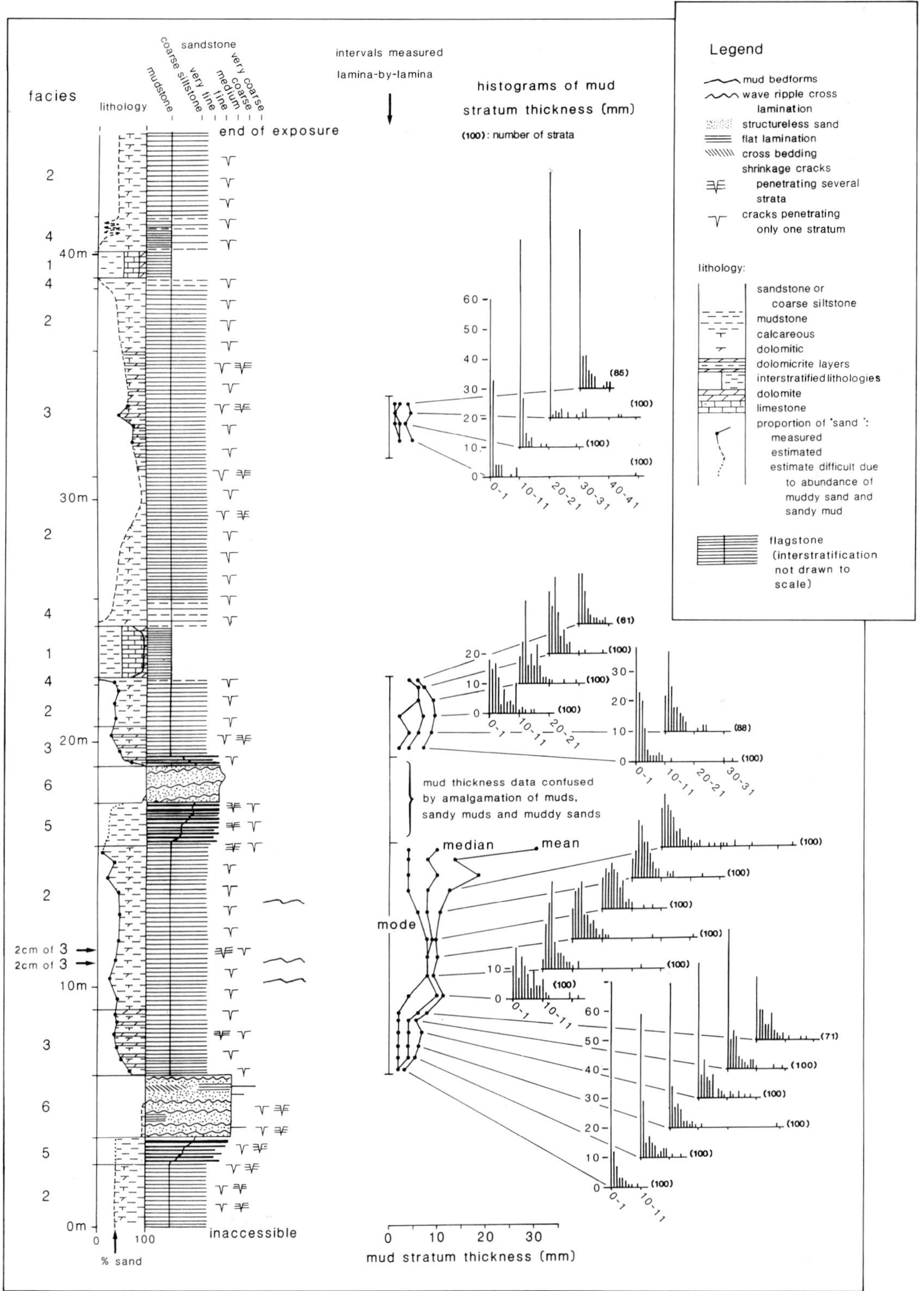

Fig. 3. Section measured at South Head quarry, Wick, Caithness.

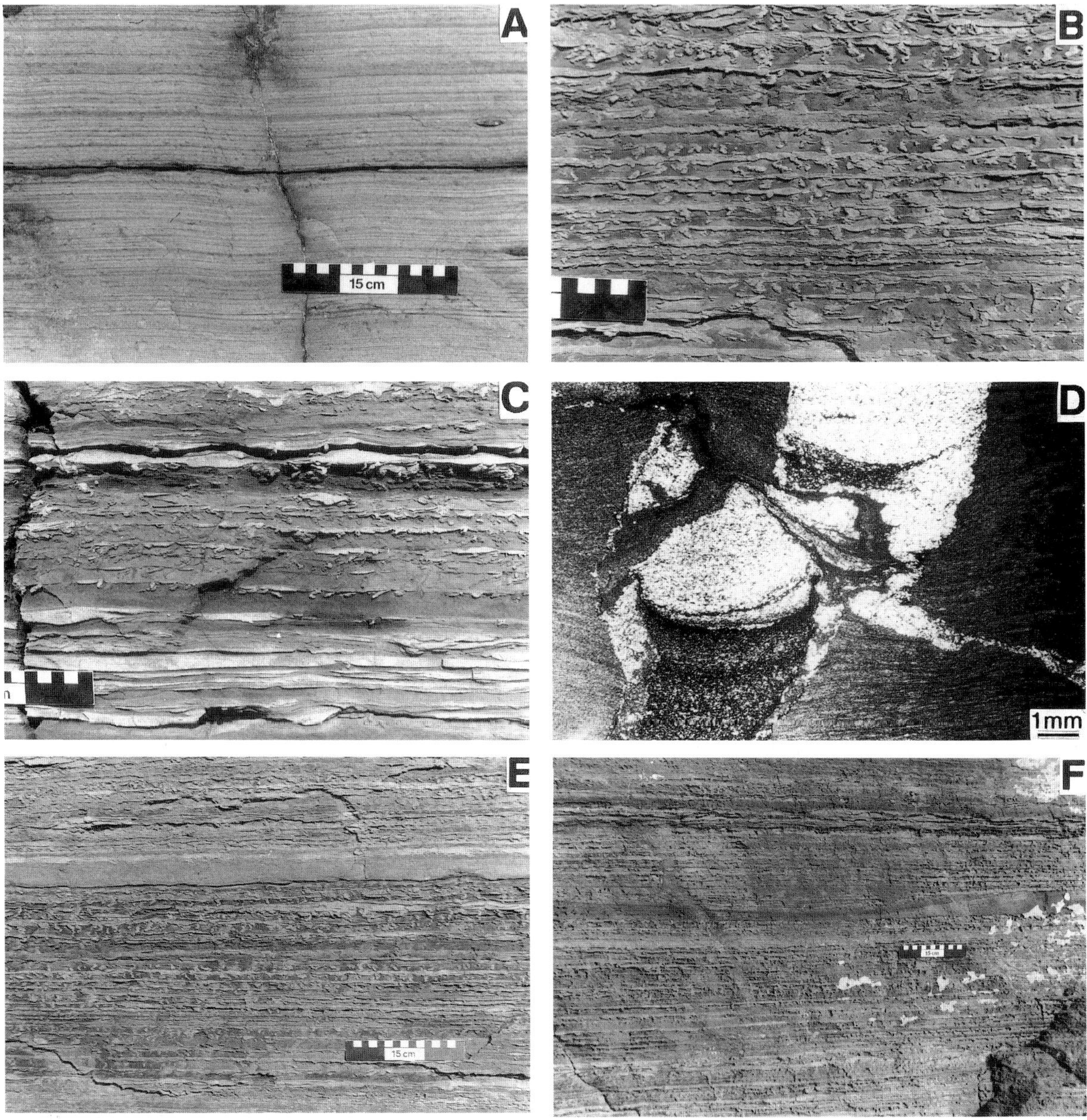

Fig. 4. (A) The lower permanent lake laminite bed (facies 1). The simple flagstone facies: (B) typical interstratification of shrinkage-cracked mud and wave-ripple cross-laminated and flat laminated coarse silt to very fine sand; (C) flagstone with some 'sand' strata reduced to trains of lenses; (D) shrinkage crack infill offset by microfault associated with injection structure (clastic dyke) in surrounding mud; (E) interval with a prominent thick clastic mudstone (underflow?) bed; (F) the lowermost bed of clastic mud with bedforms and cross-stratification (possible wind-blown mud pellet bedforms — see also Fig. 7).

1971,1978,1980; Plimmer, 1974; Trewin, 1986), and laminites are sometimes interbedded with thick turbidite or storm mudstone beds (Fannin, 1970; Trewin, 1986).

The chemistry of these permanent lakes seems to have been variable. Janaway & Parnell (1989) reported the presence of primary and secondary calcite and secondary dolomite throughout a number of Orcadian laminites, including the lower one at Wick. They found that the primary calcite in the centre of beds includes large polyhedra resembling 'suspendoids' precipitated slowly in fresh or dilute waters, whereas towards the margins it consists of micrite, which they attribute to shallower, warmer, evaporatively concentrated waters. Near the margins, they report unrecrystallized dolomicrite laminae, which may be precipitates of more saline lake waters. In support of these interpretations, they cite increasingly heavy oxygen isotope ratios and higher sodium concentrations in these components towards the margins of the laminite beds. Orcadian laminite isotope data and C/S ratios which spread between those expected for deposits of fresh waters and those with marine salinities are reported also by Duncan & Hamilton (1988).

Facies 2: simple flagstone

The simplest flagstones (Figs 4, 5, 6) consist of finely interbedded to coarsely interlaminated mudstone and siliciclastic 'sand' (coarse silt or very fine sand grade). From *every* 'sand' stratum, shrinkage cracks, 0.5–10 mm wide, penetrate down into the underlying mud layer and usually no further (Fig. 3). Many of the crack fills have bulbous cross-sections attributable to loading and compaction, but in most a downward taper can be seen. They are compacted into ptygmatic folds (Figs 4B,C, 5C) and the 'sand' strata are deformed around them. Less frequent small clastic dykes are distinguishable in section by their widths (<2 mm), their penetration upward or downward from the source stratum, often crossing several strata and/or crack fills, and their frequent passage up or down into normal or reverse microfaults with offsets of up to a few millimetres (Fig. 4D).

The mud strata are dark blue-grey to brown, weathering brown to ochreous, and have thicknesses ranging from less than 1 mm to 15 mm, with modes of 2–4 mm (Fig. 3). They are micaceous, clayey siltstones, or occasionally silty claystones, and also contain variable proportions of pyrite, organic matter, and carbonate, the latter ranging from 0 to over 50% of the volume. The silt fraction ranges from very fine to medium grade in different mud strata and is mostly quartz. The muds are structure-

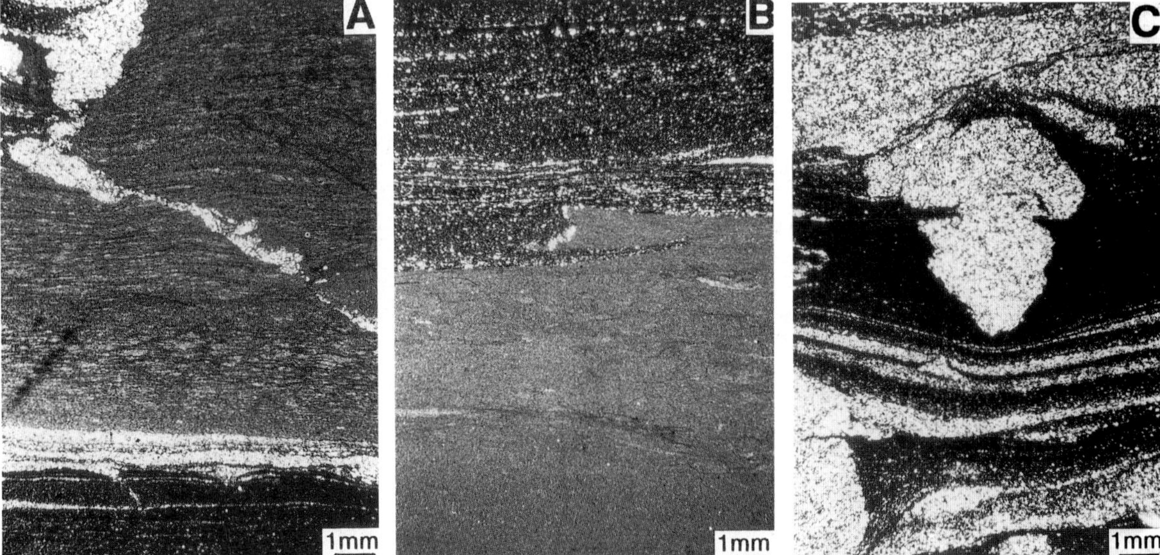

Fig. 5. The simple flagstone facies: (A) laminated carbonate-rich thin mud overlain by thick, normally graded, flat laminated clastic mud (underflow?) bed (same bed as in Fig. 4D) with thin interval of silt–clay flat lamination at base, cut by crack infill and clastic dyke; (B) upper portion of thick, structureless, fine, siliciclastic mud (underflow?) bed with sharp, exfoliating (desiccated?) top, overlain by laminated organic- and silt-rich, carbonate-poor thin mudstone; (C) 'sand' strata, ptygmatically folded crack-fills, and organic-rich, silt- and carbonate-poor, thin muds with occasional, thin, silt laminae.

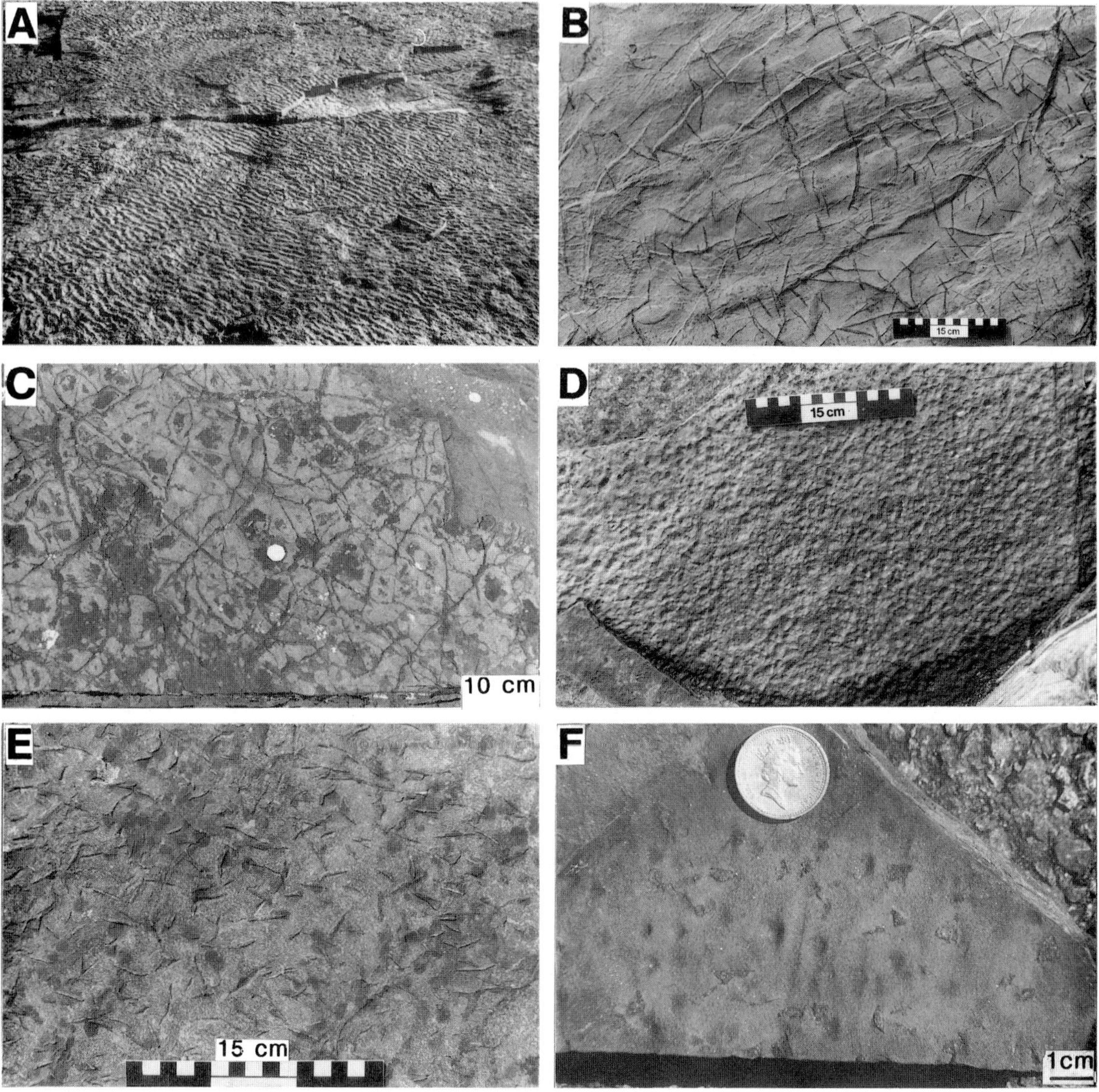

Fig. 6. The simple flagstone facies in plan view: (A) wave ripples; (B) lenticular and (C) polygonal shrinkage cracks; (D) rain drop imprints; (E) gypsum pseudomorphs (note quadruple junctions) of coarse silt in mud; (F) halite pseudomorphs, with lenticular cracks propagating from some (loose block on tarmac).

less, or have flat laminae up to 0.5 mm thick defined either by variable proportions of their major constituents, or by laminae, one or two grains thick, of medium or coarse quartz silt (Figs 4D, 5A,B). Such grains also occur scattered through some of the finer muds. Staining indicates that ferroan dolomite is the dominant carbonate mineral. In analyses of similar facies from Orkney (Fannin, 1970; Plimmer, 1974), ferroan calcite was also found, but usually in subordinate proportions.

The 'sand' (coarse silt to very fine sand) strata are 1–20 mm thick. They are well sorted, angular to subangular, micaceous subarkoses, with up to 5% dolomite and calcite clasts. The range of grain size is remarkably consistent through the section. Most strata have sharp, non-erosive bases, which were planar before compaction (Fig. 4B,C,E,F). Very occasionally (five examples), the bases are irregular and erosive, cutting down a few millimetres. The majority of strata show wave-ripple-generated subhorizontal, undulatory and cross-lamination (cf. Allen, 1981a; de Raaf *et al.*, 1977), sometimes with mudstone lenses in ripple troughs. Some of these show ripple profiles and most have lateral thickness variations attributable to rippling (Fig. 4A,B). Some 'strata' consist of trains of discrete ripple-form lenses on an otherwise 'sand'-starved surface (Fig. 4C). The tops of all these wave-rippled strata are sharp, or normally graded over less than 1 mm.

The remaining, non-rippled strata have sharp subhorizontal tops. They pinch and swell subtly over distances of decimetres to tens of metres and may pinch out completely for some distance. They are massive, or have indistinct flat laminae. The laminae are generally laterally persistent, but occasionally truncate against both the tops and bases of strata.

Where a 'sand' layer of any type pinches out, the 'stratum' can be followed laterally as a horizon of crack fills until it reappears (Fig. 4C). Apart from such interruptions, all types of stratum in this facies persist at least the width of the quarry (200–350 m).

Bedding planes of this facies are commonly wave rippled (Fig. 6A). A few show rain-drop pitting (Fig. 6D). In plan, the shrinkage cracks are mostly lenticular and centimetres to decimetres long (Fig. 6B,E), but a minority are joined into polygons (Fig. 6C). Quite a few surfaces have 'sand'-filled triangular or square depressions up to 15 mm across, which we interpret as halite pseudomorphs (Fig. 6F). Some of these are rather rounded, probably as a result of loading and/or compaction.

The few relatively thick (> 15 mm) mud beds (Fig. 3) are different. Most are laterally persistent, tabular, dark green, and poor in carbonate and organic matter (Figs 4E, 5A,B). Within the flagstone intervals their abundance increases up towards the thick sandstone bodies, as recorded by the increasing mean mud thicknesses (Fig. 3). They are finer than the typical carbonate-bearing thin muds, consisting mostly of medium to well sorted, very fine or fine, micaceous, siliciclastic silt, but with up to 6 mm of coarse silt above their sharp, erosive bases

(Fig. 5A). They occur above both carbonate-bearing muds and 'sands'. A few overlie cracks filled by coarse silt, but none were seen filling cracks themselves. They are either erosively overlain by 'sand', or pass up sharply or gradationally into calcareous mud. The sharp tops sometimes appear to exfoliate (Fig. 5B). Some beds are composite — amalgamated by erosion. Many are apparently structureless (Fig. 5B), or show normal grading and/or indistinct flat lamination (Fig. 5A). A few have convolute lamination. The coarse bases often have distinct c. 0.1 mm silt–clay flat lamination (Fig. 5A).

Three clastic mud beds in the intervals of detailed measurement (Fig. 3) are distinct from the rest in having planar bases, but tops showing periodic undulations with apparent wavelengths (in two-dimensional section) of 1–1.3 m (Figs 4F, 7). They are draped by the overlying muds and 'sands', which thicken into the hollows. They overlie and infill shrinkage cracks and are themselves cracked and filled from above. Above the lowest bed, the drapes are planed off by a horizontal erosion surface. The undulations are most pronounced in this bed; in the others the amplitudes average c. 1 cm. The uppermost bed is apparently structureless, but the others show low-angle cross-beds, which parallel certain of the undulatory surfaces and are tangential to the bases of the beds.

Interpretation

The thin mudstone strata. These are attributed to deposition from suspension in lakes, in a similar manner to the laminite facies. The clastic component was probably supplied as wind-blown dust and river-generated inter-, or overflows, whereas the carbonate precipitated from the lake either directly, perhaps aided by plankton productivity altering water chemistry, or as the tests of microorganisms. Occasionally, fish fragments are recorded from this facies (Fannin, 1970; Plimmer, 1974; Donovan, 1978,1980), but the absence of intact fish skeletons and infrequent lamination of these muds suggests that deposition was usually from relatively shallow lakes which were for most of the time unstratified (Fig. 8). The dolomite component probably formed either by proto-dolomite precipitation from saline lake waters, or by early diagenetic alteration of calcite to dolomite on dried-up lake floors, perhaps as the result of evaporative concentration of groundwaters.

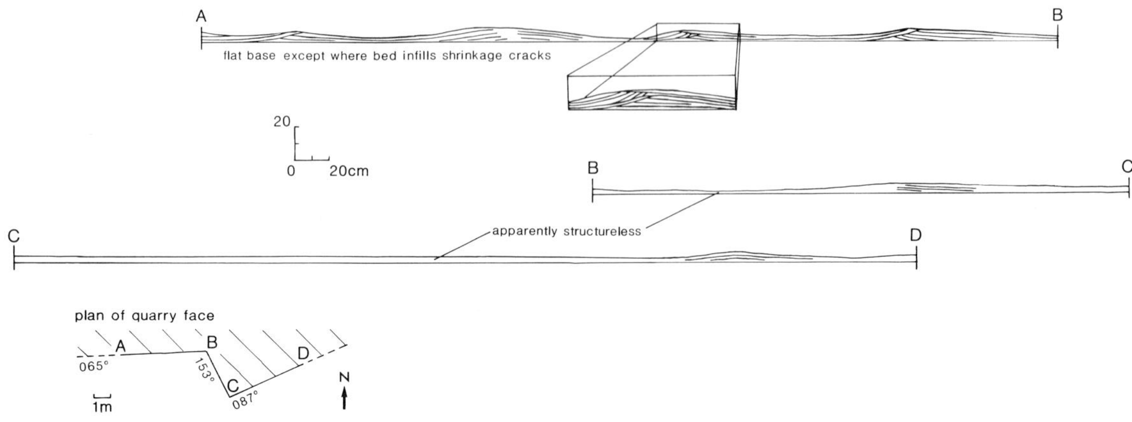

Fig. 7. Measured profile of the lowermost bed of clastic mudstone with bedforms and cross-bedding: possible wind-blown mud pellet bedforms. The bed continues east and west at an average thickness of 3.5 cm, with a smoothly undulose top and low-angle cross-beds with apparent dips in both directions.

Wind transport of the coarse silt to very fine sand strata. We propose that the 'sand' laminae were supplied by wind across emergent, desiccated lake floors (Fig. 8). No other mechanism, such as wave transport, turbidity currents, storm- or river-generated flows within a lake, or subaerial fluvial sheet-floods can explain all their characteristics and their wide distribution. Flagstone sequences similar to those at Wick have been found to extend laterally over large distances (24 km and 58 km; Fannin, 1970; Astin, 1985) and dominate the Middle Devonian of Caithness and Orkney. The mud component of the flagstones must have been deposited in very extensive, unstratified shallow lakes with near flat floors. Yet laterally persistent siliciclastic 'sand' strata supplied from the lake margin are common in all exposed outcrops of this facies. Obviously most 'sand' strata were affected by at least occasional wave activity and many record its waning in their graded tops. Allen (1981a) used measurements from Southeast Shetland (Fig. 1) to estimate that similar ripples formed in water depths of 5 m or less in the

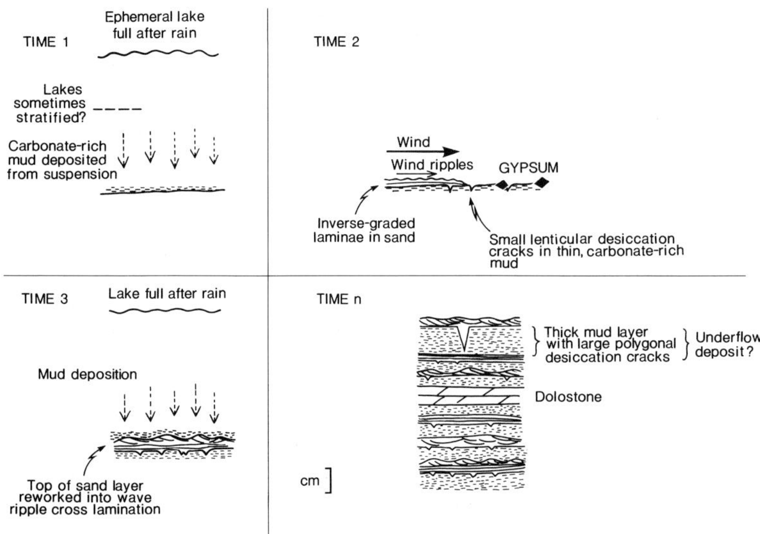

Fig. 8. Proposed mode of deposition of the flagstone facies.

large Orcadian lakes. However, waves by themselves are relatively inefficient at transporting sand far from a shoreline and a significant proportion of the 'sand' layers do not show wave rippling, so wave rippling was a modifying process, not the main transport mechanism.

Aqueous current mechanisms are ruled out by the sharp tops of the unreworked strata and their lack of current ripples, normal grading, and parting lineation. Wind transport across a dried-up lake floor, on the other hand, explains the sorting, consistent grain size and widespread distribution of the 'sands'. Indeed, large aeolian sand accumulations are found within marginal Middle Devonian alluvial fans in the basin (Astin, 1985; Rogers, 1987). So long as the lake floor remained a transport zone for wind-blown sand, the amount of sand present in any one place would be small, giving rise only to thin sand patches when the wind dropped. The planed-off tops of some of the wave ripples may have been deflated. The sharp, non-erosive bases and tops, lack of overall grading, occasional flat lamination, and large-scale patchiness of the unreworked 'sand' strata (Fig. 4) are all consistent with an origin as wind-rippled or aeolian plane-bedded (cf. Hunter, 1977), low-amplitude sand patches (Fig. 8).

Thus each mud−sand couplet is interpreted as the result of an episode of lake desiccation and aeolian supply of sand, which often was reworked by waves as the lake refilled, before being blanketed by mud (Fig. 8). The cohesive mud would have been resistant to wave reworking as the lake dried up and perhaps had the added protection of an organic film.

Desiccation rather than subaqueous 'syneresis'. Previously, cracks with lenticular geometries both in the Orcadian Basin (Fannin 1970; Donovan & Foster, 1972) and elsewhere (Picard, 1966,1969; Fuchtbauer & Tisljar, 1975; Clemmey, 1978) have been attributed to subaqueous shrinkage, or 'syneresis'. Astin & Rogers (1991) review Orcadian crack geometries and the subaqueous shrinkage literature and reinterpret such cracks as gypsum pseudomorphs and desiccation cracks. The Orcadian evidence can be summarized as follows. All gradations are seen (Fig. 6) between surfaces with lenticular crack arrays and those with polygons of undisputed subaerial origin. Often both types of crack occur on a single surface. The lenticular cracks are shallow, usually penetrating only a single mud layer, whereas polygonal cracks may be deeper. Both types of crack are overlain and infilled by a

similar range of 'sand' strata of inferred aeolian origin. We infer a common origin for all cracks, by subaerial desiccation, and argue that the unjoined crack pattern and absence of curled mud flakes resulted not from some different process, but from the limited potential for shrinkage of the thin, single layers of mud which is poor in clay. A high water table is suggested to have inhibited the formation of deeper cracks.

We contend that the lenticular geometry is so common because of crack nucleation upon gypsum crystals. Many of the cracks (Fig. 6E) closely resemble such crystals in plan, both in their basic shape and their quadruple junctions reminiscent of interpenetrating crystals, but atypical of desiccation cracks (Allen, 1987). Lenticular cracks of similar size occur with very variable densities on different surfaces. This can be explained by variable crystal density, but not by different degrees of contraction, which is normally accommodated by variations of crack width and elongation, rather than of crack density. The occasional aligned patterns of cracks (Donovan & Foster, 1972) can be explained by preferred directions of crystal elongation in response to concentration gradients. We suggest that flat-lying gypsum crystals grew in the tops of the mud layers during the last stages of lake desiccation, or following emergence of the damp surface sediments (Fig. 8). In some cases, halite crystals (Fig. 6E) were precipitated as well. The limited volumes of evaporite implied by the pseudomorphs reflect the low salinity of the lake waters prior to evaporation. Pseudomorphing was achieved by incorporation of 'sand' into the crystals during growth, or by infill following dissolution either at the surface, owing to local rain or ground-water seepage, or during burial.

Clastic mud event deposits. The occasional thick, clastic mud beds (Fig. 4E) were subject to shrinkage cracking and exfoliation (Fig. 5B), both probably the result of desiccation. They represent exceptional events of clastic supply to this point in the basin. Those with silt−clay laminated bases (Fig. 5A) can be attributed to traction in dilute, muddy turbidity currents within the lake, or fluvial sheet-floods (cf. Stow & Shanmugam, 1980). The remainder may be under-, inter-, or overflow deposits. The graded beds probably represent rapid events of waning flow, but the structureless ones may have been deposited over several weeks by river-generated flows. Similar beds in Orcadian permanent lake facies have been mentioned above. River influx was both the likely source of the sediment and the generating mechanism for the flows, whether in the lake or across its dried-up bed, though storms might be invoked. Fluvial generation is supported by the increasing frequency of such beds up towards the major sand bodies (Fig. 3).

Wind-blown clay dunes. The three undulatory topped thick mudstones (Figs 4F, 7) are probably further evidence for wind transport on the lake floor. The cross-bedding present in two of them indicates that their topography is depositional rather than being erosionally sculpted. The unusual phenomena of bedforms made of mud can be explained if they were produced from sand-grade mud pellets. An analogy may be drawn with the 'clay dunes' formed from wind-blown pellets of evaporite-impregnated mud (with $>8\%$ clay) under restricted environmental conditions in modern arid and semiarid areas (Bowler, 1973). Such pellets are created by disruption of a mud surface by efflorescent evaporite crystals, especially gypsum, by mechanical disintegration of wind-blown mud curls, and by collapse of blistered, evaporite-cemented crusts, and are held together by internal damp (Price, 1963). Given the context of the beds (Fig. 3), it is very unlikely that the pellets were of the type formed by fluvial reworking of deeply and repeatedly desiccated vertisols (Rust & Nanson, 1989). Most descriptions of aeolian clay bedforms concentrate on the foredunes ('lunettes'), up to tens of metres high, stabilized in the lee of clay/salt pans by vegetation and by hygroscopic dampening and rain rendering the pellets plastic and adhesive. However, Price (1963) describes salt flats with mobile ripples and small dunes analogous to those formed in sand, as well as stabilized fields of giant (<1 m high) slip-faceless 'desert ripples'. The Wick dunes are probably aeolian bedforms of this type, but could possibly be the product of flood reworking of aeolian pellets, though the pellets would have disaggregated if immersed for very long.

Facies 3: flagstone with dolostones

This facies (Fig. 9A,B) has gradational transitions to the simple flagstones. Its muds tend to be richer in carbonate (ferroan dolomite) and there are grey-black ferroan dolomicrite strata weathering a distinctive yellow. It differs also in that deep shrinkage

cracks, which penetrate more than one mud stratum, are present in a few layers (eight in the lower unit measured in detail — Fig. 3), and the 'sand' strata occasionally partly comprise dolomicrite pellets of up to medium sand grade (Fig. 9C). The modal mud stratum thickness is consistently small and there are very few thick, clastic muds (three in the lower unit measured in detail and one in the upper), as reflected by the reduced mean thickness (Fig. 3). The majority of thick mud beds on the histograms for this facies are dolostones. These occur in all possible situations: between mud laminae, between sand strata, or separating the two. Their early lithification is shown by the brittle deformation undergone by some, while surrounding layers remained ductile. Both boudins and thrusts occur, in variable orientations. Such beds in this part of Caithness contain $0-15\%$ calcite (Donovan, 1971; Janaway & Parnell, 1989). Janaway (pers. comm.) reports microscopic pseudomorphs after evaporites in dolostones from Wick. Plimmer (1974) states that the calcite component of similar rocks in Orkney is dedolomite.

Interpretation

This facies is attributed to a similar range of processes as the simple flagstones, but in a slightly drier setting and with a reduced clastic input by rivers; hence the increased proportion of carbonate in the muds and the reduced number of underflow/ sheet-flood muds. The water table occasionally fell low enough for shrinkage cracks to penetrate deeper than one mud stratum. The reduced modal mud thickness (Fig. 3) probably records shorter ephemeral lake durations. Janaway & Parnell (1989) report stable isotope and other geochemical data from the dolostone strata. Janaway (pers. comm.) suggests an origin for such beds, with their microscopic pseudomorphs, by carbonate mud precipitation on exposed mud flats as a result of evaporative concentration of ground-water, with lithification into crusts following repeated desiccation. This would be consistent with the drier depositional setting; the dolostones probably represent relatively prolonged periods of exposure of the lake floor with the water table at, or close to the surface. The dolomicrite pellets in some of the 'sand' strata (Fig. 9C) can be attributed to wind and wave reworking of crusts brecciated by desiccation.

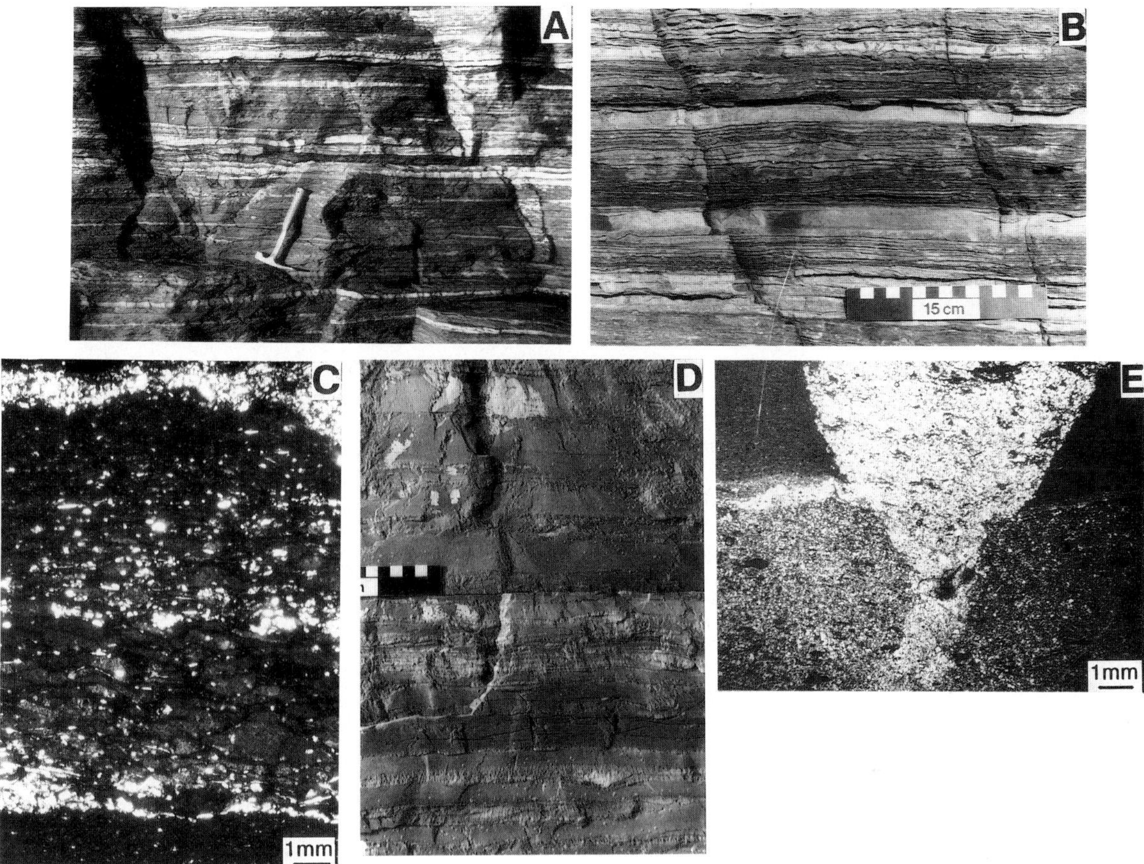

Fig. 9. (A) Flagstones with dolostones (facies 3). (B) Facies 3 close-up. (C) Facies 3, 'sand' strata composed largely of dolomicrite pellets. (D) Thick mudstones in facies 5. (E) Facies 5, mudclast-rich 'sand' stratum and crack infill.

Facies 4: disrupted laminite and 'sand'-starved flagstone

This facies (Fig. 10B,C,D) forms the gradational transition between simple flagstone and laminite. It differs from the latter in that its carbonate is mostly dolomite, in having less carbonate, in the presence of intervals of apparently unlaminated mudstone, in that some of its laminae have a disrupted, even brecciated, appearance, and in having a variable proportion of 'sand' (in the form of millimetre-thick strata, series of lenses, or isolated crack fills, gradational in form to those of the flagstones). The facies may be a relatively smooth transition from laminite to flagstone, or may comprise a centimetre to decimetre-scale alternation of laminite and 'sand'-poor flagstone. Bedding planes display lenticular and occasionally polygonal cracks as well as 'sand'-filled halite pseudomorphs.

Interpretation

Orcadian laminites are typically bounded by this transitional facies, formerly attributed to shallow permanent lakes with restricted sediment supply (Donovan *et al.*, 1974; Plimmer, 1974; Donovan, 1980). For similar reasons to those discussed above, it is interpreted here as the product of lakes that were ephemeral, but which desiccated less frequently than the flagstone lakes and were more often stratified. The brecciation of mud laminae is probably the

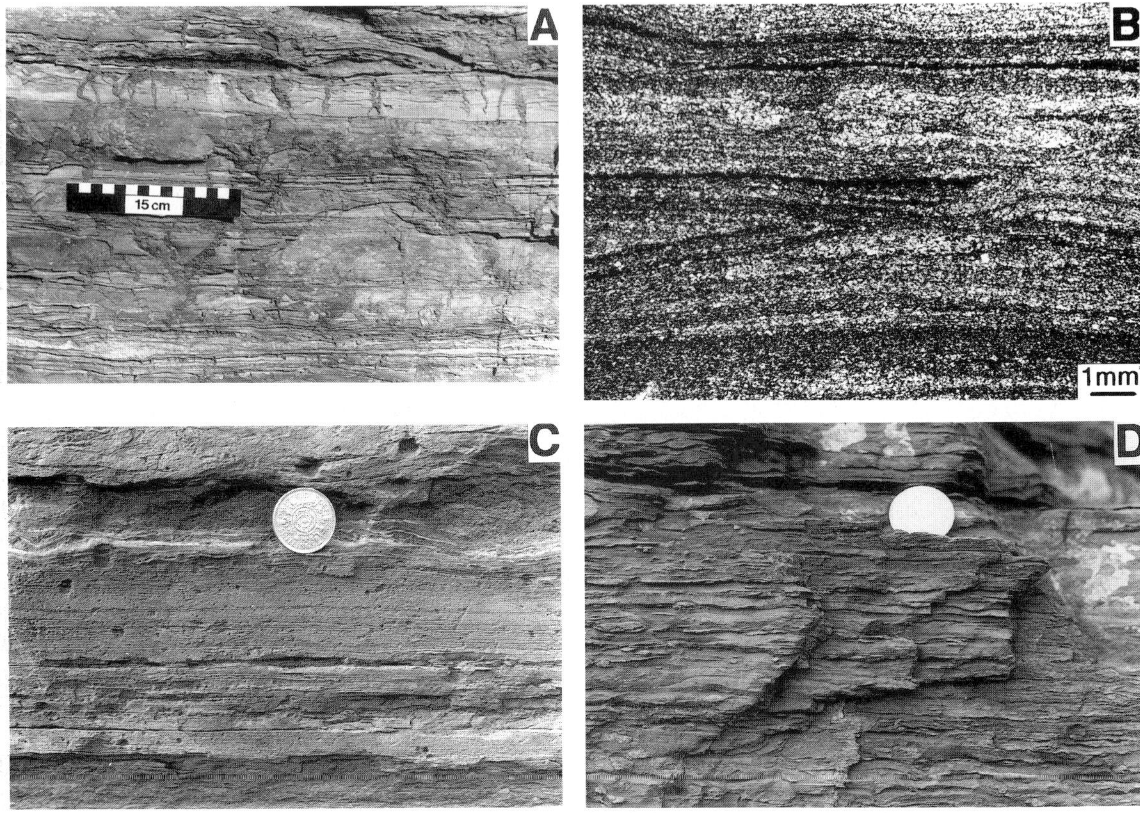

Fig. 10. (A) Facies 5: thick, turbate sands with deeply penetrating cracks. (B) Facies 4, disrupted laminite in section. (C) Upward transition from laminite to disrupted laminite. (D) Facies 4, 'sand'-starved flagstones.

result of desiccation. The reduced amount of 'sand' can be explained by less frequent and shorter desiccation events, by the concentration of sand in beaches at the edges of the longer-lived lakes and, above the laminites, by their blanketing of any previous coarse silt or sand accumulations on the lake bed.

This interpretation is supported by the carbonate petrography and geochemistry of Janaway & Parnell (1989). In such transitional facies, they describe shrinkage-cracked dolomicrite crusts attributed to evaporative pumping, fenestral dolomicrite formed as algal mats, and pelleted dolomites produced by reworking of mats or crusts, as well as the types of carbonate strata seen in the true laminite facies. They quote geochemical data that support evaporative concentration of the dolomitizing lake or ground-waters. The data of Duncan & Hamilton (1988), who found consistent $\delta^{18}O$ enrichment of

dolomite laminae relative to calcite in laminites, seem to fit the same model.

Facies 5: flagstones with turbate sandy muds and muddy sands

Below the sand bodies, the simple flagstones grade up into this transitional facies by (i) an increase (Fig. 3) in the number and thickness of thick, dark green, carbonate-poor muds (Fig. 9D), (ii) a reduced frequency of thin, dark blue-grey, carbonate-bearing mud laminae, (iii) increasing 'sand' stratum thicknesses, and (iv) a slight coarsening of the 'sand' grain-size range (to coarse silt to fine sand). Deep, polygonal shrinkage cracks, often penetrating several beds, become much more abundant than lenticular cracks (Fig. 10A). Mudstone intraclasts appear (Fig. 9E) and comprise 50% of some sands. For intervals of several centimetres it is impossible

Fig. 11. Facies 6: the upper (A) and lower (B) thick sandstone bodies (folded 1-m ruler for scale); (C) erosive base of the upper body; (D) the lower body (from base up: wave rippled sands; convoluted sands and sandy muds with traces of wave-ripple cross-lamination; flat laminated sand lens below scale; wave-rippled, mud-draped sand, grading to structureless sandstone).

to distinguish thin strata within turbate, poorly sorted sandy mud or muddy sand, apparently produced by repeated shrinkage and infill by mudclast-bearing sand (Fig. 10A).

Interpretation

The increasing frequency and thickness of river-generated muddy underflow or sheet-flood beds upwards into this facies (Figs 2, 3) probably indicates the progradation of a low-gradient terminal fan into the ephemeral lake basin. Fluvial processes possibly contributed to the transport and reworking of sand, though the continued presence of wave ripples, wind-blown 'sand' strata, and occasional carbonate-bearing muds demonstrates the continuation of the range of processes responsible for simple flagstone deposition. The presence of deep mud cracks, of

mud intraclasts, and of turbate sand attributable to repeated desiccation is probably the result mainly of deeper desiccation, but may partly reflect increased mud bed thicknesses and clay content. The water table may have been able to fall further because of increased aridity and/or because of relief generated by the terminal fan.

Facies 6: thick sandstone bodies

These two sandstone sheets (Fig. 11) have sharp, locally erosional contacts with the underlying transitional facies (Fig. 11C). The lower one has an abrupt top, whereas the upper is overlain by a thin interval of facies 5. In many places, sedimentary structures are difficult to discern in the sand bodies because the sand is well sorted and indurated.

Apart from some shrinkage-cracked mud drapes

in ripple troughs near its base, the upper sand body (Fig. 11A,C) consists of very fine to fine, or in one place fine to medium sandstone. It is divided up by wave-rippled surfaces, one to tens of centimetres apart, which over 50–100 m in an east-northeast direction descend from high in the body to near its base. Between these surfaces the sands are either wave-ripple cross-laminated, or apparently structureless. Thus the body is made up of a series of sand lenses that accreted northeastward.

The lower body (Fig. 11B,D) is dominantly fine–medium sand, with lenses and laminae reaching very coarse sand grade and with muddy partings a few millimetres thick in its lower half. Most sections are structureless or show smooth-based scours, 0–10 cm deep and 0.5–1 m across, filled by low-angle, poorly defined laminae, or weakly defined undulatory lamination with low-angle discontinuities. Discernible in places are wave-rippled surfaces and cross-lamination, convolute lamination, flat lamination, and isolated cross-sets up to 5 cm thick.

Interpretation

The poorly defined structures in these beds make their interpretation difficult. Like other sand bodies in Orcadian lake cycles (Fannin, 1970; Astin, 1990), these are systematically positioned at the top of successions that coarsen up and show increasing proportions of beds supplied directly by rivers or river-generated flows (Figs 2, 3). This suggests that a fundamental control on sand body origin was a combination of terminal fan progradation and subsequent abandonment or retreat. However, in contrast to the fluvial sand bodies seen in some cycles, the two at Wick show structures recording wave action rather than unidirectional current structures. The upper sand body was periodically reworked by lake waves to give the well-defined partings, but was not subject to continuous wave activity. Probably it was easily drowned so that each wave-rippled surface records a lake filling episode. Between such episodes, sand accreted on the east-northeast side of the sand body — its lee relative to the prevailing wind and also the direction of the probable centre of the basin (Astin, 1985; Rogers, 1987). The sedimentary structures do not point clearly to any particular process of accretion, but likely mechanisms, given the bed's context, are aeolian accretion and beach-face reworking. A similar origin is likely for the lower body; the flat lamination (Fig. 11D) is probably swash lamination (Clifton, 1969), while the thin

cross-beds may record berm or ridge migration. At different states of lake filling, the sand bodies were probably barrier beaches, nearshore bars, and exposed sand patches subject to aeolian reworking.

DISCUSSION

Comparison with previous interpretations

We attribute the flagstone facies, whatever its mud crack geometries or scale of lamination, to deposition of mud in ephemeral lakes, sometimes stratified, which dried out and refilled at intervals of one to tens of years in the manner of Lake Eyre in central Australia (Wells & Callen, 1986), and of sand and coarse silt blown across the dried-up lake bed (Fig. 8). Our interpretation contrasts with those (see Table 1) of Fannin (1970), Donovan (1971,1978,1980), Foster (1972), and Plimmer (1974), followed by Mykura (1976), Duncan & Hamilton (1988), and Hamilton & Trewin (1988). These models subdivided Orcadian flagstones according to whether both desiccation polygons and 'subaqueous shrinkage cracks', or just the latter were present, into facies deposited, respectively, in ephemeral lakes and shallow permanent water. The 'sand' strata were attributed to either, or both, turbidity flows or wave transport. Plimmer, Donovan, and Foster attributed the cracking and interstratification to short-term, probably seasonal variations of water cover, salinity, wave activity, and coarse sediment supply, as opposed to the rather longer term changes we propose.

We suggest that our reinterpretation applies to flagstones (thinly interstratified shrinkage-cracked muds and sands or coarse silts) throughout the basin. Unlike the previous model, it explains the intergradational nature of the various types of flagstone and the similarity of their coarse strata. In many Orcadian sections, flagstones are complicated by the addition of one or more of: thin fluvial sheet-flood sands, stromatolites, gutter casts, and flagstone-filled channels (Fannin, 1970; Astin, 1990), all of which are consistent with the ephemeral lake interpretation.

Controls of cyclicity and origin of the facies sequence at Wick

Our reinterpretation leads to a revised model for the origin of the Orcadian cycles. Previous models (listed in the introduction) have invoked tectonic, autocyclic, climatic, and mixed controls. The sequence of facies at Wick can be interpreted best when it is compared with other central Orcadian sequences described by Fannin (1970), Donovan (1971,1978, 1980), Foster (1972), Plimmer (1974), and Astin (1990). In all these successions, permanent-lake laminites occur as regularly spaced beds of consistent thickness (Table 1), showing rapid gradational transitions at their margins as at Wick. In those successions that have been correlated over significant distances, the Stromness and Rousay Flagstones (24 km N−S, 15 km E−W: Fannin, 1970; Astin, 1990) and the Eday Flags (58 km N−S: Astin, 1985), only a few of the laminite beds die out laterally. The rest must represent deep permanent lakes that occupied much of the basin, so their occurrence cannot be attributed to autocyclicity (e.g. the progradation, lateral migration, or avulsion of river systems). In the minority of cycles where a laminite is absent, it is often 'replaced', as in the lower cycle at Wick, by an unusually thick unit of simple flagstones, implying that such units are also the result of a non-autocyclic control. This is confirmed by the fact that the laminites that do die out laterally do so by passage into such extra-thick flagstones (Fannin, 1970; Astin, 1990). Neither was autocyclicity the control of successions *between* laminites or their equivalents, for the order of facies in such sequences is not random. Rather, except where there are erosively based sand bodies, the facies are arranged, as at Wick, into symmetrical cycles of facies representing increasing then reducing ephemerality. Local tectonic control of the cyclicity, by episodic fault movements, also can be discounted because the cycles can be correlated over such large distances. However, regional tilting clearly affected cycle thickness; the correlations of Fannin (1970) and Astin (1990) (Table 1) show consistent thickening of groups of cycles from one section to another.

The sequence of facies at Wick, and in Orcadian lake cycles generally, is therefore the result mainly of a cyclic, climatically controlled variation in lake ephemerality, with river systems able to prograde further into the basin as lakes became more ephemeral. Some at least of the facies transitions probably represent the consequent migration of facies belts, but it is uncertain whether all the environments represented in the section were in existence at any one time (e.g. a permanent lake may or may not have been continuously present distally). Avulsion may have had a subsidiary role,

by controlling some of the non-laminite facies transitions. Models of the end-member states of the lake basin are given in Fig. 12. The inferred climate varied across part of the range from hot arid to hot subhumid. This is consistent with the subtropical to tropical latitude (c. 15°S) implied by Orcadian palaeowind directions (Allen & Marshall, 1981; Astin, 1985; Rogers, 1987) and by palaeomagnetic data (Tarling, 1985).

The Wick succession is therefore interpreted as follows. The dolostone-bearing flagstones formed during the most arid periods, with the lowest accumulation rates. Their passage up into simple flagstones represents a reduction in aridity, with thicker mudstone strata reflecting some combination of higher carbonate productivity in the lake and slightly longer lake durations. In the upper two cycles at Wick and in most Orcadian cycles, deep, permanent lakes formed next, owing to an excess of runoff over evaporation during the most humid climatic conditions, before a return to slightly more arid climate restored ephemeral lake conditions. The progradation of ephemeral rivers in the form of terminal fan(s) then caused a progressive increase in the abundance and grain size of terrigenous sand and mud beds, supplied subaerially and subaqueously respectively. Frequent lake-level variations allowed both the lake-floor flats and the terminal fans to be shaped by wind during exposure and by waves at low lake levels. With yet greater aridity, river progradation ceased, and the terminal fans became abandoned and covered by flagstones with dolostones. In the Wick section, sand bodies formed at the point of fan abandonment, with sedimentary structures mostly recording wave reworking.

The increased abundance of dolomite immediately above the sand bodies suggests progressive concentration of salts in the lake basin water over a climatic cycle. During long periods of lake ephemerality, salts would be added during each river flood and slowly concentrated by evaporation into the residual lake and the interstitial water of surface sediments. The only sink for magnesium would have been direct dolomite precipitation or the dolomitization of existing calcite; hence dolomite was deposited most abundantly towards the end of long periods of ephemerality.

Sulphate was temporarily deposited in each desiccation event as gypsum. Its build up over the long term would have been prevented by conversion to pyrite by sulphate-reducing bacteria in the lake

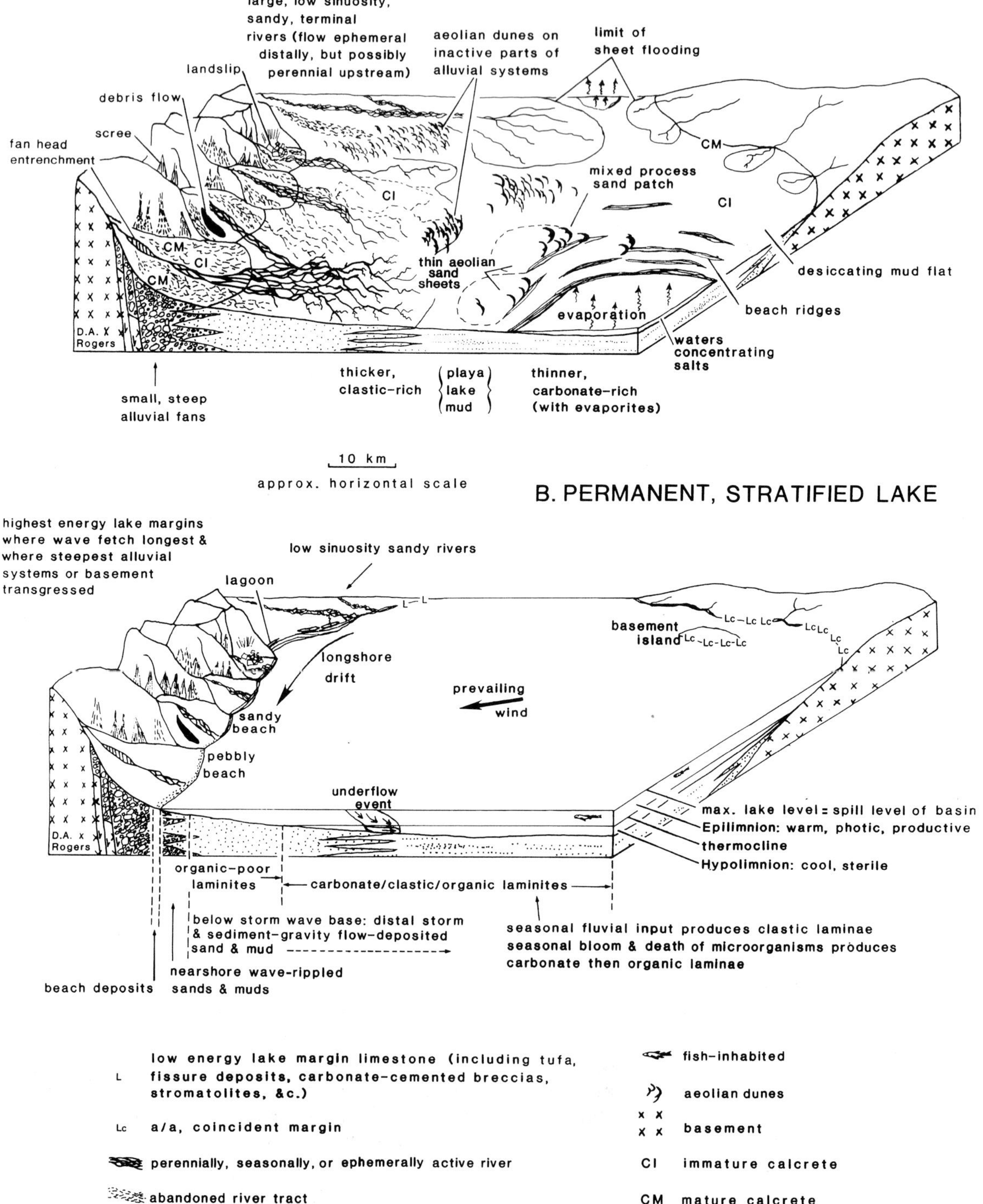

Fig. 12. Palaeoenvironmental models for the two end-member states of the Orcadian lake and surrounding environments. These diagrams have no directional significance.

sediment during periods when the ephemeral lake was full. When the lake filled up more permanently, depositing laminites, the lake waters tended to freshen owing to river inflow. If the lake reached a basin-spill point during these episodes, then salts could have been removed by overflow, but external drainage was not necessarily achieved at these times (see below).

Time-scales and controls of Orcadian climatic cyclicities

The discussion above has demonstrated (i) approximately annual cyclicity recorded by lamination within the laminites and some of the ephemeral lake muds (see also Rayner, 1963; Trewin, 1986), (ii) long-term climatic variations responsible for the c. 10 m cycles, and (iii) the intermediate period cyclicity of wet versus dry conditions in the ephemeral lakes and, at times, of water levels in the permanent lakes. The possibility that intermediate orders of cyclicity are recorded by systematic variations in flagstone mud strata thickness is still being investigated.

Figure 13 illustrates how the abrupt transitions between the end-member states of the Orcadian lake could possibly have been produced by quite gradual climatic variations, though the model proposed here does not require the climatic cyclicity to have been so smooth. The time axis is labelled to reflect the suggestion that the c. 10 m cycles record the Milankovitch obliquity cycle of about 20 000–25 000 years, possibly modulated by longer term orbital cycles (Hamilton & Trewin, 1988; Astin, 1990). This hypothesis is not essential to the model proposed here, but it is consistent with the data from Wick in a number of ways, as follows.

The laminite bed is 1.2 m thick (typical for laminites in the basin generally — Table 1). Its laminae average between 0.25 and 0.5 mm thick, implying a permanent lake duration of between 2400 and 4800 years — figures of the order of magnitude required by the hypothesis. The carbonate-rich mud strata in the flagstones are typically 1–4 mm thick (Fig. 3). Using the same assumed depositional rate, this suggests typical ephemeral lake durations of 4–16 years.

The completely ephemeral cycle, between the two sandstone bodies, has a total mudstone thickness of 7.04 m, whilst that of the half-cycle between the upper sand body and the laminite is 1.47 m. At similar depositional rates to the laminites, these muds would take totals of around 14–28 and 2–6 thousand years to form, respectively. These are overestimates because a significant proportion of the thicknesses comprises thick, carbonate-poor muds (Fig. 3) inferred to have had much higher rates of deposition. However, they are of the predicted magnitude.

An alternative approach is to use the *number* of

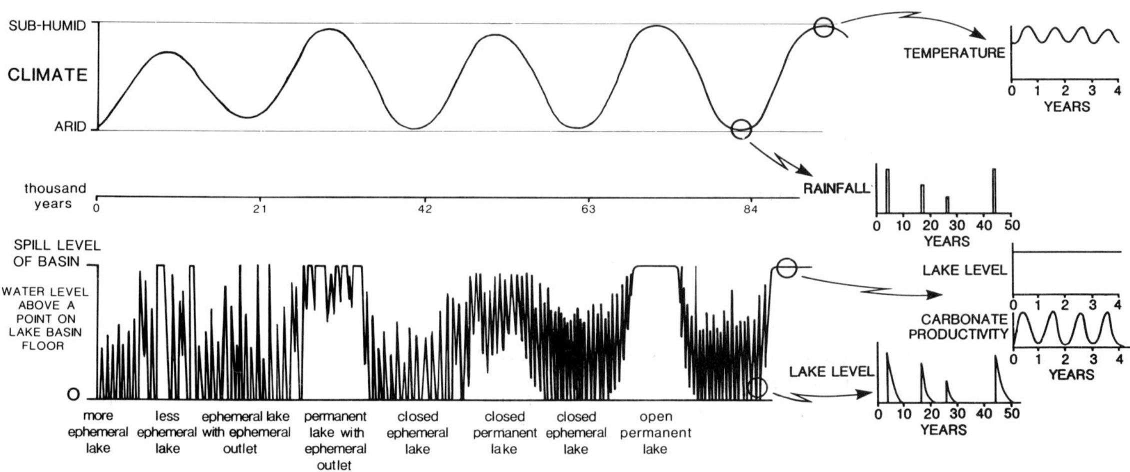

Fig. 13. Proposed mechanism for control of Orcadian lake level, ephemerality and chemistry by cyclicities of climate and weather. Note that permanent lakes may possibly have persisted for longer, perhaps continuously, in topographically lower locations.

thin mud strata, assuming each represents an ephemeral lake-filling event. The two flagstone intervals analysed contain 1314 and 363 mud strata, respectively. If the complete cycle represents about 20 000−25 000 years of an obliquity cycle and the flags directly beneath the laminite 7500−10 500 years, this would indicate, respectively, about 15−19 and 20−28 years on average between lake-filling events. Some mud strata probably represent more than one such event and the deposits of some lakes have no doubt been eroded, so these average flooding spacings may be slight overestimates. Nevertheless, they are consistent with the lake durations suggested by the modal thicknesses and are of the same order as the frequency (15 years) of filling events of Lake Eyre, Australia since 1945 (Wells & Callen, 1986).

Lake levels and outflows

Obviously the ephemeral lakes would have been hydrologically closed as they dried up (Fig. 12A), but some may have had outlets when at their deepest. The lateral passage of a few laminites into flagstones is demonstrable in Orkney (Fannin, 1970; Astin, 1990), so some at least of the permanent lakes were surrounded by areas of ephemeral standing water and therefore were closed or had only ephemeral outlets (Fig. 13). This may not have been true, though, of some of the other laminite-depositing lakes; some may have reached the basin's spill level and had *fluvial* outlets for most of their duration (models involving a connection *direct* to the sea during high lake levels require special pleading given the great thickness of the main lake succession and its lack of marine microfossils). However, the spill level of such a large basin is likely to have varied only gradually, so we suspect that the basin had a permanent outflow only during the two exceptional episodes when permanent lakes were most widespread, transgressing high on to alluvial systems, during the deposition of the Achanarras and John o'Groats Fish Beds (Fig. 1).

The depths of the various types of lakes and the spill level that limited them must remain the subject of speculation. They are difficult to constrain using wave-base estimates in the absence of reliable estimates of fetch and wind speeds (Allen, 1981a,b; Duncan & Hamilton, 1988); the topography is uncertain even of the lake that deposited the most widely correlated Orcadian lake unit, the Achanarras Fish Bed (Fig. 1). Astin (1985) used the areal dis-tribution of beach intercalations in the alluvial Lower Eday Sandstone, with an estimate of fan gradient, to show that it was very rarely transgressed by lakes more than 20 m deep; probably most of the ephemeral lakes were shallower than this even at their fullest.

Implications of the model

The proposed model of geologically rapid lake level fluctuation provides a basis for analyses of likely water budgets, changing lake chemistry, and sources and sinks of salts during the lake's history. Understanding of the pattern of lake level variations above the flat basin centre is also the key to interpretation of fluvial−lacustrine interactions in more marginal Orcadian successions complicated by the effects of basement and alluvial relief. Permanent lake marginal limestones adjacent to the basement (Donovan, 1973,1975; Janaway & Parnell, 1989) and intercalated with alluvial/aeolian successions (Allen, 1981b; Allen & Marshall, 1981; Rogers, 1987) are obviously the correlatives of basin centre laminites, but beach sands and gravels found in such settings (Astin, 1982,1985; Rogers, 1987) are equivocal; they could represent permanent or ephemeral lake margins. Conversely, the great frequency of ephemeral lake transgressions predicted by our model implies that there may be many subtle lacustrine phenomena (thin beach sands, wave ripples, muds) yet to be detected in apparently fluvial Orcadian sequences. The model also implies that marginal mixed lacustrine/fluvial successions (Allen & Marshall, 1981; Rogers, 1987) cannot be attributed to gradual transgressions and regressions. Walther's Law must be used with great caution in their interpretation; for instance, even if the only lake intercalation in an otherwise fluvial succession is of a *permanent* lake facies, it is most likely that the rivers which deposited the surrounding sediment ran into an *ephemeral* lake.

The greater lake ephemerality recognized between laminite 'events' can be added to the environmental stresses that have been invoked to explain the low abundance of trace fossils in Orcadian lake deposits and the distribution of fish fossils in the succession, particularly the distinctness of faunas found at different horizons (Donovan *et al.*, 1974; Donovan, 1980; Hamilton & Trewin, 1988). Fish are abundant in some laminite beds, but the flagstones have only rare and scattered bones. The prolonged episodes of desiccation probably excluded most soft-bodied organisms and killed off the majority of fish after a permanent lake episode. Some species perhaps aestivated or recolonized the ephemeral lakes from rivers supplying the basin or from short-lived outlets; any residual standing water would have been very saline. However, major recolonization probably had to await the next permanent lake event with a permanent or semi-permanent fluvial outflow.

CONCLUSIONS

1 The most abundant Orcadian lake facies, 'flag-stone' (shrinkage-cracked, thin interbeds or inter-laminations of mud and coarse silt or very fine sand) is reinterpreted as an interstratification mostly of lake suspension deposits and of coarser sediment blown across dried-up lake floors. Thus over the present onshore outcrop at least, the Orcadian lake was ephemeral for most (c. 90%) of its history, probably filled, to a few metres only, every few to tens of years.

2 The frequently cited examples of subaqueous (or 'syneresis') cracks in the flagstones are reinter-preted as gypsum pseudomorphs and desiccation cracks.

3 Halite pseudomorphs, occasional wind-blown mud pellet bedforms, and rain prints provide further new evidence in the flagstones for evaporites and sub-aerial exposure.

4 Only during the deposition of the volumetrically subordinate 'laminite' facies (less than 10% of most cycles — Table 1) did standing water cover much of the basin floor permanently (for hundreds or thousands of years).

5 Within the ephemeral lake facies, variations of mud stratum thickness and composition and of desiccation crack shape record increasing and de-creasing aridity.

6 Heralded by muddy underflows or sheet floods, rivers prograded across the ephemeral lake beds as terminal fans, but these sometimes retreated or were abandoned during the most arid conditions.

7 The lake's ephemerality allowed extensive aeolian and lake-margin reworking of sediment across the lake floor and on the terminal fans (Fig. 12A). Thus it is possible that reservoir-scale sand bodies may have accumulated in central parts of the basin, where previously they have not been expected. The key to predicting such sand bodies will be an understanding of the palaeowind pattern in the basin and its possible interaction with tectonically controlled topography.

8 The Orcadian lake deposits record cyclicities of climate and weather on at least three scales: (i) approximately annual temperature and rainfall variations responsible for lamination in laminites and in some ephemeral lake mudstones; (ii) the c. 10–30 year cyclicity that controlled the filling of the ephemeral lakes and, at times, the water levels of the permanent lakes; (iii) the long-term cyclic climatic variations, which may or may not have been smooth, responsible for the abrupt alternation of ephemeral and permanent lake facies on the 10 m scale, probably controlled by the obliquity cycle of c. 20000–25000 years and possibly modulated by longer term orbital cycles (Fig. 13). Further detailed analyses, especially of 'sand'–mud couplet thick-nesses and desiccation features may yield a more complete understanding of climatic change on time-scales less than 20000 years.

9 The revised model of lake level fluctuation has significant implications for understanding of the lakes' chemistry, of the palaeoenvironmental con-trols of more marginal Orcadian sequences, and of the distribution of fossils in the succession.

ACKNOWLEDGEMENTS

We are grateful for the help and advice of Drs P.A. Allen and N.H. Trewin, who reviewed the manuscript, of Drs P. Anadón, J.E.A. Marshall, P.F. Friend, S.J. Hillier, J. Parnell, T.M. Janaway, and J.E. Tyler, and of Professor J.R.L. Allen. DAR thanks the Royal Commission for the Exhibition of 1851 for a Research Fellowship and the Department of Earth Sciences, Oxford University, for use of facilities. This is Reading University P.R.I.S. Con-tribution No. 049.

REFERENCES

ALLEN, J.R.L. (1987) Desiccation of mud in the temperate intertidal zone: studies from the Severn estuary and eastern England. *Philos. Trans. R. Soc. London, Ser. B* **315**, 127–156.

ALLEN, P.A. (1981a) Wave-generated structures in the Devonian lacustrine sediments of south-east Shetland and ancient wave conditions. *Sedimentology* **28**, 369–379.

ALLEN, P.A. (1981b) Devonian lake margin environments and processes, SE Shetland, Scotland. *J. Geol. Soc. London* **138**, 1–14.

ALLEN, P.A. & MARSHALL, J.E.A. (1981) Depositional environments and palynology of the Devonian southeast Shetland Basin. *Scott. J. Geol.* **17**, 257–273.

ASTIN, T.R. (1982) *The Devonian geology of the Walls Peninsula, Shetland.* Unpublished PhD thesis, University of Cambridge, 250 pp.

ASTIN, T.R. (1985) The palaeogeography of the Middle Devonian Lower Eday Sandstone, Orkney. *Scott. J. Geol.* **21**, 353–375.

ASTIN, T.R. (1990) The Devonian lacustrine sediments of Orkney, Scotland; implications for climate cyclicity, basin structure and maturation history. *J. Geol. Soc. London* **147**, 141–151.

ASTIN, T.R. & ROGERS, D.A. (1991) 'Subaqueous shrinkage' or 'syneresis' cracks in the Devonian of Scotland reinterpreted. *J. Sediment. Petrol.* **61**, in press.

BOWLER, J.M. (1973) Clay dunes: their occurrence, formation and environmental significance. *Earth Sci. Rev.* **9**, 315–338.

CLEMMEY, H. (1978) A Proterozoic lacustrine interlude from the Zambian Copperbelt. In: *Modern and Ancient Lake Sediments* (Eds Matter, A. & Tucker, M.E.), Spec. Publ. Int. Assoc. Sediment. **2**, 259–278.

CLIFTON, H.E. (1969) Beach lamination: nature and origin. *Mar. Geol.* **7**, 553–559.

CRAMPTON, C.B. & CARRUTHERS, R.G. (1914) *The geology of Caithness.* Memoir, Geological Survey of Scotland (sheet 110 & 116, & parts of 109, 115 & 117), HMSO, Edinburgh, 194 pp.

DE RAAF, J.M., BOERSMA, J.R. & GELDER, A. (1977) Wave-generated structures and sequences from a shallow marine succession, Lower Carboniferous, County Cork, Ireland. *Sedimentology* **24**, 451–483.

DONOVAN, R.N. (1971) *The geology of the coastal tract near Wick, Caithness.* Unpublished PhD thesis, University of Newcastle-upon-Tyne.

DONOVAN, R.N. (1973) Basin margin deposits of the Middle Old Red Sandstone at Dirlot, Caithness. *Scott. J. Geol.* **9**, 203–211.

DONOVAN, R.N. (1975) Devonian lacustrine limestones at the margins of the Orcadian Basin, Scotland. *J. Geol. Soc. London* **131**, 489–510.

DONOVAN, R.N. (1978) Caithness. In: *A field guide to selected outcrop areas of the Devonian of Scotland, the Welsh borderland and South Wales* (Eds Friend, P.F. & Williams, B.P.J.), pp. 38–53. Palaeontological Association, London.

DONOVAN, R.N. (1980) Lacustrine cycles, fish ecology and stratigraphic zonation in the Middle Devonian of Caithness. *Scott. J. Geol.* **16**, 35–50.

DONOVAN, R.N. & FOSTER, R.J. (1972) Subaqueous shrinkage cracks from the Caithness Flagstone Series (Middle Devonian) of north east Scotland. *J. Sediment. Petrol.* **42**, 309–317.

DONOVAN, R.N., FOSTER, R.J. & WESTOLL, T.S. (1974) A stratigraphical revision of the Old Red Sandstone of north-eastern Caithness. *Trans. R. Soc. Edinburgh* **69**, 167–201.

DUNCAN, A.D. & HAMILTON, R.F.M. (1988) Palaeolimnology and organic geochemistry of the Middle Devonian in the Orcadian Basin. In: *Lacustrine Petroleum Source Rocks* (Eds Fleet, A.J., Kelts, K. & Talbot, M.R.) Geol. Soc. London Spec. Publ. **40**, 173–201.

FANNIN, N.G.T. (1970) *The sedimentary environment of the Old Red Sandstone of western Orkney.* Unpublished PhD thesis, University of Reading.

FOSTER, R.J. (1972) *The solid geology of north-east Caithness.* Unpublished PhD thesis, University of Newcastle-upon-Tyne.

FUCHTBAUER, H. & TISLJAR, J. (1975) Peritidal cycles in the Lower Cretaceous of Istria (Yugoslavia). *Sediment. Geol.* **14**, 219–233.

HAMILTON, R.F.M. & TREWIN, N.H. (1988) Environmental controls on fish faunas of the Middle Devonian Orcadian Basin. *Can. Soc. Petrol. Geol. Mem.* **14**, 589–600.

HUNTER, R.E. (1977) Basic types of stratification in small aeolian dunes. *Sedimentology* **24**, 361–387.

JANAWAY, T.M. & PARNELL, J. (1989) Carbonate production within the Orcadian Basin, northern Scotland: a petrographic and geochemical study. *Palaeogeogr. Palaeoclimatol. Palaeoecol.* **70**, 89–105.

MARSHALL, J.E.A., BROWN, J.F. & HINDMARSH, S. (1985) Hydrocarbon source rock potential of the Devonian rocks of the Orcadian Basin. *Scott. J. Geol.* **21**, 301–320.

MYKURA, W. (1976) *Orkney and Shetland. Institute of Geological Sciences British Regional Geology Handbook.* HMSO, Edinburgh, 149 pp.

PARNELL, J. (1985) Hydrocarbon source rocks, reservoir rocks and migration in the Orcadian Basin. *Scott. J. Geol.* **21**, 321–336.

PICARD, M.D. (1966) Oriented, linear shrinkage cracks in Green River Formation (Eocene), Raven Ridge area, Uinta Basin, Utah. *J. Sediment. Petrol.* **36**, 1050–1057.

PICARD, M.D. (1969) Oriented, linear shrinkage cracks in Alcova Limestone Member (Triassic), southern Wyoming. *Contrib. Geol.* **8**, 1–7.

PLIMMER, R.S. (1974) *The sedimentology and stratigraphy of the Middle Old Red Sandstone Rousay Group of the Orkney Islands.* Unpublished PhD thesis, University of Newcastle-upon-Tyne, 343 pp.

PRICE, W.A. (1963) Physicochemical and environmental factors in clay dune genesis. *J. Sediment. Petrol.* **33**, 766–778.

RAYNER, D.H. (1963) The Achanarras Limestone of the Middle Old Red Sandstone, Caithness, Scotland. *Proc. Yorks. Geol. Soc.* **34**, 117–138.

ROGERS, D.A. (1987) *Devonian correlations, environments and tectonics across the Great Glen Fault.* Unpublished PhD thesis, University of Cambridge, 252 pp.

ROGERS, D.A., MARSHALL, J.E.A. & ASTIN, T.R. (1989) Devonian and later movements of the Great Glen Fault system, Scotland. *J. Geol. Soc. London.* **146**, 369–372.

RUST, B.R. & NANSON, G.C. (1989) Bedload transport of mud as pedogenic aggregates in modern and ancient rivers. *Sedimentology* **36**, 291–306.

STOW, D.A.V. & SHANMUGAM, G. (1980) Sequence of structures in fine-grained turbidites: comparison of modern deep sea and ancient flysch sediments. *Sediment. Geol.* **25**, 23–42.

TARLING, D.H. (1985) Palaeomagnetic studies of the Orcadian Basin. *Scott. J. Geol.* **21**, 261–273.

TREWIN, N.H. (1986) Palaeoecology and sedimentology of the Achanarras Fish Bed of the Middle Old Red Sandstone, Scotland. *Trans. R. Soc. Edinburgh, Earth Sci.* **77**, 21–46.

WELLS, R.T. & CALLEN, R.A. (Eds) (1986) The Lake Eyre Basin — Cainozoic sediments, fossil vertebrates and plants, landforms, silcretes and climatic implications. *Australasian Sedimentologists Group Field Guide Series*, vol. 4. Geological Society of Australia, Sydney, 178 pp.

WILSON, G.V., EDWARDS, W., KNOX, J., JONES, R.C.B. &

STEVENS, J.V. (1935) *The geology of the Orkneys*. Memoir, Geological Survey of Scotland (sheets 117–122). HMSO, Edinburgh, 205 pp.

Spec. Publs Int. Ass. Sediment. (1991) **13**, 223–243

Wave-dominated lacustrine facies and tectonically controlled cyclicity in the Lower Carboniferous Horton Bluff Formation, Nova Scotia, Canada

A.T. MARTEL[1] *and* M.R. GIBLING

Department of Geology, Dalhousie University, Halifax, Nova Scotia, B3H 3J5, Canada

ABSTRACT

The Lower Carboniferous (Tournaisian) Horton Bluff Formation was deposited in a tectonically subsiding basin with a major bounding fault (Cobequid Fault precursor) to the north and with onlapping relationships to the south. Excellent exposure of the Middle and Upper Members along the basin axis shows that the members thicken northward and contain repeated shallowing upward cycles formed within a hydrologically open lacustrine system. An ideal cycle shows from base to top: (1) grey clayshale ('deep' lake); (2) alternating clayshale/wavy-bedded siltstone and sandstone (lake shoreline), composed of three subfacies (a) lenticular hummocky cross-stratified siltstone (sediment-starved, transitional zone), (b) wave-rippled sandstone (shoaling wave zone), and (c) planar-laminated siltstone (attenuated wave zone); (3) green rooted mudstone (marsh); and (4) tabular and nodular dolomite (early diagenetic subaqueous and pedogenic horizons).

Four types of lacustrine cycle are described: offshore dominated (sediment-starved), shoreline dominated, marsh dominated, and delta dominated. Cycles are thicker and contain fewer marsh-dominated cycles toward the northern basin margin. These trends in cycle type and thickness can be explained by more rapid subsidence toward the fault-bounded, northern basin margin. Cycle thickness decreases upward, reflecting a decreasing rate of tectonic subsidence within the basin.

INTRODUCTION

Lacustrine deposits occur in both the upper Middle Member and the Upper Member of the Tournaisian Horton Bluff Formation of the Minas Basin area, Nova Scotia. They consist predominantly of repeated sequences of clayshale overlain by wave-dominated shoreline deposits (including hummocky cross-stratification), which in turn are overlain by marsh deposits. The cyclic nature of the succession indicates that basinal filling involved repeated progradation of shoreline over offshore deposits. Whereas the Upper Member lacustrine cycles are interbedded with deltaic deposits, the upper Middle Member cycles lack deltaic strata and are inferred to have formed by longshore drift from distant (deltaic?) sources.

In this paper we document the facies of the cyclic sequences, identify four types of cycle based on facies proportions, and investigate the relationship of cycle type to geographical location within the fault-controlled, asymmetric Windsor sub-basin.

REGIONAL SETTING

The Horton Bluff Formation, up to 1025 m thick, is overlain by the fluvial Cheverie Formation and together they comprise the Horton Group within the Windsor sub-basin area of Nova Scotia (Fig. 1). Bell (1929, 1960) recognized three gradational members of the Horton Bluff Formation (Fig. 2), which broadly conform to depositional environments: (i) Lower Member, fluvial; (ii) Middle Member, fluvial/lacustrine below to open lacustrine above; and (iii) Upper Member, lacustrine/deltaic (Fig. 2). The age of the formation is Tournaisian as determined from plant fossils and invertebrates (Bell, 1960), and from miospores (Playford, 1963; Hacquebard, 1972; Utting, 1987; Utting *et al.*, 1989).

[1] Present address: Department of Earth Sciences, Parks Road, Oxford OX1 3PR, UK

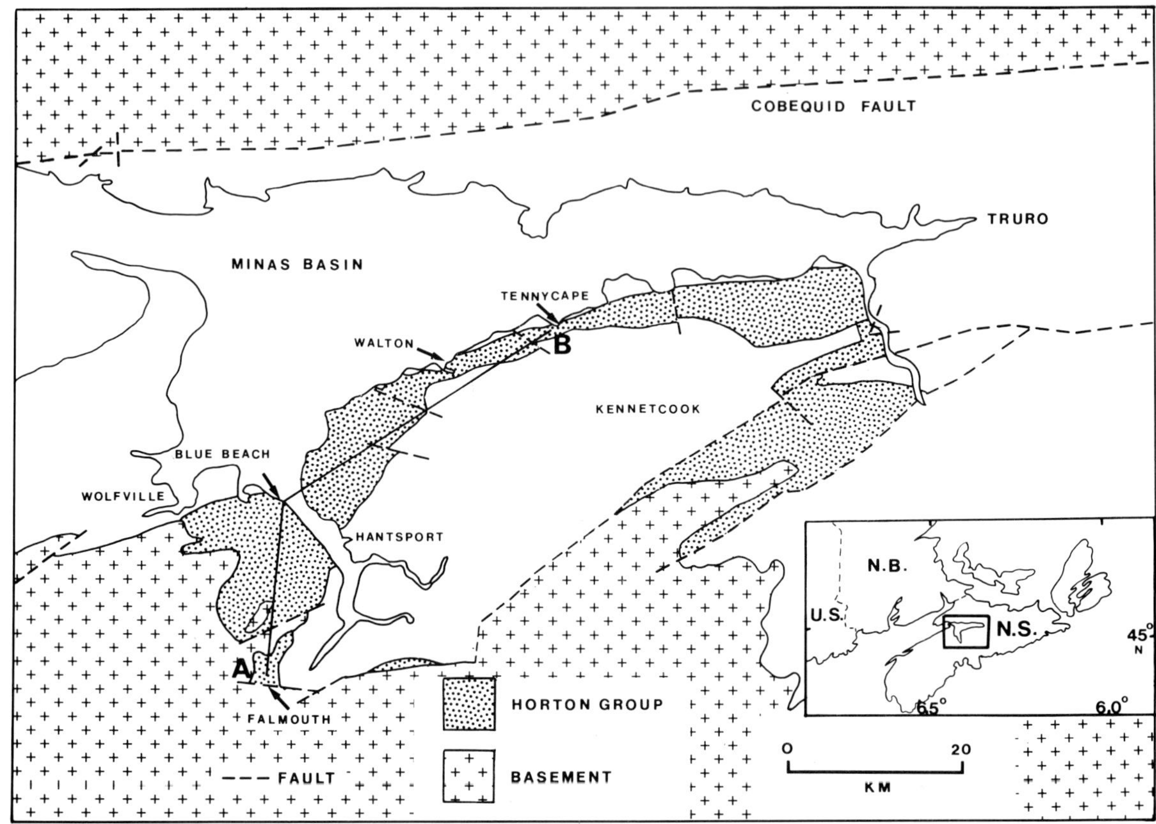

Fig. 1. Geological map of the study area (Windsor sub-basin). The basement is composed of undifferentiated Meguma Group and granitic rocks south of the Cobequid Fault (Meguma terrane), and granitic and metamorphic rocks of the Avalon terrane to the north of the Cobequid Fault. The Horton Group is composed of undifferentiated Horton Bluff and Cheverie Formations. The blank areas onshore indicate younger Carboniferous and Mesozoic rocks. Inset shows location of map in Nova Scotia (N.S.) and the relationship to New Brunswick (N.B.) and the United States of America (U.S.). Cross section A−B is shown in Fig. 2.

The Cobequid Fault forms the present-day northern margin of the Windsor sub-basin (Fig. 1.). The fault is known to have been a major zone of dextral strike-slip during the Late Carboniferous (Eisbacher, 1969; Mawer & White, 1986; Yeo & Gao, 1986; Nance, 1987). The Meguma and Avalon terranes were juxtaposed along the line of the Cobequid Fault during the Devonian Acadian Orogeny (Schenk, 1981; Keppie, 1982) and Tournaisian volcanics are present in the Cobequid Hills north of the present-day fault trace (Utting et al., 1989). Mooney (1987) concluded that basin development near the Chedabucto Fault (an eastward continuation of the Cobequid Fault) in southeastern Nova Scotia indicated activity along the fault zone between the latest Devonian and mid-Carboniferous. These data collectively suggest that the Cobequid Fault was active during deposition of

the Horton Bluff Formation. Seismic profiles and lithologic correlation of Horton Group strata show that stratigraphic units thin southward (Fig. 2) with onlap over the Meguma Group, from which it is inferred that subsidence adjacent to the Cobequid Fault resulted in an asymmetric basin during Horton times.

Hesse & Reading (1978) described well-developed clastic dykes and collapse structures from the Horton Bluff Formation, and interpreted them as the result of sediment extrusion associated with contemporaneous earthquake activity. However, evidence presented below indicates that these structures were generated by downward injection of coarser grained sediment into mud, possibly as a result of cyclic wave-loading.

The Horton Bluff strata were deposited in lacustrine, rather than marine, waters as inferred

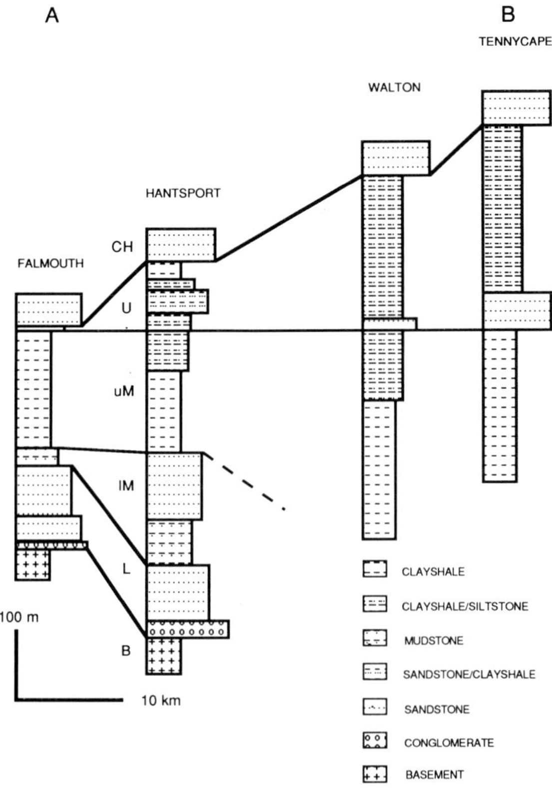

Fig. 2. Southwest (A) to northeast (B) cross-section of the southern part of the Windsor sub-basin. Both Falmouth and Hantsport sections are composites of a number of measured sections and drill holes. The Hantsport composite section includes the Blue Beach section. B, basement rocks; L, Lower Member; lM, lower Middle Member; uM, upper Middle Member; U, Upper Member; CH, Cheverie Formation. Note thickening of the Upper Member and probably the upper Middle Member to the northeast. A−B on Fig. 1 denotes line of cross-section.

from the lack of marine fossils (Bell, 1960) and an endemic amphibian and reptile population (Carroll *et al.*, 1972; Sarjeant & Mossman, 1978). Evidence for a low-salinity, hydrologically open lake includes the lack of evaporite minerals and casts within sub-aerial facies; the diverse fish population (Lambe, 1910; Carroll *et al.*, 1972), not expected within saline lakes where ecological stress reduces diversity (Picard & High, 1972); the presence of amphibians (almost exclusively freshwater in the modern although possibly tolerant of brackish waters early in their history (Milner, 1987)); and the *in situ* casts of trees, which rarely border modern saline lakes (E. Gierlowski-Kordesh, pers. comm., 1989).

FACIES

The lacustrine deposits contain four facies that are commonly superimposed in upward sequence (facies 1−4; Fig. 3) within the component cycles of the sections represented in Fig. 2. Major criteria for facies identification are lithology, the assemblage of sedimentary structures, and form and continuity of beds and bedsets. The classification scheme of Potter *et al.* (1980) has been used to describe shale lithology.

Grey clayshale and claystone (Facies 1)

Grey clayshale and claystone varies in colour from medium grey to black, and contains less than 1.5% organic matter ('poor to good' source rock potential (McMahon, 1988)). Units commonly begin with intensely bioturbated (*Planolites*) medium to coarse-grained lags containing sandstone grains, fish debris, and reworked nodule and duricrust fragments. The shale is predominantly clay-rich, but coarsens upward to mudshale in some cycles, locally with isolated siltstone and sandstone lenses (accompanied by clastic dykes and convolute lamination), dolomitic septarian nodules, and plant material.

Evenly splitting, fissile clayshale predominates at Blue Beach (Hantsport), and is typically rich in fossil material, including ostracods and Conchostraca (Bell, 1960; Bless & Jordan, 1971), fish and amphibian material (Carroll *et al.*, 1972), and plant material. Coprolites of varied size (maximum 2.5 cm), are also common, whereas visible trace fossils are rare to uncommon. Thin (< 3 cm) layers mainly composed of fish and ostracod material are common near unit tops, but dispersed fossils are scarce.

Platy, or irregularly splitting claystone (with minor mudstone) predominates at Tennycape, and is fossil-poor, containing some ostracods, Conchostraca, robust fish-teeth, *Glyptolepis* scales, and impressions 4 cm wide of large fish scales or bivalve shells. Moderate to intense bioturbation is inferred from the abundance of trace fossils on basal surfaces of sandstone and siltstone beds (see facies 2) and from the random orientation of mica grains.

Interpretation

Units of this facies were deposited from suspension in a quiet environment, mainly below wave base. A shallowing upward trend within units is demonstrated by the upward increase in plant debris, silt

Fig. 3. Thin lacustrine cycle from the Blue Beach section consisting of: clayshale facies (base not seen) (1); lenticular HCS siltstone (2a); wave-rippled sandstone (2b); planar-bedded siltstone (2c); green mudstone (3); carbonate duricrust (4c); and overlain by clayshale facies again (1). Scale is 1 m long.

content, and siltstone and sandstone lenses. The generally low organic content of the fissile clay-shales at Blue Beach, and the presence of ostracods and disarticulated fish remains, suggests that, periodically at least, bottom waters were aerobic and supported a benthic fauna. However, the fissile, non-bioturbated nature of the clayshales suggests predominantly anaerobic or dysaerobic interstitial waters (Ekdale & Mason, 1988). Substantial winnowing and/or transport formed the fossil-rich layers at the unit tops, and the scarcity of dispersed fossils in associated beds suggests that periodic events (storms?) redeposited and concentrated fossils in shallow water.

The uneven-splitting, platy nature of the Tennycape claystones, with randomly oriented mica grains, is evidence of intense bioturbation. The general absence of visible bioturbation reflects lack of lithological contrast. Relatively intense bioturbation implies oxygenated interstitial waters (Ekdale & Mason, 1988), and the poor preservation of all but the most robust body fossils may reflect the high level of biological activity.

Alternating sandstone, siltstone and clayshale (Facies 2)

Facies 2 gradationally overlies facies 1 in most cycles, and is composed predominantly of siltstone and sandstone beds 0.5−12 cm thick, which in turn commonly comprise composite bedsets (Collinson & Thompson, 1982) up to 60 cm thick. The beds are separated by erosion surfaces and/or thin clayshale layers, whereas bedsets are separated by thicker, locally silty, clayshale layers that contain thin, discontinuous to continuous siltstone and sandstone lenses (de Raaf *et al.*, 1977). The siltstones and sandstones (the coarsest are fine grained) are moderately to well-sorted and quartzose. Septarian nodules are less common than in facies 1; in one case roots are preserved in an uncompacted state within a nodule, suggesting early post-depositional nodule formation (Raiswell, 1971; Astin & Scotchman, 1988).

Subaerial and shallow-water features are common. These include rhizoliths (Klappa, 1980a), mudcracks, rain prints, mini-ripples (Singh & Wunderlich, 1978), ladder ripples, planed-off ripples, asymmetrical oscillation ripples (Davis, 1965), wave ripples with rounded crests and pointed troughs (Reineck & Singh, 1980, p. 365; Bang, 1987), water-level marks within ripple troughs (Hantzschel, 1939), scratch circles (Prentice, 1962), and amphibian and reptile footprints (Sarjeant & Mossman, 1978). A greenish weathering colour (olive-grey on fresh surfaces) is typical of pedoturbated (disrupted by rooting (Hole, 1961)) and mudcracked clayshales.

The facies is divided into three intergradational subfacies, which, although not all are present in every unit of the facies, tend to occur in upward succession from 2a to 2c (Fig. 3).

Hummocky cross-stratified siltstone (Subfacies 2a)

Siltstone and, less commonly, sandstone beds occur singly and as thin bedsets deposited within low-angle, elongate scours. Bedsets are separated by relatively thick, silt-free clayshale beds. Most beds and bedsets contain hummocky cross-stratification (HCS) (Harms *et al.*, 1975). Hummocky beds and bedsets are 2–40 cm thick and hummock wavelengths range between 30 and 150 cm. Whereas a lower limit of 1 m for hummock wavelength was suggested by Harms *et al.* (1975), other workers, such as Campbell (1966), Brenchley (1985) and the present authors, find a gradation into small-wavelength hummocks (micro-HCS of Dott & Bourgeois, 1982; Eyles & Clark, 1986).

Dott & Bourgeois (1982) and Walker *et al.* (1983) described an ideal HCS sequence that shows two orders of scouring, and from base to top, massive or faintly laminated sediment, hummocky laminae, flat laminae, and wave ripples. Horton Bluff subfacies 2a sequences conform, in general, to this model. The sequences are underlain by first-order scours that show tool marks, loads, minor lags of fossil material, and rare gutter casts (Whitaker, 1973). Massive or faintly laminated zones are rare. The hummocky zone contains laminae that typically thicken into depressions (Fig. 4a), but locally thicken into antiforms (Brenchley, 1985). Laminae onlap on to the first-order scours and locally are normally graded, and separated by thin partings of claystone or mica. Flat laminae are rare and wave ripples, in places irregularly shaped, cap some sequences. Climbing wave-ripples (Kreisa, 1981) occur very rarely in the uppermost strata.

The HCS beds and bedsets typically contain regularly spaced hummocks at the same stratigraphic level, isolated within clayshale (Figs 3, 4b). At most, thin upper layers connect with adjacent hummocks. Hummocky lenses are usually linear (rarely circular) in plan, and between 30 and 300 cm wide and lengths of 600 cm are noted. Lens elongation direction invariably parallels the strike of associated wave ripples and clastic dykes, and lies at right angles to tool marks and gutter casts (Fig. 5).

Rare, fine grained, planar-laminated sandstone beds up to 10 cm thick are interbedded with the HCS units, and comprise the most continuous layers and coarsest sediments of subfacies 2a.

Tool marks (groove and prod casts) are most common beneath small hummocks and lenses and on the shallow extremities of deep scours. They show consistent orientation on individual surfaces, but associated asymmetrical prod marks (Dzulynski & Walton, 1965) commonly show opposing flow directions on a single surface. V-shaped grooves (Fig. 6a) further suggest reversing flow. Gutter casts within and associated with HCS units commonly show groove marks on their walls, which decrease in concentration downward. Bidirectional prod marks associated with HCS were noted by Bloos (1982) and Duke (1987).

Clastic dykes are very common within the upper Middle and Upper Members and invariably occur within subfacies 2a. Size varies from 1 cm wide and 5 cm deep (depth of penetration measured vertically) to 40 cm wide and 40 cm deep; most show ptygmatic folding (Kuenen, 1968) due to preferential compaction of the surrounding muds, and their original depth of penetration was estimated as up to 80 cm. Larger dykes usually underlie loads or downfolds (cf. Eyles & Clark, 1985), and all dykes display a downward thinning. Dykes in each layer are aligned parallel to one another and from one layer to the next (Fig. 5), and show a semi-regular spacing (Hesse & Reading, 1978).

Siltstone and sandstone beds rarely show bioturbation, but their basal surfaces commonly display trace fossils in convex hyporelief: 'small stuffed burrows' and *?Margaritichnus* (Bromley & Asgaard, 1979), and both the resting and locomotion traces of *Isopodichnus* (Trewin, 1976: see also the *Rusophycus* and *Cruziana* described by Bromley & Asgaard (1979)). Subaerial exposure features are very rare.

Interpretation. Most authors agree that HCS forms under storm conditions, where waves interact with the bottom (Dott & Bourgeois, 1982; Harms *et al.*, 1982). Walker (1984) suggested that HCS is best preserved below fair-weather wave base, but Duke (1985, 1987) considered that preservation is likely in very shallow lacustrine environments, and Greenwood & Sherman (1986) recorded HCS from the surf zone of Lake Huron.

The HCS of subfacies 2a formed under conditions of (storm related?) high-energy flow (indicated by the presence of scours and tool marks), associated with and/or followed by rapid deposition (indicated by load casts, convolute bedding, rare climbing wave ripples, and lack of bioturbation within the siltstone and sandstone beds). Normally graded laminae probably represent suspension fall-out from passing waves (Dott & Bourgeois, 1982). A setting below fair-weather wave base is suggested by the thick

Fig. 4. (A) Hummocky cross-stratified lens showing laminae thickening into swales. (B) Hummocky cross-stratified lens with underlying clastic dyke. Hammer is 31 cm long.

interbeds of clayshale, interpreted as suspension deposits. Lack of subaerial and pedoturbation structures suggests a setting below low-stand lake level.

Lenticular morphology of the hummocky units probably reflects sediment starvation. Isolated hummocky cross-stratified beds occur in the ancient marine record (Kreisa, 1981; Dott & Bourgeois, 1982; Myrow *et al.*, 1988), and in lacustrine deposits (Eyles & Clark, 1986). Hummocky cross-stratification has been described from ancient lake facies (van Dijk *et al.*, 1978 (as interpreted by Allen & Collinson (1986)); Duke, 1985; Eyles & Clark, 1986; Hamblin, 1988) and from modern lakes (Greenwood & Sherman, 1986).

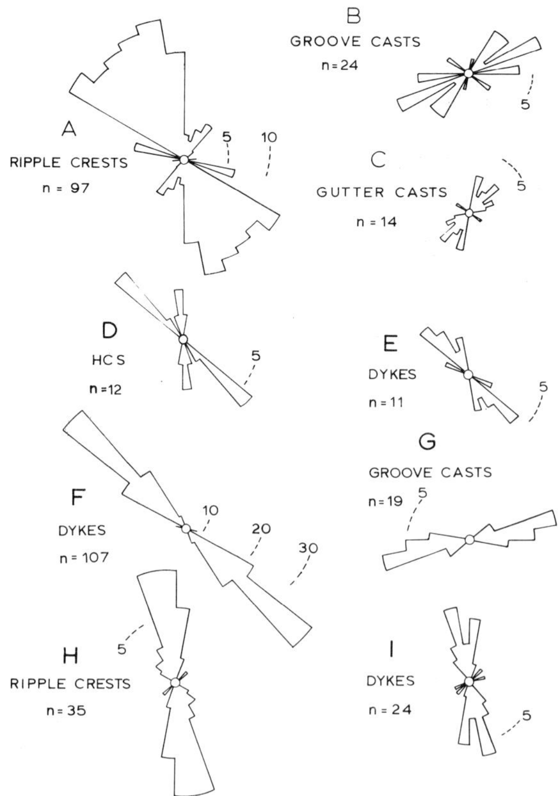

Fig. 5. (A−F) Rose diagrams for the Blue Beach section: (A) wave-ripple crest orientations; (B) groove cast orientations; (C) gutter cast orientations; (D) elongation direction of HCS lenses; (E) mean clastic dyke orientations of separate layers; (F) clastic dyke orientations from a single layer (Upper Member). (G−I) Rose diagrams from the Tennycape section: (G) groove cast orientations; (H) wave-ripple crest orientations; (I) clastic dykes from separate layers. Dashed lines denote number of readings within each 10° division; *n* equals total number of readings.

The associated planar-laminated sandstones probably reflect relatively high-energy events, with deposition from suspension (Reineck & Singh, 1972) or during upper regime flow (Harms *et al.*, 1982). The lack of unidirectional current structures and parting lineation within this subfacies suggests deposition from suspension clouds generated by wave activity (Brenchley, 1985).

Tool marks and gutter casts at the base of HCS units have been cited as indications of strong unidirectional flow prior to HCS deposition (Hamblin & Walker, 1979; Walker, 1984). The presence of prods that show opposed flow directions and v-shaped grooves oriented at right angles to wave-ripple crests, together with a lack of unidirectional indicators, suggest that tool marks were formed by either dominantly oscillatory flows (e.g. Duke & Leckie, 1986; Leckie, 1988) or from combined flows with a strong oscillatory component (Swift *et al.*, 1983; Allen, 1985; Greenwood & Sherman, 1986). Lateral offset of the v-shaped grooves (Fig. 6A) seems to favour the latter. Gutter casts are commonly associated with wave-dominated shorelines and HCS (Goldring & Bridges, 1973; Kreisa, 1981) and may reflect sand abrasion during wave action (Norrman, 1964; Plint *et al.*, 1988). The concentration of tool marks in the shallow parts of scours may reflect the greater strength of the deeper, better compacted mud: Zullig (1956; in Muller, 1967) recorded a significant decrease in water content and porosity within the top few centimetres of Swiss lake sediments, and McCave (1985) noted a large strength increase in muds at depths of only a few centimetres.

Clastic dykes of the upper Middle and Upper Members have been interpreted previously as eruption features originating from dewatering in response to earthquake activity (Hesse & Reading, 1978). However, the dykes taper downward below hummocky lenses, and undisturbed fish-scale layers at dyke margins can be followed downward into the dyke centres, where they become deformed. Rose diagrams (Fig. 5) show the close relation of dyke orientation to the orientation of wave ripples and hummocky lenses. We believe that the orientation of wave-generated hummocky lenses controlled the orientation of the dykes, which were injected downward. Injection probably occurred during the storms that deposited the lenses, perhaps as a result of cyclic wave-loading (Seed & Lee, 1966; Poulos, 1988).

Wave-rippled sandstone (Subfacies 2b)

Subfacies 2b consists of very fine to fine-grained sandstone and siltstone bedsets, 5−20 cm thick, that are continuous within outcrop, but vary laterally in thickness and number of beds. Individual beds are wavy to lenticular bedded, moderately continuous to discontinuous, and 1−5 cm thick. Erosive bases of beds record numerous truncations and amalgamations indicative of a complex depositional history. Siltstone and sandstone bedsets alternate with clay-shale-rich bedsets 2−10 cm thick. Bioturbation, pedoturbation (perhaps subaqueous in part), and subaerial features are common, and tend to increase in abundance upward within units of this subfacies.

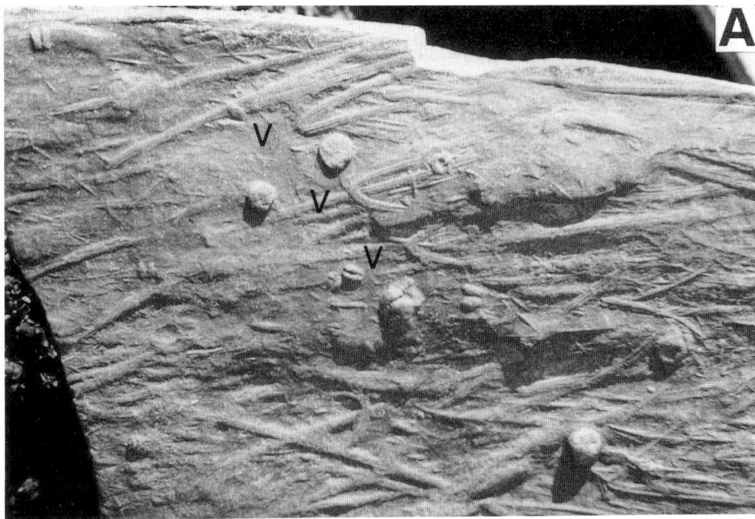

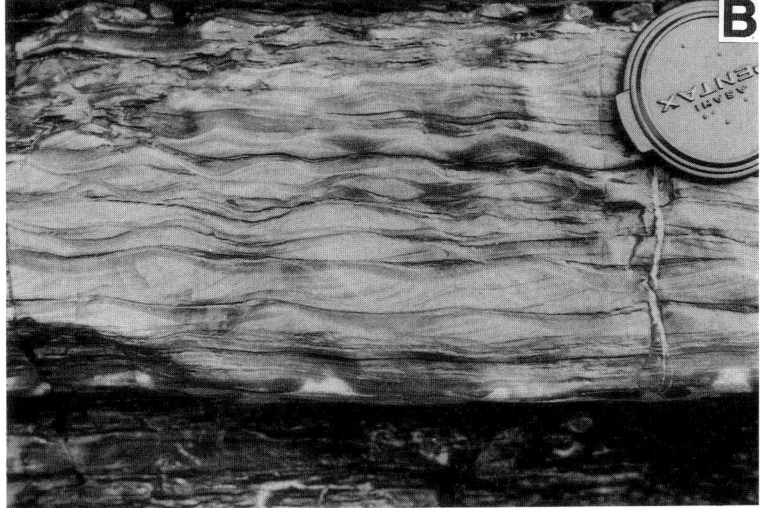

Fig. 6. (A) Basal surface of a subfacies 2a bed showing oscillatory-formed v-shaped (V) groove casts. Burrows are *Isopodichnus*; width of photograph is 12 cm. (B) Single bedset of wave-rippled sandstone and shale (subfacies 2b). Note the slightly asymmetrical to symmetrical form of ripple crests in cross-section, the apparent reversals in asymmetry direction between layers, and the preferred dip-direction of cross-laminae to the left. Set thickness and ripple wavelength decrease upward as shale increases. Lens cap is 5 cm in diameter.

Trace fossils include *Pelecypodichnus, ?Margaritichnus*, 'small stuffed burrows' (Bromley & Asgaard, 1979), *Isopodichnus* (Trewin, 1976), *Palaeophycus*, and *Planolites* (Pemberton & Frey, 1982).

Sandstone and siltstone bedsets contain symmetrical (rarely asymmetrical) wave ripples that are long-crested and irregular in plan (Clifton, 1976). Ripple wavelength averages 5.7 cm (range 1.5−11 cm). Most ripple sets have a bidirectional cross-lamination orientation, with one direction predominant. Individual rippled sets commonly fine upward through clayshale interlaminae to clayshale, and some bed-

sets show ripple wavelength and set thickness decreasing upward (Fig. 6B). Maximum ripple wavelength within a cycle also tends to decrease upward. As in subfacies 2a, prod marks with opposing flow directions, groove casts, including some v-shaped examples, and gutter casts are oriented at right-angles to ripple crests.

Single beds of planar-laminated, fine-grained sandstone with erosional bases occur rarely in subfacies 2b. They are up to 10 cm thick and as such are the thickest beds within units of this subfacies.

Clayshale-rich bedsets separate siltstone and sandstone bedsets and generally contain very thin

(<0.5 cm) siltstone lenses and layers. Mudcracks are common in the upper portion of most subfacies units.

Interpretation. The wave-rippled sandstone beds with clayshale drapes are inferred to record waning episodes of bedload deposition associated with moderate to high-energy waves (Harms *et al.*, 1982), followed by periods of quiescence (suspension deposition). The deposits resemble in their fine grain-size and ripple morphology deposits formed in a restricted body of water (Clifton, 1976, Fig. 15). The constant dip direction of the cross laminae within wave-rippled units probably reflects deposition beneath shoaling waves (Clifton, 1976) and indicates the direction of wave propagation (Allen, 1979). Associated subaerial and shallow-water structures indicate deposition within the zone of lake-surface fluctuation commonly at levels a few metres below the cycle top. Clayshale interbeds within coarser grained bedsets show that wave energy was frequently low enough for clays to settle.

The planar-laminated sandstone resembles that of subfacies 2a, but units are thicker, coarser-grained and show deeper basal scours. Upper plane-bed and suspension deposition are both possibilities, and probably reflect storm processes in shallow water (Clifton, 1976) or on beaches.

Clayshale-rich bedsets represent prolonged periods of quiet-water suspension deposition (Harms *et al.*, 1982). Mudcracks and other subaerial features within the upper bedsets of some subfacies units suggest shallow-water deposition.

Planar-bedded siltstone (Subfacies 2c)

Tabular siltstone and subordinate sandstone beds, $1-15$ cm thick, are generally continuous, being traceable for hundreds of metres. Slight lateral thickness variation is the result of local thickening in low-angle scours. Sole marks are rare or absent. The beds form thick ($10-60$ cm), laterally continuous bedsets separated by clayshale-rich bedsets $2-5$ cm thick (Fig. 7A). There is much less shale, and generally less sandstone within subfacies 2c than within either 2a or 2b.

Beds commonly show planar laminae or fining upward couplets of sandstone to siltstone, and less commonly coarsening upward trends. A typical couplet shows a thin wave-rippled, fine-grained sandstone layer overlain by a thick siltstone layer (Fig. 7B). Siltstone layers are massive or show faint,

normally graded planar laminae. Some incipient cross laminae (de Raaf *et al.*, 1977) are present. The dip direction of ripple cross-laminae and the crestal orientation closely resemble those of subfacies 2b ripples. Water escape structures, especially small dish structures (Lowe, 1975), and disrupted bedding are commonly observed, and dewatering may explain the structureless appearance of some beds. Bedding surfaces are planar, or show small-wavelength (>3 cm) ripples or mini-ripples ($0.5-3$ cm wavelength: Singh & Wunderlich, 1978) (Fig. 8A). A few beds grade up into bioturbated clayshale, in a sequence reminiscent of the 'tempestite' beds of Brenchley (1985, fig. 13), here associated with shallow-water structures. Rare beds of planar-stratified sandstone ($3-15$ cm thick) are similar to those of subfacies 2b. Rare trace fossils include *Palaeophycus* and *Planolites* (Pemberton & Frey, 1982).

The top few beds of a bedset contain mudcracks, mini-ripples, tree casts, pedoturbation structures, and, in one case, flat linguoid ripples (Reineck & Singh, 1980, p. 365), indicating shallow water and exposed conditions. Some upper surfaces also display mudchip and plant-debris layers. Vertical casts and associated rhizoliths of *Archaeocalamites* (synonymous with *Asterocalamites*, the generic name used by Bell (1960): Walton, 1958) commonly occur in the upper beds of a bedset (Fig. 8B). Where *Archaeocalamites* is absent, bed surfaces display small pits (less than 0.5 cm wide) that rarely lead into 'grasslike', flow-oriented impressions. These are interpreted as the base and stems of subaqueous plants. One bedding surface shows a lateral transition from pits to 1 cm diameter root casts, to 20 cm diameter root casts surrounded by mudcracks. A clayshale-rich bedset (Fig. 8) commonly overlies these bedsets.

The clayshale-rich bedsets are similar to those of subfacies 2b, but contain more evidence of shallow water and subaerial exposure, including rhizoliths and abundant impressions and fossils of aligned tree trunks, branches, and cones.

Interpretation. The planar-bedded siltstones are interpreted as the deposits of storm-flows in shallow water. Shoaling waves deposited thin rippled layers with preferred cross-lamination orientation, after which suspension clouds deposited planar-laminated silt. Thin sand streaks and graded silt laminae probably reflect the variable carrying capacity of the wave trains. Rapid deposition from suspension clouds has been documented at current velocities below those required for the genesis of ripples

Fig. 7. (A) A single cycle (younging to the right) with an unusually thick unit of subfacies 2c. Note tabular, continuous beds and siltstone bedsets separated by clayshale bedsets. Knapsack is 40 cm high, and rests approximately on the subfacies 2b−2c contact. (B) Sandstone to siltstone couplet (subfacies 2c) with wave rippled sandstone on the base overlain by a thick layer of planar-laminated siltstone. Lens cap is 5 cm in diameter.

(Reineck & Singh, 1972). Sediment suspensions can be generated by shoaling waves (Reineck & Singh, 1972; Clarke *et al.*, 1982), and by high winds in shallow lakes (Kenney, 1985). Rare planar-stratified sandstones represent deposits of high-energy flows, possibly those of severe storms or of a beach zone.

Shallow water and subaerial structures and colonization by small plants and *Archaeocalamites* at the top of the bedsets indicate a shallowing upward trend. *Archaeocalamites* flourished in very shallow water conditions, similar to present-day reed marshes (Scott, 1980). Plants may grow far into lakes where shallowness permits (Pulan, 1968). The plant community probably baffled wave energy, ac-

counting for the predominance of clayshale-rich bedsets following plant colonization.

Green mudstone (Facies 3)

The facies is composed of olive-grey (5Y3/2) (fresh surface) to greenish grey (5GY6/1) and dusky yellow (5Y6/4) (weathered surface) mudstone and silty mudstone. Many units contain calcareous siltstone and sandstone beds that display wave cross-lamination, horizontal lamination, and, rarely, normal grading. Many siltstone beds show moulds of roots and tree trunks with diameters up to 40 cm (Fig. 9). Most are probably *Lepidodendron* (Bell, 1960), with

Fig. 8. (A) Upper portion of a subfacies 2c bedset. Note small-wavelength ripples (above lens cap) and mini-ripples (M) from reworking of bed tops, and rhizoliths (R) that together indicate shallow water to exposed conditions. Lens cap is 5 cm in diameter. (B) *In situ Archaeocalamites* in the upper strata of a subfacies 2c bedset. Note the overlying shale-rich strata. Portion of pen in view is 8 cm long.

Archaeocalamites. The intensity of pedoturbation and mudcracking increases upward within units, which are also lighter coloured upward.

Interpretation

The facies was deposited in a lake-fringing marsh (marsh is defined here as both forested and non-forested wetlands where peat does not accumulate (Moore, 1987)) that was periodically covered by siltstone and sandstone (storm?) layers. The abundance of pedoturbation structures and mudcracks, especially at unit tops, indicates a shallow, vegetated, and routinely subaerially exposed environment characterized by progressive shallowing or shoreline progradation. The greenish weathering colour is characteristic of subaerially exposed surfaces in the Horton Bluff Formation. Lake shorelines and delta tops of the British Upper Carboniferous were dominated by the reed-like *Calamites* lakeward (*Archaeocalamites* probably occupied a similar niche (Scott, 1980)), and were replaced landward by the larger, probably marsh dwelling, *Lepidodendron* (Scott, 1979).

Dolomite (Facies 4)

Three dolomite subfacies (4a, 4b, and 4c) are recognized within the lacustrine cycles, not including minor nodular and lenticular dolomite. Whereas

Fig. 9. Moulds of trees preserved by a siltstone layer within facies 3.

siltstones and sandstone of facies 2 and 3 are calcite-cemented, this facies is typified by dolomite that is dark grey (5N3) on fresh surfaces and dusky yellow (5Y6/4) on weathered surfaces. Thin sections of all subfacies show uniform-sized (5–10 μm) anhedral dolomite crystals that appear to have displaced a lesser percentage (10–15%) of clay minerals, and in some cases silt or sand grains.

Subfacies 4a consists of continuous, planar, 5–10 cm thick beds of vertically splitting dolomite. Although commonly structureless, some beds contain peloids and planar laminated organic-rich layers. Subfacies 4a usually overlies facies 1 or caps incomplete cycles that terminate with subfacies 2a.

Subfacies 4b consists of continuous, sharp-based dolomite layers 15–40 cm thick that are massive, well-indurated, and, in contrast to subfacies 4a, split randomly. Rhizoliths are present locally. Most strata are planar bedded and some show polygonal patterns of mud flames on their bases, whereas others display buckling. A subplanar laminar zone (Gile *et al.*, 1966; Lattman, 1973) occurs on the upper (and in one case lower) portion of some units. This subfacies is associated either with facies 1 or slightly pedoturbated and/or mudcracked shallow-water sediments of facies 2.

Subfacies 4c consists of nodular dolomite, locally coalesced into irregular but continuous buff-weathering, massive layers (Fig. 3). These typically occur near the top of severely pedoturbated and mudcracked strata of facies 3. Rhizoliths are common to abundant. Rare laminar zones are wavy and have an algal-like morphology (Lattman, 1973). Dolomite nodules 1–2 cm in diameter and vertical cylinders up to 20 cm in diameter (Hubert, 1977) (probably tree casts (Klappa, 1980a; Kraus, 1988)) are present locally.

Interpretation

Units of facies 4, with the possible exception of laminated and peloidal units of facies 4a, formed through carbonate precipitation within the pores of water-rich mud. Early precipitation is inferred from the large percentage of dolomite relative to the host mud (although displacive growth may have been a factor (Raiswell, 1971)), the occurrence of dolomite clasts in associated sandstones, and the non-compacted nature of some horizontal roots within the beds. The fine crystal size may reflect the fine grain size of the host material (Matsumoto, 1978).

Subfacies 4a is associated predominantly with offshore deposits, and pedogenic structures are absent. Subfacies 4b is associated with slightly pedoturbated and mudcracked beds suggesting a shallow water depositional setting. Although substantial pedoturbation is lacking, buckled strata, probably owing to displacive crystal growth (Watts, 1978) or rooting (Klappa, 1980b), and laminar zones are reminiscent of caliches (Lattman, 1973; Watts, 1978). Subfacies 4b may be analogous to 'non-pedogenic' calcrete formed in the phreatic (ground-water) zone (Mann & Horwitz, 1979).

Subfacies 4c occurrences are interpreted as pedogenic duricrusts, based upon the occurrence of rhizoliths and their intimate association with intensely mudcracked and pedoturbated soil zones (facies 3), nodules, and tree casts. These structures are typical of pedogenic horizons formed in the vadose zone (Goudie, 1983), and closely resemble the 'paludine' or lacustrine marsh carbonates of Freytet (1973). The gradation from small nodules to coalesced nodular layers reflects varying degrees of duricrust development (Reeves, 1976).

Other facies

Distinctive units of fine to very coarse-grained sandstone form the tops of some cycles, especially at Tennycape. Up to 10.5 m thick, they overlie units of facies 1 and, rarely, facies 2 with erosional or gradational contact, and generally coarsen upward. Sedimentary structures include trough and planar cross-bed sets up to 35 cm thick, flaser bedding, wave and current-ripples. The facies are interpreted as deltaic distributary channel and mouth bar deposits on the basis of their exceptionally coarse grain size, the large scale of the cross-bedding, and the coarsening upward pattern (Coleman & Prior, 1980). These units occur in the Upper Member and distinguish it from the upper Middle Member and are more common in the Tennycape section than the Blue Beach section.

SHORELINE FACIES PATTERNS

Figure 10 is a schematic diagram for the lacustrine cycles within the upper Middle and Upper Members, showing the main sediment types, and inferred wave-energy and processes. Not every cycle shows all these features and facies. Following Walther's Law, the upward succession of facies found in the most complete cycles is taken to represent an offshore to onshore transition within the Horton Bluff Lake.

Several features of Fig. 10 require comment. Hummocky cross-stratified siltstone (subfacies 2a) was deposited under sediment-starved conditions as storms carried sediment to the transition zone between facies 1 and 2. Offshore sand transport was thus limited to the coastal area (Thomas *et al.*, 1972). Storm waves scoured the muddy bottom, and collected coarser sediment into lenses oriented predominantly parallel to the wave crests (Fig. 10). Although wave-ripple orientation reflects the wave orientation, it does not necessarily reflect the shore-

line orientation (Boyd *et al.*, 1988) as shown in Fig. 10.

Tool marks within subfacies 2a and 2b were formed by oscillatory waves. This is significant for two reasons: (i) wave-produced tool marks are not necessarily indicative of net sediment transport direction; and (ii) oscillatory processes were apparently dominant *prior* to HCS deposition. Johnson & Baldwin (1986) stated that basal erosional surfaces, sole marks, and gutter casts associated with HCS indicate the presence of strong initial unidirectional flows, although they also noted that the oscillatory origin of HCS is inferred from the rarity of associated current-formed structures. In subfacies 2a, however, erosive structures are apparently the result of *oscillatory* (wave) flows. We envisage below fair-weather wave base deposition by combined flows where the oscillatory component was dominant (e.g. Greenwood & Sherman, 1986).

Subfacies 2b represents the highest energy environment of facies 2, as indicated by the scoured bases of beds and coarse grain size. The increase in bed continuity compared with subfacies 2a indicates a more reliable sediment source. Water depth was commonly below the influence of suspension-capable waves, but the sediment surface was within the range of short-term climatic (seasonal?) lake-level variation.

Subfacies 2c contains more siltstone (less sandstone) and shows fewer sole marks and scoured surfaces than subfacies 2b, but contains thicker beds and bedsets and much less clayshale (see Fig. 11b). These data collectively suggest that although flow energy was less for subfacies 2c, sediment load was relatively abundant, rapidly deposited, and background energy was enough to prevent clay settling. We propose that subfacies 2c was deposited by storms in shallow water areas protected from high-energy waves (Fig. 10). Wave energy was probably attenuated shoreward over subfacies 2b deposits (offshore) and during storms, the rapidly deposited suspended load resulted in planar-laminated siltstone with water-escape structures. The subfacies 2c deposits differ from commonly described marine (Heward, 1981; Elliott, 1986), and lake (Davis *et al.*, 1972) shoreline deposits in that: (i) peak wave energy occurred over an offshore area rather than a beach; and (ii) storms deposited sediment in shallow water whereas storms commonly remove material along sandy shorelines (e.g. Cook & Gorsline, 1972).

Sandy beaches are notably absent from the Horton Bluff lacustrine cycles. This probably reflects wave

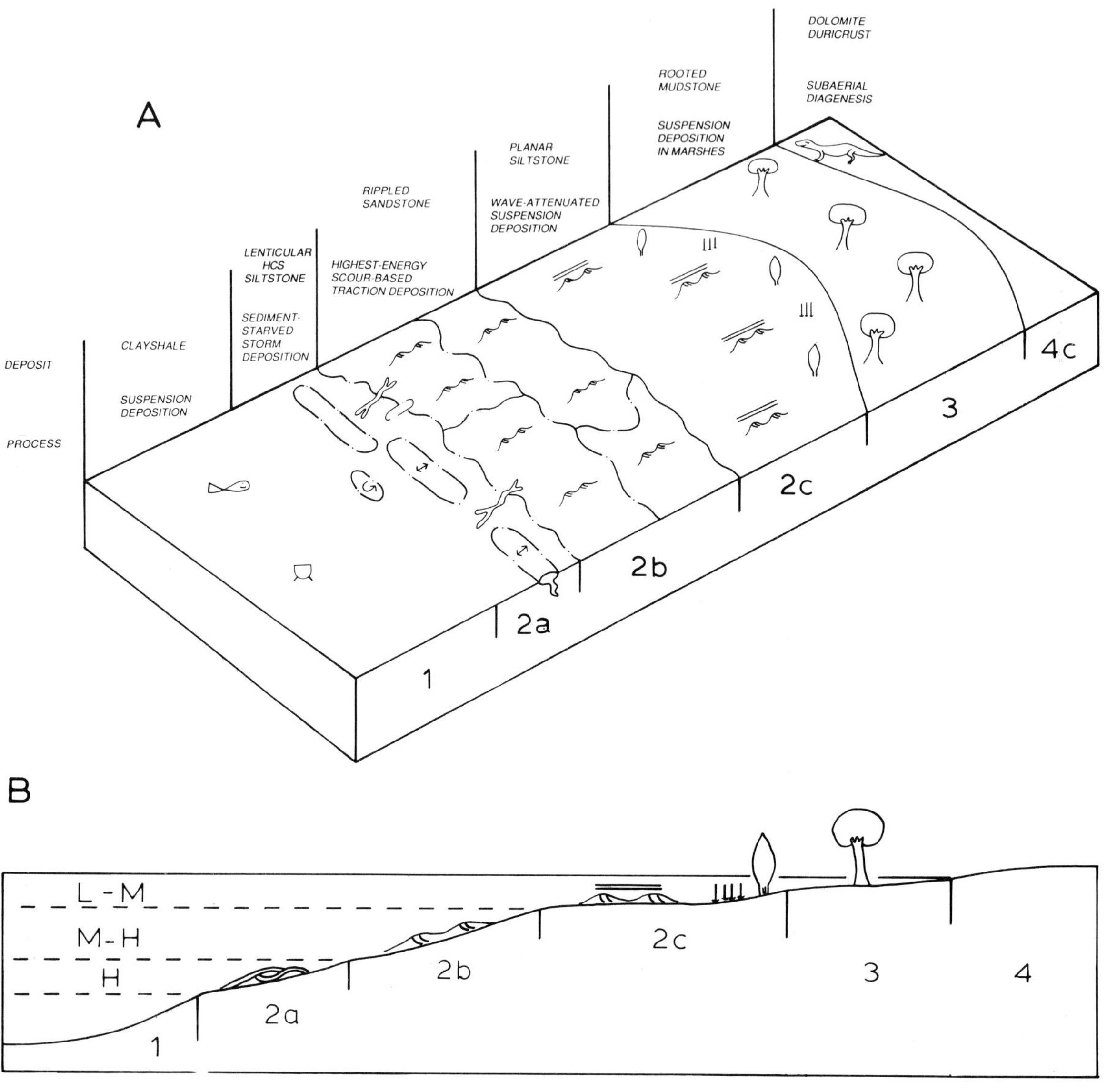

Fig. 10. (A) Model of a Horton Bluff lacustrine shoreline. Dashed lines in subfacies 2a and 2c denote scoured bases to beds. Palaeocurrents in the model suggest an orientation at right-angles to the shoreline, but this is not necessarily the case. (B) Schematic shoreline profile. L, M, and H represent low, medium, and high wave-energy conditions respectively, at the lake bottom. Note that attenuation of waves over zone 2b allows only medium and low-energy waves to affect zone 2c. See Fig. 11 for legend.

attenuation and/or colonization by vegetation. Wave power is commonly reduced across extensive low-gradient shore zones (Hutchinson, 1957; Wells & Coleman, 1981; Kemp, 1986), resulting in an onshore decrease in wave shear stress, which can deposit, for example, laminated and graded silt and shell and mud beds from fluid mud suspensions (Kemp, 1986). Wave attenuation allows the establishment of shoreline vegetation even where lakes are of considerable size (Hutchinson, 1957). A dampened wave regime in the Horton Bluff lake (and perhaps lake lowstands (Carmouze & Lemoalle, 1983)) would have allowed plant colonization, which would have further dampened wave power, creating the facies 3 marshes. Ancient lacustrine shoreline deposits lacking beach sediments have been documented by Smith (1986), and shoreline siltstones similar to subfacies 2c were described by Donnelly & Jackson (1988). Lake to marsh transitions are common in many African lakes (Debenham, 1947; Beadle, 1974, p. 248; Carmouze *et al.*, 1983; Cohen *et al.*, 1986).

The absence of deltaic or fluvial deposits within the upper Middle Member of the Horton Bluff Formation suggests that sediment was transported by wave-driven longshore currents from unknown (deltaic?) sources. Longshore transport, and associated longshore fining, is common in modern lakes (Cohen *et al.*, 1986; Frostick & Reid, 1987), and has been inferred in ancient lakes (Allen, 1981). Deltaic deposits occur within the Upper Member locally.

CYCLES

Cycle types

Three common facies associations (cycle types A, B, C, Fig. 11) illustrate depositional variation within the Horton Bluff Formation. The facies occur in repeated asymmetrical, shallowing upward cycles, commonly with abrupt tops. Type D cycles contain the deltaic strata described within 'other facies'. Figures 12 and 13 show the distribution of facies and cycle types within the Blue Beach and Tennycape sections.

Type A cycles

Type A cycles consist mainly of grey clayshale (facies 1), with facies 2 and 3 sediments thin to absent (Figs 11, 12). Sediment-starved conditions are indicated by the thin facies 2 deposits and layers of reworked fossil material.

Type B cycles

Type B cycles are the most common, and vary the most in thickness and facies association (Fig. 12).

Whereas Tennycape type B cycles contain approximately equal percentages of subfacies 2b and 2c, the Blue Beach cycles are typically dominated by subfacies 2b. Marsh (facies 3) and associated duricrust (subfacies 4c) deposits vary from non-existent to moderately thick, and are best developed at Blue Beach. Both Tennycape and Blue Beach type B cycles display a decrease in thickness up-section (Fig. 13), and, on average, type B cycles are thicker at Tennycape (8.2 m) than at Blue Beach (3.9 m).

Type C cycles

Type C cycles (Fig. 11) are dominated by marsh deposits (facies 3), with facies 1 and 2 poorly developed. They represent low-energy shorelines in which well-developed marshes prograded over shallow, submerged platforms. Only three type C cycles occur at Tennycape whereas 18 occur at Blue Beach. In addition, of these 21 cycles, 16 occur in the Upper Member and only five in the upper Middle Member (Fig. 13).

Type D cycles

Type D cycles (Figs 12, 13) contain deltaic distributary channel and mouth bar deposits described under 'other facies'.

Distribution and interpretation of cycle types

Both the Blue Beach and Tennycape sections display an up-section thinning of cycles (Figs 12, 13). In lakes with a stable, outlet-controlled water level, cycle thickness should reflect the amount of syndepositional subsidence. Increased water depth (and hence potential sediment aggradation) in such lakes implies a corresponding increase in the elevation difference between the basin floor and the outlet, caused by raising the outlet or lowering the basin floor, both commonly tectonic in origin. In consequence, the upward decrease in cycle thickness in both sections (Fig. 13) is interpreted to reflect a progressive, basinwide decrease in tectonic activity, probably in subsidence rate. Steel & Gloppen (1980) explained 5–20 m thick, coarsening upward cycles in alluvial deposits by periodic downward movement of the basin floor. In the Windsor sub-basin, compactional lowering of the basin floor was probably of minor significance, at least for the thicker lowermost cycles described, because the underlying Lower and lower Middle Members are thin and composed

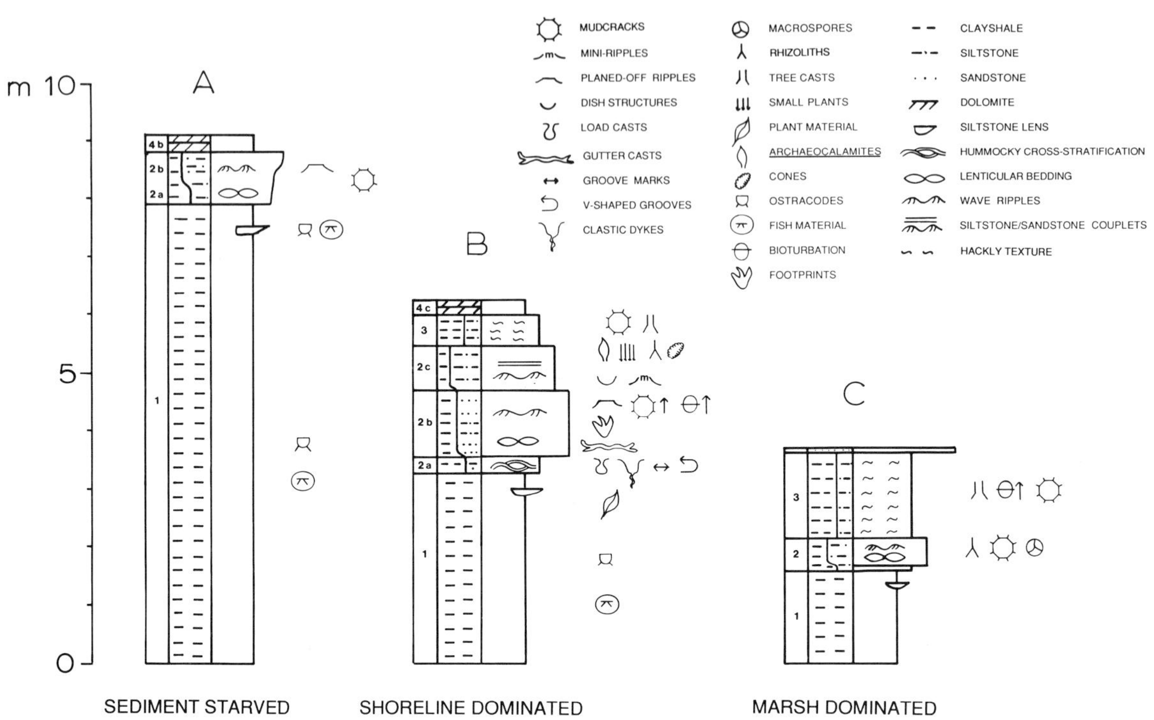

Fig. 11. Schematic sections of cycle types A, B, and C. The left-hand column denotes facies and subfacies. Type A cycles contain greater than 80% facies 1. Type B and C cycles contain less than 80% facies 1, more facies 2 than facies 3 (type B), and more facies 3 than facies 2 strata (type C). Facies (and cycle) thicknesses are averaged from cycles of each type (A = 16, B = 51, C = 21) from both Tennycape and Blue Beach sections. Note in cycle B, clayshale decreases from subfacies 2b to subfacies 2c, while relative grain-size also decreases.

mainly of relatively non-compactable sandstone and siltstone (Fig. 2).

Type A cycles are most common in the lower part of both sections (upper Middle Member), type B cycles occur throughout but thin up-section, and type C cycles are most abundant in the upper part of both sections (Upper Member). Deltaic deposits (type D) occur only in the upper portions of the sections, above the Middle–Upper Member boundary (Fig. 13). Thus, there is a progressive change up-section both at Blue Beach and Tennycape from offshore-dominated cycles to cycles with a higher proportion of nearshore facies to cycles with abundant marsh and associated deltaic deposits. A progressive decrease in subsidence rate would have permitted coastal deposits to prograde over much of the basinal area, eventually becoming the predominant component of the cycles. Several authors have inferred that lacustrine, especially deeper lacus-

trine, conditions prevailed during the most rapid periods of basin subsidence (Steel & Gloppen, 1980; Manspeizer, 1981; Li Sitian *et al.*, 1984; Blair, 1987; Martel, 1987; Blair & Bilodeau, 1988).

Both the upper Middle and Upper Members are thicker at Tennycape than at Blue Beach (Fig. 2). Cycles are, on average, thicker at Tennycape also (Fig. 13). These observations can be explained by the location of the Tennycape section in the central portion of the asymmetrical Windsor sub-basin and closer to the Cobequid Fault (Fig. 1), adjacent to which greater subsidence and sediment thickening would be expected. At Tennycape, type B cycles are especially common whereas type C cycles are relatively uncommon (Fig. 13). We infer that the relatively high subsidence rate at and the basinward location of Tennycape promoted prolonged periods of nearshore sedimentation (type B cycles), but precluded prolonged periods of subaerial sedimen-

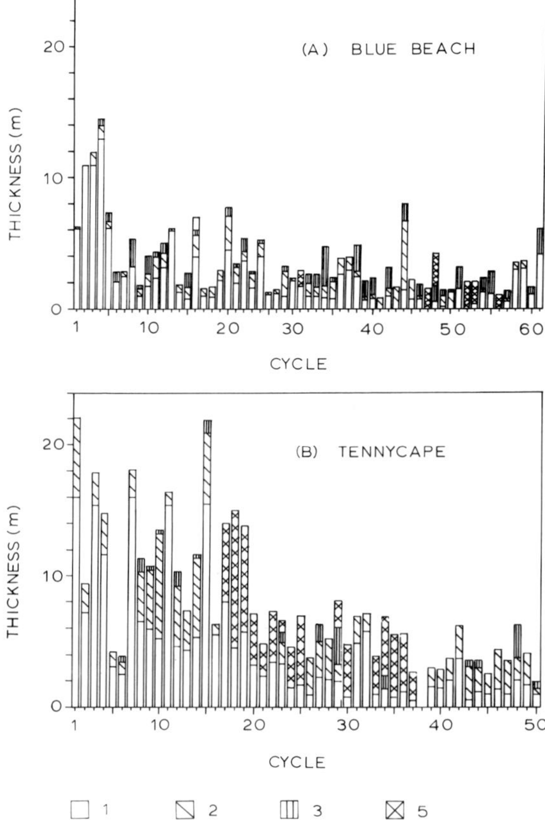

Fig. 12. Thickness and facies proportions of each cycle from the Blue Beach (A) and Tennycape (B) sections. Cycles are numbered from base to top of the section. Facies 4 strata are too thin to be included. The incoming of facies 5 ('other facies') deposits marks the contact between the upper Middle and Upper Members (cycle 31 at Blue Beach and cycle 17 at Tennycape).

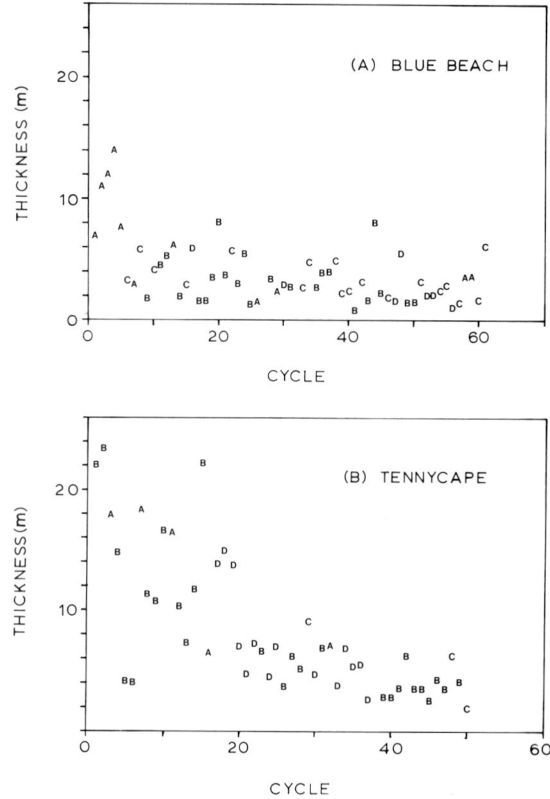

Fig. 13. Cycle thickness and type from the Blue Beach (A) and Tennycape (B) sections. A, sediment starved; B, shoreline dominated; C, marsh dominated; D, delta dominated. Cycles are numbered from base to top of section.

tation (type C cycles). The better development of type A cycles at Blue Beach and type D deltaic deposits at Tennycape probably reflects the distribution of local sediment sources within the basin.

Interpretation of a hydrologically open lake system for the Horton Bluff Formation is based upon evidence indicative of lake chemistry (see Regional Setting). A second line of evidence for hydrologically open conditions comes from the consistently shallowing upward cycles. Although small-scale fluctuations in lake level are inferred, their effects are superimposed on consistent regressive trends. An outlet would prevent the lake from rising above a certain level and periodic downward movement of

the basin floor would result in abrupt deepening followed by a gradual shallowing upward. In contrast, lakes without an outlet to regulate the maximum water level (i.e. hydrologically closed) should show periodic increases in water volume resulting from climatic variation, and hence deepening upward trends in their deposits. Gibling *et al.* (1985) noted both shallowing and deepening upward trends in carbonate-rich oil shale deposits in a small, probably internally drained, Tertiary lake basin.

Although it is impossible to completely discount the effects of climate in generating the Horton Bluff cyclicity, we believe that the evidence favours a tectonic cause. Such evidence includes the asymmetrical nature of the cycles, and evidence presented earlier for local and regional tectonic activity during deposition. The upward trend of decreasing cycle thickness is more consistent with tectonic than climatic variation. Finally, climate would be expected to exercise more significant control on cyclicity for a hydrologically closed lake, whereas the Horton Bluff lake apparently was hydrologically open throughout its life.

CONCLUSIONS

1 The upper Middle and Upper Members of the Horton Bluff Formation were deposited in an hydrologically open lacustrine system. Sediment was probably derived from local deltaic sources and transported by longshore drift.

2 An idealized lacustrine cycle commences with offshore, low-energy deposits of grey clayshale (facies 1), overlain by nearshore deposits of inter-bedded clayshale/wavy bedded siltstone and sandstone (facies 2). The latter is composed, in upward sequence, of: (a) hummocky cross-stratified siltstone; (b) wave-rippled sandstone; and (c) planar-laminated siltstone. Shoaling over subfacies 2a and 2b probably caused wave attenuation and deposition of the relatively low-energy subfacies 2c deposits. Colonization by marsh vegetation produced green rooted mudstones (facies 3). Dolomite beds (facies 4) represent early diagenetic subaqueous and pedogenic horizons.

3 Groove, prod, and gutter casts and scours from the hummocky cross-stratified zone were produced by predominantly oscillatory flows. The orientation of clastic dykes was related to the orientation of the wave-crests, and dyke generation probably resulted from cyclic wave-loading of the sediments.

4 Three main types of lacustrine cycle are defined: (A) offshore-dominated, (B) shoreline-dominated, and (C) marsh-dominated. Cycles (and members) are thicker, and strata contain fewer type C cycles, toward the northern basin margin — trends that can be explained by more rapid subsidence toward the fault bounded, northern basin margin.

5 The presence of freshwater flora and fauna and the absence of chemical sediments suggests that the lacustrine system was hydrologically open, and a maximum lake-level was imposed by an outlet. Maximum water depth and cycle thickness would have been controlled by the elevation difference between the outlet and the basin floor. A trend of decreasing cycle thickness upward is inferred to reflect decreasing tectonic subsidence within the basin. The transition from type A cycles at the base of the sections to type C cycles at the top supports this inference.

ACKNOWLEDGEMENTS

The manuscript was greatly improved from reviews by Philip Allen and Elizabeth Gierlowski-Kordesh. The authors thank Rob Naylor, Guy Plint, Ron Boyd, Yasu Makino, and Fran Hein for stimulating discussions. Support (for ATM) was provided by Dalhousie University and an NSERC Student Fellowship, and by NSERC Operating Grant A-8437 (for MRG).

REFERENCES

ALLEN, J.R.L. (1979) A model for the interpretation of wave ripple-marks using their wavelength, textural composition, and shape. *J. Geol. Soc. London* **136**, 673–682.

ALLEN, P.A. (1981) Devonian lake margin environments and processes, SE Shetland, Scotland. *J. Geol. Soc. London* **138**, 1–14.

ALLEN, P.A. (1985) Hummocky cross-stratification is not produced purely under progressive gravity waves. *Nature* **313**, 562–564.

ALLEN, P.A. & COLLINSON, J.D. (1986) Lakes. In: *Sedimentary Environments and Facies* (Ed. Reading, H.G.), pp. 63–94. Blackwell Scientific Publications, Oxford.

ASTIN, T.R. & SCOTCHMAN, I.C. (1988) The diagenetic history of some septarian concretions from the Kimmeridge Clay, England. *Sedimentology* **35**, 349–368.

BANG, T.H. (1987) *Lacustrine sedimentation in the Lower Carboniferous Horton Bluff Formation, at Rainy Cove, Nova Scotia.* BSc thesis, Dalhousie University, 79 pp.

BEADLE, L.C. (1974) *The Inland Waters of Tropical Africa: an Introduction to Tropical Limnology.* Longman, London, 365 pp.

BELL, W.A. (1929) Horton–Windsor district, Nova Scotia. *Geol. Surv. Can. Mem.* **155**, 269 pp.

BELL, W.A. (1960) Mississippian Horton Group of type Windsor–Horton district, Nova Scotia. *Geol. Surv. Can. Mem.* **314**, 58 pp.

BLAIR, T.C. (1987) Tectonic and hydrologic controls on cyclic alluvial fan, fluvial, and lacustrine rift-basin sedimentation, Jurassic–Lowermost Cretaceous Todos Santos Formation, Chiapas, Mexico, *J. Sediment. Petrol.* **57**, 845–862.

BLAIR, T.C. & BILODEAU, W.L. (1988) Development of tectonic cyclothems in rift, pull-apart, and foreland basins: sedimentary response to episodic tectonism. *Geology* **16**, 517–520.

BLESS, M.J.M. & JORDAN, H. (1971) The new genus Copelandella from the Carboniferous — the youngest known Beyrichiacean Ostracodes. *Lethaia* **4**, 185–190.

BLOOS, G. (1982) Shell beds in the Lower Lias of south Germany — facies and origin. In: *Cyclic and Event Stratigraphy* (Eds Einsele, G. & Seilacher, A.), pp. 223–239. Springer-Verlag, Berlin.

BOYD, R., FORBES, D.L. & HEFFLER, D.E. (1988) Time-sequence observations of wave-formed sand ripples on an ocean shoreface. *Sedimentology* **35**, 449–464.

BRENCHLEY, P.J. (1985) Storm influenced sandstone beds. *Modern Geol.* **9**, 369–396.

BROMLEY, R. & ASGAARD, U. (1979) Triassic freshwater Ichnocoenoses from Carlsberg Fjord, East Greenland. *Palaeogeogr. Palaeoclimatol. Palaeoecol.* **28**, 39–80.

CAMPBELL, C.V. (1966) Truncated wave-ripple laminae. *J. Sediment. Petrol.* **36**, 825–828.

CARMOUZE, J.P. & LEMOALLE, J. (1983) Physical and chemical characteristics of the waters. In: *Lake Chad: Ecology and Productivity of a Shallow Tropical Ecosystem* (Eds Carmouze, J.P., Durand, J.R. & Leveque, C.), Dr W. Junk, The Hague, 575 pp.

CARMOUZE, J.P., DURAND, J.R. & LEVEQUE, C. (1983) *Lake Chad: Ecology and Productivity of a Shallow Tropical Ecosystem.* Dr W. Junk, The Hague, 575 pp.

CARROLL, R.L., BELT, E.S., DINELEY, D.L., BAIRD, D. & MCGREGOR, D.C. (1972) Vertebrate paleontology of Eastern Canada. *XXIV Intern. Geol. Congr. Montreal, Quebec, Excursion A59.*

CLARKE, T.L., LESHT, B., YOUNG, R.A., SWIFT, D.J.P. & FREELAND, G.L. (1982) Sediment resuspension by surface wave action: an examination of possible mechanisms. *Mar. Geol.* **49**, 43–59.

CLIFTON, H.E. (1976) Wave-formed sedimentary structures — a conceptual model. In: *Beach and Nearshore Sedimentation* (Eds Davis, R.A., Jr. & Ethington, R.L.), Spec. Publ. Soc. Econ. Paleontol. Mineral. **24**, 126–148.

COHEN, A.S., FERGUSON, D.S., GRAM, P.M., HUBLER, S.L. & SIMS, K.W. (1986) The distribution of coarse-grained sediments in modern Lake Turkana, Kenya: implications for clastic sedimentation models of rift lakes. In: *Sedimentation in the African Rifts* (Eds Frostick, L.E., Renaut, R.W., Reid, I. & Tiercelin, J.-J.), Geol. Soc. London Spec. Publ. **25**, 127–139.

COLEMAN, J.M. & PRIOR, D.B. (1980) Deltaic sand bodies. *Am. Assoc. Petrol. Geol. Cont. Educ. Course Note Ser.* **15**, 171 pp.

COLLINSON, J.D. & THOMPSON, D.B. (1982) *Sedimentary Structures.* George Allen & Unwin, London, 194 pp.

COOK, D.O. & GORSLINE, D.S. (1972) Field observations of sand transport by shoaling waves. *Mar. Geol.* **13**, 31–55.

DAVIS, R.A., JR. (1965) Underwater studies of ripples, southeastern Lake Michigan. *J. Sediment. Petrol.* **35**, 857–866.

DAVIS, R.A., JR., FOX, W.T., HAYES, M.O. & BOOTHROYD, J.C. (1972) Comparison of ridge and runnel systems in tidal and non-tidal environments. *J. Sediment. Petrol.* **42**, 413–421.

DE RAAF, J.F.M., BOERSMA, J.R. & GELDER, VAN A. (1977) Wave generated sequences from a shallow marine succession, Lower Carboniferous, County Cork, Ireland. *Sedimentology* **24**, 451–483.

DEBENHAM, F. (1947) The Bangweulu Swamp of Central Africa. *Geogr. Rev.* **37**, 351–368.

DONNELLY, T.H. & JACKSON, M.J. (1988) Sedimentology and geochemistry of a mid-Proterozoic lacustrine unit from northern Australia. *Sediment. Geol.* **58**, 145–169.

DOTT, R.H., JR. & BOURGEOIS, J. (1982) Hummocky stratification: significance of its variable bedding sequences. *Bull. Geol. Soc. Am.* **93**, 663–680.

DUKE, W.L. (1985) Hummocky cross-stratification, tropical hurricanes, and intense winter storms. *Sedimentology* **32**, 167–194.

DUKE, W.L. (1987) Hummocky cross-stratification, tropical hurricanes, and intense winter storms: reply. *Sedimentology* **34**, 344–359.

DUKE, W.L. & LECKIE, D.A. (1986) Origin of hummocky cross-stratification: Part 2, paleohydraulic analysis indicates formation by orbital ripples in the wave-formed flat bed field. In: *Shelf Sands and Sandstones* (Eds Knight, R.J. & McLean, J.R.), Can. Soc. Petrol. Geol. Mem. **11**, 339.

DZULYNSKI, S. & WALTON, E.K. (1965) Sedimentary features of flysch and greywackes. *Dev. Sedimentol.* **7**, 274 pp.

EISBACHER, G.H. (1969) Displacement and stress field along part of the Cobequid Fault, Nova Scotia. *Can. J. Earth Sci.* **6**, 1095–1104.

EKDALE, A.A. & MASON, T.R. (1988) Characteristic trace-fossil associations in oxygen-poor sedimentary environments. *Geology* **16**, 720–723.

ELLIOTT, T. (1986) Siliciclastic shorelines. In: *Sedimentary Environments and Facies* (Ed. Reading, H.G), pp. 63–94. Blackwell Scientific Publications, Oxford.

EYLES, N. & CLARK, B.M. (1985) Gravity-induced soft-sediment deformation in glaciomarine sequences of the Upper Proterozoic Port Askaig Formation, Scotland. *Sedimentology* **32**, 789–814.

EYLES, N. & CLARK, B.M. (1986) Significance of hummocky and swaley cross-stratification in late Pleistocene lacustrine sediments of the Ontario basin, Canada. *Geology* **14**, 679–682.

FREYTET, P. (1973) Petrography and paleo-environment of continental carbonate deposits with particular reference to the Upper Cretaceous and Lower Eocene of Languedoc (Southern France). *Sediment. Geol.* **10**, 25–60.

FROSTICK, L.E. & REID, I. (1987) Tectonic control of desert sediments in rift basins ancient and modern. In: *Desert Sediments: Ancient and Modern* (Eds Frostick, L.E. & Reid, I.), Geol. Soc. London Spec. Publ. **35**, 53–68.

GIBLING, M.R., TANTISUKRIT, C., UTTAMO, W., THANASUTHIPITAK, T. & HARALUCK, M. (1985) Oil shale sedimentology and geochemistry in Cenozoic Mae Sot Basin, Thailand. *Bull., Am. Assoc. Petrol. Geol.* **69**, 767–780.

GILE, L.H., PETERSON, F.F. & GROSSMAN, R.S. (1966) Morphological and genetic sequences of carbonate accumulation in desert soils. *Soil Sci.* **101**, 347–360.

GOLDRING, R. & BRIDGES P. (1973) Sublittoral sheet sandstones. *J. Sediment. Petrol.* **43**, 736–747.

GOUDIE, A.S. (1983) Calcrete. In: *Chemical Sediments and Geomorphology* (Eds. Goudie, A.S. & Pye, K.). Academic Press, London, 439 pp.

GREENWOOD, B. & SHERMAN, D.J. (1986) Hummocky cross-stratification in the surf zone: flow parameters and bedding genesis. *Sedimentology* **33**, 33–45.

HACQUEBARD, P.A (1972) The Carboniferous of Eastern Canada: *Compte Rendu, 7th International Carboniferous Congress,* Krefeld, 1971, tome I, pp. 69–90.

HAMBLIN, A.P. (1988) A preliminary report on the sedimentology, tectonic control and resource potential of the Upper Devonian–Lower Carboniferous Horton Group, Cape Breton Island. *Geol. Surv. Can., Curr. Res. Part B. Pap.* **88–1B**, 17–22.

HAMBLIN, A.P. & WALKER, R.G. (1979) Storm-dominated

shelf deposits: the Fernie–Kootenay (Jurassic) transition, southern Rocky Mountains. *Can. J. Earth Sci.* **16**, 1673–1690.

HANTZSCHEL, W. (1939) Brandungswalle, Rippeln, und Fliebfiguren am Strande von Wangeroog. *Nat. Volk.* **69**, 141–148.

HARMS, J.C., SOUTHARD, J.B., SPEARING, D.R. & WALKER, R.G. (1975) Depositional environments as interpreted from primary sedimentary structures and stratification sequences. *Soc. Econ. Paleontol. Mineral., Short Course* **2**, 161 pp.

HARMS, J.C., SOUTHARD, J.B. & WALKER, R.G. (1982) Structures and sequences in clastic rocks. *Soc. Econ. Paleontol. Mineral., Short Course* **9**, 249 pp.

HESSE, R. & READING, H.G. (1978) Subaqueous clastic fissure eruptions and other examples of sedimentary transposition in the lacustrine Horton Bluff Formation (Mississippian), Nova Scotia, Canada. In: *Modern and Ancient Lake Sediments* (Eds Matter, A. & Tucker, M.), Spec. Publ. Int. Assoc. Sediment. **2**, 241–257.

HEWARD, A.P. (1981) A review of wave-dominated clastic shoreline deposits. *Earth Sci. Rev.* **17**, 223–276.

HOLE, F.D. (1961) A classification of pedoturbations and some other processes and factors of soil formation in relation to isotrophism and anisotrophism. *Soil. Sci.* **19**, 375–377.

HUBERT, J.F. (1977) Paleosol caliche in the New Haven arkose, Newark Group, Connecticut. *Palaeogeogr. Palaeoclimatol. Palaeoecol.* **24**, 151–168.

HUTCHINSON, G.E. (1957) *A Treatise on Limnology*, vol. III, *Limnological Botany*. Wiley, New York, 660 pp.

JOHNSON, H.D. & BALDWIN, C.T. (1986) Shallow siliciclastic seas. In: *Sedimentary Environments and Facies* (Ed. Reading, H.G.), pp. 229–282. Blackwell Scientific Publications, Oxford.

KEMP, G.P. (1986) *Mud deposition at the shoreface: wave and sediment dynamics on the Chenier Plain of Louisiana.* PhD thesis, Louisiana State University and Agricultural and Mechanical College, 163 pp.

KENNEY, B.C. (1985) Sediment resuspension and currents in Lake Manitoba. *J. Great Lakes Res.* **11**, 85–96.

KEPPIE, J.D. (1982) The Minas Geofracture. In: *Major Structural Zones and Faults of the Appalachians* (Eds St-Julian, P. & Beland, J.), Geol. Assoc. Spec. Pap. **24**, 263–280.

KLAPPA, C.F. (1980a) Rhizoliths in terrestrial carbonates: classification, recognition, genesis and significance. *Sedimentology* **27**, 613–629.

KLAPPA, C.F. (1980b) Brecciation textures and tepee structures in Quaternary calcrete (caliche) profiles from Eastern Spain: the plant factor in their formation. *Geol. J.* **15**, 81–89.

KRAUS, M.J. (1988) Nodular remains of Early Tertiary forests, Bighorn Basin, Wyoming. *J. Sediment. Petrol.* **58**, 888–893.

KREISA, R.D. (1981) Storm-generated sedimentary structures in subtidal marine facies with examples from the Middle and Upper Ordovician of Southwestern Virginia. *J. Sediment. Petrol.* **51**, 823–848.

KUENEN, PH.H. (1968) Origin of ptygmatic features. *Tectonophysics* **6**, 143–158.

LAMBE, L.M. (1910) Palaeoniscid fishes from the Albert shales of New Brunswick. *Geol. Surv. Can., Contrib. Can. Paleontol., Mem.* **3**, 69 pp.

LATTMAN, L.H. (1973) Calcium carbonate cementation of alluvial fans in southern Nevada. *Geol. Soc. Am. Bull.* **84**, 3013–3028.

LECKIE, D.A. (1988) Wave-formed, coarse-grained ripples and their relationship to hummocky cross-stratification. *J. Sediment. Petrol.* **58**, 607–622.

LI SITIAN, LI BAOFANG, YANG SHIGONG, HUANG JIAFU & LI ZHEN (1984) Sedimentation and tectonic evolution of the Late Mesozoic faulted coal basins in north-eastern China. In: *Sedimentology of Coal and Coal-bearing Sequences* (Eds Rahmani, R.A. & Flores, R.M.), Spec. Publ. Int. Assoc. Sediment. **7**, 387–406.

LOWE, D.R. (1975) Water escape structures in coarse-grained sediments. *Sedimentology* **22**, 157–204.

MANN, A.W. & HORWITZ, R.C. (1979) Groundwater calcrete deposits in Australia: some observations from western Australia. *J. Geol. Soc. Aust.* **26**, 293–303.

MANSPEIZER, W. (1981) Early Mesozoic basins of the Central Atlantic passive margins. In: *Geology of Passive Margins*, Am. Assoc. Petrol. Geol. Educ. Course Note Ser. **19**, 4-1 to 4-60.

MARTEL, A.T. (1987) Seismic stratigraphy and hydrocarbon potential of the strike-slip Sackville Sub-basin, New Brunswick. In: *Sedimentary Basins and Basin-Forming Mechanisms* (Eds Beaumont, C. & Tankard, A.J.). *Can. Soc. Petrol. Geol. Mem.* **12**, 319–334.

MATSUMOTO, R. (1978) Occurrence and origin of authigenic Ca–Mg–Fe carbonates and carbonate rocks in the Paleogene coalfield regions in Japan. *J. Fac. Sci. Univ. Tokyo* **5**, 335–367.

MAWER, C.K. & WHITE, J.C. (1986) Sense of displacement on the Cobequid fault system, Nova Scotia, Canada. *Can. J. Earth Sci.* **24**, 217–223.

MCCAVE, I.N. (1985) Recent shelf clastic sediments. In: *Sedimentology: Recent Developments and Applied Aspects* (Eds Brenchley, P.J. & Williams, P.J.), Geol. Soc. London Spec. Publ. **18**, 49–65.

MCMAHON, P.G. (1988) Petroleum source rock study, onshore Nova Scotia: a progress summary. *Nova Scotia Dept. Mines Energy, Rep. Activities* **88-3**, 3–7.

MILNER, A.R. (1987) The Westphalian tetropod fauna; some aspects of its geography and ecology. *J. Geol. Soc. London* **144**, 495–506.

MOONEY, S.J. (1987) Stratigraphy and tectonic evolution of two mid-Paleozoic basins: implications for the timing of final emplacement of the Meguma Terrain. *Geol. Assoc. Can., Miner. Assess. Can., Joint Ann. Meet. Prog. Abstr.* **12**, 74.

MOORE, P.D. (1987) Ecological and hydrological aspects of peat formation. In: *Coal and Coal-Bearing Strata: Recent Advances* (Ed. Scott, A.C.), Geol. Soc. London Spec. Publ. **32**, 7–15.

MULLER, G. (1967) Diagenesis in argillaceous sediments. In: *Diagenesis in Sediments* (Eds Larsen, G. & Chilingar, G.V.), Dev. Sedimentol. **8**, 127–177.

MYROW, P., NARBONNE, G. & HISCOTT, R. (1988) Storm-shelf and tidal deposits of the Chapel Island Group and Random Formations, Burin Peninsula: facies and trace fossils. *Geol. Assoc. Can., Miner. Assess. Can., Can. Soc. Petrol. Geol., Joint Ann. Meet. Prog. Abstr.* **13**, 27–28.

NANCE, R.D. (1987) Dextral transpression and late Carboniferous sedimentation in the Fundy coastal zone of southern New Brunswick. In: *Sedimentary Basins and Basin-Forming Mechanisms* (Eds Beaumont, C. & Tankard, A.J.). Can. Soc. Petrol. Geol. Mem. **12**, 363–377.

NORRMAN, J.O. (1964) Lake Vattern: investigations on shore and bottom morphology. *Geogr. Annar.* **46A**, 1–238.

PEMBERTON, S.G. & FREY, R.W. (1982) Trace fossil nomenclature and the *Planolites–Paleophycus* dilemma. *J. Paleontol.* **56**, 843–881.

PICARD, M.D. & HIGH, L.R., JR. (1972) Criteria for recognizing lacustrine rocks. In: *Recognition of Ancient Sedimentary Environments* (Eds Rigby, J.K. & Hamblin, W.K.), Spec. Publ. Soc. Econ. Paleontol. Mineral. **16**, 108–145.

PLAYFORD, G. (1963) Miospores from the Mississippian Horton Group, Eastern Canada. *Geol. Surv. Can. Bull.* **107**, 47 pp.

PLINT, A.G., WALKER, R.G. & DUKE, W.L. (1988) An outcrop to surface correlation of the Cardium Formation in Alberta. In: *Sequences, Stratigraphy, Sedimentology: Surface and Subsurface* (Eds James, D.P. & Leckie, D.A.), Can. Soc. Petrol. Geol. Mem. **15**, 167–184.

POTTER, P.E., MAYNARD, J.B. & PRYOR, W.A. (1980) *Sedimentology of Shale*. Springer-Verlag, New York, 306 pp.

POULOS, H.G. (1988) *Marine Geotechnics*. Unwin Hyman, London, 473 pp.

PRENTICE, J.E. (1962) Some sedimentary structures from a Weald Clay sandstone at Warnham brickworks, Horsham, Sussex. *Proc. Geol. Assoc. London* **73**, 171–185.

PULAN, R.A. (1968) Lake Chad. In: *The Encyclopedia of Geomorphology* (Ed. Fairbridge, R.W.), pp. 607–611. Reinhold, New York.

RAISWELL, R. (1971) The growth of Cambrian and Liassic concretions. *Sedimentology* **17**, 147–171.

REEVES, C.C. (1976) *Caliche: Origin, Classification, Morphology and Uses*. Estacado Books, Lubbock, 233 pp.

REINECK, H.E. & SINGH, I.B. (1972) Genesis of laminated sand and graded rhythmites in storm-sand layers of shelf mud. *Sedimentology* **18**, 123–128.

REINECK, H.E. & SINGH, I.B. (1980) *Depositional Sedimentary Environments*. Springer-Verlag, Berlin, 549 pp.

SARJEANT, W.A.S. & MOSSMAN, D.J. (1978) Vertebrate footprints from the Carboniferous sediments of Nova Scotia: a historical review and description of newly discovered forms. *Palaeogeogr. Palaeoclimatol. Palaeoecol.* **23**, 270–306.

SCHENK, P.E. (1981) The Meguma Zone of Nova Scotia — a remnant of western Europe, South America, or Africa? *Can. Soc. Petrol. Geol. Mem.* **7**, 119–148.

SCOTT, A.C. (1979) The ecology of Coal Measure floras from northern Britain. *Proc. Geol. Assoc. London* **90**, 97–116.

SCOTT, A.C. (1980) The ecology of some Upper Paleozoic Floras. In: *The Terrestrial Environment and the Origin of Land Vertebrates* (Ed. Panchen, A.L.), Systematics Assoc. Spec. Vol. **15**, 87–115.

SEED, H.B. & LEE, K.L. (1966) Liquefaction of saturated sands during cyclic loading. *Proc. Am. Soc. Civil Eng.,*

J. Soil Mech. Fdns. Div. **92**(SM6), 105–134.

SINGH, I.B. & WUNDERLICH, F. (1978) On the terms wrinkle marks (Runzelmarken), millimetre ripples, and miniripples. *Senckenbergiana Marit.* **10**, 75–83.

SMITH, J.D. (1986) Depositional environments of the Tertiary Colton and Basal Green River Formations in Emma Park, Utah. *Brigham Young Univ. Geol. Stud.* **33**(1), 135–174.

STEEL, R. & GLOPPEN, T.G. (1980) Late Caledonian (Devonian) basin formation, western Norway: signs of strike-slip tectonics during infilling. In: *Sedimentation in Oblique-Slip Mobile Zones* (Eds Ballance, P.F. & Reading, H.G.), Spec. Publ. Int. Assoc. Sediment. **4**, 79–103.

SWIFT, D.J.P., FIGUEIREDO, JR. A.G., FREELAND, G.L. & OERTEL, G.F. (1983) Hummocky cross-stratification and megaripples: a geological double standard? *J. Sediment. Petrol.* **53**, 1295–1317.

THOMAS, R.L., KEMP, A.L.W. & LEWIS, C.F.M. (1972) Distribution, composition and characteristics of the surficial sediments of Lake Ontario. *J. Sediment. Petrol.* **42**, 66–84.

TREWIN, N.H. (1976) *Isopodichnus* in a trace fossil assemblage from the Old Red Sandstone. *Lethaia* **9**, 29–37.

UTTING, J. (1987) Palynostratigraphic investigation of the Albert Formation (Lower Carboniferous) of New Brunswick, Canada. *Palynology* **11**, 73–96.

UTTING, J., KEPPIE, J.D. & GILES, P.S. (1991) Palynology and stratigraphy of the Lower Carboniferous Horton Group, Nova Scotia. In: *Contributions to Canadian Paleontology*. Geol. Surv. Can. Bull. **396**, 117–143.

VAN DIJK, D.E., HOBDAY, D.K. & TANKARD, A.J. (1978) Permo-Triassic lacustrine deposits in the Eastern Karoo Basin, Natal, South Africa. In: *Modern and Ancient Lake Sediments* (Eds Matter, A. & Tucker, M.), Spec. Publ. Int. Assoc. Sediment. **2**, 223–239.

WALKER, R.G. (1984) Shelf and shallow marine sands. In: *Facies Models* (Ed. Walker, R.G.), Geosci. Can. Reprint Ser. **1**, 141–170.

WALKER, R.G., DUKE, W.L. & LECKIE, D.A. (1983) Hummocky cross-stratification: significance of its variable bedding sequences: discussion. *Geol. Soc. Am. Bull.* **94**, 1245–1249.

WALTON, J. (1958) *An Introduction to the Study of Fossil Plants*. Adams & Charles Black, London, 201 pp.

WATTS, N.L. (1978) Displacive calcite: evidence from recent and ancient calcretes. *Geology* **6**, 699–703.

WELLS, J.T. & COLEMAN, J.M. (1981) Physical processes and fine-grained sediment dynamics, coast of Surinam, South America. *J. Sediment. Petrol.* **51**, 1053–1068.

WHITAKER, J.H. McD. (1973) 'Gutter casts', a new term for scour-and-fill structures with examples from Llandoverian of Ringerike and Malmoya, Southern Norway. *Norsk. Geol. Tidsskr.* **53**, 403–406.

YEO, G.M. & GAO, R.X. (1986) Late Carboniferous dextral movement on the Cobequid–Hollow fault system, Nova Scotia: evidence and implications. *Geol. Surv. Can., Curr. Res. Part A. Pap.* **86-1a**, 399–410.

ZULLIG, H. (1956) Sedimente als Ausdruck des Zustandes eines Gewassers. *Schweiz. Z. Hydrol.* **18**, 5–143.

Spec. Publs Int. Ass. Sediment. (1991) **13**, 245–256

Rhythmically laminated lacustrine carbonates in the Lower Cretaceous of La Serranía de Cuenca Basin (Iberian Ranges, Spain)

J.C. GÓMEZ FERNÁNDEZ *and* N. MELÉNDEZ

Departamento de Estratigrafía, Universidad Complutense, 28040 Madrid, Spain

ABSTRACT

Lacustrine carbonate deposits are well represented in the Lower Cretaceous of the Iberian Ranges, eastern Spain. Previously, only shallow lacustrine carbonate deposits have been described from the Iberian Basin, and interpreted as deposits of hard water, shallow lakes and wide, ponded alluvial plains. This paper deals with some of these carbonates located in La Serranía de Cuenca, in the northwestern part of the Iberian Ranges.

A notable feature of the lacustrine Upper Hauterivian? to Lower Barremian La Huerguina Formation, in La Serranía de Cuenca Basin, is the presence of rhythmically laminated limestones interpreted as lacustrine varves. The existence of these lacustrine varves and the preservation of different organisms (especially fish), suggest that thermal stratification of an open lake produced anoxic bottom conditions. Other evidence to support this interpretation includes the absence of bioturbation and high organic matter content. Irregular slabby limestones were deposited below the euphotic zone. Calcarenite beds and algal limestones represent shallower areas of the lake. This lake was surrounded by palustrine areas and a distal alluvial carbonate plain.

INTRODUCTION

Rhythmically laminated carbonate sediments (non-glacial varves) are a common feature of lakes situated in the temperate zone (e.g. Kelts & Hsü, 1978; Hsü & Kelts, 1978; Eugster & Kelts, 1983). This paper analyses rhythmically laminated carbonate deposits formed in a lake system during a warm and semi-arid climate. These lacustrine deposits form La Huerguina Formation, part of the continental Lower Cretaceous of the Iberian Ranges in east-central Spain. The research area is located in the southern part of La Serranía de Cuenca, about 20 km east of the province capital, Cuenca (Fig. 1).

Open and marginal lacustrine deposits of the palaeolake Las Hoyas are associated with distal alluvial and paludal sediments. These deposits have been the focus of attention in the last few years because of their palaeontological interest. This interest has increased with the discovery (Sanz *et al.*, 1988 a,b) of one of the earliest fossil birds.

Data from logged sections and other outcrops, sedimentary petrography, and geological mapping are integrated. The observations include: lithology, grain size variations, sedimentary structures, fossils, and diagenetic features. Many samples were slabbed and many thin-sections were made. The marls were sieved to obtain charophytes and ostracods. X-ray diffraction and chemical analysis were obtained for some samples of special interest.

GENERAL GEOLOGICAL AND PALAEOGEOGRAPHIC SETTING

La Serranía de Cuenca is part of the Iberian Ranges, which is considered by Alvaro *et al.* (1979) to be an aulacogen. This aulacogen started to evolve during the Early Triassic and sedimentation continued, practically uninterrupted, until the Late Cretaceous (Alvaro *et al.*, 1979). The Early Cretaceous was a time of extension and vertical movement of fault blocks. Vilas *et al.* (1983) proposed an evolutionary model for the Iberian Ranges throughout the

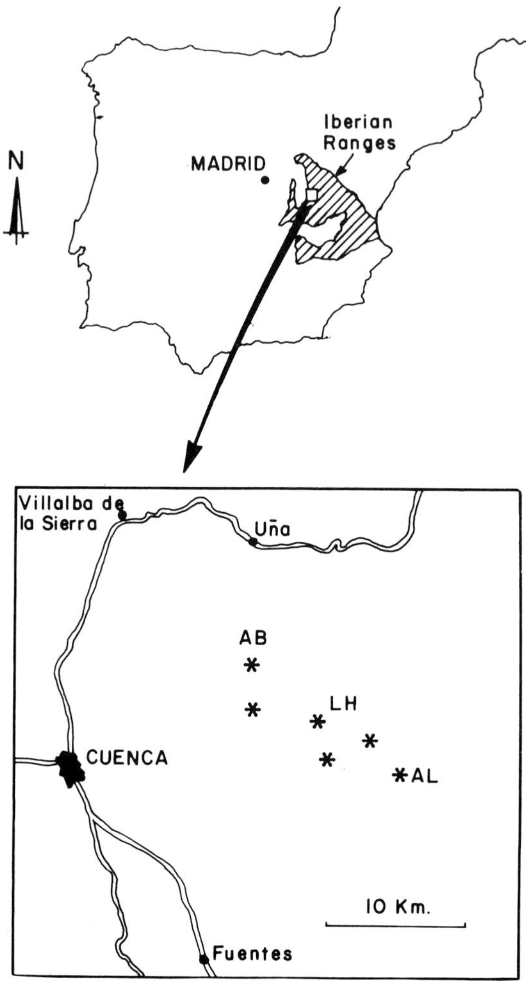

Fig. 1. Location map of the research area showing logged sections and location of outcrops.

Cretaceous as a complete aulacogen development, from the Early Cretaceous early graben stage to the beginning of the folding stage in the latest Cretaceous.

Early Cretaceous sedimentation in the Iberian Basin records infilling of an elongated NW–SE oriented trough resulting from active block movements during the rifting phase. The subsequent basins were filled by continental sequences of variable thickness. There were several depocentres, located on differentially subsiding blocks, which were preferentially oriented along the trough axis. In the study area, two groups of continuous fractures of NW–SE and WNW–ESE directions (F. Meléndez, 1972; F. Meléndez *et al.*, 1974; Gómez Fernández, 1988) controlled subsidence. The research area is located on one of these subsiding blocks; it contains a thick record of continental sediments (about 200 m), most of them lacustrine in origin.

The Lower Cretaceous of the Iberian Ranges has been the focus of many sedimentological and stratigraphical papers (García, 1977; Mas, 1981; N. Meléndez, 1983). Mas *et al.* (1982) reviewed the stratigraphical and sedimentological development of the Lower Cretaceous continental rocks in a zone of the Iberian Basin located towards the southeast of our study area. Mas *et al.* (1982) report the existence of a NW–SE directed trough in which large coastal, alluvial, and deltaic plains developed toward the southeast. These systems lose their marine influence towards the northwest, passing into vast alluvial plains laterally contiguous with systems of shallow carbonate lakes. Deposits of open lakes (deep enough to allow the formation and preservation of lacustrine varves) have not yet been described. The exact location of our study area is in the northwestern region of the Iberian Ranges (Fig. 1).

There are numerous works that interpret the palaeoclimate of the Iberian Plate during the Early Cretaceous. Menéndez Amor (1970), based on palynological analyses, concluded that the prevailing climatic conditions were tropical. Rat (1982), applied sedimentary and palaeogeographic data to suggest that the climate was warm, semi-arid, and seasonal. Ziegler *et al.* (1983) placed the Iberian Peninsula during that time at latitude 30°N, which today would be a dry tropical zone. Ziegler *et al.* (1987) suggested that the climatic zones had a different distribution from today because the polar fronts were weaker and continental mass distribution was different. Sanz *et al.* (1988b) reported the presence of warm-temperate and subtropical floras in the deep lake deposits of Lake Las Hoyas. Sedimentological studies of Gómez Fernández (1988) and Gierlowski-Kordesch *et al.* (this volume) postulated that the continental sediments of the La Serranía de Cuenca were deposited under warm climatic but seasonal conditions, and it is possible to differentiate seasons (humid and dry), as well as longer climatic cycles.

A climate with these characteristics seems to have acted as a main factor in the dynamics of the Early Cretaceous lacustrine system. Long-term climatic cycles may have influenced processes such as the lake margin exposure and the paludal area development, formation of paleosols, etc. On the other hand, short-term cycles (seasons) probably controlled the ambient air and lake water temperatures.

STRATIGRAPHIC FRAMEWORK

The Lower Cretaceous continental rocks rest unconformably on a marine calcareous series of Middle Jurassic (Dogger) age. The contact is erosive (Upper Jurassic (Malm) strata crops out towards the south-east) and shows a well-developed paleokarst. The paleokarst has cavities lined with calcite crystals and filled with red clays in recrystallized limestones.

Above this unconformity, the lacustrine carbonate dominated La Huerguina Formation is found over the entire research area (Fig. 2). Data indicate that this formation is laterally equivalent to the El Collado Formation (Vilas *et al.*, 1982), which does not crop out in this area. The El Collado Formation is mainly clastic and alluvial in origin. The La Huerguina Formation (lacustrine) and El Collado Formation (fluvial) constitute a depositional sequence *sensu* Mitchum *et al.* (1977). The age of these formations ranges from uppermost Hauterivian to Lower Barremian (N. Meléndez, 1983).

Another unconformity occurs above these two

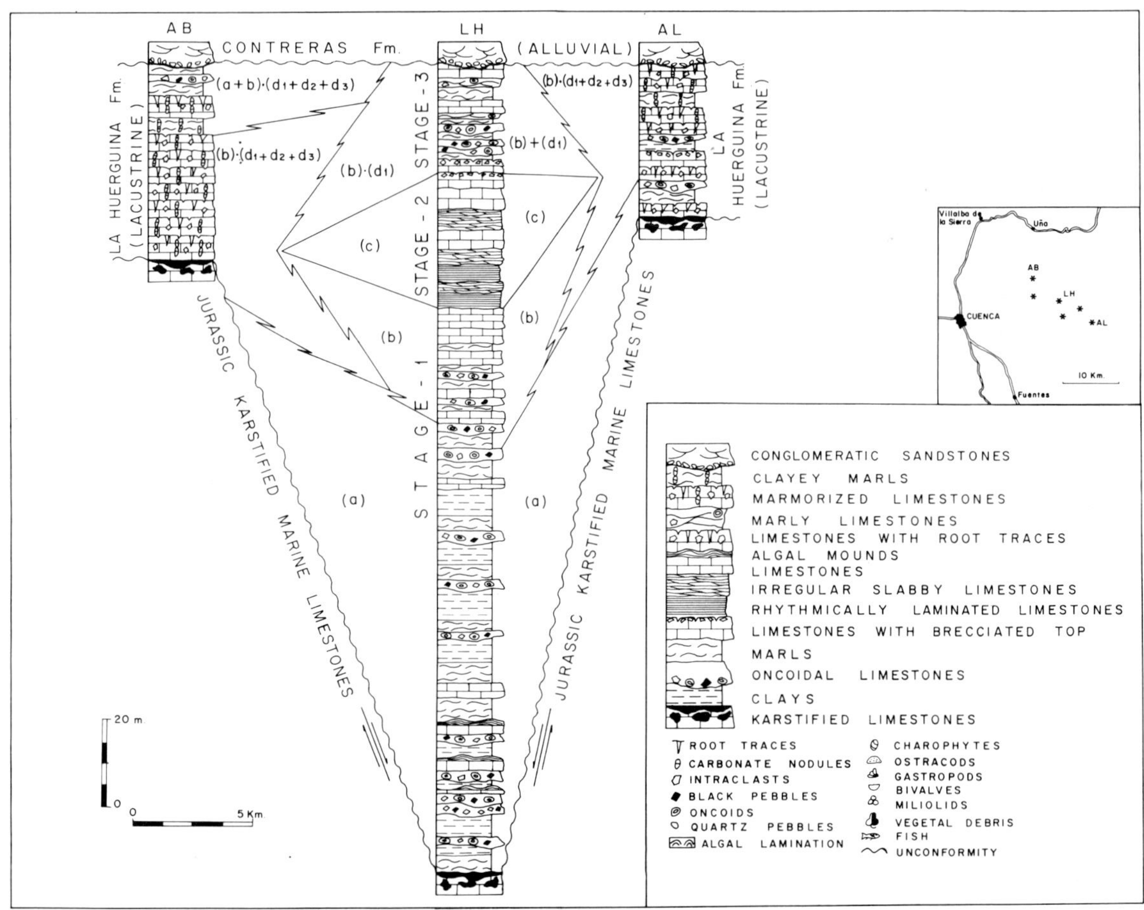

Fig. 2. Logged sections and stratigraphic framework. General legend. Key to sections: AB, Arroyo Bonilla; LH, Las Hoyas; AL, Arroyo del Lavadero. Letters designate facies (see Fig. 3).

formations (Fig. 2). Finally, the alluvial siliciclastics of the Contreras Formation overlie this continental sequence. This formation contains carbonate clasts reworked from the underlying lacustrine formation.

In the research area, a series of control points (three measured sections and other outcrops) distributed over the whole area were used as base points (Fig. 1). The lacustrine deposits are mainly carbonates (limestones and marls). Clastics, mainly mudstones, are less common. There are large changes in the thickness of the lacustrine unit across the study area (Fig. 2), from that in the peripheral zones (Arroyo Bonilla (52 m) and Arroyo del Lavadero (32 m) sections) to a maximum thickness of 208 m in the central areas (Las Hoyas section). The main lithological features in this unit can be represented in a short description from the northwest to the southeast.

The materials outcropping in the Arroyo Bonilla section (towards the northwest) are mainly variegated and brecciated limestones and contain many root traces. Charophytes and ostracods are present. Clayey marls also occur; these display many vertically arranged carbonate nodules.

In the central part, the thicker Las Hoyas section (Fig. 2) consists of three separate units:
1 The Lower Unit is composed of red silts; massive limestones (wackestones) up to 1 m thick with large lateral continuity, containing abundant charophytes and ostracods; and grey-greenish marls with charophytes, ostracods, and gastropods. Also limestones (rudstones and grainstones) crop out in lenticular bodies, 0.50−0.75 m thick; they have erosive bases, and contain many oncoids (up to 3 cm) and lithoclasts, as well as some charophytes, ostracods, bone fragments, and plant debris.
2 The lower part of the Middle Unit consists of finely laminated limestones of dark colour, whose laminae are laterally continuous (a single lamina can be followed along the total length of an outcrop). These laminated limestones alternate with irregularly laminated limestone lenses, which pinch out over 1 m or less. The upper part of this unit consists of metre-scale packstone and grainstone beds displaying low-angle planar cross-stratification and rare algal lamination.
3 The Upper Unit in the Las Hoyas section consists mostly of limestones (wackestones) in beds up to 1 m thick with great lateral continuity. Many layers show a brecciated top. Limestones (rudstones) in lenticular bodies and grey-greenish marls also occur.

The Arroyo del Lavadero section (in the southeast) includes variegated and brecciated limestones with widespread root traces (Fig. 2). It is possible to recognize charophytes and ostracods. Lenticular oncoidal limestones (rudstone) and grey-greenish marls also occur.

In summary, the main features of the La Huerguina lacustrine limestones are:
1 Presence of early diagenetic features (brecciation, root marks, marmorization), which are developed with a greater intensity in the Arroyo Bonilla and Arroyo del Lavadero sections. These features are very rare and poorly developed (only brecciated tops) in the central area (Las Hoyas).
2 Three well-defined sedimentological units are recognizable in the central area (Las Hoyas), not recorded in peripherical areas (Fig. 2).
3 Finely laminated limestones and other carbonate facies showing irregular layers only occur in the middle unit of the Las Hoyas section. The presence of these limestones coincides with the maximum thickness measured in the La Huerguina Formation in the study region.
4 The total absence of clastics in the middle unit of Las Hoyas section.

FACIES AND FACIES ASSOCIATIONS: INTERPRETATION AND DISCUSSION

Four lacustrine−palustrine facies associations have been recognized (Fig. 3): three are depositional (a, b, and c) and one is early diagenetic (d). These facies associations are important clues for the interpretation of the environments of sedimentation.

Association a

This association is widely distributed over the study area; however, it is easier to recognize in the central zone (in the Upper Unit and, especially, in the Lower Unit of the Las Hoyas section). It consists of four facies: clayey marls, marls, marly limestones, and oncoidal rudstones (Fig. 4A). This association has been widely recognized in the La Huerguina Formation in zones close to the southeast of the present study area (Mas, 1981; Mas *et al.*, 1982; N. Meléndez, 1983), and in nearby zones to the north (Gierlowski-Kordesch *et al.*, this volume). This association is interpreted to be carbonate alluvial plain deposits: the oncoidal rudstones represent channel deposits and the clayey marls, marls, and marly limestones represent floodplain sediments characterized by significant pedogenesis.

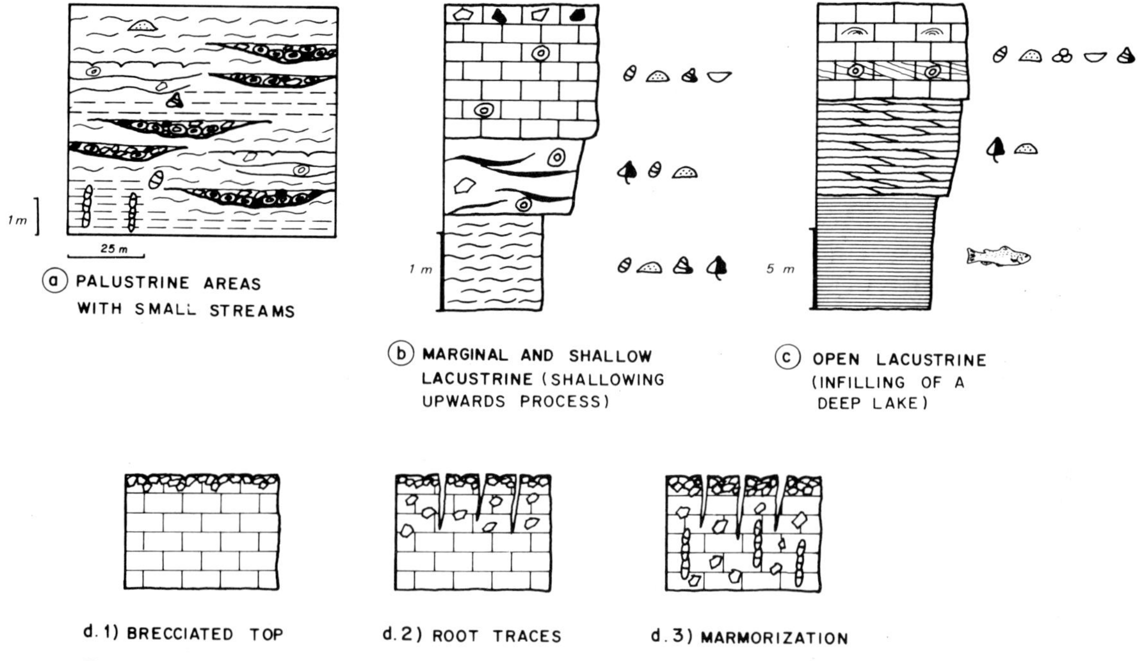

Fig. 3. Facies associations (see general legend in Fig. 2).

Association b

This association consists of three facies (marls, marly limestones, and charophyte micrites). It appears to be widely distributed over the entire study area, especially in the Las Hoyas zone (mainly in the Upper Unit). Like the previous association, this one also has been described from other areas in which the La Huerguina Formation crops out (Mas, 1981; Mas *et al.*, 1982; N. Meléndez, 1983). This association has also been identified north of the study area (Gierlowski-Kordesch *et al.*, this volume) and is interpreted as deposits of shallow hardwater lakes and ponds.

Association c

It is only represented in the Middle Unit of the Las Hoyas section in the central zone of the study area. It is composed of three facies.

Rhythmically laminated limestones. These are dark grey with a shaley appearance (Fig. 4B). The laminae are laterally continuous, uniform and parallel, with a thickness ranging from 0.7 to 1.5 mm. They consist entirely of low-magnesium calcite (LMC) and contain approximately 5% organic matter by volume (visual estimation). In these layers, fish remains occur (Fig. 4C), locally in clusters with ostracods and plant impressions. Sanz *et al.* (1988b) described a rich faunal and floral assemblage in this facies.

In thin-section (Fig. 4D), each layer contains the following couplet:

1 0.3 mm brown micrite with an abrupt contact with the couplet below.

2 0.7 mm — calcite crystal size increases gradually upward. The largest ones are found in the upper part of the couplet. The content of organic matter increases upwards as well. The upper contact with the next couplet is abrupt, representing a break in the development of this clear lamina.

The accumulation of these couplets can be truncated by the presence of some thicker layers (up to 1 cm thick), which contain coarser components (lithoclasts and bioclasts of ostracods and bivalves). Beneath these clast layers the couplets appear slightly deformed and eroded.

J.C. Gómez Fernández and N. Meléndez

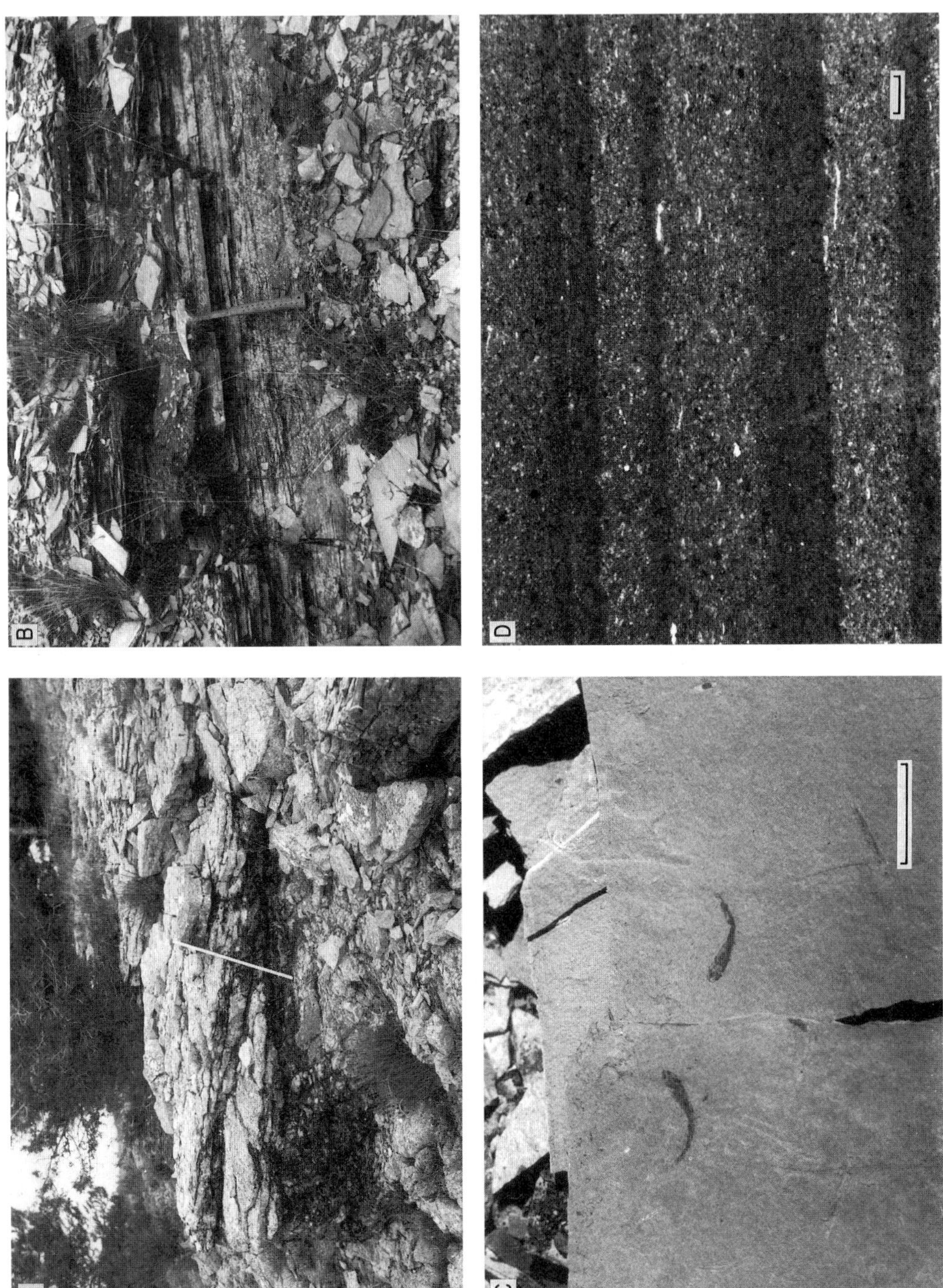

Fig. 4. (A) Oncoidal rudstone showing cross-bedding and erosive base (Lower Unit, Las Hoyas section). Scale 60 cm in length. (B) Common field aspect of the rhythmically laminated limestones (Middle Unit, Las Hoyas section). Hammer for scale. (C) Fish remains from the rhythmically laminated limestones (Middle Unit, Las Hoyas section). Scale bar 3 cm in length. (D) Thin-section of the rhythmically laminated limestones interpreted as lacustrine varves; each varve is formed by a couplet containing a dark micrite lamina and a light lamina with larger crystals. Note the persistence of parallel laminae and the sharp contact at the base of the dark lamina, while the contact between dark lamina and light lamina is gradual (Middle Unit, Las Hoyas section). Scale bar 0.25 mm.

Fig. 5. (A) Transition from the irregular slabby limestones to the bioclastic limestones (Middle Unit, Las Hoyas section). Hammer for scale. (B) Detail of the bioclastic limestones showing alternation of bioclastic intervals and algal laminated intervals, reflecting periods of high- and low-water agitation periods (Middle Unit, Las Hoyas section). Key for scale. (C) Brecciation at the top of a shallow lacustrine limestone (charophyte limestones) (Arroyo del Lavadero section). Hammer for scale. (D) Thin-section of a marmorized limestone. See root trace filled by calcite (arrow) (Arroyo Bonilla section). Scale bar 1 cm.

Irregular slabby limestones. Grey limestones (wackestones) occur in thin slabs or plates. These slabs are several millimetres thick, and pinch out over less than 1 m; they have an irregular morphology (Fig. 5A). Their main components are irregular peloids, but they also contain abundant ostracods, some bivalves, and plant debris (mainly leaves).

Bioclastic limestones. These are light brown packstones and grainstones that show low-angle planar cross-bedding. Cosets are about 0.30 m thick. They almost exclusively consist of bioclasts (charophytes, ostracods, bivalves, gastropods, and miliolids). Centimetre-thick layers of small oncoids (about 2 mm diameter) also occur. In addition, intervals containing an alternation of cross-stratification and irregular laminae occur with possible aligned algal filaments (Fig. 5B).

The *rhythmically laminated limestones* record quiet sedimentation of carbonate particles out of suspension in an environment without agitation or currents, over a flat lake floor. Burrowing organisms were absent, since otherwise these fine laminations would not have been preserved with great lateral continuity. On the other hand, the variety of depositional components in the laminae that form each couplet suggests some variations in the palaeoenvironmental conditions. These were probably periodic, maybe seasonal. The rhythmically laminated limestones are equivalent to non-glacial varves, which form in lakes located in temperate climate zones (Kelts & Hsü, 1978; Hsü & Kelts, 1978; Eugster & Kelts, 1983). Anoxic bottom conditions, developed in thermally stratified lakes, allow preservation of organic matter. These conditions of quiet sedimentation were occasionally interrupted by turbiditic currents bringing bioclasts and lithoclasts reworked from shallow lake areas.

The *irregular slabby limestones* originate from sedimentation in relatively deep zones of the lake, where heterotrophic organisms were present, but photosynthesizing organisms were excluded, possibly owing to dim light. Similar sediments are reported from deep permanently oxic lake waters (Treese & Wilkinson, 1982), and from situations where oxygenation occurs only during or shortly after thermal overturn (Murphy & Wilkinson, 1980).

The depositional components and other sedimentological features of the *bioclastic limestones* point toward sedimentation in a shallow environment with stronger agitation. These conditions occur in lacustrine littoral zones (Ryder *et al.*, 1976; Treese & Wilkinson, 1982; Dean & Fouch, 1983). The presence of layers of small oncoids and algal laminations shows that there were frequently suitable conditions for the growth of blue-green algae. Algal balls form in agitated zones and algal mats in quiet zones.

In short, the sequential arrangement of Association c represents a typical infill of a meromictic lake with an open zone characterized by thermal stratification of the water column and a permanently anoxic bottom (monimolimnion). An annual overturn of epilimnion and hypolimnion took place in this lake. This overturn was a partial overturn, and did not involve the monimolimnion. The large size of the lake permitted occasional wave agitation in littoral zones during storms.

Varve formation: discussion

The formation of non-glacial varves is clearly linked to climatic parameters, especially seasonal changes. As mentioned above, during the Early Cretaceous this area was located in a tropical climate zone with distinct humid and dry seasons, with temperatures probably remaining relatively high throughout the year.

In present-day lakes located in low latitudes, water stratification can be established with little difference in temperature (1 or 2°) between the epilimnion and the hypolimnion (Burgis & Morris, 1987). A slight fall in temperature can disrupt water stratification. Temperature changes of small magnitude can be produced by insolation decrease during the rainy season and may result in annual overturn of the water column during the humid season, resulting in mixing of the epilimnion and hypolimnion of such meromictic lakes.

Considering the above-mentioned points, and taking into account that the water of this lake contained a great quantity of carbonate in solution, a mechanism can be proposed for the formation of these lacustrine varves.

In a thermally stratified lake the hypolimnion would contain nutrients and carbonates in solution. With an overturn (in the rainy season), these components reach the surface. The temperature increase would contribute to calcite saturation. Nutrients would contribute to the organic productivity, which drives calcite supersaturation (Kelts & Hsü, 1978). Under these conditions, calcite precipitation would be very quick, so the crystals would be of small size

(Folk, 1974; Kelts & Hsü, 1978). These conditions would induce the formation of the brown micrite laminae during the humid season, and probably also at the beginning of the dry season.

During the dry season, all dissolved ions (carbonates and nutrients) supplied by the hypolimnion would be slowly depleted. With carbonate depletion, organic productivity would also diminish along with biochemical precipitation of calcite. This is a gradual process, which would slow the rate of crystal growth and result in an increase in crystal size (Folk, 1974; Kelts & Hsü, 1978). Planktonic organism mortality would also increase and organic matter would accumulate in the upper part of the second layer of the couplet.

Association d

This association contains three early diagenetic facies, which are transitional to the limestone facies of Associations a and b. Original depositional features (generally bioclasts) record the original sedimentary facies prior to early diagenetic modification. These facies are represented mainly in the marginal zones of the study area (Arroyo Bonilla and Arroyo del Lavadero), although they are also occasionally developed in the Upper Unit of the Las Hoyas section. Important diagnostic features include:

Brecciated top (Fig. 5C). This occurs at the top of limestone units. The clasts have not been transported and they are fractured, resembling mud cracks. The matrix appears slightly recrystallized with a secondary vug porosity (less than 10%), filled by sparry calcite.

Root traces. These occur in limestones with a brecciated top as tube-like traces normal to the bedding surface, with curving morphologies and different sizes (from 1 to 4 cm in diameter, and up to 40 cm in length). The matrix is normally microsparite. The secondary porosity (between 15 and 20%) is vug and channel type and appears to be filled by sparry calcite crystals with clay and iron-oxide impurities.

Marmorization (Fig. 5D). This occurs in brecciated limestones, which are variegated in colour (red, pink, yellow, grey, and white), with a great quantity of vertical traces filled with marly material and lithoclasts or large calcite crystals. The matrix is microsparite. The secondary porosity can reach 30%, as vug and channel type, and is filled by sparry calcite with clay impurities and iron oxide. In thin-section small nodules (up to 2 cm) are marked with iron oxide coatings and the matrix is recrystallized.

These facies show a spectrum of diagenetic alteration from marmorization to brecciation at the tops of beds, reflecting the duration of exposure and the extent of the pedogenetic modification in wet vegetated palustrine areas (Duchaufour, 1960; Freytet & Plaziat, 1982).

PALAEOGEOGRAPHIC EVOLUTION

Lithological and sedimentological data show that sedimentation took place during the uppermost Hauterivian and Lower Barremian, in three well-defined episodes, marked by three different lithological units described above, which are preserved only in the central area. This is probably owing to a more intense subsidence: the sediments are thicker here and the three major units are clearly differentiated. These three episodes are the consequence of the block tectonic structure that prevailed during the Early Cretaceous (Mas *et al.*, 1982; Vilas *et al.*, 1983). The extensional tectonics provoked a relative displacement of blocks, each with a particular subsidence rate. The prevailing fracture directions in this area were NW−SE and WNW−ESE.

Related to these block movements and to the differential subsidence, a general lacustrine system with different areas or palaeogeographic domains can be recognized. In these domains a range of depositional processes and sedimentary features are reflected in the sedimentary record.

The Early Cretaceous lacustrine system contained an open lake of considerable size and depth, at least during part of the time of continental sedimentation in the Iberian Basin. This lacustrine system reached approximately 150 km^2. When the lacustrine system was completely developed, three palaeogeographic domains with individual characteristics resulted. This setting corresponds with the second stage or episode (Figs 2, 6).

Marginal zone. Situated along the shores of the lake, this zone was dominated by a vegetated palustrine environment where early diagenetic (pedogenetic) processes affected the carbonates. This palustrine zone was drained by small streams carrying, almost exclusively, calcareous reworked material.

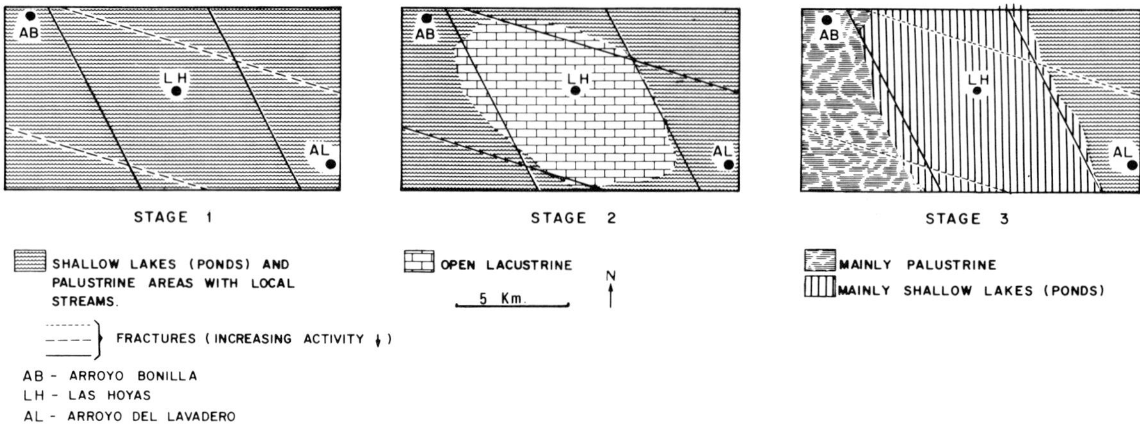

Fig. 6. Palaeogeographic setting of La Huerguina Formation in three evolutionary stages. These three stages are a consequence of relative displacement of tectonic blocks. Diverse palaeogeographic domains are recognized related to the differential subsidence rate.

Lacustrine littoral zone. This shallow zone was inhabited by a diversified biological community (charophytes, blue-green algae, gastropods, ostracods, bivalves). Sedimentation of high-energy deposits reflects lake shoreline dynamics.

Open and deeper lacustrine zone. The lake was meromictic with a water mass large enough to produce thermal stratification during the dry season, which was broken in the rainy season with a partial overturn. This partial overturn did not affect the lower part of the water column, which was permanently anoxic (monimolimnion).

The spatial distribution of these three domains changed during continental sedimentation. This evolution can be represented in three stages within the study area (Fig. 6).

Stage 1

Sedimentation started in this basin in small shallow carbonate lakes (ponds). Pure carbonate micrites were deposited in these hard waters. These lakes were surrounded by palustrine zones, which were located in a wide alluvial plain drained by small streams. During this stage the sedimentary environment was similar throughout the research area, as is shown by the facies distribution. However, thickness changes across the area suggest that sedimentation was principally controlled by fractures in a NW−SE direction (Fig. 6).

Stage 2

The transition from the first to the second stage is only recorded in the central area by a gradual change from sedimentation in an alluvial carbonate plain to a shallow lacustrine and, later, a deeper lacustrine environment. This change seems to be the result of increased subsidence with respect to the sedimentation rate in this central area. This increase in the subsidence rate could have been in response to reactivation of WNW−ESE fractures while the NW−SE fractures were still active. In the central area, a lake with open waters resulted from this strong differential subsidence. It is difficult to estimate its depth, but it was deep enough to maintain thermal stratification of the water column and a permanent anoxic bottom. These conditions allowed the production and preservation of lacustrine varves. Around this central zone, charophytic carbonates were deposited in shallower areas of the lake. Palustrine areas with small lakes or ponds similar to those of stage 1 were located in these outer zones.

Stage 3

The transition from the second to the third stage was gradual, and is characterized by sedimentation in shallow environments, the infilling of the Las Hoyas Lake with the establishment of shallower environments. Three different zones can be recognized: a western zone with palustrine development; a central zone with shallow lakes, and an eastern zone of palustrine character with small streams and shallow hardwater lakes. This environmental zonation probably reflected differing subsidence rates across the basin, controlled by the NW−SE fractures, while the WNW−ESE fractures became inactive.

CONCLUSIONS

A lacustrine system from the Lower Cretaceous continental phase in the Iberian Basin contained a large lake (named Las Hoyas Lake) with different palaeogeographic domains: a central region of open water; a littoral area; and a marginal zone, with an associated fringing alluvial plain.

The most interesting facies are rhythmically laminated limestones interpreted as non-glacial lacustrine varves. They are composed of carbonate laminae couplets: dark micritic and clear sparite laminae. These varves formed in a lake within a warm and semi-arid climatic zone. The alternation of dry and humid seasons controlled the thermal stratification and water overturn. Varve preservation indicates that the lake was thermally stratified with an anoxic bottom (monimolimnion). The couplets within the varves show an alternation between pure carbonate laminae and more organic-rich laminae. Good preservation of fauna (e.g. fish remains) in the varve facies supports this interpretation.

From the palaeogeographic point of view, three stages of sedimentation, controlled by NW−SE and WNW−ESE fractures, are recognized in this continental sequence. An open lake only appeared during the second stage. This lake was surrounded by palustrine marginal zones. During the first and third stages, sedimentation took place in alluvial−palustrine areas and shallow carbonate lakes. The sedimentation in the marginal areas shows minor changes over time. In the central area the sedimentation starts with an alluvial plain with carbonate ponds; later, owing to an increase in the subsidence rate, the central part of the basin became deeper than the surrounding areas, and a lake (Las Hoyas Lake) was established. Finally, the infilling of the lake and a change in the subsidence patterns led to a new configuration of the basin, with shallow-lake sedimentation.

ACKNOWLEDGEMENTS

The authors would like to thank the CAICYT ID-452 Project for providing the financial support. Thanks are extended to A. Arche, M.E. Arribas, E. Gierlowski−Kordesch, J. Lopez, N. Platt and M.E. Tucker for their constructive reviews of the manuscript.

REFERENCES

ÁLVARO, M., CAPOTE, R. & VEGAS, R. (1979) Un modelo de evolución geotectónica para la Cadena Celtibérica. *Acta Geol. Hisp.* **14**, 172−177.

BURGIS, M.J. & MORRIS, P. (1987) *The Natural History of Lakes.* Cambridge University Press, Cambridge, 218 pp.

DEAN, W.E. & FOUCH, T.D. (1983) Lacustrine environment. In: *Carbonate Depositional Environments* (Eds Scholle, P.A., Bebout, D.G. & Moore, C.H.), Mem., Am. Assoc. Petrol. Geol. **33**, 97−130.

DUCHAUFOUR, P. (1960) *Précis de pedologie.* Masson et Cie, Paris, 426 pp.

EUGSTER, H.P. & KELTS, K. (1983) Lacustrine chemical sediments. In: *Chemical Sediments and Geomorphology* (Eds Goudie, A.S. & Pye, K.), pp. 321−368. Academic Press, London.

FOLK, R.L. (1974) The natural history of crystalline calcium carbonate: effect of magnesium content and salinity. *J. Sediment. Petrol.* **44**, 40−53.

FREYTET, P. & PLAZIAT, J. (1982) Continental carbonate sedimentation and pedogenesis. *Contrib. Sedimentol.* **11**, 216 pp.

GARCÍA, A. (1977) Jurásico terminal y Cretácico inferior en la región central de la provincia de Valencia y noreste de la provincia de Albacete. *Seminarios de Estratigrafía* **1**, 334 pp.

GÓMEZ FERNÁNDEZ, J.C. (1988) *Estratigrafía y Sedimentología del Cretácico inferior en 'Facies Weald' de la Región meridional de La Serranía de Cuenca.* Tesis de Licenciatura, Universidad Complutense de Madrid, 228 pp.

HSÜ, K.J. & KELTS, K. (1978) Late Neogene chemical sedimentation in the Black Sea. In: *Modern and Ancient Lake Sediments* (Eds Matter, A. & Tucker, M.E.), Spec. Publ. Int. Assoc. Sediment. **2**, 129−145.

KELTS, K. & HSÜ, K.J. (1978) Freshwater carbonate sedimentation. In: *Lakes: Chemistry, Geology, Physics.* (Ed. Lerman, A.), pp. 295−323. Springer-Verlag, Berlin.

MAS, R. (1981) El Cretácico inferior de la región noroccidental de la provincia de Valencia. *Seminarios de Estratigrafía* **8**, 408 pp.

MAS, R., ALONSO, A. & MELÉNDEZ, N. (1982) El Cretácico Basal 'Weald' de la Cordillera Ibérica Suroccidental (NW de la provincia de Valencia y E de la de Cuenca). *Cuad. Geol. Ibérica* **8**, 309−335.

MELÉNDEZ, F. (1972) *Estudio geológico de la Serranía de Cuenca en relación a sus posibilidades petrolíferas.* Tesis Doctoral, Universidad Complutense de Madrid, 245 pp.

MELÉNDEZ, F., VILLENA, J., RAMÍREZ, J., PORTERO, J.M., OLIVÉ, A., ASSENS, J. & SÁNCHEZ, P. (1974) Síntesis del Cretácico de la Zona Sur de la Rama Castellana de la Cordillera Ibérica. *I Symposium Cretácico de la Cordillera Ibérica*, pp. 241−252.

MELÉNDEZ, N. (1983) El Cretácico de la Región de Cañete–Región de Ademuz (provincias de Cuenca y Valencia). *Seminarios de Estratigrafía* **9**, 242 pp.

MENÉNDEZ AMOR, J. (1970) Primera contribución al estudio esporo-polínico de los lignitos de Uña (Cuenca). *Bol. R. Soc. España. Hist. Natural. (Geol.)* **68**, 275–281.

MITCHUM, R.M., VAIL, P.R. & THOMPSON, S. (1977) Part two: the depositional sequence as a basic unit for stratigraphic analysis. In: *Seismic Stratigraphy — Applications to Hydrocarbon Exploration* (Ed. Payton, C.E.). Mem., Am. Assoc. Petrol. Geol. **26**, 53–62.

MURPHY, D.H. & WILKINSON, B.H. (1980) Carbonate deposition and facies distribution in central Michigan marl lake. *Sedimentology* **27**, 123–135.

RAT, P. (1982) Factores condicionantes en el Cretácico de España. *Cuad. Geol. Ibérica* **8**, 1059–1076.

RYDER, R.T., FOUCH, T.D. & ELISON, J.H. (1976) Early Tertiary sedimentation in the western Uinta Basin, Utah. *Geol. Soc. Am. Bull.* **87**, 496–512.

SANZ, J.L., BONAPARTE, J.F. & LACASA, A. (1988a) Unusual Early Cretaceous birds from Spain. *Nature* **331**, 433–435.

SANZ, J.L., WENZ, S., YÉBENES, A., ESTES, R., MARTÍNEZ-DELCLÓS, X., JIMÉNEZ-FUENTES, E., DIÉGUEZ, C., BUSCALIONI, A.D., BARBADIJO, J. & VIA, L. (1988b) An Early Cretaceous faunal and floral continental assemblage: Las Hoyas fossil site (Cuenca. Spain). *Geobios* **21** (5), 611–635.

TREESE, K.L. & WILKINSON, B.H. (1982) Peat marl deposition in a Holocene paludal–lacustrine basin. Sucker Lake, Michigan. *Sedimentology* **29**, 375–390.

VILAS, L., MAS, A., GARCÍA, A., ARIAS, C., ALONSO, A., MELÉNDEZ, N. & RINCÓN, R. (1982) Ibérica Suroccidental. In: *El Cretácico de España*. Universidad Complutense de Madrid, pp. 457–513.

VILAS, L., ALONSO, A., ARIAS, C., MAS, R., RINCÓN, R. & MELÉNDEZ, N. (1983) The Cretaceous of the Southwestern Iberian Ranges (Spain). *Zitteliana* **10**, 245–254.

ZIEGLER, A.M., BARRETT, S.F. & SCOTESE, C.R. (1983) Mesozoic and Cenozoic paleogeographic maps. In: *Tidal Friction and Earth's Rotation II* (Eds Brosche, P. & Sundermann, J.), pp. 240–252. Springer-Verlag, Berlin.

ZIEGLER, A.M., RAYMOND, A.L., GIERLOWSKY, T.C., HORRELL, M.A., ROWLEY, D.B. & LOTTES, A.L. (1987) Coal, climate and terrestrial productivity: the present and the early Cretaceous compared. In: *Coal and Coal-bearing Strata. Recent Advances* (Ed. Scott, A.C.) Geol. Soc. London Spec. Publ. **32**, 25–49.

Spec. Publs Int. Ass. Sediment. (1991) **13**, 257–275

Sequential arrangement and asymmetrical fill in the Miocene Rubielos de Mora Basin (northeast Spain)

P. ANADÓN*, LL. CABRERA[†], R. JULIÀ* and M. MARZO[†]

**Institut de Ciències de la Terra 'Jaume Almera', C.S.I.C., c. Martí i Franqués s.n., 08028-Barcelona, Spain*
[†]Facultat de Geologia, Dept. G.D.G.P., Universitat de Barcelona, 08028-Barcelona, Spain

ABSTRACT

The Rubielos de Mora Basin is an early–middle Miocene synclinal basin developed in an extensional setting. The basin fill consists of over 800 m of alluvial and lacustrine deposits, which can be divided into three main units: Lower (alluvial dominated), Middle (alluvial and lacustrine) and Upper (lacustrine-dominated) units.

The depositional framework of the Upper Unit, which is given particularly detailed attention in this paper, was a perennial meromictic lake influenced by contributions from surrounding alluvial systems. The deposits of the lake system are characterized by a striking asymmetric facies distribution and by a hierarchical sequential arrangement.

The asymmetry of the lacustrine system is shown by the distribution of facies assemblages. In the east, relatively steep gradients occurred, with subaqueous gravitational breccia rim and carbonate–terrigenous progradational benches grading from the lake margins into the inner lake zones, where laminated organic-rich deposits developed under highly persistent anoxic bottom conditions. In the west the depositional gradient was gentler, and extensive terrigenous sublacustrine ramps developed. The changing bottom conditions in these zones was enhanced by the low gradient, allowing the development of the striking cyclical (anoxic–oxic) successions. From this point of view the morphometry of the ancient lacustrine system emerges as a major feature to take into account during exploration for oil shale deposits.

The hierarchical arrangement of the basin fill resulted from the evolution of the lacustrine system. A major vertical trend of rising water level and expansion of the lacustrine environment defines a first-order sequence that encompasses the whole basin fill. This can be subdivided, on the basis of major lacustrine flooding episodes, into three second-order sequences that represent the Lower, Middle, and Upper stratigraphic units. The Upper Unit can be subdivided further into third and fourth-order sequences that resulted from minor oscillations in water level. Finally, at the smallest scale, rhythmite deposits probably record seasonal palaeoclimatic variations.

The tectonic sedimentation balance is invoked as the main factor responsible for the generation of the first and second-order sequences. Third and fourth-order sequences may also record climatically forced processes.

INTRODUCTION

The establishment of three-dimensional boundaries for genetically related deposits leads to the definition of stratigraphic units that record the major changes produced in the sedimentary setting. As a consequence of their genetic character, each unit should reflect a major reorganization of the basin depositional framework. Genetic and sequential stratigraphic analyses have produced useful approaches to the definition of stratigraphic units (Payton, 1977; Bally, 1987; Wilgus *et al.*, 1988; Galloway, 1989a,b). Although these concepts have largely evolved from the study of marine successions it can be accepted that, in a general sense, the recognition of genetically related units is valid for any palaeotectonic, palaeoclimatic, and palaeogeographic scenario. From this point of view, the feasibility of applying the genetic approach to the analysis of non-marine basin deposits is evident, despite the fact that limitations may result from a scarcity of subsurface data and/or correlation problems (Reynolds *et al.*, 1989). An initial step in the application of sequence stratigraphy to lacustrine rifts has been shown by Scholz *et al.* (1990).

The Miocene Rubielos de Mora Basin provides an opportunity for carrying out a detailed facies analysis, as well as the application of some of the general aspects of the genetic stratigraphic approach to unravelling the depositional history of a non-marine basin. This small, fault-bounded basin displays a large variety of alluvial and lacustrine facies assemblages, which show a complex sequential arrangement related to an overall expansion of the lake, punctuated by several fluctuations in lake level.

GEOLOGICAL SETTING

The main structures of the northeastern Iberian Plate resulted from the Paleogene continental collision between Iberia and Europe (Fig. 1): the Pyrenean thrust-fold belt and its late southern foreland Ebro Basin, the Iberian Range, the Catalan Coastal Range strike-slip fault system, and the Linking Zone (Guimerà, 1984; Anadón *et al.*, 1985). Several Neo-

gene rift basins are superimposed upon these Paleogene structures. The structural relationships of some of the fault-bounded Neogene basins and the previously formed 'compressional' structures often record the development of a tectonic inversion.

The early−middle Miocene Rubielos de Mora Basin is located in the southernmost part of the Linking Zone (Fig. 1), which is made up of Hercynian basement unconformably overlain by a thick Mesozoic and Cenozoic cover. The Mesozoic cover was strongly affected by Neogene normal faults, with vertical displacements ranging from some hundred metres to 1500 m.

GEOLOGICAL FEATURES OF THE RUBIELOS DE MORA BASIN

The Rubielos de Mora sedimentary basin fill is early−middle Miocene in age (Crusafont *et al.*, 1966) and is unconformably overlain by Upper Miocene to Lower Pleistocene alluvial deposits.

From present exposures it can be inferred that the basin was at least 10 km long by 3 km wide (Fig. 2). The basin origin was closely related to the activity of major ENE−WSW to NE−SW striking faults (Figs 2, 3). The basin shows a structural asymmetry, with a gentle northern margin opposing a steeper southern one. The basin structure is explained as a result of early−middle Miocene reactivation of the basin-bounding fractures as normal faults. Two small-scale tablelands with similar altitude occur to the north and south of the eastern basin area, providing a reference level in relation to which a vertical slip of up to 1500 m can be deduced (Guimerà, 1990).

The basin-fill sequence is made up of over 800 m of terrigenous and minor carbonate deposits of lacustrine and alluvial origin. This sequence shows an overall synclinal structure, with its axis close to the southern margin. No large faults have been detected within the basin fill (Fig. 3).

Thick Lower Cretaceous formations, characteristic of the Linking Zone, make up most of the outcrops that surround the Rubielos de Mora Basin (Moissenet & Gautier, 1971). These Lower Cretaceous successions consist mainly of a thick red-bed sequence ('Wealdian facies' of Valanginian−Barremian age) overlain by a thick marine limestone sequence with minor interbedded dolostones, carbonate mudstones, and sandstones (Barremian−Aptian). This carbonate-dominated succession is

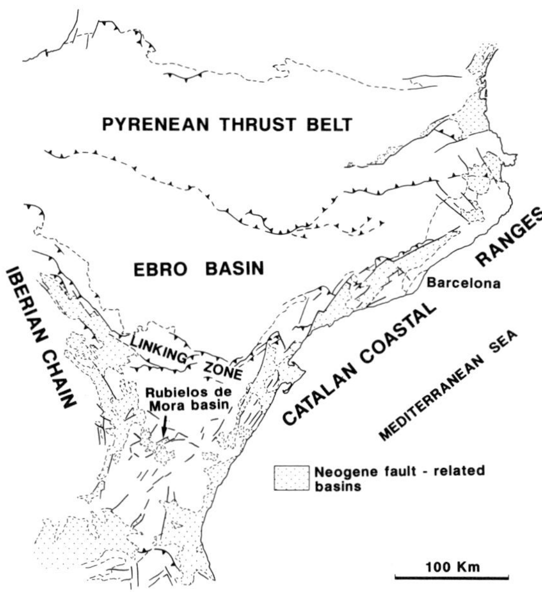

Fig. 1. Location map of the Rubielos de Mora Basin.

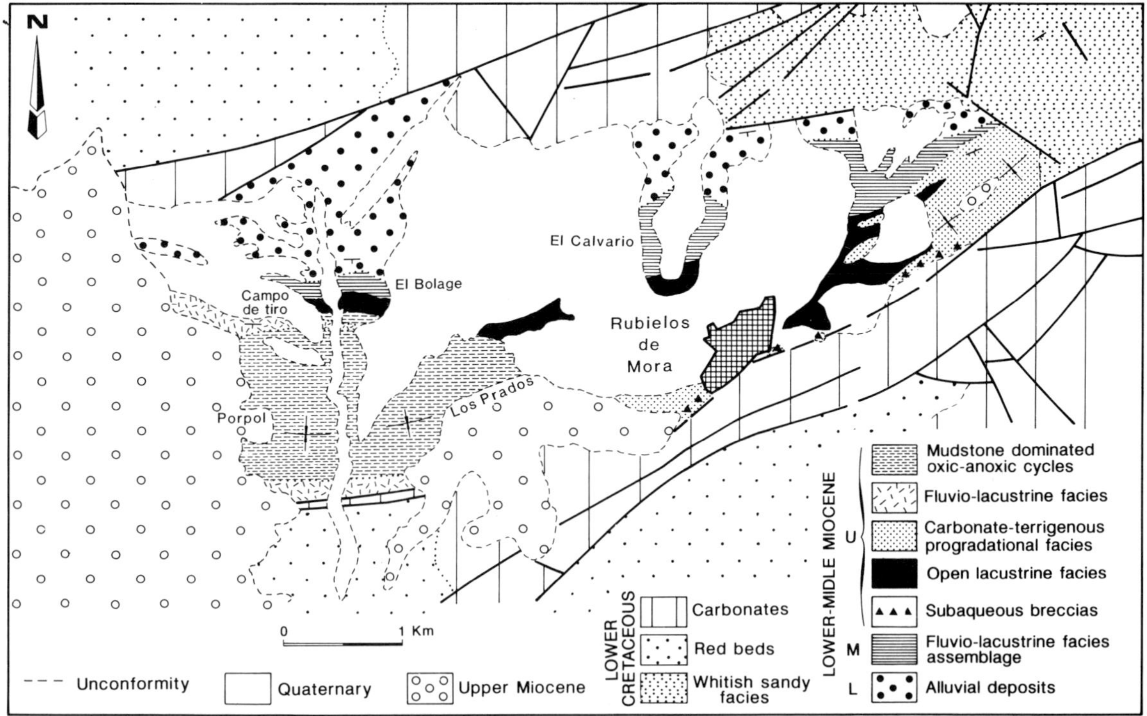

Fig. 2. Geological map of the Rubielos de Mora Basin. L, M, and U refer to Lower, Middle, and Upper Units, respectively (modified after Anadón *et al.*, 1988a).

unconformably overlain by an Albian whitish arkosic sandstone unit ('Utrillas facies'), which in turn is overlain by Cenomanian–Turonian carbonate units.

The diverse lithological composition of the Lower Cretaceous substratum and the above-mentioned structural asymmetry resulted in the development of two kinds of source areas during basin evolution:

1 Areally restricted carbonate source areas, located along the southern margin.

2 More widespread terrigenous source areas, located mainly on the Wealden red beds and whitish Utrillas sandy sequences.

The very distinct size of the clasts fed into the basin, as well as the diverse volume of terrigenous contributions delivered from the different source areas, gave rise to essentially quite distinct deposits.

STRATIGRAPHIC FRAMEWORK

The basin fill stratigraphy has been established on the basis of the study of laterally extensive and well-developed outcrops in the eastern and western parts of the basin (Fig. 2). Some well-log data from coal

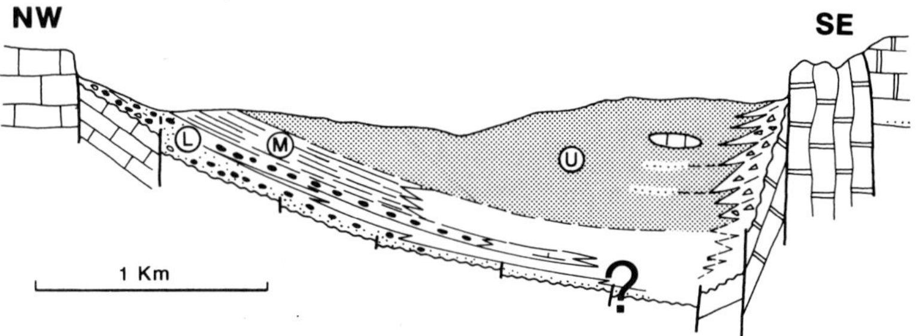

Fig. 3. Schematic cross-section of the Rubielos de Mora Basin showing the overall synclinal arrangement of the basin fill as well as its structural asymmetry. L, M, and U refer to the Lower, Middle, and Upper Units, respectively.

and oil shale exploration (IGME-CGS, 1980) are also available, and have been useful in the basinwide correlations of the various units. The overall basinward megasequential arrangement observed in the eastern and western outcrop areas clearly records an areal expansion of the inner lacustrine facies as well as a general deepening of the lacustrine system (Anadón, 1983; Anadón *et al.*, 1989). Taking into account these trends, it is possible to define three main genetic stratigraphic units (Figs 3, 4), which reflect major rearrangements of the basin depositional framework (Anadón *et al.*, 1988a,b).

Lower Unit

This unit is up to 300 m thick and consists of sandstones with minor interbedded mudstones. This unit unconformably overlies the Lower Cretaceous limestones and sandstones. The Lower Unit cannot be traced as far as the southern basin margin because of the lack of subsurface data. The bulk of this unit is formed of alluvial deposits. At the eastern end of the basin, the unit consists of proximal to middle alluvial fan sandy facies, supplied by Upper Cretaceous sandy source areas. Widespread diagenetic

iron remobilization and local pedogenetic features occur in these facies. In the western zones the Lower Unit consists of fluvial red mudstones and sandstones interbedded with some thin lacustrine limestones. Colour and clast composition of these deposits reflect Lower Cretaceous red bed source areas.

To the north of Rubielos de Mora this unit is represented by fluvial channel sands and floodplain deposits with interbedded thin, organic rich, lacustrine shales. These deposits overlie a few metres of limestone breccias with minor lenses of reworked lacustrine gastropods.

Middle Unit

This unit is up to 100 m thick. Outcrop and well-log data show that in the eastern basin sector it consists mainly of lacustrine deposits formed of bioclastic limestones with interbedded grey and yellowish mudstones, sandstones, and bituminous brown coals (Fernández-Navarro, 1914; Gavala, 1921).

In the western area and to the north of Rubielos the unit is mainly terrigenous. Reddish fluvial mudstones and sandstones are predominant, but thin beds, and packets up to a few metres thick of lacustrine bioclastic carbonates and mudstones are widespread. Lignites have not been recorded in this zone. The Middle Unit shows that shallow carbonate and coal lacustrine sedimentation spread into the eastern basin areas. This spread might have been accompanied by shrinkage of the fringing alluvial zones owing to a relative rise and stabilization of lake water level. It is not certain whether relatively deeper lacustrine facies exist close to the southern basin margin (Fig. 5).

Outcrop and well-log data clearly record a noticeable change between eastern basin areas and those to the west, where alluvial red bed facies are dominant and lacustrine deposits become scarce.

Upper Unit

This unit is up to 400 m thick and displays a large variety of facies. Alluvial, marginal lacustrine, and open lacustrine facies occur in close proximity (Figs 4, 5, 6).

The bulk of the alluvial contributions were fed into the basin from its east-northeast and west-southwest ends, from where they were later axially distributed. From the southern margin, terrigenous contributions were minor, coarser, and restricted to subaqueous debris flows, sometimes accompanied

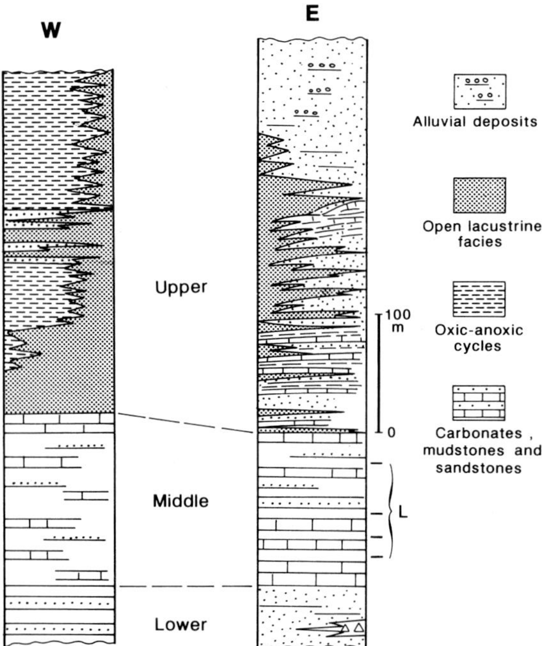

Fig. 4. Simplified representative stratigraphic logs of the Rubielos de Mora Basin showing the diverse sedimentary features in the eastern and western basin zones. L = lignite seams.

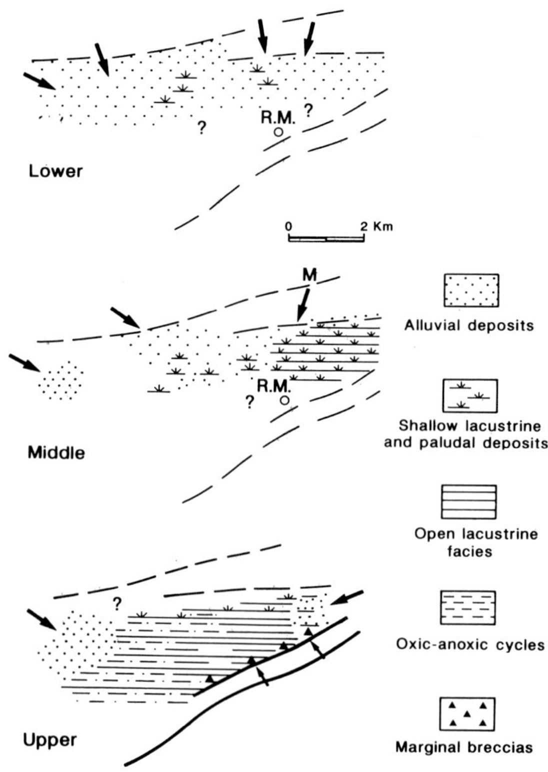

Fig. 5. Simplified palaeogeographic maps of the Rubielos de Mora Basin showing the main evolutionary stages of the basin.

by olistolith emplacements (Anadón, 1983).

In the eastern zone (Figs 4, 5, 6) the alluvial and the subaqueous gravitational deposits grade into other facies assemblages: (i) transitional alluvial–lacustrine sandstones and mudstones, (ii) marginal lacustrine carbonates, and (iii) open lacustrine, very thinly laminated, shales and carbonate–clay rhythmites. Small-scale slumps are common in the open lacustrine facies. In the western zone, cyclically arranged massive and laminated lacustrine facies occur (Fig. 6). Several fluvial and deltaic sandstone beds, which interfinger with the lacustrine sequences, record the influence of the southwest alluvial system.

The Upper Unit records the maximum extent of a lacustrine system with a 'deep' persistently stratified water column. During the deposition of this unit the lacustrine system attained its greatest environmental diversity, as indicated by the variety of facies assemblages and the complexity of their relationships (Figs 4, 5, 6). This fact, as well as the good and

laterally extensive exposures make the Upper Unit particularly suitable for detailed facies analysis and the comprehension of their sequential relationships.

FACIES ANALYSIS OF THE UPPER UNIT

The Upper Unit records the maximum extent of the open lacustrine facies. Several facies associations have been recorded (Fig. 6), ranging from alluvial to open lacustrine deposits.

Marginal subaqueous breccias

A distinctive coarse-grained facies belt occurs along the fault-related southeastern margin of the basin (Figs 2, 5, 6). This belt, up to 100 m wide, shows a tectonically deformed wedge-shaped cross-section and an overall depositional pattern involving a rapid basinward transition from marginal breccias to open lacustrine facies.

The coarse-grained proximal facies is represented by unstratified and highly disorganized limestone 'megabreccias'. This facies reflects a direct provenance from the highly fractured Cretaceous carbonate succession, exposed along the southern basin margin (Fig. 2). The breccias are polymodal, clast to matrix supported and characterized by the presence of large boulders, up to several metres in diameter. The matrix is a poorly sorted mixture of chalky mud, sand, and granules.

These unstratified 'megabreccias' generally merge basinward into a well-stratified alternation of breccias, sand-rich conglomerates, and pebbly sandstones. Thin layers (0.2–0.4 m) of calcareous mudstone and limestone also occur locally. Basinwards, two prominent 'megabreccia' units (4–10 m thick) are intercalated between fine-grained open lacustrine facies, 500 m away from the southern margin of the basin (Fig. 2).

The stratified breccias appear as sheet-like to broadly lenticular beds ranging in thickness from 0.2 to 3.5 m and showing slightly erosive to non-erosive bases. These breccias are characterized by a polymodal and disorganized clast to matrix support fabric with frequent 'outsized' cobbles and/or boulders (up to 5 m in maximum diameter). The matrix contains variable amounts of mud, sand, and granules.

The sand-rich conglomerates and pebbly sandstones display a preferred subhorizontal clast orien-

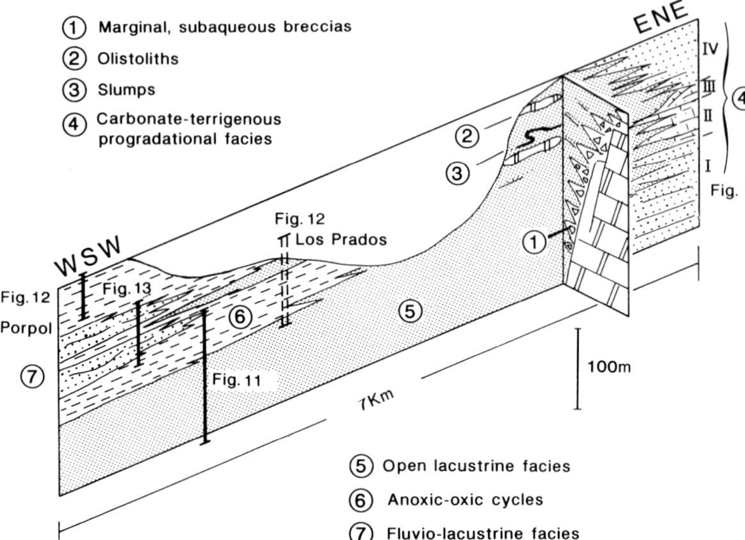

Fig. 6. Stratigraphic relationships of the main facies assemblages in the Upper Unit. Note the location of the sedimentological logs from Figs 11, 12, 13, and sketch of Fig. 7.

tation and vague, horizontal to inclined, internal stratification. The clasts are poorly rounded, ranging in size from pebbles to cobbles. The mud content in the matrix is apparently negligible. The beds, 0.35–1.5 m thick, are sheet-like to lenticular and show weakly erosive bases.

The proximal, unstratified 'megabreccias' are interpreted as olistoliths and debris flow deposits that originated by gravity processes along a tectonically active, steep, and unstable lacustrine margin. Similar coarse-grained deposits have been described along the steep, faulted margins of Lake Tanganyka (Mondeguer *et al.*, 1986; Cohen, 1989) and along the San Gabriel fault margin of the Pliocene Ridge Basin, California (Link & Osborne, 1978). According to Lowe (1982) and Nemec & Steel (1984), stratified breccias of this sort can be attributed to cohesionless debris flow deposits, and the sand-rich conglomerates and pebbly sandstones to heavily sediment-laden fluidal flows. The subaqueous emplacement of all the above described deposits is proved by their lateral and vertical association with lacustrine mudstones, limestones, and rhythmites.

Carbonate–terrigenous progradational facies

A mixed carbonate–terrigenous marginal succession occurs in the eastern part of the basin, overlying the limestone and lignite beds of the middle unit (Figs 2, 4). The bulk of the marginal facies inter-

fingers and merges to the southwest into laminated mudstones, whereas the lateral equivalents of the sand-dominated uppermost part of the marginal succession are not preserved (Fig. 6).

Four principal subunits are recorded in this marginal facies belt on the basis of facies development and geometric relationships (Figs 6, 7).

Subunit I. The lowermost subunit (120 m thick) is mostly represented by stacked beds of structureless, moderately to well-sorted, very fine to medium grained, micaceous quartz arenites. Silty to sandy mudstone interbeds, as well as amalgamated lenticular layers of cross-bedded quartzarenites and small pebble conglomerates are less frequent. Some beds of calcisiltites, calcarenites, and laminated mudstones are locally observed in the upper part of the subunit.

This sand-dominated succession is poorly exposed and passes basinward into an alternation of mudstones, sandstones, minor thin limestones, and rare lignite beds. The uppermost two mudstone–sandstone couplets (Fig. 8a) delineate poorly defined coarsening upward sequences (8–12 m thick). A lacustrine fauna (Sphaeriidae, ostracods) is associated with the laminated mudstones.

The assemblage is interpreted as the lateral transition from mainly subaerial fluvio-deltaic facies to subaqueous, lacustrine deltaic facies. The terrigenous progradations were punctuated by some lacus-

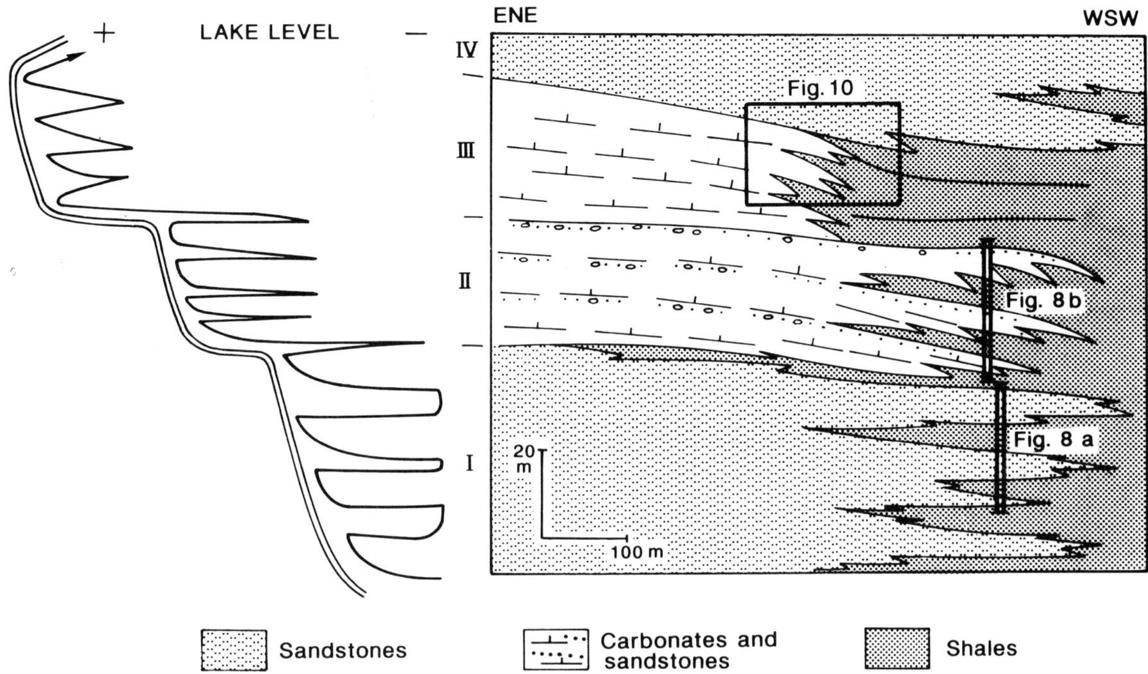

Fig. 7. Sketch of the relationships between the terrigenous and the mixed carbonate–terrigenous progradational facies and their transition into the inner lacustrine facies in the easternmost basin zone. Note the interpretation of the overall sequential arrangement of the facies assemblages depending on the relative lake-level variations. The bars show the location of the sedimentological logs and sketches in Figs 8, 10.

trine expansions, which allowed the deposition of laminated mudstones in relatively marginal basin areas.

Subunit II. Near the northeastern margin of the basin, the second subunit consists of a succession 20–25 m thick of thinly bedded calcarenites and calcisiltites arranged in several poorly defined, up to 10 m thick, thickening upwards sequences. These sequences include intercalations of calcareous mudstones towards the base and are capped by base-scoured thin pavements of quartzarenite and small pebble conglomerates (Fig. 7). The calcarenites and calcisiltites contain abundant ostracods, charophytes, gastropods, and bivalve debris. In some cases they are parallel-laminated but typically display a low-relief wavy stratification with cosets of hummocky and swaley cross-bedding. Soft sediment deformation structures, such as overturned laminae and convolute lamination are common.

This carbonate-dominated succession grades basinward into four stacked carbonate–terrigenous coarsening upward sequences (Fig. 8b). From base to top, these sequences (up to 6 m thick) comprise three terms: (i) laminated calcareous mudstones, (ii) parallel or ripple cross-laminated calcarenites and calcisiltites, and (iii) cross-bedded quartzarenites recording a northeast provenance and bioclastic sandstones.

The proximal carbonate-dominated sequences in subunit II are thought to have formed on the platform of a wave-worked carbonate bench, which evolved under relatively high energy conditions. The more distal stacked carbonate–terrigenous coarsening upward sequences are interpreted as bench platform break deposits (Fig. 9).

The depositional setting of subunit II was a high-gradient carbonate bench rimming a shore-attached terrigenous belt. The occurrence of thicker terrigenous deposits at the outer platform edge than in the inner platform zones is a feature which suggests that the quartzarenites and well-rounded conglomerates were deposited by bypass mechanisms from the northeast source areas. Both the petrographical and the palaeocurrent criteria enable the southeast provenance to be rejected as a source for these terrigen-

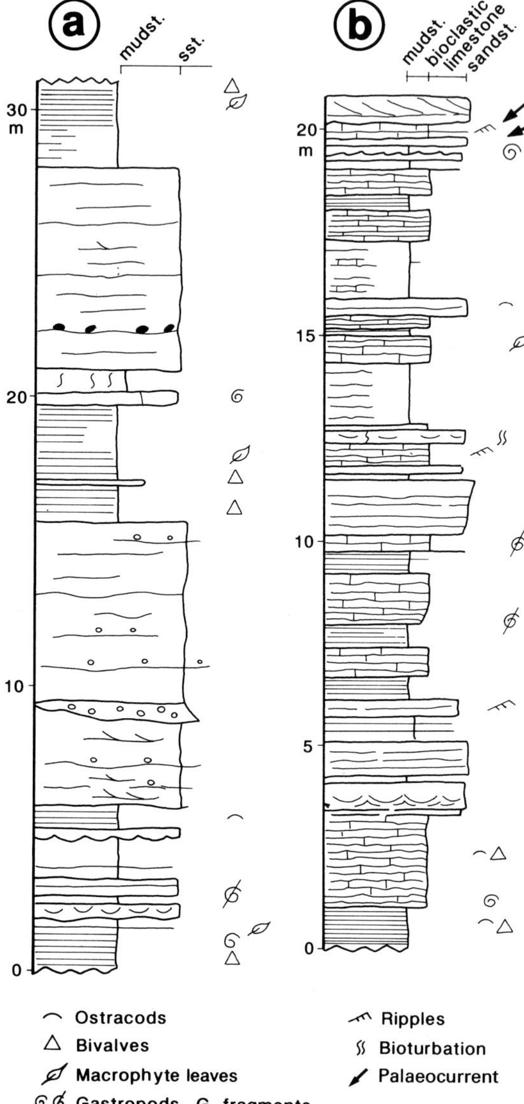

Fig. 8. Sedimentological logs of the terrigenous (a) and mixed carbonate−terrigenous (b) progradational facies in the eastern basin zone. See location of logs on Fig. 7.

ous deposits. From this point of view it has been inferred that the progradational carbonate−terrigenous bench developed under changing lacustrine water level conditions. During high-stands, most of the terrigenous input was presumably trapped in proximal areas, favouring high carbonate accumulation and progradation of the carbonate-platform−talus over the laminated mudstones, which were deposited below wave base. On the other hand, during low-stand conditions, the terrigenous inputs mostly bypassed the carbonate bench platform, probably through a series of (not observed, hypothetical) proximal channels whose distal equivalents would be the base-scoured thin pebbly pavements. The bypassed sediment accumulated as mainly terrigenous platform-break aprons (Figs 7, 9). It must be stressed that no evidence of subaerial exposure (pedogenesis, karstification) has been detected in the outcrops studied. Therefore the suggested water level oscillations did not give rise to subaerial erosive surfaces.

Subunit III. This subunit consists of a 25-m-thick alternation of sandstones, biocalcarenites, biocalcisiltites, and rare mudstone layers. This alternation merges distally into laminated mudstones, which contain lacustrine faunas, and intercalate with rhythmites and limestone beds (Fig. 7). Basinwards dipping sigmoidal clinoforms (Fig. 10), as well as synsedimentary extensional listric faults and compressional thrusts are recorded locally. The assemblage is interpreted as a terrigenous−carbonate talus that prograded in successive pulses into open lacustrine facies, represented here by shales and carbonate-dominated rhythmite beds.

The tentatively restored geometry of these deposits (Fig. 10b) records the development of successive small-scale progradational pulses into open lacustrine zones.

The outcrop quality and lack of accessibility do not allow a detailed reconstruction of the geometry of these prograding deposits. However, their development was caused by an overall subsiding situation combined with changes in the depositional equilibrium level (cf. Moore & Curray, 1964; Dietz, 1963, 1964; Mougenot *et al.*, 1983). It is assumed here that these changes were influenced by tectonics, sediment contribution, and water level oscillations.

Subunit IV. The poorly exposed uppermost subunit is mainly formed by stacked beds of cross-bedded fine to very coarse-grained sandstones, interpreted as fluvio-deltaic deposits. No lateral equivalents of this subunit have been preserved in the inner basin zones.

In conclusion, the successive subunits of the progradational eastern marginal succession (Fig. 7) record: (I) several low-stand terrigenous expansions punctuated by minor risings; (II) the development

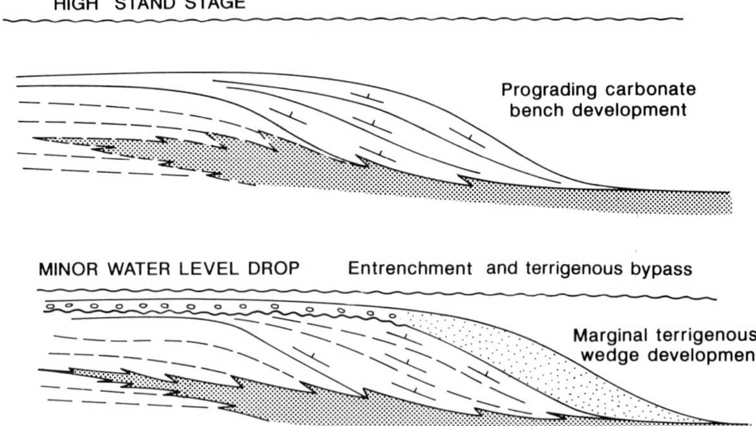

Fig. 9. Interpretative sketches showing the successive development of the prograding carbonate benches and the platform edge terrigenous aprons resulting from minor lacustrine level changes. This development is deduced from the sequences shown in Fig. 8b.

of a high-gradient carbonate bench under a rising to high-stand situation, but with noticeable fluctuations of the lacustrine water level; (III) renewed rising and high-stand water level, perhaps leading to the 'drowning' and/or to the landward retrogradational displacement of the former high-gradient carbonate bench and their substitution by a prograding mixed carbonate–terrigenous bench margin — the development of this margin would also have been punctuated by minor risings and drops of the water level; and (IV) the beginning of a general lowering trend with the generation of low-stand terrigenous expansion.

Thus, the accumulation of the successive basinward progradational margin deposits could take place under an overall rising water level trend punctuated by transitional high-stand and minor lowering stages. This happened prior to the definitive lowering recorded by the transition from subunit III to IV.

It must be emphasized that no major entrenchment surfaces (i.e. type 1 unconformities in the sense of Haq *et al.*, 1987) have been recorded that affected the above-described carbonate-bench–talus deposits. This fact leads to the conclusion that the rises and falls in lacustrine water level remained restricted to a relatively narrow range. Moreover the vertical passage from unit III to IV is quite transitional too, suggesting that, at least during its earlier stages, the fall in water level was gentle and could have resulted in the development of a marginal

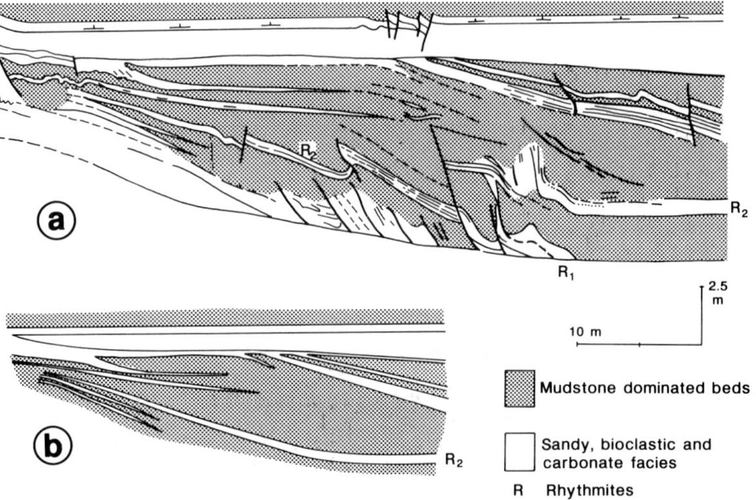

Fig. 10. (a) Basinward progradational clinoforms observed in subunit III (see location on Fig. 7), showing the development of a prograding terrigenous carbonate-bench–talus grading into inner lacustrine facies. Note the development of synsedimentary extensional listric faults. The lower part of the section has been affected by small-scale thrusting of uncertain origin. (b) Tentative restored section.

prográdational wedge (i.e. type 2 unconformities of Haq *et al.*, 1987).

Open lacustrine facies

The open lacustrine facies are formed by sequences up to 250 m thick, consisting of laminated grey mudstones (oil shales) and interbedded rhythmite beds up to 2.5 m thick (Fig. 11). Sparse sandstone and pebble conglomerate beds up to 0.1 m thick are also present.

The laminated grey mudstones (oil shales) display lack of bioturbation and thin lamination. In some cases the latter is not so evident and is only marked by the arrangement of fossil leaves, displaying a poor fissibility. Macrophyte leaves, insects, and amphibian skeletons have been recorded in the grey mudstones. Sparse lacustrine molluscs and ostracods also occur.

The rhythmites consist of alternating couplets of carbonate (aragonite, calcite) and clay laminae. The relative amount and thickness of both clay and carbonate laminae in the rhythmite beds is highly variable. They range from very thin, sparse carbonate laminae interbedded in laminated mudstones to laminated carbonates with interbedded very thin clay partings. Nevertheless, in the intervals where the laminae couplets are equally developed, a mean lamina thickness of 0.22 mm has been recorded. This figure agrees well with the thickness ranges recorded by several authors in modern and ancient varved lacustrine deposits (see Picard & High (1981) for a summary; Boyer, 1981).

In the oil shales the oil yield ranges from 5 to 55 l/t, with a mean value of 20 l/t (IGME-CGS, 1980). Recent studies of the organic-rich rocks of the Rubielos Basin (Anadón *et al.*, 1988c; unpublished data) show that the oil shales from the open lacustrine facies have consistently highly reactive kerogen contents (Hydrogen Index (HI) values from 244 to 564 mg HC/g TOC (total organic carbon)) and TOC contents from 4 to 14.8%. Pyrolysis gas chromatography and visual kerogen analysis show the organic matter to be of mixed higher plant and algal origin, with the former dominant. These data agree with the results of preliminary studies of the extractable organic matter (X. de las Heras, in Anadón *et al.*, 1988b). The nature of the organic matter also has been established according to visual analysis in oil shale samples from several shallow boreholes (Prado *et al.*, 1988). A lacustrine origin is indicated by significant quantities of two types of

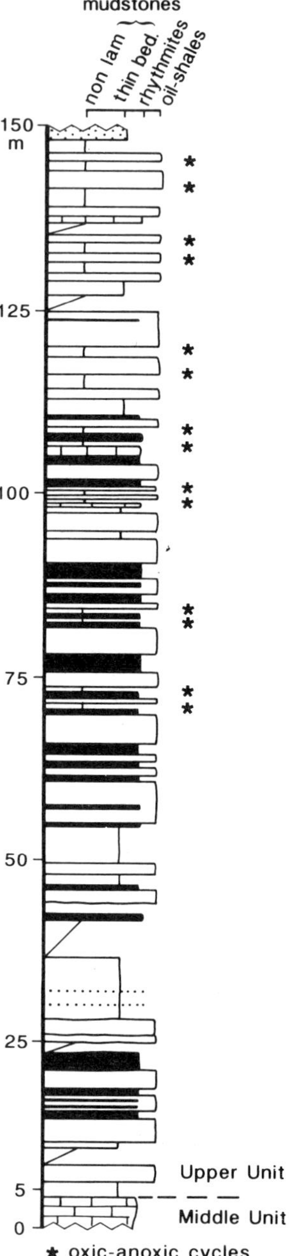

Fig. 11. Transition from the Middle to the Upper Unit in the Bolaje section (western basin zone). The open lacustrine facies assemblage (mainly shales and rhythmites) sharply overlies the upper shallow lacustrine carbonates and is covered in turn by a cyclically arranged oxic—anoxic succession. See location on Fig. 6.

Botryococcus sp. and other types of algae. In some intervals, *Botryococcus* occurs concentrated in thin laminae of the laminated mudstones and rhythmites.

The thick successions described above have been interpreted previously as open lacustrine deposits' that formed beneath the oxycline of a meromictic lake (Anadón *et al.*, 1988b). The frequent varve-like pattern of the carbonate−clay couplets was caused by the alternation of carbonate precipitation and very fine terrigenous contributions. This rhythmic facies emphasizes the development of short-term cyclical oscillations of aqueous ionic equilibria that controlled the carbonate precipitation. These oscillations were probably caused by successive algal blooms. This interpretation is supported by the presence of *Botryococcus* accumulations at the base of the carbonate laminae (Prado *et al.*, 1988). Laminated mudstones bearing *in situ* lacustrine bivalves (Sphaeriidae), are common in the transition from the thick laminated successions (open lacustrine deposits) to the above-described marginal successions (Fig. 8). These laminated mudstones are interpreted as being formed under weakly oxic bottom-waters.

Anoxic−oxic cyclical lacustrine ramp deposits

These lacustrine cycles occur as minor features in the northern basin margin although they are very well developed in the western basin zones (Fig. 2), where a succession more than 200 m thick of lacustrine deposits of the Upper Unit includes cyclically arranged sequences (Figs 4, 11, 12). A detailed description of these cyclical sequences has been provided by Anadón *et al.* (1988b). The cycles consist of an alternation of laminated and non-laminated facies (Fig. 12a). An erosive, but not deeply incised surface, separate the two kinds of facies. The laminated facies (b) comprise sandy mudstones (b_1), bioclastic laminae (b_2), oil shales (b_3), rhythmites (b_4, carbonate−clay varve-like couplets), and crudely laminated white marls (b_5). The non-laminated facies comprise massive white marls (a_1) and green, bioturbated mudstones (a_2). Magnesium calcite, aragonite, and dolomite have been recorded in the laminated facies, whereas low-Mg calcite and dolomite are the carbonate minerals in the non-laminated facies (Fig. 12a). The oil shales (b_3) reach a mean oil yield of 30 l/t, with a maximum of 70 l/t. whereas the non-laminated facies attain 5 l/t. The oil shales have high reactive kerogen contents (up to 850 mg HC/g TOC) and are of similar composition to the oil shales from the open lacustrine

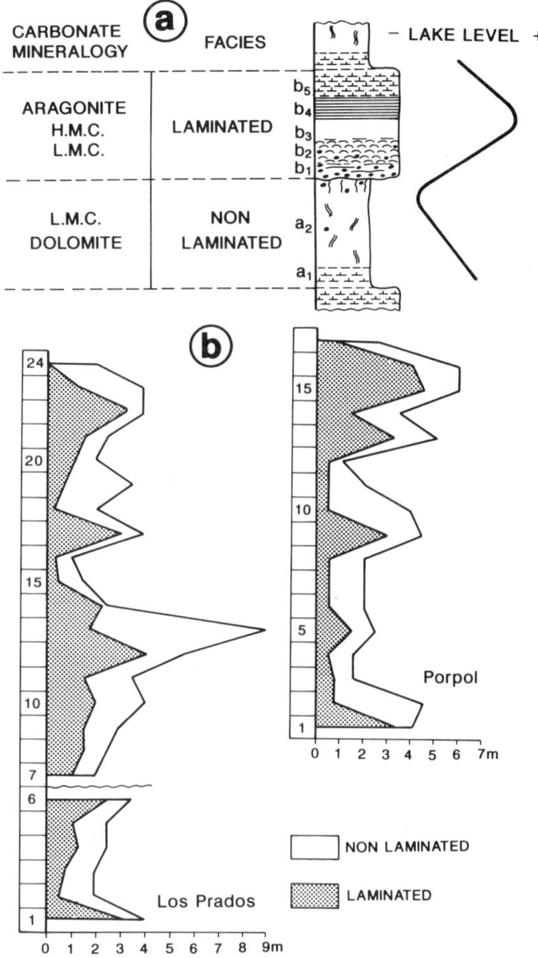

Fig. 12. (a) Ideal cyclical oxic−anoxic sequence resulting from alternating oxic−anoxic bottom conditions linked to variations in lake water levels. Note the mineralogical changes of the primary carbonate minerals from the b3 to b4 facies, recording possible changes in lake solute concentration. (b) Trends of thickness changes in the oxic−anoxic cyclically arranged successions of the Porpol and Prados sections (see Figs 2, 6 for location).

facies (Anadón *et al.*, 1988c; unpublished data).

Cycles range from 0.5 to 9 m in thickness, with maximum frequencies centred around 1.75 m and 4 m (Fig. 12b). In some cases individual cycles can be traced for up to 1 km. Anadón *et al.* (1988b) interpreted the cyclical arrangement to be a result of cyclical changes in lake water volume causing alternating oxic−anoxic bottom conditions in marginal zones of a meromictic lake. The relation of these repeated oxic–anoxic cycles to minor water level changes is shown well by the carbonate mineralogical evolution recorded in the transition from the oil shale facies (b_3) to the rhythmites (b_4) and to the crudely laminated marl facies (b_5). The primary carbonate mineralogy (aragonite and/or high magnesium calcite) of some rhythmite beds is interpreted to be a result of evaporative concentration of solutes in an alkaline lake (Müller *et al.*, 1972). Increasing water concentration suggests a correlative gradual decrease in lake volume and, in turn, a lowering of the water level. Conversely, the major carbonate mineralogy of the b_1–b_3 laminated facies can be linked to conditions of more dilute, higher levels of lake water.

These water volume changes caused transgressive and regressive pulses, which are recorded by the cyclically arranged sequences. The sandy and bioclastic laminated facies (b_1, b_2) represent the transgressive deposits, laid down on a transgressive surface. This surface could be coincident, in (not observed) more marginal zones, with erosive surfaces developed during the low-level stages. Transgressive deposits were covered by the high-stand level deposits, consisting of laminated organic-rich deposits (b_3). The later reduction of water volume and the subsequent initial lowering of the water level would be recorded by the deposition of aragonite-bearing carbonate–clay rhythmites and marls. This fact suggests that the required conditions for aragonite precipitation were attained during the first stages of lake level fall. Further lowering of the water led to the establishment of oxic conditions, which resulted in the deposition of massive dolomitic marls. Later, continued base-level fall enhanced the progradation of the marginal terrigenous facies on the inner lake zones, precluding the formation of dolomitic marls and leading to the deposition of the overlying massive pale green mudstones.

Fluvial–lacustrine facies in the western area

A sequence of fluvial and open lacustrine deposits up to 45 m thick (Figs 6, 13) is interbedded in a thick succession of anoxic–oxic cycles in the western area. These fluvial and fluvio-lacustrine deposits consist of red non-laminated mudstones, yellow sandstones, and grey mudstones. They pass laterally eastwards into the anoxic–oxic cycles and the laminated mudstones of the open lacustrine facies (Fig. 6).

The non-laminated red mudstones, up to 1.2 m

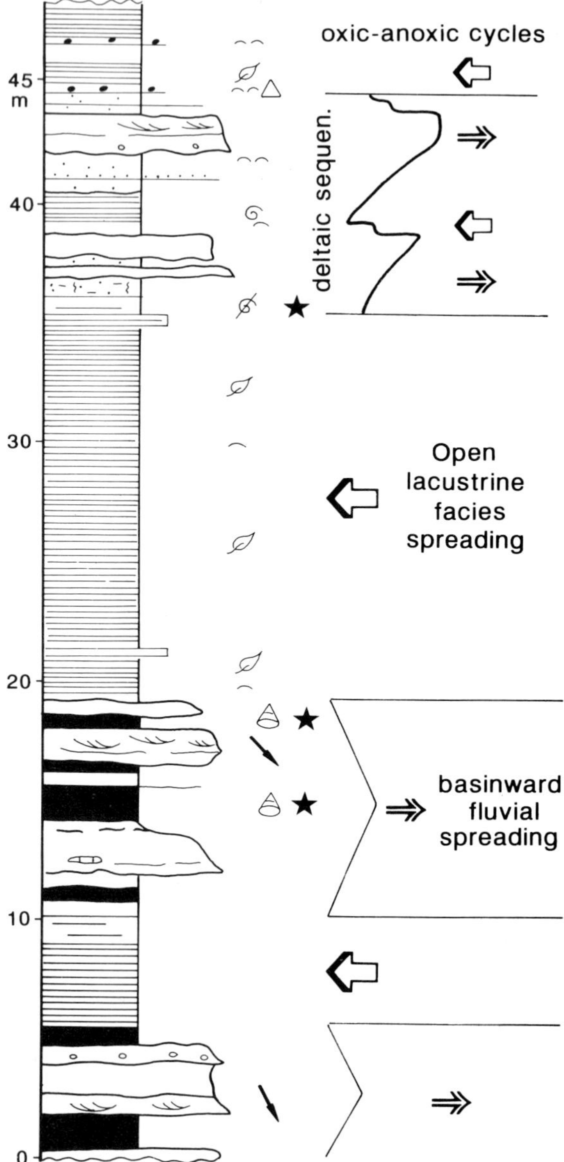

Fig. 13. Fluvial and fluvio-lacustrine deltaic facies alternating with open lacustrine facies in the Campo de Tiro section (see Fig. 6 for location). Note that several distinctive oil shale packets separate the successive episodes of sandstone deposition.

thick, contain terrestrial fossil gastropods (Helicidae) and vertebrate bone fragments. These red mudstones can be interpreted as fluvial overbank deposits. The red colour in the mudstones is inherited from the Cretaceous red mudstones. The fluvial channel deposits are formed by fine to coarse-grained sandstones, up to 3 m thick, which show trough cross-bedding and ripple cross-lamination. Quartz granules and pebbles occur through the sandstone beds and in some places make up conglomerate lenses.

Interbedded lacustrine facies are formed of grey, brown, and black laminated mudstones up to 16 m thick and green non-laminated mudstones up to 0.8 m thick. The latter display widespread bioturbation and, locally, abundant ostracods and vertebrate bone fragments. The laminated mudstones contain macrophyte leaves, ostracods, and sparse lacustrine gastropods (Planorbidae). The laminated mudstones include carbonate—clay rhythmite beds up to 0.3 m thick.

The fluvial—lacustrine facies assemblage records basinward spreading of alluvial tongues from the system that rimmed the western lacustrine basin zones. The rapid vertical transitions between the open lacustrine facies and the coarse-grained fluvial deposits suggest rapid changes in lake water levels, which would account for the observed sharp sequential arrangement. It must be emphasized again, however, that the lowering of water level that resulted in the observed sequence (Fig. 13) was not so pronounced as to give rise to fluvial channel incisions into the deep-water deposits.

SEQUENTIAL DEPOSITIONAL FRAMEWORK AND ARRANGEMENT OF THE BASIN FILL

Two major features characterize the (Rubielos de Mora) basin fill:

1 the asymmetrical depositional framework;
2 the hierarchical sequential arrangement.

The depositional framework asymmetry

The facies changes observed between the diverse basin zones show that the lacustrine assemblages of the Middle and Upper Units resulted from an asymmetrical depositional system (Figs 4, 6). Depositional asymmetry is a feature often observed in tilt-block/half-grabens as a consequence of the asymmetrical subsidence vectors developed across the basin (Leeder & Gawthorpe, 1987; Leeder *et al.*, 1988).

A clear distinction may be established in the Rubielos de Mora Basin between (i) an eastern zone, where a lacustrine carbonate and lignite depocentre developed, and (ii) a western zone, with a clear dominance of subaerial alluvial red beds interbedded with minor lacustrine carbonate deposits and lacking lignites (Figs 4, 5). The nearly exclusive occurrence of well-developed lacustrine assemblages in the eastern sector could be considered the prior announcement of the overall later evolution of the basin: the preferential location of deeper, and more persistent lacustrine environments in the eastern sector of the basin.

The Upper Unit facies assemblage distribution also shows striking differences, in both transverse and longitudinal basin sections (Figs 5, 6, 14).

In short, a major environmental and depositional asymmetry existed between the eastern basin zones (persistently deep and anoxic, with high-gradient margins, Fig. 15) and the western zones (from shallow to moderately deep, displaying a rather low gradient and changing bottom conditions). A similar distinction can be proposed for the southern margin and the northern margin (with steep and gentle bottom gradients, respectively) on the basis of a clearly distinctive facies development (Fig. 14).

The hierarchical sequential arrangement

The major trends of lake level rise and of the lacustrine environment are clearly recorded by the vertical superposition in the inner basin zones of alluvial dominated deposits (Lower Unit), followed by shallow lacustrine facies (Middle Unit) and deeper lacustrine deposits (Upper Unit). This vertical trend defines a 'first order' sequential arrangement of the whole basin fill, which is up to 800 m thick (Figs 3, 4).

Apart from this major deepening trend other minor sequences can be established from the recognition of lake level oscillations recorded in the basin fill.

Thus, second-order sequences have been defined on the basis of major basinwide lacustrine flooding episodes. These episodes caused dramatic palaeogeographic rearrangements in the depositional pattern. The 'second order' sequences are represented by each one of the above-described informal stratigraphical units, which range from 100 to 400 m in thickness.

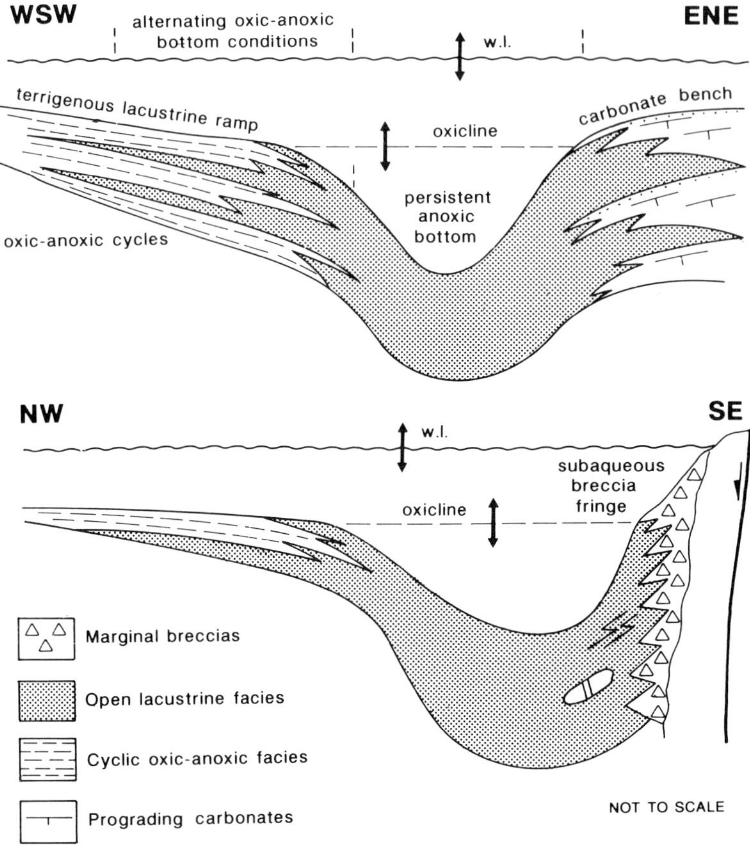

Fig. 14. Schematic cross-section models showing the relationships between the marginal and inner zones of the Rubielos lacustrine system and emphasizing its asymmetrical arrangement. Top: longitudinal cross-section displaying the relationships of the eastern progradational mixed carbonate benches and the western terrigenous lacustrine ramps to the inner lacustrine zones. Bottom: transverse cross-section showing the relationships of the high-gradient subaqueous breccia fringe and the low-gradient sublacustrine terrigenous ramp to the inner lacustrine zones. Note the potential sensitivity of the low-gradient sublacustrine ramp zones in relation to even minor changes of the oxycline.

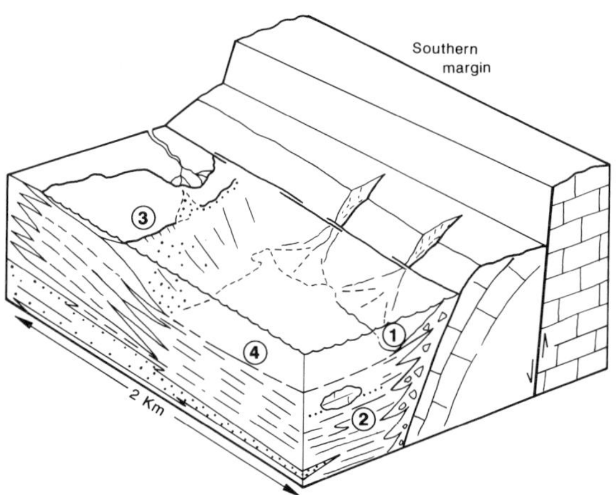

Fig. 15. Schematic three-dimensional representation of the depositional framework of the Upper Unit in the eastern basin zones. (1) Subaqueous gravitational breccia rim, (2) olistoliths and slumps, (3) progradational mixed carbonate—terrigenous benches, (4) inner lacustrine zones.

Third-order sequences are bounded by lacustrine flooding episodes recorded by the expansion of oil shale accumulation (Fig. 16). These sequences are up to some tens of metres thick and have been recognized only in the Upper Unit, thanks to its more continuous and laterally extensive outcrops. The third-order sequences record palaeogeographic rearrangements that caused marked environmental variations along the basin margins.

In the easternmost basin area, third-order sequences are represented by successive development of clastic deltaic wedges and mixed carbonate—terrigenous progradational benches (Figs 7, 8b, 16a).

In the western basin areas the third-order sequences are recorded in diverse ways, depending on the lacustrine zone where the sedimentary record developed. In the Bolajes section the third-order sequences are up to 30 m thick and consist of the alternation of relatively thicker laminated packets with oxic—anoxic cyclical episodes (see Figs 11, 16b;

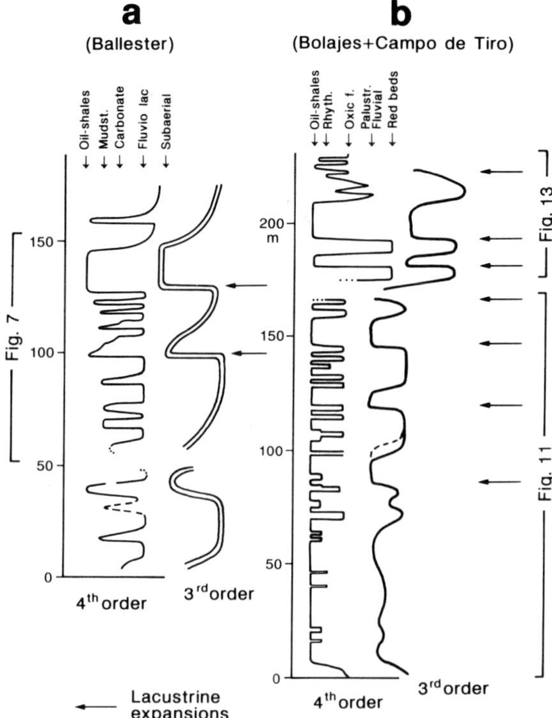

Fig. 16. Relationships between facies and the third and fourth-order sequences as recorded in the eastern (a) and western (b) basin zones. See Figs 7, 11, 13 for reference sedimentological logs.

from 60 to 160 m). In the oxic–anoxic dominated successions (Fig. 16b), a general upward increase in cycle thickness characterizes each third-order sequences. They are up to 25 m thick and include from seven to nine cycles (Fig. 12b).

Finally, in the fluvial–lacustrine facies (Figs 13, 16b) the third-order sequences (ranging from 15 to 25 m in thickness) consist of alternations of relatively thick shale packets and fluvial and fluvio-lacustrine deposits.

The fourth-order sequences, up to 10 m thick, are also bounded by transgressional episodes. In the western zones they are especially well recorded by the oxic–anoxic cycles related to oxicline oscillations, which are in turn linked to minor water level changes (Anadón *et al.*, 1988b). In the eastern progradational basin margin the only well-exposed sections that may be equivalent to the oxic–anoxic cycles, are mudstone–carbonate sequences (up to 15 m thick) recording water level changes (Figs 7, 8b). These sequences resulted from the depo-

sition of open lacustrine facies (rising and high-stand stages) followed by bench progradation (steady high-stand) and by a later gentle entrenchment of the terrigenous deposits (fall in water level). Because of the steeper gradient of the eastern progradational margins, the influence of minor water level changes in the sedimentary record was not so striking as in the western margins.

Finally, apart from the above-described hierarchical sequential arrangement, a smaller scale cyclicity is recorded by the superposition of the carbonate–clay couplets (rhythmites). As stated previously this rhythmic pattern was probably the result of seasonal changes.

Considering all the above-mentioned facts, lake level changes dominated the basin-fill evolution, and these changes have been used here as a major criterion in order to subdivide its fill (Fig. 17).

The establishment of first and second-order sequences and their basinwide correlation can be realized easily in the Rubielos de Mora Basin. In contrast, the possibility of correlating third or fourth-order sequences over the entire basin is not so straightforward. The limited subsurface data and the outcrop gaps between the eastern and western basin zones make this objective difficult (Figs 2, 4). Nevertheless, the potential usefulness of the analysis of the sequential hierarchy in non-marine basin fills that include well-developed lacustrine deposits must be stressed.

DISCUSSION: TECTONIC AND PALAEOCLIMATIC CONTROL ON THE LACUSTRINE BASIN EVOLUTION

Changes in base level, subsidence rate, and sediment supply are claimed as the main factors that influence the evolution of any basin. Tectonics and climate provide the fundamental control (Reynolds *et al.*, 1989). The evolution of the Rubielos de Mora Basin reflects the long-term deepening and expansion of a lacustrine system that was punctuated by repeated oscillations in lake level. Moreover the basin displayed a persistent asymmetry, which exerted a decisive influence upon the final sedimentary facies distribution. The causes of changing lake levels, taking into account its palaeotectonic and palaeogeographic setting, could be:

1 Changes in basin geometry caused by variations in the relative subsidence and sedimentation rates. The resulting variations in basin volume would change

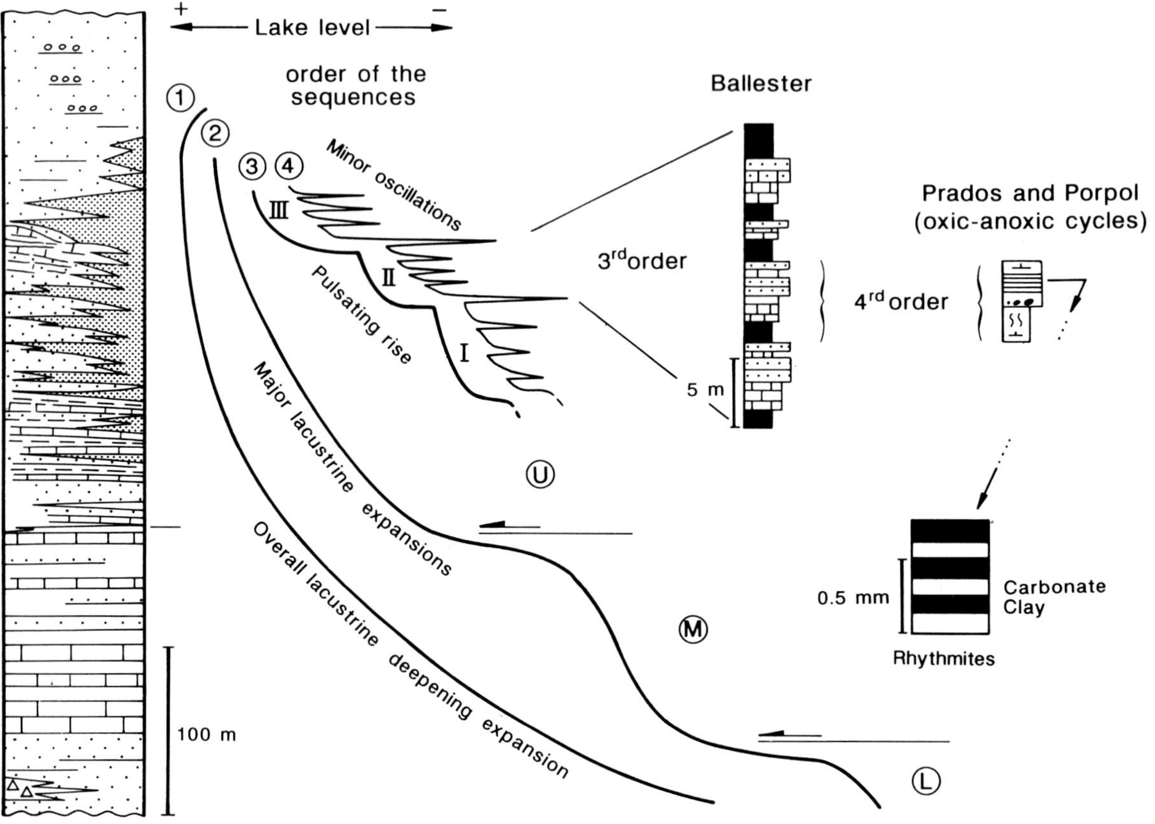

Fig. 17. Schematic diagram showing the diverse order of the sequences and rhythms recognized in the Rubielos de Mora Basin. L, M, and U refer to Lower, Middle, and Upper Units, respectively. The logs for fourth-order sequences are a simplification of the respective sequences in Figs 8b, 12a. Therefore this part of the figure only portrays the fourth-order cyclicity reported in two different zones of the basin.

the water and sediment capacities.

2 Variations in water balance caused by climatic change (i.e. evaporation/precipitation balance) and/ or by watershed modification (i.e. by river diversion owing to evolution of the surrounding drainage network through blockage, avulsion, or stream capture).

Although it is difficult to assess the exact importance of each of the above-mentioned factors, some general ideas can be suggested.

Tectonic influence

The early alluvial and lacustrine deposits in the Lower and Middle Units record the onset of differential tectonic subsidence along the basin-bounding faults. Nevertheless, evidence for significant tectonic

activity is most apparent in the Upper Unit, where slumps and gravitational megabreccias are widespread in some facies assemblages. The synsedimentary tectonic activity recorded by these features may have resulted in a higher subsidence rate that outstripped sedimentation rates. This feature, with or without drainage network evolution, resulted in the expansion of the lacustrine system that is recorded by the first and second-order sequences.

As a result of the final tectonic sedimentation adjustment, a nearly continuous anoxic lacustrine facies deposition (linked to the high persistence of a deep lacustrine depositional setting) took place in the eastern zones, where rather narrow, high-gradient marginal zones developed.

In contrast, anoxic and oxic facies alternated in the western (and northern) zones, characterized by wider, low-gradient terrigenous-dominated sub-lacustrine ramps.

Changes in water balance

Apart from the above-mentioned major deepening trend of the lake system, small-scale fluctuations of the lake level also exerted a strong influence on the system. These minor oscillations resulted in cyclical transgressive–regressive third and fourth-order sequences.

It is not possible to eliminate a tectonic-subsidence–sedimentation-rate control on the third and fourth-order sequences observed in the basin. This uncertainty is especially striking for the third-order sequences. However, the palaeoclimatic setting and sedimentological features of the cyclical fourth-order sequences emphasize the possible influence of water balance changes on basin-fill evolution.

A warm, subtropical–tropical climatic regime was well established in the northwest Mediterranean region by the early to middle Miocene (Bessedik, 1985). The palynological and macrophyte fossil remains recorded in the Upper Unit agree with this fact (Fernández-Marrón & Alvarez-Ramis, 1988). Moreover the present Salamandridae related to those recorded in the Rubielos de Mora sequences (*Chelotriton paradoxus*, Sanchís, pers. comm.) are restricted to southeast Asia, where warm and humid climatic conditions occur. Such a climate would have favoured thermal stratification of the water column in the Rubielos lake. In addition, the palaeolatitudinal position of the basin (close to its present location) probably made it sensitive to possible cyclical palaeoclimatic changes (i.e. from rainy to drier conditions).

The limited total thickness of the fourth-order cyclical sequence of the western Rubielos de Mora Basin and the absence of similar synchronic sequences in neighbouring basins, preclude definitive statistical testing of the periodicity of these deposits (cf. Olsen, 1986). Nevertheless it must be stressed that the range in fourth-order cycle thickness (0.5–9 m) is similar to that observed in some cyclic Phanerozoic lacustrine sequences that have been interpreted as climatically forced (van Houten, 1964; Eugster & Hardie, 1975; Donovan, 1980; Olsen, 1986; Gore, 1988,1989).

This fact could support the idea that a similar forcing process gave rise to the oxic–anoxic cyclical sedimentation in Rubielos. Van Houten (1964) and Olsen (1986) have established periods of a few tens of thousands of years (20 000–30 000 years) for the generation of some of the most important and frequent cycles in the Newark Supergroup. Berger (1984) established that the Earth's precession cycle, with a periodicity of about 25 000 years, has a direct impact upon precipitation evolution. A similar control could be suggested for the Rubielos lacustrine oxic–anoxic cycles, taking into account their thickness and magnitude range (i.e. a few metres).

The occurrence in the recent sedimentary record of several modern lakes (Tanganyka, Victoria, Kivu) of minor transgressive–regressive sequences similar to those recognized in ancient Phanerozoic lakes, has been attributed to palaeoclimatic changes (Talbot, 1988; Talbot & Livingstone, 1989). On the other hand, the influence of watershed modifications upon basin hydrology has been emphasized recently (Gore, 1989). The palaeocurrent trends and the alluvial facies distribution observed in the Rubielos de Mora Basin (Fig. 5), do not contradict a possible closing of the basin, despite some uncertainty resulting from the absence of outcrops in the westernmost basin zones. Anyway, a combination of both factors (palaeoclimatic water balance change and repeated transition from open to closed conditions) are not necessarily contradictory.

CONCLUDING REMARKS

1 The Miocene Rubielos lacustrine system evolved in an extensional tectonically influenced setting that allowed the establishment and subsequent spreading of a lacustrine environment.

2 As a consequence of the half-graben basin structure the arrangement of the lacustrine depositional framework was clearly asymmetrical. A southeastern, deep, anoxic lacustrine depocentre persisted all along the Upper Unit successions (Figs 5, 15).

3 The major, deepening evolutionary trend of the lacustrine basin, recorded by first and second-order sequences, was controlled mainly by tectonics. In addition, minor oscillations of the water level also influenced the lacustrine sedimentation and resulted in third and fourth-order sequences. These minor oscillations were probably climatically forced and modulated the overall deepening trend of the lake system (Fig. 17).

4 The genetic stratigraphy approach arises as a useful tool for lacustrine basin analysis owing to the fact that water level variations will lead to the generation of striking bounding surfaces, which may be traced in some cases throughout the basin.

5 The water level oscillations in the Rubielos de Mora Basin remained in a relatively narrow range and no dramatic falls in water level took place. This fact is shown clearly by the absence of deeply entrenched erosive surfaces (type 1 unconformities) affecting-either the carbonate benches or the low-gradient, terrigenous lacustrine ramp (Fig. 14).

6 The sedimentary record of the propounded water level oscillations changed depending on the lacustrine floor gradient in the diverse basin zones (Figs 14, 17).

7 The morphometry of the ancient lacustrine systems arises as a major feature to take into account during exploration of oil shale deposits, since the relationship between oxicline oscillations and the gradient of the lake bottom is the main factor that controls the areal extent of the anoxic lacustrine floor zones (Fig. 14).

ACKNOWLEDGEMENTS

This work was partially supported by the Comisión Asesora de Investigación Científica y Técnica and Consejo Superior de Investigaciones Científicas (CAICYT-CSIC, ID: 851). We are grateful to Michael Leeder and Michael Talbot for constructive criticism and helpful suggestions. The manuscript benefited from their reviews.

REFERENCES

ANADÓN, P. (1983) Características generales de diversas cuencas lacustres terciarias con pizarras bituminosas del NE de la Península Ibérica. *Com. X Congr. Nacional de Sedimentología, Menorca* 1, 9–12.

ANADÓN, P., CABRERA, LL., GUIMERÀ, J. & SANTANACH, P. (1985) Paleogene strike-slip deformation and sedimentation along the southern margin of the Ebro Basin. In: *Strike-slip Deformation, Basin Formation and Sedimentation* (Eds Biddle, K.T. & Christie-Blick, N.), Soc. Econ. Paleontol. Mineral. Spec. Publ. 37, 303–318.

ANADÓN, P., CABRERA, LL., INGLES, M., JULIÀ, R. & MARZO, M. (1988a) The Miocene lacustrine basin of Rubielos de Mora. *International Workshop on 'Lacustrine Facies Models in Rift Systems and Related Natural Resources', Barcelona–Rubielos de Mora. Excursion Guidebook*, International Association of Sedimentologists, 32 pp.

ANADÓN, P., CABRERA, LL. & JULIÀ, R. (1988b) Anoxic–oxic cyclical lacustrine sedimentation in the Miocene Rubielos de Mora Basin. In: *Lacustrine Petroleum Source Rocks* (Eds Fleet, A.J., Kelts, K. & Talbot, M.R.), Geol. Soc. London Spec. Publ. 40, 353–367.

ANADÓN, P., CAWLEY, S.J. & JULIÀ, R. (1988c) Oil source rocks in lacustrine sequences from Tertiary grabens, Western Mediterranean Rift System, Northeast Spain. *Bull., Am. Assoc. Petrol. Geol.* 72, 983.

ANADÓN, P., CABRERA, LL., JULIÀ, R., ROCA, E. & ROSELL, L. (1989) Lacustrine oil-shale basins in Tertiary Grabens from NE Spain (Western European Rift System). *Palaeogeogr. Palaeoclimatol. Palaeoecol.* 70, 7–28.

BALLY, A.W. (Ed.) (1987) Atlas of seismic stratigraphy. *Am. Assoc. Petrol. Geol., Stud. Geol.* 27, 124 pp.

BERGER, A. (1984) Accuracy and frequency stability of the Earth's orbital elements during the Quaternary. In: *Milankovitch and Climate.* (Eds Berger, A.L., Imbrie, J., Hays, J., Kukla, G. & Saltsman, B.). *Nato Advanced Science Institutes Series* 126, 3–39.

BESSEDIK, M. (1985) *Réconstitution des environnements miocènes des régions Nord-ouest Mediterraneennes à partir de la Palynologie.* Thèse Université des Sciences et Techniques du Languedoc, Montpellier, 162 pp.

BOYER, B.W. (1981) Tertiary lacustrine sediments from Sentinel Butte, North Dakota and the sedimentary record of ectogenic meromixis. *J. Sediment. Petrol.* 51, 429–440.

COHEN, A.S. (1989) Facies relationships and sedimentation in large rift lakes and implications for hydrocarbon exploration: examples from lakes Turkana and Tanganyika. *Palaeogeogr. Palaeoclimatol. Palaeoecol.* 70, 65–80.

CRUSAFONT, M., GAUTIER, F. & GINSBURG, L. (1966) Mise en évidence du Vindobonien inférieur continental dans l'Est de la province de Teruel (Espagne). *C. R. Somm. Soc. Geol. France* 1966, 30–32.

DIETZ, R.S. (1963) Wave-base, marine profile of equilibrium, and wave-built terraces: a critical appraisal. *Geol. Soc. Am. Bull.* 74, 971–990.

DIETZ, R.S. (1964) Wave-base, marine profile of equilibrium, and wave-built terrace: reply. *Geol. Soc. Am. Bull.* 75, 1275–1282.

DONOVAN, R.N. (1980) Lacustrine cycles, fish ecology and stratigraphic zonation in the Middle Devonian of Caithness. *Scott. J. Geol.* 16, 35–50.

EUGSTER, H.P. & HARDIE, L.A. (1975) Sedimentation in an ancient playa-lake complex: the Wilkins-Peak member of the Green River formation of Wyoming. *Geol. Soc. Am. Bull.* 86, 319–334.

FERNÁNDEZ-MARRÓN, M.T. & ÁLVAREZ-RAMIS, C. (1988) Note preliminaire sue l'étude paléobotanique du gisement de Rubielos de Mora (Teruel), Espagne. *Colloque de l'Organisation Française de Paléobotanique (OFP), Lille (31 May 1988)*, 6 pp.

FERNÁNDEZ-NAVARRO, L. (1914) La cuenca petrolífera de Rubielos de Mora. *Rev. Acad. Cienc.* 13, 237–255.

GALLOWAY, W.E. (1989a) Genetic stratigraphic sequences in basin analysis. I: Architecture and genesis of flooding-surface bounded depositional units. *Bull. Am. Assoc. Petrol. Geol.* 73, 125–142.

GALLOWAY, W.E. (1989b) Genetic stratigraphic sequences in basin analysis. II: Application to northwest Gulf of Mexico Cenozoic Basin. *Bull. Am. Assoc. Petrol. Geol.* 73, 143–154.

GAVALA, J. (1921) Nota acerca de los yacimientos de lignitos y pizarras bituminosas de Rubielos de Mora (Teruel). *Bol. Inst. Geol. Minero España* 42, 263–302.

GORE, P.J.W. (1988) Lacustrine sequences in an early Mesozoic rift basin: Culpeper Basin, Virginia, USA. In:

Lacustrine Petroleum Source Rocks (Eds Fleet, A.J., Kelts, K. & Talbot, M.R.) Geol. Soc. London Spec. Publ. **40**, 247–278.

GORE, P.J.W. (1989) Toward a model for open- and closed-basin deposition in ancient lacustrine sequences: the Newark Supergroup (Triassic–Jurassic), Eastern North America. *Palaeogeogr. Palaeoclimatol. Palaeoecol.* **70**, 29–51.

GUIMERÀ, J. (1984) Paleogene evolution of deformation in the northeastern Iberian Peninsula. *Geol. Mag.* **121**, 413–420.

GUIMERÀ, J. (1990) Formación de una cubeta sinclinal en un contexto extensivo: la cuenca miocena de Rubielos de Mora (Teruel). *Geogaceta* **8**, 33–35.

HAQ, B.V., HANDERBOL, J. & VAIL, P.R. (1987) Chronology of fluctuating sea levels since the Triassic. *Science* **235**, 1156–1167.

IGME-CGS (1980) *Investigación geológica y geofísica de la cuenca terciaria de Rubielos de Mora (Teruel).* Report Instituto Geológico Minero de España, Madrid (unpublished).

LEEDER, M.R. & GAWTHORPE, R.L. (1987) Sedimentary models for extensional tilt-block/half-graben basins. In: *Continental Extensional Tectonics* (Eds Coward, M.P., Dewey, J.F. & Hancock, P.L.). Geol. Soc. London Spec. Publ. 28, 139–152.

LEEDER, M.R., ORD, D.M. & COLLIER, R. (1988) Development of alluvial fans and fan deltas in neotectonic extensional settings: implications for the interpretation of basin fills. In: *Fan Deltas: Sedimentology and Tectonic Settings* (Eds Nemec, W. & Steel, R.J.), pp. 173–185. Blackie, Glasgow.

LINK, M.M. & OSBORNE, R.M. (1978) Lacustrine facies in the Pliocene Ridge Basin, California. In: *Modern and Ancient Lake Sediments* (Eds Matter, A. & Tucker, M.E.), Spec. Publ. Int. Assoc. Sediment. 2, 169–187.

LOWE, D.R. (1982) Sediment gravity flows. II. Depositional models with special reference to the deposits of high-density turbidity currents. *J. Sediment. Petrol.* **52**, 279–297.

MOISSENET, E. & GAUTIER, F. (1971) La region de Rubielos de Mora. Contribution a l'étude geologique et morphologique. *Mel. Casa Velázquez* **7**, 5–28.

MONDEGUER, A., TIERCELIN, J.-J., HOFFERT, M., LARQUE, PH., LE FOURNIER, J. & TUCHOLKA, P. (1986) Sédimentation actuelle et récente dans un petit basin en contexte extensif et décrochant: La baie de Burton, fosé nord-Tanganyika, rift est-african. *Bull. Cent. Rech. Exp. Prod. Elf-Aquitaine* **10**, 229–247.

MOORE, D.G. & CURRAY, J.R. (1964) Wave-base, marine profile of equilibrium, and wave-built terraces: discussion. *Geol. Soc. Am. Bull.* **75**, 1267–1274.

MOUGENOT, D., BOILOT, G. & REHAULT, J.P. (1983) Pro-grading shelfbreak types on passive continental margins: some European examples. *Soc. Econ. Paleontol. Mineral. Spec. Publ.* **33**, 61–77.

MÜLLER, G., IRION, G. & FÖRSTNER, U. (1972) Formation and diagenesis of inorganic Ca–Mg carbonates in the lacustrine environment. *Naturwissenchaften* **59**, 158–164.

NEMEC, W. & STEEL, R.J. (1984) Alluvial and coastal conglomerates: their significant features and some comments on gravelly mass-flow deposits. In: *Sedimentology of Gravels and Conglomerates* (Eds Koster, E.M. & Steel, R.J.). *Can. Soc. Petrol. Geol. Mem.* **10**, 1–31.

OLSEN, P.E. (1986) A 40-million year lake record of Early Mesozoic orbital climatic forcing. *Science* **234**, 842–848.

PAYTON, C.E. (Ed.) (1977) Seismic stratigraphy — applications to hydrocarbon exploration. *Mem., Am. Assoc. Petrol. Geol.* **26**, 516 pp.

PICARD, M.D. & HIGH, L.R. (1981) Physical stratigraphy of ancient lacustrine deposits. In: *Recent and Ancient Non-marine Depositional Environments: Models for Exploration.* (Eds Ethridge, F.G. & Flores, R.M.) Soc. Econ. Paleontol. Mineral. Spec. Publ. 31, 233–259.

PRADO, J.G., SUAREZ-RUIZ, I., BORREGO, M.A. & GARCÍA, A.M. (1988) The nature of organic matter in oil shales from Rubielos de Mora (abstr.) *International Workshop 'Lacustrine Facies Models in Rift Systems and Related Natural Resources', Barcelona–Rubielos de Mora.* International Association of Sedimentologists.

REYNOLDS, S.A., GALLOWAY, W.E. & GLASFORD, J.L. (1989) Genetic stratigraphic sequences in nonmarine basins. *Abstr. XXVIII Int. Geol. Congr.* **2**, 691–692.

SCHOLZ, C.A., ROSENDAHL, B.R. & SCOTT, D.L. (1990) Development of coarse grained facies in lacustrine rift basins: examples from East Africa. *Geology* **18**, 140–144.

TALBOT, M.R. (1988) The origins of lacustrine oil source rocks: evidence from the lakes of tropical Africa. In: *Lacustrine Petroleum Source Rocks* (Eds Fleet, A.J., Kelts, K. & Talbot, M.R.). Geol. Soc. London Spec. Publ. **40**, 29–43.

TALBOT, M.R. & LIVINGSTONE, D.A. (1989) Hydrogen index and carbon isotopes of lacustrine organic matter as lake level indicators. *Palaeogeogr. Palaeoclimatol. Palaeoecol.* **70**, 121–137.

VAN HOUTEN, F.B. (1964) Cyclic lacustrine sedimentation, Upper Triassic Lockatong Formation, Central New Jersey and adjacent areas. *Kans. State Geol. Surv. Bull.* **169**, 497–531.

WILGUS, C.K., HASTINGS, B.S., KENDALL, C.G.St.C., POSAMENTIER, H.W., ROSS, C.A. & VAN WAGONER, J.C. (Eds) (1988) Sea-level changes: an integrated approach. *Soc. Econ. Paleontol. Mineral. Spec. Publ.* **42**, 407 pp.

Geochemistry and Organic Remains in Lacustrine Deposits

Spec. Publs Int. Ass. Sediment. (1991) **13**, 279–290

Evolution of lacustrine systems in the Tertiary Narbonne Basin, northern Pyrenean foreland, southeast France

J. SZULC*, PH. ROGER[†], M.P. MOULINE[†] *and* M. LENGUIN[†]

**Institute of Geological Sciences. Jagellonian University 30-063, Oleandry str. 2a, Cracow, Poland*
[†]Institut de Géodynamique, Université Bordeaux III Av. des Facultes, 33 405 Talence Cedex, France

ABSTRACT

The Narbonne Basin was formed within an irregular depression located on the Eastern Corbières thrust sheet. During late Oligocene and early Miocene (Aquitanian) times, this piggyback basin was filled with non-marine sediments including lacustrine and palustrine carbonates, evaporites (gypsum), alluvial fine-grained siliciclastics, and fanglomerates. Minor components of the basin fill include cherts, lignites, and bituminous matter. Three lithostratigraphical units are distinguished. These units comprise facies assemblages characteristic of fluctuating, evaporitic–non-evaporitic lacustrine episodes. The lacustrine suecession is attributed to subtropical–tropical climate fluctuation, as it is interpreted here that syndepositional tectonics was of minor significance and only influenced local sedimentation. Some geochemical investigations (including $\delta^{18}O$, $\delta^{13}C$ and $\delta^{34}S$), are used to reconstruct brine evolution.

INTRODUCTION

During late Oligocene and early Miocene (Aquitanian) a small foreland basin on the northern Pyrenean eastern thrust sheet was filled with continental deposits, including fanglomerates, fine-grained siliclasts, lacustrine and palustrine carbonates and sulphates. The lithology of the basin fill changed during Oligocene–Miocene times owing to environmental variations.

This paper examines the fluctuations of lacustrine systems (*sensu lato*) from non-evaporitic through to evaporitic sedimentary conditions. Sedimentological studies enabled overall environmental reconstruction, whilst isotope data refined hydrological and climatological interpretation.

GENERAL SETTING

The Oligocene–Miocene Narbonne Basin belongs to the meridional system of foreland depressions that originated within the northeastern bend of the Pyrenean overthrust (Figs 1, 2). This bend is postulated to be the result of the strike-slip motion of a basement fault; the Cevennes Fault (Arthaud & Mattauer, 1972; Ellenberger, 1980) or the Nimes Fault (Gottis, 1967). This part of the Pyrenean orogen is known as the Eastern Corbières Range. The rocks of the autochthonous basement and the allochthonous units comprise: red Keuper (Upper Triassic) marls, limestones, dolomites and shales of the Middle and Upper Jurassic, as well as the sandstones, argillites and carbonates of the Upper Cretaceous. In its western part, the outer unit of Corbières has overthrusted the Eocene molasse.

Lack of knowledge of the tectonics of this part of the Pyrenean overthrust hinders structural interpretation of the area (Ellenberger, 1980; Viallard, 1987). It may be accepted, however, that the Narbonne depression is a piggyback basin carried on the northeast thrust sheet of the Pyrenees (Fig. 3).

The western boundary of the Narbonne Basin is formed by an elevated front of the main Corbières overthrust (Fig. 2). A compressional–tensional regime was important in the western margin and resulted in a strong variation of the sedimentary processes. Thus fining upward alluvial sequences —

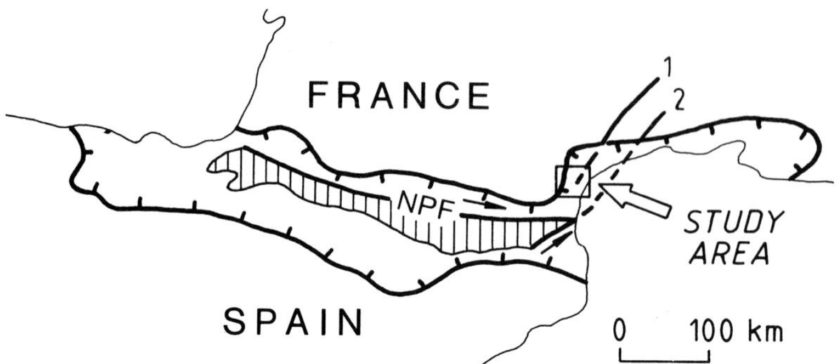

Fig. 1. Location map and structural framework of the Pyrenean orogen. Hatched area: Hercynian (pre-Upper Carboniferous) rocks; NPF, North Pyrenean Fault; 1, Cevennes Fault; 2, Nîmes Fault.

beginning with fault-derived breccias followed by fine clastics and then by lacustrine and palustrine sediments — have been observed very close to synchronous, coarsening upwards sequences, which include basal palustrine sediments overlain by fanglomerates often overthrusted by basement slices or nappes. Such rapid, lateral changes (every 1–2 km) fit well the en échelon faulting style of the western basin boundary (Fig. 2B; Gottis, 1967).

The eastern basin margin is bounded by a compact fault-block called 'la Clape Massif' (Fig. 2A). Sediments linked directly to the uplifted basement rocks show fairly progressive internal unconformities (cumulative wedge system of Riba (1976)). The sediment arrangement and facies changes indicate the punctuated, minor synsedimentary uplift of the

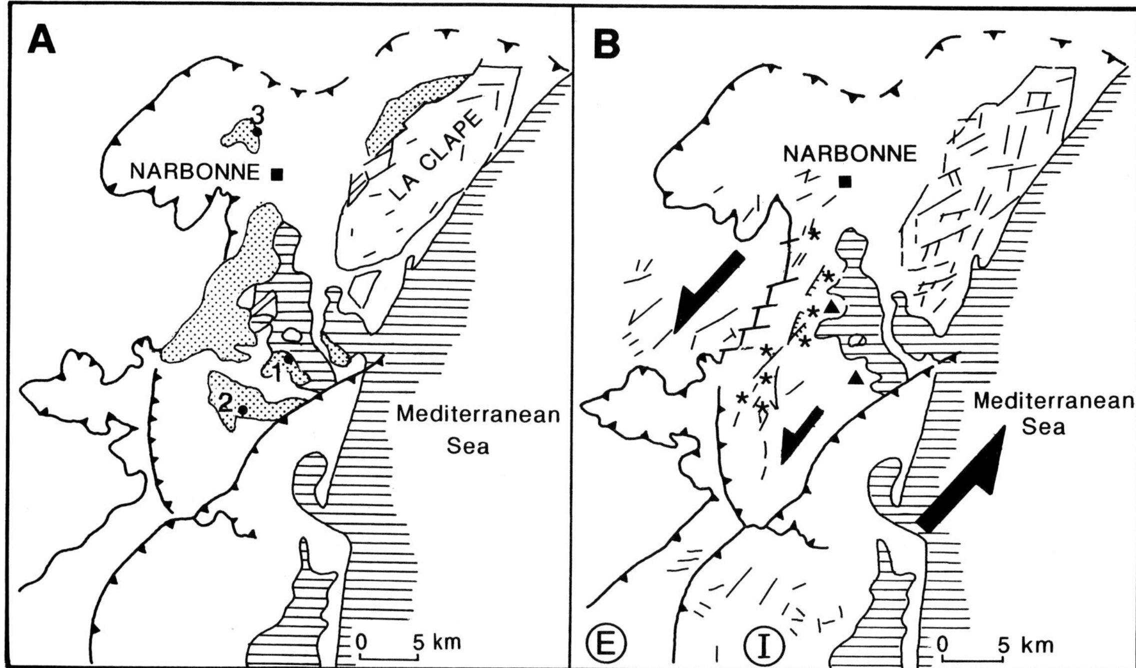

Fig. 2. Sketch maps of the Narbonne Basin. (A) Location map of the more important outcrops (numbered points). Dotted area: Oligocene–Miocene outcrops. Sites: 1, Port Mahon; 2, Trois Moulins; 3, Malvézy. (B) Tectonic map of study area. Arrows indicate movement directions. Size of the arrows reflects the relative scale of motion. I and E, internal and external units of the Corbieres nappe; ▲, diapiric ring structures; *, upfaulted basement blocks and/or olistholites. (Modified after Gottis, 1967; Viallard, 1987).

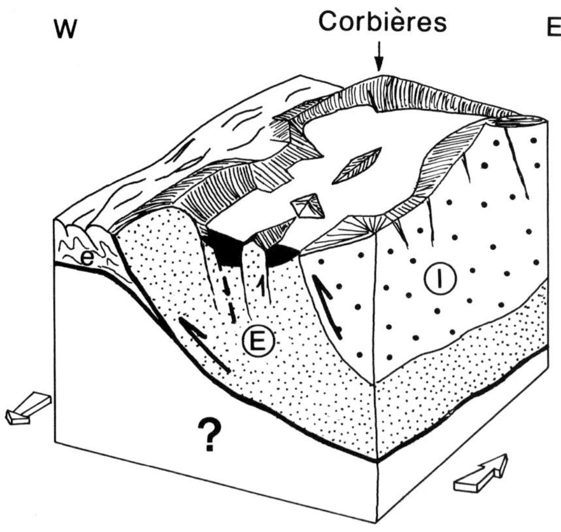

Fig. 3. Schematic diagram of the Narbonne Basin, showing structural variability within the basin. See discussion in the text. I and E, internal and external units of the Corbieres nappe; e, Eocene molasse deposits. Arrows indicate movement direction.

eastern marginal block (Szulc *et al.*, in preparation).

A dense network of SW−NE oriented and ENE−WSW trending faults occurs within the internal part of the basin (Fig. 2B). Many of the faults are normal, though some of them are reverse or strike-slip faults. The results of sedimentological studies (Szulc *et al.*, in preparation) indicate that most of these faults were of synsedimentary nature, and suggest that the great transverse fault system was already active after the last phase of the Corbières thrusting, which is accepted to have been active during the early Oligocene (Puigdefàbregas & Souquet, 1986). An elongated zone of uplifted basement rocks occurs within the inner part of the basin (Fig. 2). This zone consists of Mesozoic rock slices, upfaulted owing to local compression.

Tectonic influences on sedimentation may be traced both along faulted margins as well as in the central part of the basin. Seismic events may have been the cause of some of the sediment structures observed, including small fractures, faults, and slumps, as well as kilometre-scale complexly deformed units associated with basement warping. Szulc *et al.* (in preparation) proposed that modification of the local tectonic fabric was facilited by the occurrence of deformable gypsiferous Keuper rocks within the basement.

STRATIGRAPHY AND OVERALL FACIES ARRANGEMENT OF THE BASIN FILL

Exposures of basin fill are inadequate for precise analyses of both lateral and vertical facies patterns. The data collected were used to construct a framework of facies and their stratigraphic succession. Three lithostratigraphic units have been defined for the basin fill (Fig. 4). The lower (A) and upper (C) units comprise limestones deposited in palustrine and lacustrine subenvironments, while the middle unit (B) is dominated by dolostones and sulphates characteristic of the mudflat and lacustrine subenvironments. The thickness of the basin fill varies from 300 to 500 m.

Alluvial sediments are a common component of all units. Alluvial fans surround the basin, forming a

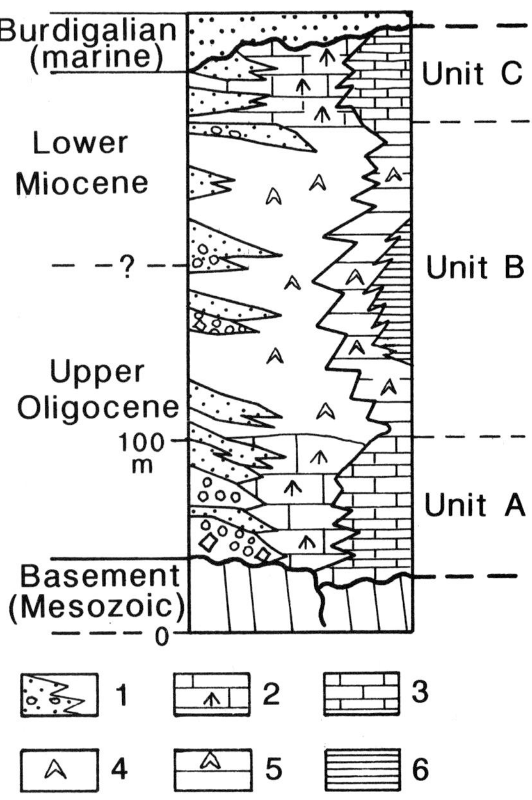

Fig. 4. Idealized lithostratigraphical sketch of the Narbonne Basin deposits. 1, alluvial deposits; 2, palustrine carbonates; 3, lacustrine limestones; 4, sandflat−mudflat deposits with gypsum; 5, lacustrine dolomites and gypsum; 6, deep lacustrine carbonates and black shales.

ring up to 300 m thick and up to 1 km wide. Texture of the fan deposits depends on the lithology of the source area. The fans adjacent to the Jurassic and Cretaceous limestone terrains are built of coarse, poorly sorted gravels with blocks up to metre size, whereas Cretaceous argillites supplied material for variegated mud and sand fans. Gypsiferous Keuper marls contributed the clayey red matrix of the coarse conglomerates, while the sulphates were either dissolved and reprecipitated within the clastics or transported to the central parts of the basin. Bipyramidal quartz crystals from the eroded Keuper rocks may be used as an indicator of the presence of Keuper material. Small talus fans also adjoined the uplifted basement ridge in the central part of the basin. They contain a significant proportion of redeposited sediments. Alluvial fan sediments grade laterally into marginal lacustrine facies, though direct contacts with inner sediments are very common.

The age of the units cannot be determined precisely. It seems, however (Rosset, 1964; Aguilar, 1977; Aguilar & Michauz, 1977) that unit A and the lower part of unit B are of late Oligocene age, while the upper part of unit B and unit C are of early Miocene age ('Aquitanian').

Geochemical investigations supplemented sedimentological analysis. Typical carbonates from units B and C were selected for carbon and oxygen isotopic analysis; $\delta^{34}S$ measurements of selected gypsum samples from unit B served to distinguish various types of gypsum in the Narbonne Basin. All samples were analysed by X-ray diffraction analysis, and a few by scanning electron microscopy.

LACUSTRINE SYSTEMS IN THE NARBONNE BASIN

As mentioned above the lithology of units A and C differs markedly from unit B. This difference is assumed to reflect the lack or presence of evaporite components within the basin fill. Based on this assumption, we divide the basinal sedimentary environments into evaporitic (unit B) and non-evaporitic (units A and C) lake systems. Table 1 summarizes the main sedimentary features of these two lake systems.

Non-evaporitic lacustrine systems (units A and C)

Marginal facies assemblage

The marginal facies consists of massive and nodular limestones. The limestones form continuous horizons and lenticular bodies, from centimetre to metre scale. Carbonates consist of bioclastic material; debris of gastropod and ostracod shells, and charophytes, as well as micrite of algal origin. The lithoclastic content varies from small amounts of dispersed siliciclasts to plane-bedded sandstones and rare, poorly sorted conglomerate sheets, up to 0.5 m thick. The limestones are highly bioturbated, mainly by root systems, which resemble the Potamogeton-type rootlets as well as burrows by insects and worms. Part of the carbonates are white, lacking non-oxidized organic components, though most of these sediments are rich in accumulated organic matter. Evaporites have not been found in this facies.

The sediments are interpreted as hardwater, poorly aerated swamp deposits, episodically prograded by clastic toes of alluvial fans.

Inner lacustrine facies assemblage

Lacustrine sequences up to 40 m thick consist of massive or faintly laminated calcarenites and calcilutites. They contain abundant gastropod and ostracod debris as well as charophyte bioclasts. Sediment colour varies from light-grey to brown, reflecting the non-oxidized organic matter content. The massive limestones pass laterally into finely laminated, fissile, organic-rich limestones. Most of the laminated beds display rhythmicity, although thin, graded ostracod-bioclasts and fine-intraclast turbidites are common. Some of the laminites also contain inconspicuous diatomite intercalations and well-preserved, carbonized leaves.

Sequential relationships and interpretation

The section Trois Moulins (site 2 in Fig. 2A) shows the sequential relationships between the carbonates of unit C. The sediment sequence at this site is twofold (Figs 5, 6). The lower part is composed of soft, grey, finely laminated limestones, rich in debris of gastropod shells (mostly Planorbidae), as well as millimetre scale lignite laminae and a subordinate amount of siliciclastics. The upper part is composed of thin-bedded indurated limestones alternating with

Table 1. Summary of the main sedimentary characteristics of evaporitic and non-evaporitic lacustrine sequences of the Narbonne Basin

	Marginal lacustrine assemblages		Inner lacustrine assemblages	
	Evaporitic lacustrine sequences: sandflat–mudflat	Non-evaporitic lacustrine sequences: hardwater swamp	Stratified saline lake	Ephemerally stratified hardwater lake
Mineralogy of the autochthonous components	Dolomites, LMC*, gypsum	LMC	Dolomite, gypsum, LMC, silica (diatomites), pyrite (abundant)	LMC, pyrite (common), dolomite (minor)
Allochthonous siliciclastic content	Dominant or substantial	Subordinate	Low	Low
Bedding	Planar, thin and medium bedded, laterally continuous horizons	Massive horizons and lenticular bodies	Well laminated, seasonally varved, thin continuous horizons (+ turbidites)	From thick, massive to rhythmically bedded thin horizons (+ turbidites)
Pedogenic fabrics	Calcrete, silcrete, teepee, desiccation cracks	Hydromorphic, nodular soils, rhizoliths, corrosion surfaces, karstic sinkholes		
Synsedimentary deformation (slumps, folds, faults, etc.)			Common	Common
Organic matter content	Low	Significant	High to very high (oil shale)	Low to high

*LMC = low-magnesium calcite.

Fig. 5. Lacustrine−palustrine limestones representing the non-evaporitic sedimentary environment (unit C, site 2, Trois Moulins section). See Fig. 2 for location. Note the karstically widened wedge-shaped cracks (rundkarren type of runnels) in the middle part of the section.

dark, homogenous marls comprising rootlet systems and subaerial karstic sinkholes. Deposits of the lower part of the section are interpreted as marls deposited in the central, deeper parts of a shallow, open lacustrine environment. The upper part of the section represents periods of relatively deep lacustrine sedimentation, which alternated with periods of sinking and emersion typical of a palustrine environment.

The Potamogeton-type reed rootlets occurring within some horizons, suggest the depth of the sedimentary environments to have been between 1 and 5 m (Fuchs & Haude, 1987) and indicate that deposition took place in hardwater lakes where massive limestones were deposited on a shallow-water marl bench (Murphy & Wilkinson, 1980).

Carbonate isotopic geochemistry

The section Trois Moulins was chosen for the geochemical isotope study. X-ray diffraction studies of this sediment have shown the dominance of low-magnesium calcite (LMC) and a small amount of aragonite (probably from shell debris), as well as traces of dolomite (Fig. 6.)

The twofold nature of the sedimentary environment deduced from the sedimentological analysis is also differentiated by stable isotope contents (Figs 6, 7). Values of $\delta^{13}C$ that group around 0‰ (-1.3‰ to 1.3‰) are characteristic of the lower part of the section. Average $\delta^{18}O$ values are negative (-2.3‰), but in general tend to be more positive than in the higher part of the section (-3.3‰). The carbonates of the upper part of the section have isotopic values depleted in ^{13}C and ^{18}O (average values $\delta^{13}C$ -2.8, and $\delta^{18}O$ 3.0).

The values of the lower part of the section are consistent with deposition of the marls in stagnant, slightly evaporitic and poorly drained swamps (long residence times). Flooding and the restoration of freshwater conditions by impinging of meteoric waters explain the recorded isotopic values in the upper part (Whelan & Roberts, 1973).

Evaporitic lacustrine system (unit B)

Marginal facies assemblage

In contrast to the marginal facies of units A and C, this facies assemblage is dominated by medium and fine-grained, variegated siliciclastic sediments. The argillites are characterized by planar bedding, large lateral continuity of individual horizons, and sporadic intercalations of conglomerates.

Carbonates, both dolostones and limestones, form thin beds (< 2 m thick) within the siliciclastic sediments. They are rich in ostracods as well as in macerated organic matter. Gastropods are common, although their specific diversity is poor. Algal laminites occur formed mainly by cyanobacterial mats. Well-preserved, primary accretion structures are disturbed by episodic dessication and exfoliation of the mats. Gypsum precipitates are common either as displacive crystals within the siliciclastic deposits or, rarely, forming thin crusts (< 1.5 cm thick).

Interpretation

The clastic sediments are interpreted as sandflat−

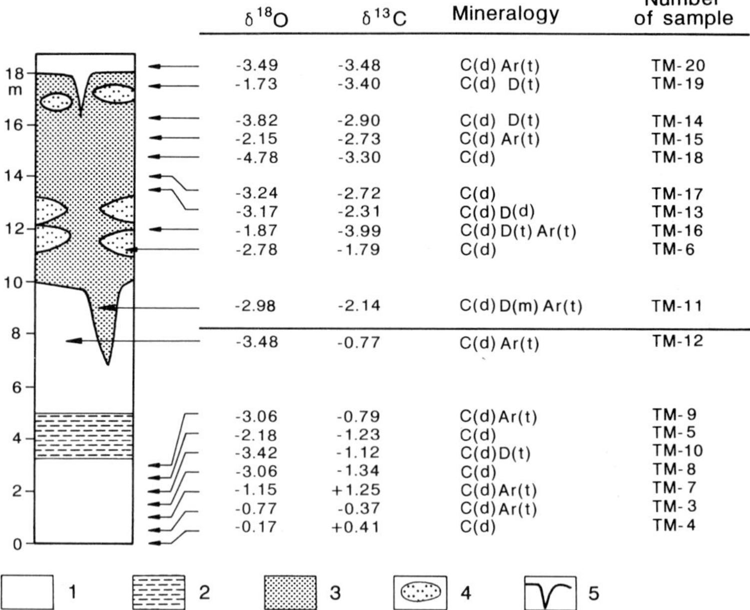

$\delta^{18}O$	$\delta^{13}C$	Mineralogy	Number of sample
-3.49	-3.48	C(d) Ar(t)	TM-20
-1.73	-3.40	C(d) D(t)	TM-19
-3.82	-2.90	C(d) D(t)	TM-14
-2.15	-2.73	C(d) Ar(t)	TM-15
-4.78	-3.30	C(d)	TM-18
-3.24	-2.72	C(d)	TM-17
-3.17	-2.31	C(d) D(d)	TM-13
-1.87	-3.99	C(d) D(t) Ar(t)	TM-16
-2.78	-1.79	C(d)	TM-6
-2.98	-2.14	C(d) D(m) Ar(t)	TM-11
-3.48	-0.77	C(d) Ar(t)	TM-12
-3.06	-0.79	C(d) Ar(t)	TM-9
-2.18	-1.23	C(d)	TM-5
-3.42	-1.12	C(d) D(t)	TM-10
-3.06	-1.34	C(d)	TM-8
-1.15	+1.25	C(d) Ar(t)	TM-7
-0.77	-0.37	C(d) Ar(t)	TM-3
-0.17	+0.41	C(d)	TM-4

Fig. 6. Stable isotope composition and mineralogy of carbonates of the Trois Moulins section (site 2, see Fig. 2 for location): D, dolomite; Ar, aragonite; C, low-magnesian calcite; Q, detrital quartz; G, gypsum moulds; d, dominant; a, abundant; m, minor; t, trace; 1, finely laminated lacustrine limestones; 2, massive marly sediments; 3, palustrine limestones; 4, lenses of rhizoliths; 5, karstic runnels.

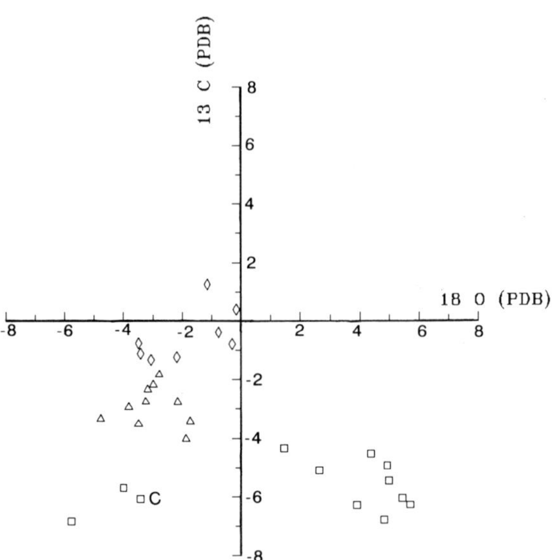

Fig. 7. Isotopic data for carbonates from the Narbonne Basin: △, Trois Moulins section, upper part (calcite); ◊ Trois Moulins section, lower part (calcite); □, Port Mahon section (dolomite predominant, except for sample labelled C, which corresponds to predominant calcite).

mudflat 'redbeds' accumulated during sheet-floods. Dolomites, of evaporative origin, formed in small, post-flood ponds, or highly evaporative playa aprons.

A high bitumen content in the algal laminites and the occurrence of 'black pebbles' indicate that deposition took place in a shallow basin with high organic productivity, probably in a sheltered, quiet bay or in hot pans. The abundant dessication cracks, teepee structures, as well as the pedogenic deposits, caliche and silcretes that are present, are consistent with this interpretation.

Gypsum precipitated from flood waters, which probably came from drained Keuper marls. The displacive gypsum crystals formed in the interstitial water of conglomerates and sandstones. Very low $\delta^{34}S$ values from this gypsum (-5.4 to -4.0‰ vs. CDT) differ from those of the Keuper source rocks ($+15.5$‰ vs. CDT). If Keuper rocks are the source of sulphate, it has been recycled from isotopically heavier Keuper gypsum through intermediate sulphide phases. Oxidation of sulphides and the reprecipitation of sulphate as isotopically light gypsum is required if this is the mechanism (Krouse, 1980).

Inner lacustrine facies assemblage

This sediment assemblage consists mainly of well-laminated bitumen-rich dolostones and limestones. Dolostones are blue-grey if the pyrite content is high. Thin-sections show that these sediments are composed of dolomicrite (< 5 μm, anhedral crystals)

Fig. 8. Slump-deformed dolomite varves from a closed deep-lake environment (unit B, site 3, Malvezy). Light laminae are dolomite. The darker laminae include a small amount of detrital quartzose silt and organic matter. Black scale-bar is 1 cm.

and were initially deposited probably as soft (or even colloidal) dolomitic oozes, similar to Quaternary analogues from Turkey (Müller & Irion, 1969), Balaton Lake (Müller & Wagner, 1978) and Southern Australia (Von der Borch & Lock, 1979; Rosen *et al.*, 1988). Some dolomite is certainly the result of early post-depositional changes from LMC deposits, such as, for example, charophytic limestones.

The deepest lacustrine facies is represented by black oil shales and clearly varved, organic-rich dolomites (Fig. 8.). Dolomite varve couplets vary in organic matter and detrital content; the dark laminae are richer in bitumen, containing *Botryococcus* (M. Gottis, pers. comm., 1989) as well as quartzose silt and clays, whereas the light-coloured bands are composed mainly of pure dolomite. The thickness of a single couplet ranges from 0.3 to 0.7 mm. Diatomites and lignites form accessory facies and reach up to 5 cm in thickness.

In contrast to the units A and C, this inner lacustrine facies commonly contains a significant amount of gypsum. Gypsum is present in two forms; either as empty moulds after dissolution of the original gypsum crystals dispersed in carbonates, or as thin, but continuous horizons, often accompanying the lignite and/or diatomite layers. The bipyramidal quartz found in the sporadic, fine-clastic beds, indicates that some parts of the basin were probably fed by waters flowing through (or from) the gypsiferous Keuper rocks and have hence become enriched in dissolved SO_4^{2-} ions.

Brine evolution in the evaporitic lacustrine system — an attempt at palaeoenvironmental interpretation

The section at Port Mahon (site 1 in Fig. 2A.), which shows a sequential relationship between palustrine and open lacustrine sediments, has been chosen for a geochemical reconstruction of the evaporitic lake system (unit B). The section comprises laminated dolomites and subordinate massive limestones. The dolomites are of both primary and early diagenetic origin. The latter commonly contains dispersed gypsum moulds and/or small amounts of detrital quartz and kaolinite. One thicker limestone layer and a lenticular horizon occur only in the middle part of the outcrop (Figs 9, 10). The limestones comprise abundant rhizoliths and are accompanied by lenticular cherts. Organic matter may be dispersed, accentuating fine laminations in dolomites, or may form separate lignite layers up to 5 cm thick. Some lignites are accompanied by diatomites and/or by thin (< 2 cm) gypsum horizons. A large amount of non-oxidized organic matter, as well as a lack of bioturbation, suggest dysaerobic or anaerobic bottom conditions.

The evolution of brines with a high original sulphate concentration was classified by Eugster & Hardie (1978) by specific pathways (IIIc). As calcite precipitates the solution is driven toward saturation with respect to gypsum. As evaporation and sulphate precipitation proceed, the Mg/Ca ratio reaches values that favour dolomite. Moreover, a decrease in the concentration of SO_4^{2-} promotes dolomite formation (Baker & Kastner, 1981). An enrichment

Fig. 9. Lacustrine dolomites with palustrine limestone horizon (middle part of the section) representing the evaporitic sedimentary environment (unit B, site 1, Port Mahon section). See Fig. 2 for location, and text for further explanation.

in $\delta^{18}O$ observed in dolomites from this unit is consistent with evaporation of basin waters and with the model of brine evolution (Figs. 7, 10). The depleted ^{13}C values of the dolostones could be explained as a result of CO_2 derived from organic matter degradation in relation to sulphate reduction processes.

In an anoxic hypolimnion, carbonate precipitates in the epilimnion settle to the bottom, while sulphate may undergo bacterial reduction to sulphides (H_2S, FeS_2). Since the gypsum moulds are rare and dispersed within the carbonates we assume the sulphates were precipitated during short oxic episodes and redissolved in the sediments during a subsequent anoxic phase.

Co-occurrence of gypsum horizons with thicker

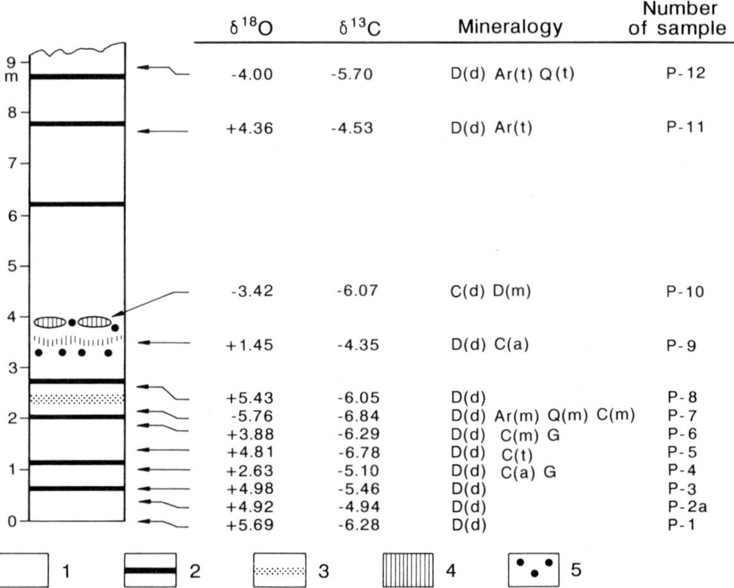

$\delta^{18}O$	$\delta^{13}C$	Mineralogy	Number of sample
-4.00	-5.70	D(d) Ar(t) Q(t)	P-12
+4.36	-4.53	D(d) Ar(t)	P-11
-3.42	-6.07	C(d) D(m)	P-10
+1.45	-4.35	D(d) C(a)	P-9
+5.43	-6.05	D(d)	P-8
-5.76	-6.84	D(d) Ar(m) Q(m) C(m)	P-7
+3.88	-6.29	D(d) C(m) G	P-6
+4.81	-6.78	D(d) C(t)	P-5
+2.63	-5.10	D(d) C(a) G	P-4
+4.98	-5.46	D(d)	P-3
+4.92	-4.94	D(d)	P-2a
+5.69	-6.28	D(d)	P-1

Fig. 10. Stable isotope composition and mineralogy of the carbonates from the section at Port Mahon (unit B, site 1). Symbols are the same as those used in Fig. 6 except: 1, finely laminated lacustrine dolostones; 2, lignite horizons; 3, horizon with conspicuous detrital admixture; 4, palustrine limestones; 5, cherts.

diatomites and lignites may indicate longer intervals with complete oxygenation of the water column. Diatom blooms indicate either water freshening (De Deckker, 1988) or nutrient pulses owing to the overturn (and upwelling?) of lake water. Freshening events may have been caused by increasing precipitation. Strongly $\delta^{18}O$ depleted dolomites (samples P-7, with $\delta^{18}O = -5.76‰$, and P-12, with $\delta^{18}O = -4‰$; Figs 7, 10) are difficult to explain. Perhaps the inflow of SO_4^{2-} poor waters could dilute SO_4^{2-} and promote the formation of dolomite. An increase in the amount of detrital quartz and aragonite from shell debris indicates a major interval with fresh water. Magaritz (1987) reports an example of precipitation of ^{18}O depleted dolomites as a consequence of a freshwater inflow and SO_4^{2-} dilution for shallow-marine evaporite basins.

Limestone horizons from the middle part of the section contain abundant traces of roots and represent palustrine deposits. The continuous or lenticular nature of the limestone may reflect the density of plant cover. Similar lenticular bodies were described by Calvo *et al.* (1985) as individual mud mounds bound around a hygrophyte root system in the ephemerally emerged nearshore zone. Silica from the diatomites was remobilized and precipitated as accompanying cherts. The occurrence of cherts in the boundary zone, between the limnic and palustrine sediments, is consistent with models of silica precipitation in the zone of mixing of alkaline ground-waters with more neutral surficial waters (Knauth, 1979; Wells, 1983). Cherts in the section are associated exclusively with the LMC horizons; they are absent from dolomitic sediments.

HYDROCARBONS

The presence of oil is a common feature of the Narbonne Basin sediments (particularly in unit B), although not yet studied from an economic point of view. The source of hydrocarbons is mostly from the abundant macro- and microphytic debris present. The preliminary results of the geochemical studies revealed the presence of *n*-alkanes ($C_{27}-C_{32}$ range) characteristic of terrestrial plant wax (Eglinton & Hamilton, 1967). However, a significant fraction of *n*-alkanes from the $C_{17}-C_{19}$ range also suggest algal matter as a source for part of the hydrocarbons (Schreiber, 1988).

The environmental conditions in the evaporitic lacustrine system were in all respects favourable for bitumen formation. Organic matter could accumulate in numerous anoxic traps (both deep and shallow, where it was subject to anaerobic bacterial maceration). Sonnenfeld (1987) has suggested that a decrease in the Mg ion concentration accompanying dolomite deposition could favour the dissociation of the alkanes from the brines, while the remaining LMC and gypsum could act as catalysts for homologue formation.

ENVIRONMENTS AND CLIMATE

Recent studies of various lacustrine systems demonstrate that the environmental interpretation based solely on the mineralogical composition of lake sediments may be problematic. Early diagenetic effects can mask the primary mineral phases (cf. Eugster & Hardie, 1978; Talbot & Kelts, 1986). Furthermore, syndepositional tectonics may generate sedimentary impacts similar to those that could have resulted from climatic fluctuations (cf. Anadón *et al.*, 1989).

Tectonic versus climatic fingerprints are easier to discern in marginal, shallow-water and subaerial facies, so we have therefore used these facies assemblages as monitors of general environmental changes of the basin. In Table 1 we summarize the main characteristics of the facies assemblages representing the evaporatic and non-evaporatic lacustrine systems developed within the Narbonne Basin. Because the evaporitic inner facies assemblage is associated with semi-arid sandflat−mudflat marginal facies, whereas the non-evaporitic inner facies assemblage is always accompanied by apparently wet, swamp marginal facies, we interpret climate as being the principal agent controlling the evolution of this lacustrine system. During late Oligocene to early Miocene the climatic regime changed from wet tropical (unit A) through semi-arid, subtropical (unit B) and then again to tropical conditions (unit C).

The stable isotope results document the twofold nature and match well the lithological alternations of the basin fill. Intense synsedimentary tectonism (Szulc *et al.*, in preparation) may have also affected the local drainage and thus short-term hydrochemical variations of the lake water, reflected by some of the geochemical properties of the basin fill.

CONCLUSIONS

The Narbonne Basin is a late Oligocene to early Miocene continental foreland basin of the north Pyrenean eastern thrust sheet. The basin fill comprises fanglomerates, lacustrine and palustrine carbonates, sandflat–mudflat fine-grained siliciclastics and gypsum. The lithofacies succession indicates fluctuation of the sedimentary environments between evaporitic and non-evaporitic lacustrine systems. Analyses of the overall facies systems arrangement and geochemical data suggest climatic factors to be a main controlling agent of the lacustrine system changes. Syndepositional tectonics was of minor significance and influenced local sedimentary processes. The geochemical study of carbonate samples illustrates the complex nature of the brine evolution. The study of the gypsum sediments from the basin fill reveals that most of these sediments originated from recycled sulphates of the Keuper.

Climatic conditions and basin topography favoured high organic productivity and bitumen formation during both humid and dry periods.

ACKNOWLEDGEMENTS

We are grateful to Drs S. Hałas and K. Różański for their assistance in the stable isotope measurements. We thank Drs S. Dżułyński and M. Gottis for valuable discussion and advice. Special acknowledgements are due to Drs E. Gierlowski-Kordesch, J. P. Calvo, P. Anadón, Ll. Cabrera, K. Kelts, R. Utrilla, and A. Vázquez for their critical and very constructive comments on the earlier draft.

REFERENCES

AGUILAR, J.P. (1977) Donnes nouvelles sur l'âge des formations lacustres des bassins de Narbonne — Sigean et Leucate (Aude) à l'aide des micromammifères. *Geobios* **10**, 643–645.

AGUILAR, J.P. & MICHAUX, J. (1977) Remarques sur la stratigraphie des terrains tertiaires des bassins de Narbonne — Sigean et Leucate (Aude). *Geobios* **10**, 647–649.

ANADÓN, P., CABRERA, L., JULIÀ, R., ROCA, E. & ROSELL, L. (1989) Lacustrine oil-shale basins in Tertiary grabens from NE Spain (Western European rift system). *Palaeogeogr. Palaeoclimatol. Palaeoecol.* **70**, 7–28.

ARTHAUD, F. & MATTAUER, M. (1972) Presentation d'une hypothèse sur la génèse de la virgation pyréneenne du Languedoc et sur la structure profonde du Golfe du Lion. *C.R. Acad. Sci. Ser. D* **274**, 524–527.

BAKER, P.A. & KASTNER, M. (1981) Constraints on the formation of sedimentary dolomite. *Science* **213**, 214–216.

CALVO, J.P., HOYOS, M. & GARCÍA del CURA, M.A. (1985) Mudmound structures in shallow lacustrine sediments, middle Miocene, Madrid Basin. *International Association of Sedimentologists, Abstracts of the Sixth European Regional Meeting*, pp. 539–542.

DE DECKKER, P. (1988) Biological and sedimentary facies of Australian salt lakes. In: *Paleolimnology* (Ed. Grey, J.) pp. 237–270. Elsevier, Amsterdam.

EGLINTON, G. & HAMILTON, R.J. (1967) Leaf epicuticular waxes. *Science* **156**, 1322–1335.

ELLENBERGER, F. (1980) La Zone Alpine du Bas-Languedoc. In: *Pyrénées et son avant-pays aquitan-languedocian* (Ed. Durang-Delga, M.). Cent. Rech. Exp. Prod. Elf Aquitaine **3**, 40–46.

EUGSTER, H.P. & HARDIE, L.A. (1978) Saline lakes. In: *Lakes: Chemistry, Geology, Physics* (Ed. Lerman, A.), pp. 237–293. Springer-Verlag, Berlin.

FUCHS, K. & HAUDE, H. (1987) Durchwurzelte Paläoboden in Miozän de beckens von Teruel — Ademuz (NE–Spanien). *Facies* **17**, 91–98.

GOTTIS, M. (1967) Sur certains aspects de l'architecture du Languedoc Mediterranéen. *Actes Soc. Linn. Bordeaux Ser. B* **104** (13), 3–10.

KNAUTH, L.P. (1979) A model for the origin of chert in limestones. *Geology* **7**, 274–277.

KROUSE, H.R. (1980) Sulphur isotopes in our environment. In: *Handbook of Environmental Isotope Geochemistry*, Vol. **1** (Eds Fritz, P. & Fontes, J. Ch.), pp. 435–471. Elsevier, New York.

MAGARITZ, M. (1987) A new explanation for cyclic deposition in marine evaporite basins: meteoric water input. *Chem. Geol.* **62**, 239–250.

MÜLLER, G. & IRION, G. (1969) Subaerial cementation and subsequent dolomitization of lacustrine carbonate muds and sands from Paleo-Tüz Golu ('Salt Lake'), Turkey. *Sedimentology* **12**, 193–204.

MÜLLER, G. & WAGNER, F. (1978) Holocene carbonate evolution in Lake Balaton (Hungary): a response to climate and impact of man. In: *Modern and Ancient lake Sediments* (Eds Matter, A. & Tucker, M.E.) Spec. Publ. Int. Assoc. Sediment. **2**, 57–81.

MURPHY, D.H. & WILKINSON, B.H. (1980) Carbonate deposition and facies distribution in a central Michigan marl lake. *Sedimentology* **27**, 123–136.

PUIGDEFÀBREGAS, C. & SOUQUET, P. (1986) Tectonosedimentary cycles and depositional sequences of the Mesozoic and Tertiary from the Pyrenees. *Tectonophysics* **129**, 173–203.

RIBA, O. (1976) Syntectonic unconformities of the Alto Cardener, Spanish Pyrenees: a genetic interpretation. *Sediment. Geol.* **15**, 213–233.

ROSEN, M.R., MISER, D.E. & WARREN, J.K. (1988) Sedimentology, mineralogy and isotopic analysis of Pellet Lake, Coorong region, South Australia. *Sedimentology* **35**, 105–122.

ROSSET, C. (1964) Les formation du bassin oligocène de Sigean-Portel (Aude) et leur chronologie. *C.R. Somm. Soc. Géol. Français* **10**, 415–417.

SCHREIBER, CH. (1988) Relations entre évaporites et accumulation d'hydrocarbures. In: *Evaporites et Hydrocarbures* (Ed. Busson, G.), Mem. Mus. Nat. Hist. Nat., Paris, Ser. C. **55**, 15–18.

SONNENFELD, P. (1987) Hydrocarbon prospects around Neogene evaporites. *Ann. Inst. Geol. Publ. Hung.* **70**, 625–634.

TALBOT, M.R. & KELTS, K. (1986) Primary and diagenetic carbonates in the anoxic sediments of Lake Bosumtwi, Ghana. *Geology* **14**, 912–916.

VIALLARD, P. (1987) Un model de charriage epiglyptique: la nappe des Corbières orientales (Aude, France). *Bull. Soc. Géol. France* **(8)** 3, 551–559.

VON DER BORCH, C.C. & LOCK, D.E. (1979) Geological significance of Coorong dolomites. *Sedimentology* **26**, 813–824.

WELLS, N.A. (1983) Carbonate deposition, physical limnology and environmentally controlled chert formation in Paleocene–Eocene Lake Flagstaff, Central Utah. *Sediment. Geol.* **35**, 263–296.

WHELAN, T. III & ROBERTS, H.H. (1973) Carbon isotope composition of diagenetic carbonate nodules from freshwater swamp sediment. *J. Sediment. Petrol.* **43**, 54–58.

Spec. Publs Int. Ass. Sediment. (1991) **13**, 291–311

Organic tissues in Tertiary lacustrine and palustrine rocks from the Jiyang and Pingyi rift depressions, Shandong Province, eastern China

E.I. ROBBINS*, ZHOU ZILI[†] *and* ZHOU ZHICHENG[‡]

US Geological Survey, Reston, VA, 22092, USA
[†]*Geological Research Institute of the Shengli Oilfield, Dongying, China*
[‡]*Nanjing Institute of Geology and Paleontology, Nanjing, China*

ABSTRACT

Organic tissues and acid-resistant minerals in Tertiary lacustrine and palustrine rocks from the Jiyang and Pingyi rift depressions in eastern China were analysed to determine if palynological techniques are useful where sedimentological techniques do not result in unique solutions. The study sites are two rifted, fault-block sub-basins of the North China Basin containing thick sequences of Lower Tertiary lacustrine rocks. The basins are so close to the sea today that periodic marine incursions could not be ruled out, but diagnostic marine fossils are lacking. Delicate dolomitic tubes thought to have been precipitated by marine algae, two seemingly similar coals, and an interbedded carbonaceous shale thought to have been deposited by an anoxic lake were analysed from the Shehejie Formation in the Jiyang Depression. 'Polychaetic' boundstone, thought to be the product of marine worms, carbonaceous micrite and an underlying terra rosa, and oncolites, were analysed from the Guanzhuang Formation in the Pingyi Depression.

Palynological analysis reveals that the algal filaments in dolomitic tubes are unlike modern marine algae, that the coals were produced by two different plant communities, and that the high pyrite content of the lower coal supports an influx of sulphate to the basin. The 'polychaetic' boundstone tufa contains plant cells; re-examination of a thin-section shows that the tubes were produced by reed-like plants having hollow stems and polygonal leaves. Tissues in the carbonaceous micrite and terra rosa suggest that the depositional regime was one of a wetland resting on a mottled and gleyed aquitard-forming soil. The oncolites are composed of masses of intertwined filaments, thereby supporting a biotic interpretation for their production. In all, the organic tissues show that lacustrine and palustrine conditions dominated the depositional regime in the sub-basins, but that periodic marine incursions cannot be ruled out as a source of sulphate.

INTRODUCTION

Palynology is often used for oilfield biostratigraphy, kerogen analysis, and palaeoecology; rarely, however, has it been applied to sedimentology. However, palynological studies, such as types of palynomorphs for biostratigraphy, types of kerogen for palaeoecologicai reconstructions, degradation of tissues for diagenetic changes, and changes in colour of tissues for depth of burial, may help to unravel problematic sedimentary sequences (Cushing, 1967; Staplin, 1977; Batten, 1982; Caratini *et al.*, 1983; Price, 1983). For example, in black carbonaceous shale that is seemingly uniform in the field, Robbins & Textoris (1988a) were able to differentiate marsh, swamp, and lake sedimentation on the basis of microscopic remains of herbaceous plant cells, woody plant cells, and zooplankton fecal pellets.

The North China rift basin in eastern China contains a thick sequence of Lower Tertiary rocks deposited primarily in lacustrine settings (Lee, 1989). The rocks in the Jiyang Depression (sub-basin) are in the subsurface in the Shengli oilfield and correlative rocks in the Pingyi Depression are at the surface (Shuai, 1988). The basin is so close to the sea today that periodic marine incursions could not be ruled out as a source of sulphate and invading marine organisms. Diagnostic marine fossils, however, are lacking and many sedimentary structures are ambiguous. It was thought that a palynological analysis of the organic remains within these controversial sequences might add useful microscopic information, some of which could then be reapplied in the field.

Palynologists who study organic components that are insoluble in HCl and HF use a variety of words interchangeably, primarily to reflect different analytical techniques and biases. In this paper, organic tissues — an ecological term — will be used instead of the more generally used terms organic matter, organic clasts, phytoclasts, macerals, or palynomorphs. Kerogen is the chemical term for the organic matter that is insoluble in HCl and HF, and in common organic solvents (Tissot & Welte, 1984).

Like all sediments, organic tissues are also clay, silt, and sand-size particles. Because of the limits of light microscopic equipment and the necessity for mounting particles under coverslips, palynological study focuses on the silt-size component of sedimentary rocks (Traverse, 1988). The remains of cyanobacteria, algae, fungi, plants, and microscopic animals typically occur with minerals in the silt-size fraction (Boulter & Riddick, 1986; Rudolf & Heroux, 1987). The clay-size fraction contains broken or fragmented remains of these tissues as well as sulphide minerals inferred to have been precipitated as the result of microbial metabolism. The analytical technique used to concentrate the organic tissues also preserves these and other minerals that can add such important information as direction of nutrient transport in the form of sulphur or minerals into the ecosystem in question (Robbins, 1987).

As rocks are buried, their organic tissues change in composition, colour, and reflectance because of the removal of gaseous and liquid components. Tissues that are transparent and pale yellow when fresh, become light brown as the rocks enter the temperature range 60−70°C and attain a reflectance value of around 0.6%, which defines the top of the petroleum window (Staplin, 1977). With heating, the organic remains become richer and richer in carbon, and change from medium brown to dark brown. The bottom of the petroleum window is defined as the point where all the tissues become black, which is within the temperature range of 110−130°C and with a reflectance value of around 1.2%. Thin-walled silt-size pollen grains are analysed

to observe the change in colour. Wood cells or vitrinite are analysed for reflectance changes.

This paper identifies the organic components in selected lacustrine and palustrine rocks from the Jiyang and Pingyi rift depressions and interprets palaeoecological aspects of the depositional environment. Palustrine is a general term for the wetland, swamp, mire, or marsh environment in which coal was deposited. Data about the weathering, diagenetic, and heating history of the tissues also are reported. Thin-section data are added wherever available.

GEOLOGICAL SETTING

Petroleum exploration has been the major impetus in the study of the sedimentary basins in China (Chen & Wu, 1979; Guan *et al.*, 1981; Chen *et al.*, 1982; Mao, 1983; Wan *et al.*, 1983; Wu, 1983,1986; Zhang *et al.*, 1985; Zhou & Du, 1986; Zhou & Lu, 1987; Lee & Masters, 1988). The major oilfields in eastern China lie between the Taihang-Shan highlands and the Tan-Lu fault system (Fig. 1). Rifting and basin tilting of the North China Basin, which lies between these two major structural elements, began with the Mesozoic Yanshanian orogeny (Lee, 1989). Reactivation during the Cenozoic resulted in a major episode of basin filling (Lee, 1989).

The North China Basin is composed of seven individual sub-basins, one of which is the Jiyang Depression (Lee & Masters, 1988; Lee, 1989). The Jiyang Depression is 25 000 km² in size and is filled with Mesozoic to Neogene age rocks (Fig. 2) (Shuai, 1988; Shuai *et al.*, 1988; Lee, 1989). The basement rocks of the Jiyang Depression consist of Precambrian metamorphic rocks and Palaeozoic carbonates. Block faulting in the Mesozoic was accompanied by volcanism and sedimentation; an erosional unconformity separates the underlying Cretaceous from the overlying Early Tertiary strata. Reactivation of the fault system resulted in lacustrine and fluvial deposition in the North China Basin and produced the sediments of the Kongdian Formation at the base, the Shahejie Formation in the middle, and the Dongying Formation at the top. Fluvial deposition during the Late Tertiary produced the Guantao Formation, which is overlain by the Minghuazhen Formation.

The major lacustrine unit is the Shahejie Formation, which is as much as 2750 m thick and Eocene to Oligocene in age (Shuai, 1988). The lower, or fourth,

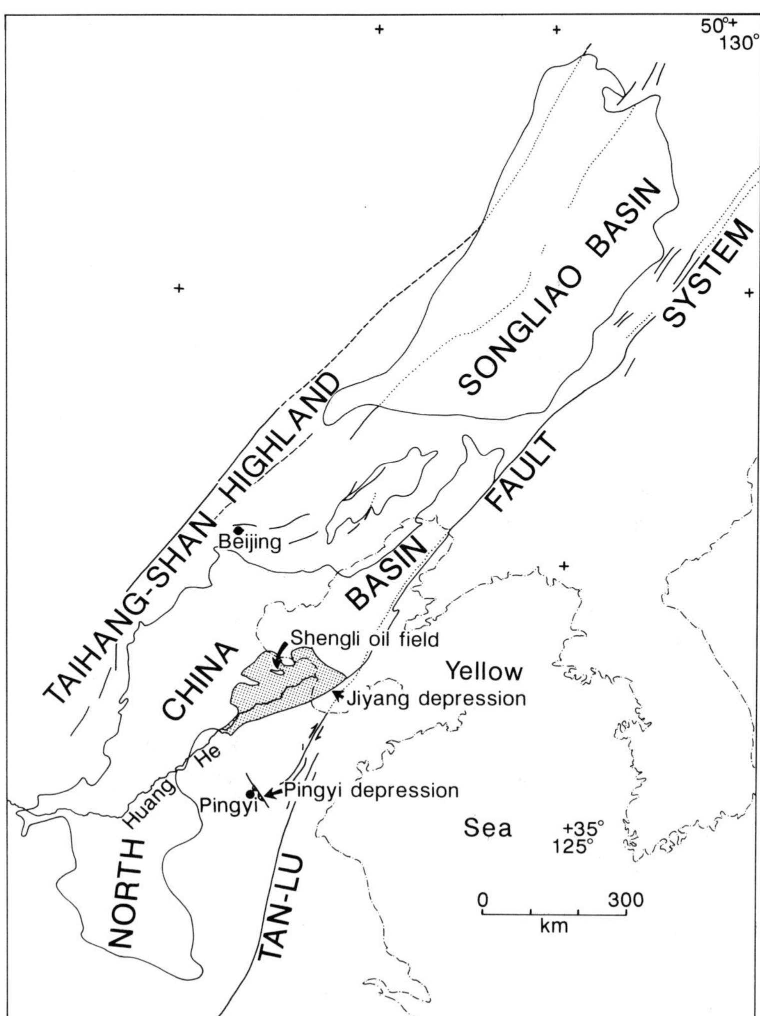

Fig. 1. Map showing major basins between the Taihang-Shan highlands and the Tan-Lu fault system and the location of Jiyang and Pingyi Depressions in eastern China (modified from Lee & Masters, 1988).

member contains anhydrite, halite, dolomite, and oil shale, which reflect an arid climate. The third member is a thick unit of dark mudstone and shale, which are the main oil source rocks in the depression. The second member contains coal and carbonaceous shale. The upper, or first, member is characterized by a dark mudstone intercalated with oil shale and dolomite.

The depositional environment of the Shahejie Formation is interpreted to be lacustrine on the basis of charophytes, ostracods, plant fossils, and unusual thin-walled dinoflagellates (Sung *et al.*, 1978) that are similar to those found in the Early Cretaceous lacustrine sequences in eastern China and western Canada (Lentin *et al.*, 1988). Beds

containing small foraminifers, dinoflagellates that were assigned to *Deflandrea* and *Lejeunia*, and glauconite are less common. The presence of these beds led Sung *et al.* (1978) and Wang *et al.* (1988) to suggest that there may have been intermittent connection to the sea during deposition of the upper part of the fourth member of the Shahejie Formation. However, support for this interpretation has been withdrawn on the basis of re-examination of the slides and transfer of all the dinoflagellate species to known or suspected lacustrine genera (Chen *et al.*, 1988; Lentin & Williams, 1989).

The Shengli oilfield in the Jiyang Depression is the second largest oilfield in China. This oilfield produces over 33 million tons of oil annually,

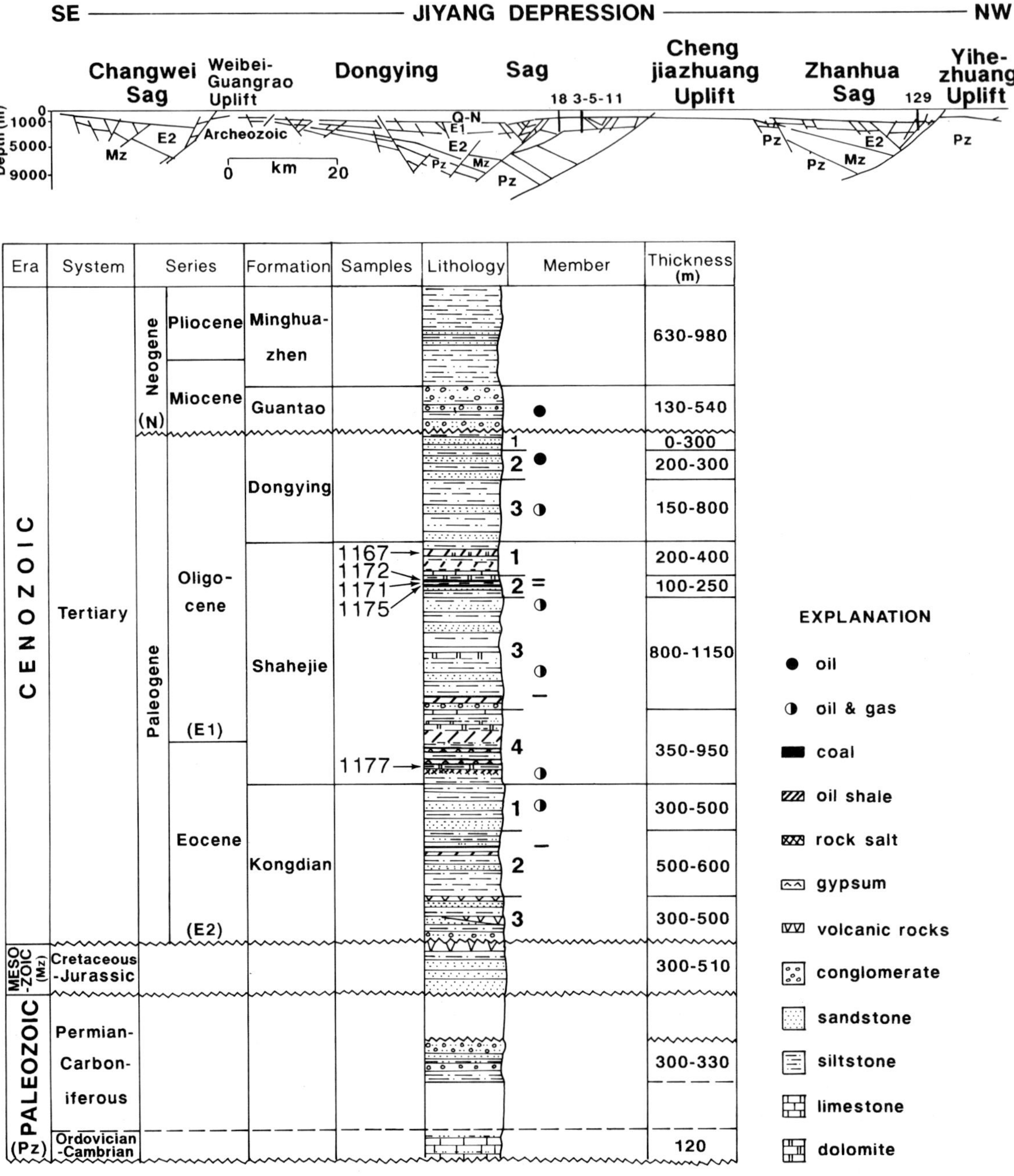

Fig. 2. Structural cross-section and composite well-log of the Jiyang Depression. Sample localities are noted (from Shuai, 1988; Lee, 1989).

primarily from Oligocene and Eocene lacustrine strata (Shuai, 1988). Potential resources may be around 10 billion tons of crude oil, and proven reserves are over three billion tons. The chemistry of the oils is unusual in that it has a preponderance of even over odd alkanes, biomarkers such as onoceranes and gammaceranes, immature pigments such as chlorins and red pigments, and Ni-bearing porphyrins (Shen *et al.*, 1980; Fu *et al.*, 1988; Wang *et al.*, 1988).

Lacking petroleum, the Pingyi Depression has not received as much study as the Jiyang Depression. The Pingyi Depression, which is 800 km² in size, lies southeast of the North China Basin (Fig. 1). It lies along one of many northwest trending faults, some of which offset and are oblique to the major north-northeast structural trend (Lee, 1989).

Sedimentological studies from outcrops show that the Pingyi Depression was filled with Jurassic, Cretaceous, and Eocene sedimentary and volcanic rocks (Gu, 1988a) (Fig. 3). The basement rocks of the Pingyi Depression consist of Precambrian metamorphic rocks and Cambrian and Ordovician limestones. The Jurassic section consists of fine to coarse-grained sedimentary rocks that are associated with plant fossils. Andesitic volcanism characterizes the Cretaceous. The Eocene section, which is cor-relative in age with rocks at depth in the Jiyang Depression, is represented by the Guanzhuang Formation, which is 2000 m thick and reflects al-luvial, fluvial, and lacustrine deposition in a rift setting.

The lower, or first, member of the Guanzhuang Formation in the Pingyi Depression is composed of sandstones and conglomerates that contain clasts of weathered andesite. The second member has a total thickness of about 435 m, and is composed primarily of lacustrine deposits having charophytes, oncolites, stromatolites, and plant fossils (Gu, 1988a). The upper, or third, member is characterized by alluvial fan conglomerates.

SAMPLING AND ANALYTICAL PROCEDURES

Cores and outcrops of the Shahejie and Guanzhuang Formations were sampled. Table 1 shows the age, formation, locality, and lithology of the rocks.

Rock samples consisted of organic-rich and organic-poor lithologies. Samples were washed and crushed to 0.5 cm or less, and 10 g were subjected to acid extraction procedures for kerogen (Robbins & Traverse, 1980). Gravity settling rather than centri-fugation was used during all procedures to keep fragile tissues from disintegrating. The rock pieces were leached by using 10 or 30% HCl to remove carbonate minerals, and by using 50% HF to remove silicate minerals. To maximize microscopic study and to approximate sizes used in sedimentary pet-rology, the residues from acid processing were sieved into three fractions. These fractions are less than 25 μm (fine or 'clay'), more than 25 μm and less than 125 μm (medium or 'silt'), and greater than 125 μm (coarse or 'sand'). The two smaller size fractions were affixed on to microscope slides by using glycerin jelly, scanned by using transmitted and reflected light microscopy, and photographed. Residues of the two coal samples and carbonaceous shale from the Shahejie Formation were further treated with Schulze's mixture (equal portions of saturated $KClO_3$ and concentrated HNO_3) for 2 h until the solution turned yellow. The solution was then washed out, and 10% KOH was added. When the solution turned a rich brown, the base was washed out. The residues were then sieved and mounted on microscope slides.

The organic matter and acid-resistant minerals were analysed by using standard palynological procedures (Traverse, 1988). Percentage recovery of insoluble residue was noted by comparing water-settled residues to a known volume. Such an analysis of volume percentage organic matter can be used to approximate weight percentage organic matter (Robbins *et al.*, 1988). The identity of organic tissues was determined by comparison with tissues from known organisms. Content, distribution, and crystal form of opaque sulphide (probably pyrite), oxide, and refractory silicate minerals were studied in acid residues.

The content of acid resistant organic and inor-ganic components in the rocks is shown in Fig. 4. Figures 5−13 show the diagnostic organic tissues in the rocks. Where measurements of tissues and minerals are given in the text as 3.5(7)24 μm, the first and last numbers are the size range and the number in parentheses is the median.

JIYANG DEPRESSION

Algal dolomite, coal, and interbedded carbonaceous shale in the Eocene and Oligocene Shahejie Forma-tion was made available from Shengli oilfield cores. The projected locations of the wells (3-5-11, Guan 18, and Yi 129) and samples are shown on Fig. 2. Table 1 provides more information about individual samples. Study of the organic matter along with the sedimentary structures in this core material is con-sidered by the Shengli Oil Corporation to be very

E.I. Robbins et al.

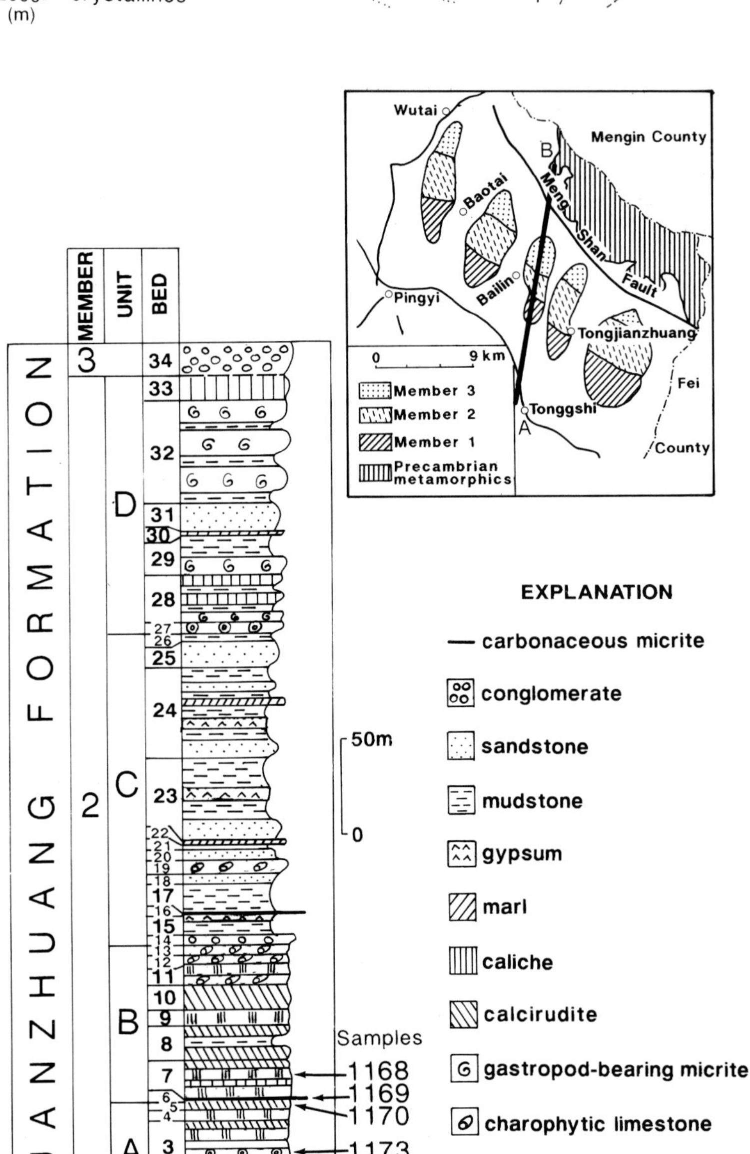

Fig. 3. Structural cross-section, locality map, and composite stratigraphic column of the Pingyi Basin. Sample localities are noted (partly from Gu, 1988a,b).

Table 1. Age, formation, locality, and lithology of rocks from the Jiyang and Pingyi rift depressions

Sample no.	Age	Formation	Locality	Lithology
Jiyang Depression, Shengli oilfield				
1167	Oligocene	Shahejie (1st member)	Well Guan 18 (1818.4−1820.5 m)	Medium−light-grey, massive, siphonaceous algal dolomite (oil stained)
1171	Oligocene	Shahejie (2nd member)	Well 3-5-11 (2271−2287 m)	Greyish-black laminated carbonaceous shale
1172	Oligocene	Shahejie (2nd member)	Well 3-5-11 (2271−2287 m)	Black massive coal
1175	Oligocene	Shahejie (2nd member)	Well 3-5-11 (2271−2287 m)	Greyish-black bedded coal
1177	Eocene	Shahejie (4th member)	Well Yi 129 (2476−2482 m)	Massive greyish-brown, siphonaceous algal dolomite (oil stained)
Pingyi Depression				
1168	Eocene	Guanzhuang 2nd member (unit B, bed 7)	Tongjianzhuang section	Greyish-yellow, massive, tube-bearing boundstone having upstanding polygonal structures
1169	Eocene	Guanzhuang 2nd member (unit B, bed 6)	Tongjianzhuang section	Medium-grey, massive, carbonaceous micrite
1170	Eocene	Guanzhuang 2nd member (unit B, bed 5)	Tongjianzhuang section	Pinkish-grey, massive, calcareous claystone having red specks
1173	Eocene	Guanzhuang 2nd member (unit A, bed 3)	Bailin section	Greyish-orange rounded oncolites

important for subsurface correlation and source potential.

Algal dolomite

Delicate, upright, tube-like structures composed of dolomite have been found in cores from a number of wells in the Shengli oilfield, and compose an oily dolomite unit, 2−10 m in thickness. The tubes are 5−10 mm in length and 0.1−0.25 mm in width (Shuai, 1988). On the basis of size, branching, and surface morphology of filaments, Zhu (1979) identified the tube structures in cleared acid residues as belonging to the siphonaceous marine algae (Order Siphonales).

Because siphonaceous algae are either uncommon, rarely reported, or rarely recognized in the fossil record, separating marine from non-marine deposition on the basis of such meagre fossil history is difficult. Other studies have interpreted the tubes to be worm tubes or stem casts (Shuai, 1988). The dominance of algal remains, the abundance of an unusual suite of ostracods, and the absence of any known marine fossils led Zhou & Du (1986) to suggest that the algae proliferated in the photic zone

of a lacustrine lagoon or algal marsh. It is not known as to whether such algae made or just lived in the tubes.

The unit is stained with oil and represents a reservoir rock that has as much as 22% porosity. Much of the porosity is growth framework, which is the result of both intra-framework and interframework variations (Zhou & Du, 1986). Peloidal grainstone interbedded in the dolomite unit has intergranular porosity, although much of this is occluded by cementation and hydrocarbon residue. It is unknown if the oil is being generated from the tissues or has migrated into the dolomite.

Observations

Samples of the algal dolomite facies were selected from the fourth member of the Shahejie Formation within the interval 2476−2482 m in Well Yi 129 (Eocene sample 1177) and from the first member of the Shahejie Formation in the interval 1818.4−1820.5 m in Well Guan 18 (Oligocene sample 1167).

Petrologically, the Oligocene dolomite shows clear millimetre-scale banding that consists of alternations of upright tubes in place and a calcarenite that is

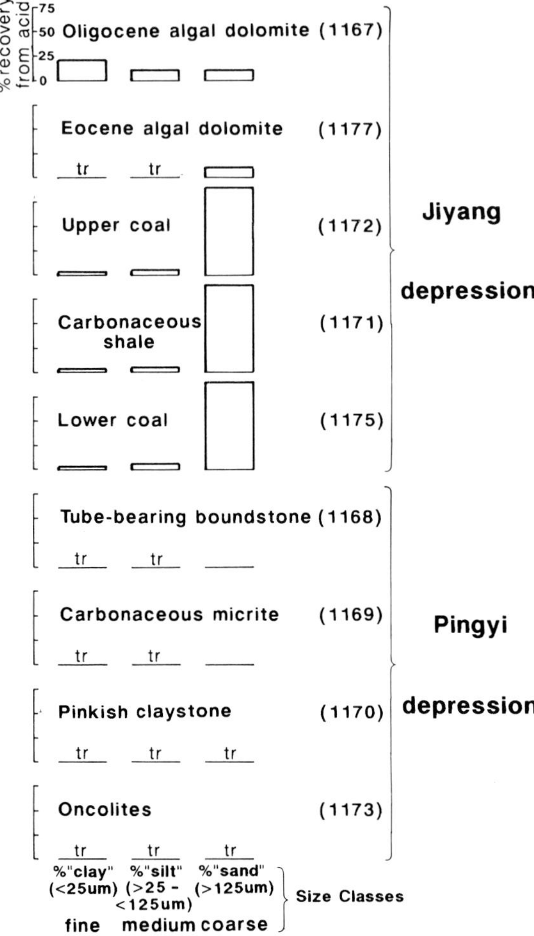

Fig. 4. Histograms showing percentage of clay, silt, and sand-size tissues that remained after acid treatment. Numbers in parentheses are sample numbers: (tr, trace).

formed of broken tubes, numerous ostracod valves, few gastropods, and very fine to fine-grained, sand-size peloids. The dolomitic tubes are filled with sparite or micrite; vugs between the tubes are lined or completely filled with sparry calcite (Fig. 5A).

The Eocene dolomite consists of concentric algal tubes that have been surrounded by radially oriented calcite spar similar to calcite crusts forming modern tufas (Irion & Müller, 1968; Risacher & Eugster, 1979). The calcite locally shows evidence of competitive growth; similar rims are formed around ostracod valves (Fig. 5B).

Even though the dolomitic tubes from the two cores appear to be similar in hand specimen, there are some differences between the organic and inorganic components in palynological preparations. In the Oligocene sample, 40% of the rock remained after acid processing, whereas only 10% remained from the more calcareous Eocene samples (Fig. 4). This difference indicates that the Oligocene sample has a higher content of organic matter. The silt and sand-size tissues in the Oligocene sample are usually long, sometimes branching, dark orange-brown, probably unsegmented filaments (Fig. 6A). Figure 6A shows that individual filaments are the same width as the holes in the tubes (Fig. 5A). The heads of some filaments are black (Fig. 6B), and budding, almost leaf-like structures are evident (Fig. 6A). Filaments appear to have an outer cuticle (Fig. 6C). The filaments are so sticky and oily that washing laboratory glassware after use was difficult.

These dark orange-brown filaments do not dominate the algal tissues in the Eocene sample. Furthermore, the Eocene algae may have had a vascular base (Fig. 7A); reproduction may have been by budding (Fig. 7D). The width of these filaments (Fig. 7B) is similar to the width of the dolomitic tubes (Fig. 5B). Small segmented filaments may be fungal in origin (Fig. 7B).

The silt and clay-size fractions of both samples are otherwise similar. Pollen and spores are rare, and the thin-walled pollen grains (Fig. 7E) are light brown. Oil globules (Fig. 7C) and rounded organic blebs that may be degraded oil globules (Fig. 6D) are present in both samples. Although most of the globules and blebs are not attached to other tissues, in one particular case, the non-descript tissue appears to be expelling oil (Fig. 7C). The Oligocene sample contains a few acritarchs that have fine, simple projections (Fig. 6E). Whereas the Eocene sample has no associated heavy minerals, the Oligocene sample has minor pyrite that occurs as particulate octahedrons and framboids. Rhombohedral holes in tissues show where the acid treatment eliminated carbonate minerals.

Interpretations

Acid residues support the sedimentological interpretation that the tube-like dolomitic structures were produced by algae. The abundant filamentous remains of algal tissues are similar in size to the openings in the tubes. Some of the small dispersed Eocene filaments are similar to *Praegongrosira* (Zhu, 1979) but none has formed colonial clumps characteristic of that genus. Other filaments in the Eocene

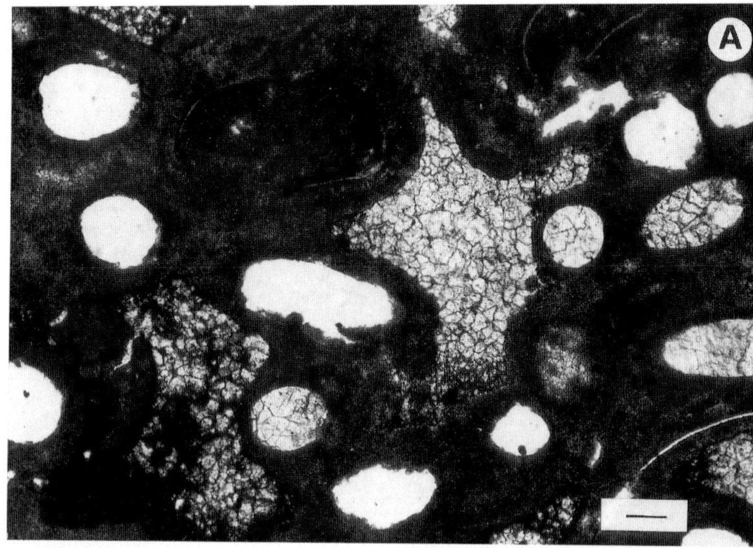

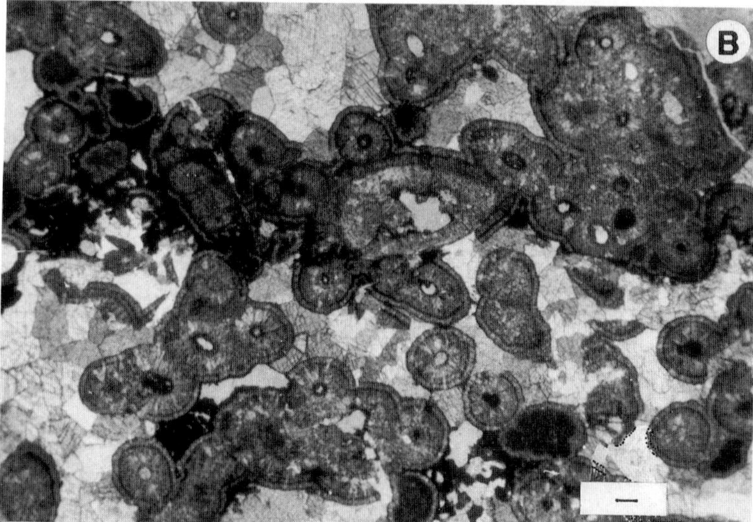

Fig. 5. Thin-section photomicrographs of siphonaceous algal dolomites. (A) Cross-sections of tubes in sample 1167. Some pores are filled with sparry calcite. (B) Algal dolomite tubes in Sample 1177 are coated with radially oriented sparite. Within the tubes are growth intra-framework pores, and between the tubes are growth interframework pores filled with sparry calcite. Scale bars are 10 μm.

samples resemble *Disparocladophora* (Zhu, 1979) in that they have elongate cells and sculptured walls, but individual cells do not intersect to form triangular basal cells characteristic of that genus. The algal remains in the Shahejie Formation are not whorled like modern Dasycladaceae or segmented in the manner of modern Codiaceae (Wray, 1977), both of which occur today in lagoonal and marine settings. More work is required on more samples to classify these remains to their proper taxon. These unusual algae may account for at least some of the unusual biomarkers.

Other components in the algal dolomite can be used to augment the known depositional and post-depositional history of the rocks. Pyrite is rare, which suggests that the environment may have been freshwater and low in sulphate rather than marine and high in sulphate (Williams & Keith, 1962). The algal tissues may be one source of the petroleum in the basin because tissues appear to be expelling oil. The rounded organic blebs that are the same size and shape as the oil globules may be crystallized or polymerized oil globules; thus there may be two populations of oil. The light brown colour of the

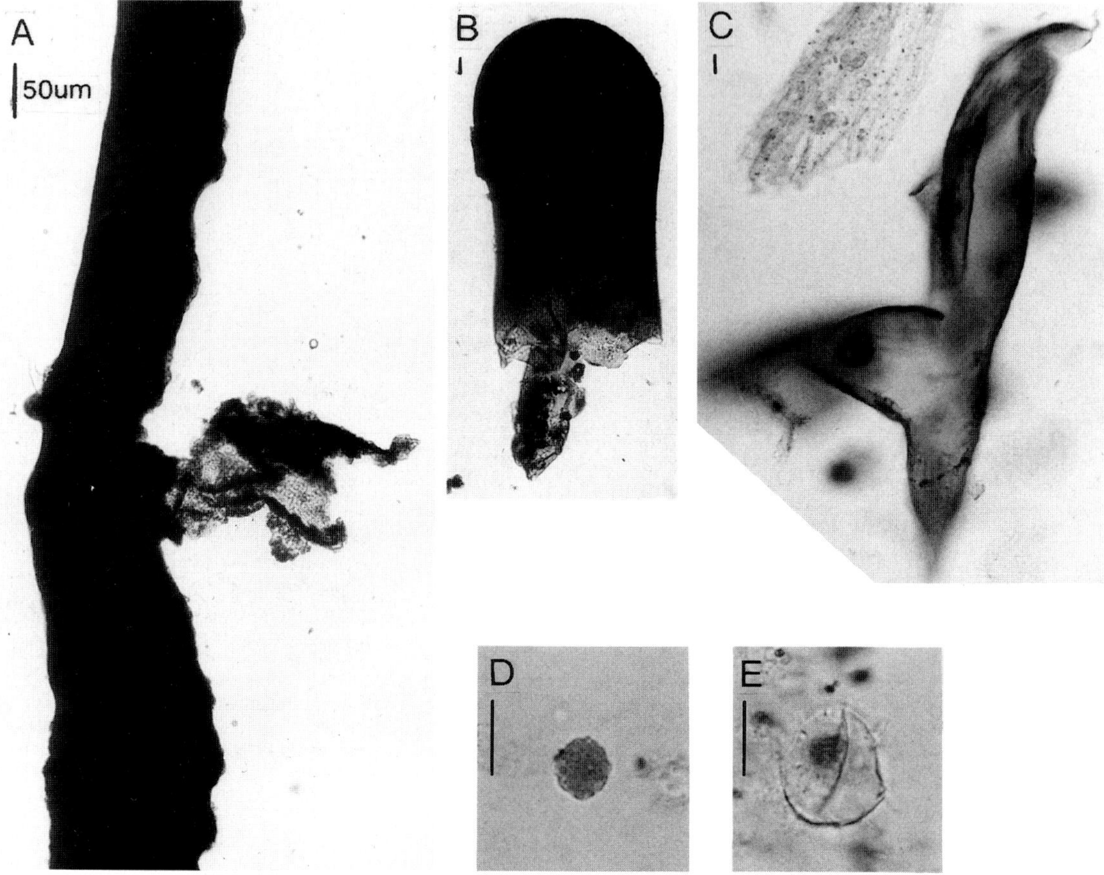

Fig. 6. Photomicrographs of organic tissues and minerals in Oligocene algal dolomite (sample 1167). (A) Algal filament showing bud structure. (B) Filament head. (C) Algal cuticle (dark particle). (D) Polymerized oil globule. (E) Spiny acritarch. Scale bars are 10 μm unless otherwise noted.

pollen grains shows that these rocks have been buried to the depths that mark the onset of the petroleum window.

The palynological observations on the softness of the algal tubes suggest a technique that can be applied in the field. Because a pointed object can be used to probe the softness of the tissues, the presence of oil-rich algal filaments can be noted upon collection.

Coal

In the Shengli oilfield, the Oligocene coal is too deep (greater than 2000 m) and too thin (less than 1 m) to mine (Shuai, 1988). Eocene and Palaeocene coal is mined in other parts of the North China Basin (Lee, 1989).

Although the coal beds had been interpreted as deposits from peat swamps (Shuai, 1988), they had never been studied petrographically or chemically. The sulphur content, rank, and nature of the vegetation that produced the coal were unknown. Another question that could be answered by the study of the organic tissues in the coal is the potential for sourcing gas or petroleum. Coal produced by herbaceous plants or trees may be a source of gas; coal rich in algal remains may be a source of petroleum (Murchison, 1987).

Observations

Grab samples from two coal beds in the Shahejie Formation were studied in a core from the interval 2271–2287 m in Well 3-5-11. The coal samples

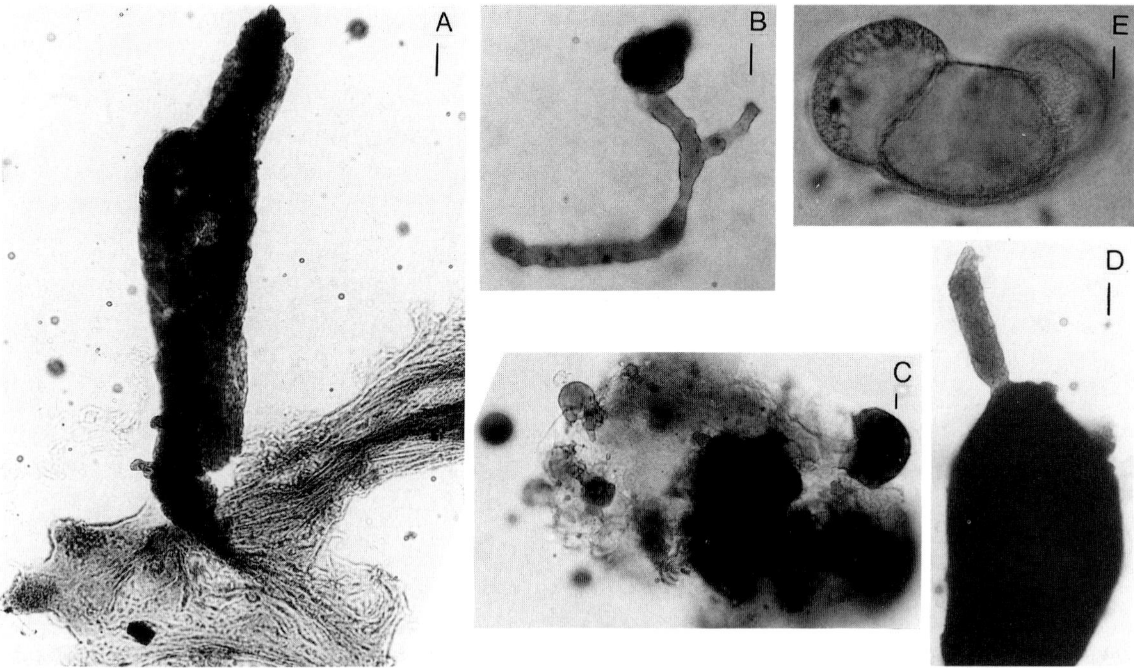

Fig. 7. Photomicrographs of organic tissues in Eocene algal dolomite (sample 1177). (A) Algal filament showing vascular-like base. (B) Segmented filament showing bud structure. (C) Organic clast that appears to be expelling oil. Note oil globule at left of photograph. (D) Algal filament showing budding structure. (E) Light-brown pollen grain of pine. Scale bars are 10 μm.

included an upper bed that is about 10 cm thick (sample 1172) and a lower bed that is about 50 cm thick (sample 1175).

Silt-size tissues in both beds are mainly coalified plant cells. Clearing with Schulze's mixture did not release many tissues in the upper coal, but in the lower coal, unusual wood cells having lignified pits are abundant (Fig. 9A). Dispersed wood cells are present in the upper coal (Fig. 8B), as well as leaf/stem cuticle of monocotyledonous plants, so classified because of the typical rectangular cells (Fig. 8C), and of dicotyledoneous plants, so classified because of the typical equant cells (Fig. 8D). Rare pollen and spores are dark yellow. Spores, sclerotia (Fig. 8E), and hyphae (Fig. 8F) of fungi are present in both coals. Whereas the sand-size fraction of the lower coal is composed of coalified woody tissues, the sand-size tissues in the upper coal have no identifiable morphologies. The clay fraction of both coals is composed of plant parts and pyrite.

In the upper coal, pyrite framboids are less than 1% of the residues and are small (3.5(7)24 μm) in size. Most are particulate, but some are enmeshed in the tissues. In the lower coal, two populations of pyrite (Fig. 9B) compose an estimated 7% by volume. Framboids are particulate, abundant, and 12(16)28 μm in size; octahedrons are particulate, rare, and 3.5(7)7 μm in size. No pyrite is enmeshed in the tissues.

Interpretations

The organic tissues show that woody vegetation probably dominated in the palustrine environment that produced the coal. Based on the abundance of cuticle, the younger wetland had more herbaceous vegetation than the older wetland. The wood cells in the lower coal are distinctive because they have lignified pits. The dark yellow colour of the pollen means that the coals are probably sub-bituminous in rank.

The lowermost coal of the Shahejie Formation has abundant pyrite. The framboids are relatively large and are not bound to the woody tissues, which

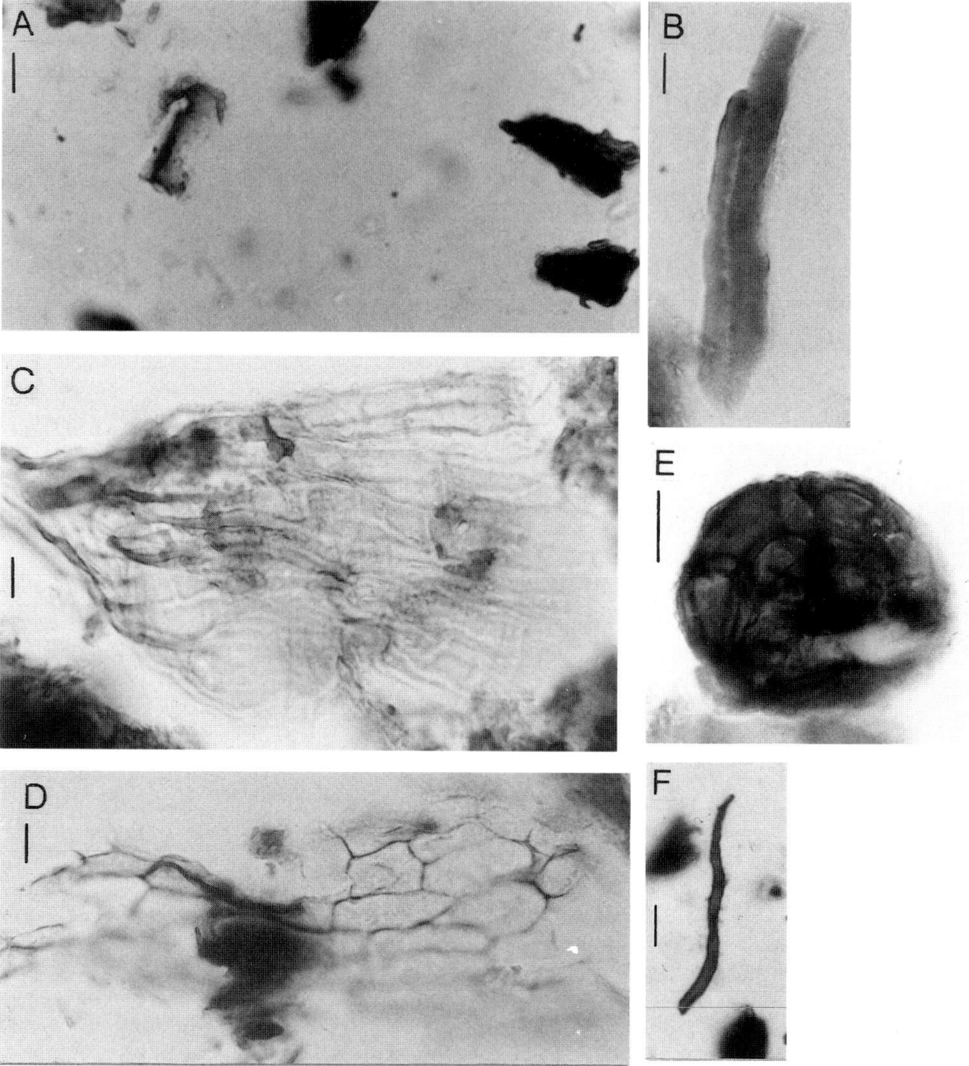

Fig. 8. Photomicrographs of organic tissues in upper coal (sample 1172). (A) Coalified wood cells. (B) Wood cell. (C) Leaf or stem cuticle of monocotyledenous plant. (D) Cuticle of dicotyledenous plant. (E) Fungal sclerotia. (F) Fungal hypha. Scale bars are 10 μm.

means the sulphur could be beneficiated from the coal. However, the coal is not economic because it is thin and occurs deep in the Jiyang Depression. A large rise in pyrite percentage is usually ascribed to marine incursions (Williams & Keith, 1962), but lakes surrounded by sulphur-bearing rocks can also have abundant pyrite (Robbins, 1987). The presence of abundant pyrite in the lower coal supports a sulphate incursion around the time of wetland deposition.

The palynological observation about the abundance of pyrite in the lower coal suggests an important observation can be noted in the field. The lower coal was noted to have a surficial white bloom. Because pyrite oxidizes to iron sulphate minerals that form yellow and white blooms, their presence in the field yields important information on the content of pyrite in a coal.

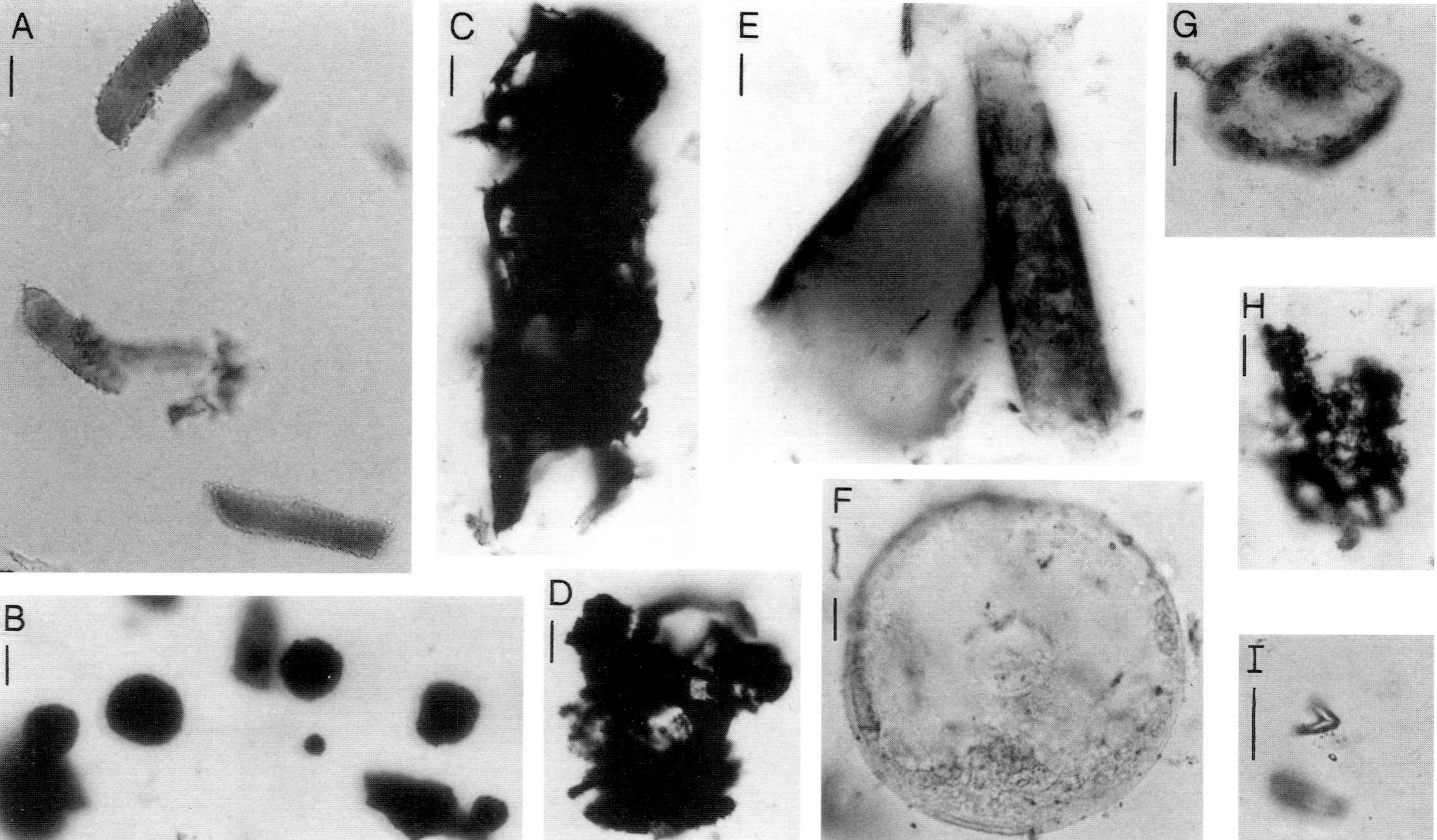

Fig. 9. Photomicrographs of organic tissues and minerals. Lower coal (sample 1175): (A) dispersed wood cells; (B) two size populations of pyrite. Carbonaceous micrite (sample 1169): (C) coalified wood cell; (D) rhombohedral holes in coalified plant cell; (E) algal cuticle; (F) organic-walled test of testicate amoeba. Pinkish micrite (sample 1170): (G) mineral grain coated with goethite; (H) rhombohedral holes in organic tissue; (I) rutile twin. Scale bars are 10 μm.

Carbonaceous shale

Between the two coal beds in the Shahejie Formation is a greyish-black laminated carbonaceous shale unit, 30–50 cm thick (Shuai, 1988). The unit is a good marker bed in the oilfield. Although the underlying and overlying coal beds probably were deposited in swamps, the depositional environment of this Oligocene shale was unknown. Carbonaceous shale is common in coal sequences, where its organic tissues may be derived from communities as diverse as non-woody herbaceous marsh vegetation that grew prior to colonization by coal-forming trees, or as the lacustrine product of algal mats. Another question that could be answered from the mineral suite in the shale concerns the direction of sediment transport, from onshore or offshore sources.

Observations

A sample of carbonaceous shale (sample 1171) was taken directly underlying coal sample 1172 in Well 3-5-11. The shale was primarily composed of organic matter (Fig. 4) rather than minerals. Tissues in the clay, silt, and sand fractions are primarily those of coalified plant cells (Fig. 10D). Clearing with Schulze's mixture revealed many limp plant cells (Fig. 10A), and leaf or stem cuticles. Some wood cells have the bordered pits typical of conifers (Fig. 10B). Pollen and spores are rare, and the palynomorphs are light brown (Fig. 10E). Some of the plant cells have strange pits; some adhere (Fig. 10C) and some detach. No dinoflaggelates or acritarchs are present. Pyrite is the only heavy mineral, and it occurs in trace amounts as octahedrons.

Interpretations

Plant cells dominate in the carbonaceous shale, which suggests that the environment producing the unit was a wetland. The cells are elongate but limp, and rigid cells are scarce, suggesting that the plants may have been of the herbaceous marsh variety and not of the woody swamp variety that dominated the overlying and underlying coals. The absence of plant remains and the lack of amorphous remains of the algal variety rules out a deep lacustrine depositional environment for the shale. The lack of heavy minerals suggests that the marsh may have been somewhat shielded from any upland sediment input.

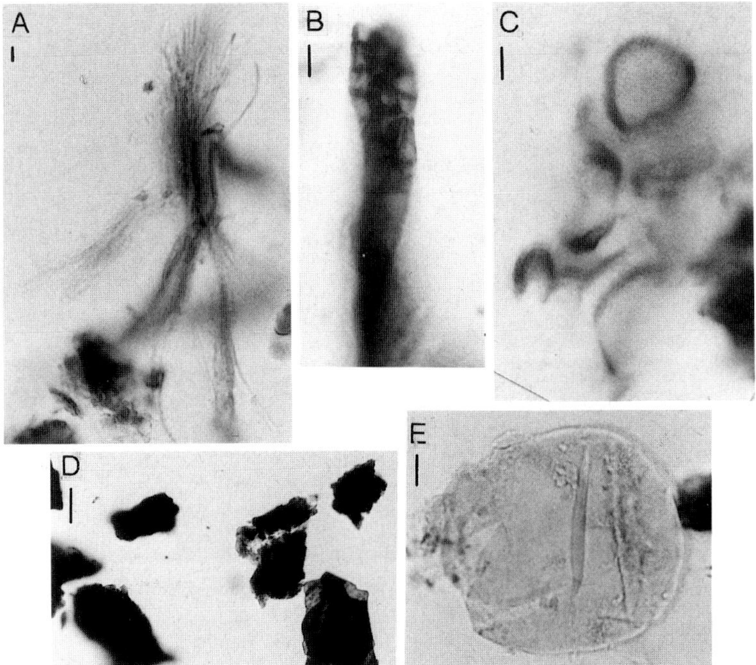

Fig. 10. Photomicrographs of organic tissues in carbonaceous shale (sample 1171). (A) Limp plant cell. (B) Conifer-type wood cell having bordered pits. (C) Radial pits of distinct but unidentified plant cell. (D) Coalified plant cells. (E) Light-brown unidentified palynomorph. Scale bars are 10 μm.

PINGYI DEPRESSION

Light-coloured calcareous rocks from the Eocene Guanzhuang Formation were collected from outcrops near Bailin and Tongjiazhuang in the Pingyi Depression (Fig. 3). Sample locations are shown on Fig. 3 and sample data are shown on Table 1. Such rocks are usually barren and therefore are rarely sampled for palynological analysis. The presence of any organic tissues in such lithologies was unexpected.

Tube-bearing boundstone

A greyish-yellow tufa crops out near Tongjiazhuang and is as much as 1 m thick. The boundstone is composed of micritic polygonal structures that stand upright and are filled with sparite (Gu, 1988a). These structures have been interpreted to be worm tubes, and the unit named 'polychaetic' boundstone (Gu, 1988a). Because the tubes are upright and parallel, they may alternatively be micrite-coated stem moulds of plants (Klappa, 1980). The study of the organic tissues might help to choose between these two different sedimentological models.

Observations

In thin-section, a central cylinder filled with a sparry cement fabric (Fig. 11B) is surrounded by curved trapezoidal structures (Fig. 11A). The trapezoidal structures are bounded by a thin micritic wall con-

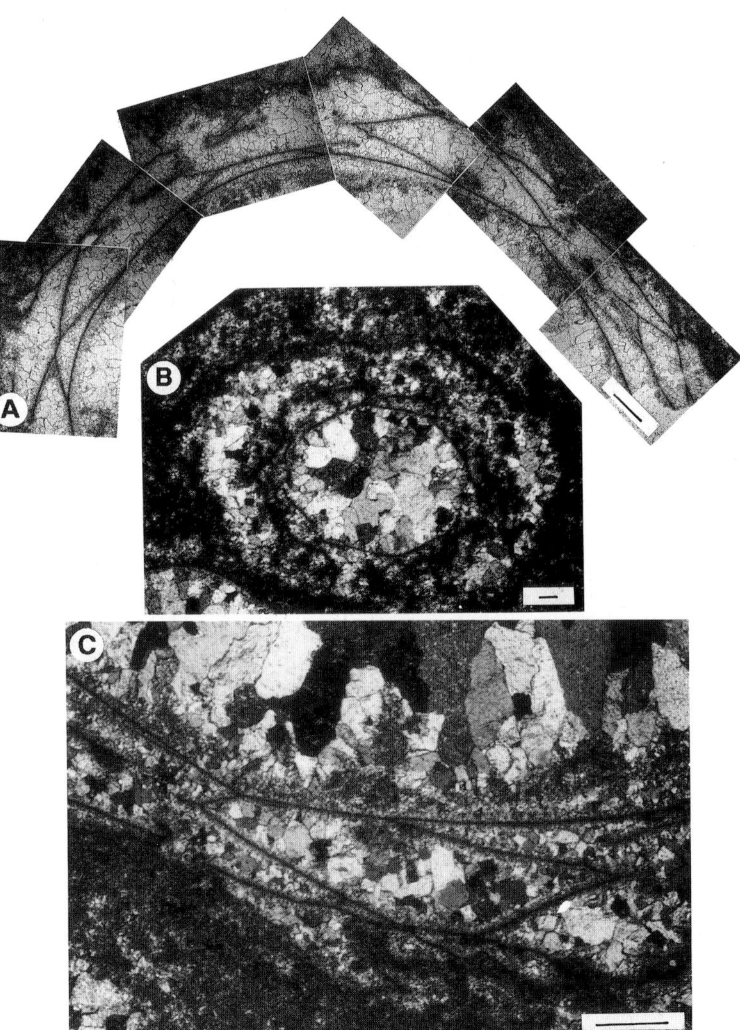

Fig. 11. Thin-section photomicrographs of tubes in calcareous boundstone (sample 1168). (A) Photographic mosaic of trapezoidal leaf cross-sections around central stem. (B) Cross-section of circular central stem surrounded by rosette of leaves. Second stem at bottom left corner of photograph. (C) Types of cements: radial alignment of sparite in plant stem, micrite along leaf margins, and sparite within leaves. Scale bars are 100 μm.

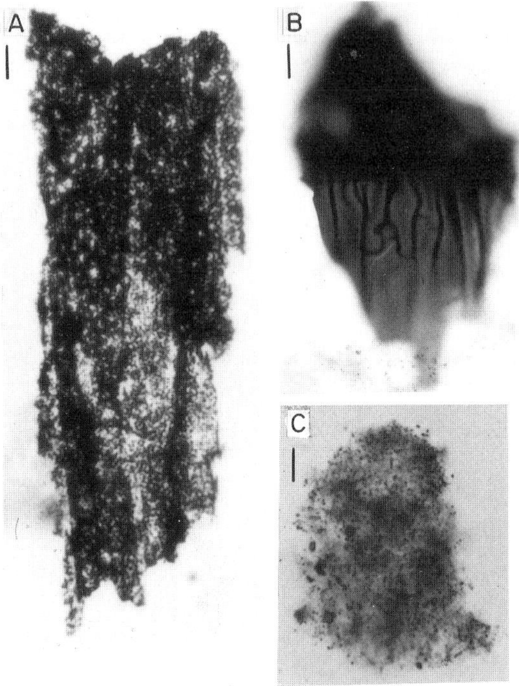

Fig. 12. Photomicrographs of organic tissues in tube-bearing boundstone (sample 1168). (A) Degraded plant cells. (B) Fungal mass (apothecium?). (C) Amorphous particle that resembles degraded fecal pellet. Scale bars are 10 μm.

taining organic material filled with similar sparry cement (Fig. 11C). The sparite centres average 2−3 mm in diameter. The polygonal structures average 5−6 mm in diameter and range in length from 3 to 6 cm. Micrite also fills the spaces between these discrete structures.

The tube-bearing boundstone (sample 1168) was highly calcareous and only trace amounts of clay and silt-size organic and mineral components remain after acid processing (Fig. 4). Minerals form the most abundant component; greyish-black, degraded plant cells are the dominant organic tissues in the clay and silt fractions (Fig. 12A). A minor amount of greyish amorphous organic matter also is present (Fig. 12C), along with thick tissue that probably is fungal in origin (Fig. 12B), so characterized because of its glassy or hyaline nature and the clamping type of cell arrangement. Pollen and spores in the unit are rare, transparent, and probably modern contaminants. Heavy minerals consist of rutile twins

and needles, tourmaline, zircon, an unidentified stubby mineral, and haematite; no pyrite was present.

Interpretations

The tissues in the tube-bearing boundstone are composed of plant cells rather than worm parts; this composition supports the stem mould interpretation. Furthermore, thin-sectioning shows that the stem of the reed-like plant, now filled with sparite, was originally round and hollow and therefore probably once filled with pith. The trapezoidal structures around the stem are the cross-sections of leaves that were basally inserted in a rosette fashion. This can be envisioned as the slice off the bottom of celery or bok choy. The waxy outer cuticles of the stem and leaves can be seen vaguely in thin-section and account for the degraded tissues seen in acid residues. These tissues have the greyish colour typical of oxidation (Masran, 1984), and the holes in the cells are consistent with fungal degradation rather than surficial weathering. The scarce amorphous component has the size, shape, and tapering terminations that resemble fecal pellets. The most likely depositional environment for this type of ecological association is a marshy wetland composed of herbaceous plants, such as rushes, sedges, and grasses.

The stem and basal leaves are suitably large and distinct in thin-section that this facies probably can be identified in the field. Identification will be facilitated if a hand specimen of the unit is cleanly broken perpendicular to the polygonal structures.

Carbonaceous micrite

In the Tongjianzhuang section, a medium grey, massive, carbonaceous micrite overlies a unit interpreted to be a terra rosa (Gu, 1988a). The depositional environment of the carbonaceous micrite and its relation to the underlying unit was unknown; the sedimentology of both might be augmented by the analysis of acid-resistant organic remains and minerals.

Observations

The carbonaceous micrite (sample 1169) was highly calcareous, and only a trace amount of clay and silt-size organic and mineral components remained after acid processing (Fig. 4). Rhombohedral holes in the tissues reflect dolomite crystal growth (Fig. 9D). The few tissues in the clay and silt fractions are predominantly of wood cells (Fig. 9C). Cuticle of siphonaceous algae also is present, so identified on the basis of size, thickness, and dark colour (Fig. 9E). Pollen and spores are absent. Although fungal spores are present, their unblemished surfaces suggest that they may be modern soil contaminants. An unusual fossil, an organic-walled test of a testicate amoeba (Fig. 9F), is present in the carbonaceous micrite. It is of an arcellid (Lee *et al.*, 1985), which is typical of wetland environments. Rutile is the only heavy mineral present.

Interpretations

The carbonaceous micrite contains plant cells, algal remains, a testicate amoeba, and abundant heavy minerals, but no pyrite. Such an association is typical of a wetland (Robbins & Textoris, 1988b). The plant cells are not distinctive enough to determine their origin. The enmeshed nature of the dolomite suggests that the organic tissues could have served as the source of carbon in the carbonate.

Pinkish-grey calcareous claystone

Underlying the carbonaceous micrite at the Tongjianzhuang section is a mottled pinkish-grey calcareous claystone that has been interpreted as a terra rosa, an old soil horizon (Gu, 1988a). The problems that could be approached by study of the acid-resistant remains were twofold. The distinctive red specks in the micrite could be weathered pyrite. The remains might clarify the genetic relationship between the overlying unit and the claystone.

Observations

The pinkish-grey claystone (sample 1170) was highly calcareous, and only a trace amount of clay and silt-size organic and heavy mineral components remained after acid processing (Fig. 4). Plant cells are present; their presence is unusual for a rock that was supposed to have been subjected to such intense weathering. However, these organic tissues are extremely degraded (Fig. 9H) and have the greyish cast that can be ascribed to oxidation. No pollen or spores are present. Pyrite is not present, and the heavy minerals include twins (Fig. 9I), needles, and blades of rutile, zircon, and a blue amphibole. Goethite occurs both as particulate grains and as coats on other mineral grains (Fig. 9G); the coatings preserved the minerals from acid dissolution.

Interpretations

The pinkish-grey claystone has few organic clasts, but the clasts that remain have rhombohedral holes that are characteristic of post-depositional recrystallization of dolomite. Heavy minerals dominate the acid residues. The scarcity of tissues, the dominance of heavy minerals, and the position of the unit below a carbonaceous micrite can be used to postulate that the terra rosa probably was a clay-rich, mottled and gleyed soil horizon that served as an aquitard or confining layer for the overlying wetland (Fanning & Fanning, 1989).

Oncolites

Greyish-orange oncolites are abundant in bedded calcareous rocks that make up unit A of member 2 of the Guanzhuang Formation at the Bailin section. The oncolite-bearing unit, of Eocene age, is 2.5 m thick at the outcrop (Gu, 1988a).

Organic matter had not been studied in association with these oncolites. Such an analysis of the acid-resistant components might reveal the biotic nature of microbial communities that may have precipitated or colonized the structures.

Observations

Oncolites that weathered directly out of the surrounding rocks were hand selected from the ground (sample 1173) because they resembled those in Onondaga Lake in New York (Dean & Eggleston, 1984). The weathered and unweathered oncolites are indistinguishable; they are concentrically layered and vary in size from 1 to 3 cm. The weathered ones are not very dense. Thin-sections of unweathered oncolites reveal sparite to be the dominant cement.

The oncolites are highly calcareous, and only a trace amount of clay and silt-sized components remain after acid processing (Fig. 4). Minerals dominate the residues and transparent lichen algal— fungal associates dominate the organic tissues. Because these lichen were probably surface contaminants, the processing technique was adjusted. A set of oncolites was bleached in commercial laundry bleach for 1 h, washed, and then processed through 10% HCl for 5 minutes to eliminate any surface contaminants. The oncolites were then washed clean, the residuum was decanted, and fresh 10% acid was added for 20 h. The lichen components were removed by this additional treatment. Intertwined, colourless masses of filaments were revealed in the residues. Segmented and unsegmented filaments, many greater than 100 μm long (Fig. 13A,B), are similar in size, shape, and habit to segmented and unsegmented filaments in the Onondaga Lake oncolites (Fig. 13C,D). Zircon, rutile, and a variety of unidentified resistant minerals also were present.

Interpretations

The organic component of the oncolites is dominated by colourless filamentous tissues, which show that the sediments of the Pingyi Depression have never been buried. The filaments resemble, in size and shape, cyanobacteria and algae in recent oncolites in Onondaga Lake. The Onondaga Lake oncolites form from biotic processes in alkaline, nearshore waters, 1 m or less in depth, where they are rolled by storms and waves (Dean & Eggleston, 1984). Such a depositional environment is suggested for the precipitation of the Eocene oncolites, although those from other environments were not available for comparison.

The microbial nature of the oncolites can be analysed in the field with 10% HCl. Most of the carbonate matrix of a crushed oncolite will dissolve in about 30 minutes. The intertwined nature of the filamentous microbes is obvious to the naked eye.

CONCLUSIONS

The study of organic tissues and their associated minerals, supports, augments, and suggests other avenues of sedimentological research in the Jiyang and Pingyi rift depressions of eastern China. In the Shahejie Formation, sedimentological analysis determined that the tubes were algal; the organic tissues support this interpretation. The coal beds were undoubtedly the result of deposition in swamps, but the organic tissues extend the interpretation to

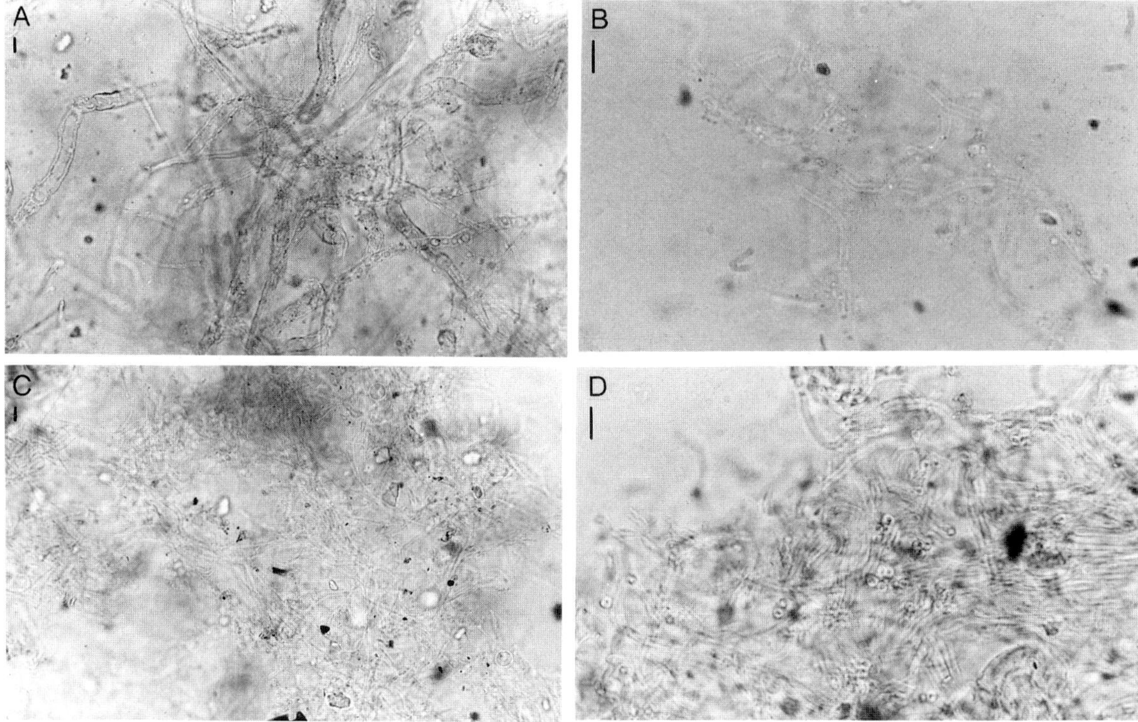

Fig. 13. Photomicrographs of microbial filaments in Eocene (sample 1173) and modern oncolites. Eocene oncolites: (A) wide filaments; (B) narrow filaments having tubular openings. Recent oncolites from Onondaga Lake, New York: (C) wide, predominantly algal filaments; (D) narrow, filamentous cyanobacterial sheaths having tubular openings. Scale bars are 10 μm.

include types of herbaceous and woody vegetation in the wetlands. The surficial weathering blooms on the lower coal suggest a high sulphur content, and abundant pyrite in the residues supports this interpretation. The source of sulphur may well have been an incursion of sea water, but no other evidence is present for marine conditions. Carbonaceous shale is generally interpreted as the remains of anoxic lakes, but abundant plant remains show the depositional regime involved herbaceous vegetation in a wetland.

In the Guanzhuang Formation, sedimentary structures interpreted to be the products of polychaetic worms were, instead, stem moulds of reed-like plants. Abundant, entwined, filamentous masses support the interpretation of oncolites as microbial products. The carbonaceous micrite, which had not been analysed before, contained wood cells, algal remains, and a testicate amoeba test, which support a palustrine depositional regime. Rhombohedral holes in the tissues suggest that the organic matter may have served as a carbon source for the carbonate. The unit that was interpreted to be a terra rosa has degraded plant tissues and mineral grains coated with goethite, supporting the sedimentological analysis. This clay-rich horizon can be interpreted as an aquitard for the overlying wetland and a product of a mottled and gleyed wetland soil.

The organic and inorganic remains suggest that although the Tertiary basins may have been near the ocean, lacustrine and palustrine processes and organisms dominated the sedimentary sequences that were deposited in the Jiyang and Pingyi rift depressions. Sedimentology coupled with palynological analysis thus produces a very powerful tool when confronted with controversial sequences.

ACKNOWLEDGEMENTS

Robbins' field work was funded by an IGCP Project 219 grant. This paper is the result of numerous discussions at the outcrops with members of Project 219, including A.S. Cohen and E. Gierlowski-Kordesch (who suggested the 'polychaetic' tubes may be plant stems that trapped carbonate mud), K. Kelts, B.F. Long, P.O. Roehl, Shuai Defu, Tang Tianfu, J.T. Teller, D.A. Textoris, P. Timofeev, and R. Yuretich. The reviews of J.P. Smoot, J.R. Eggleston, C. Poumot, and an anonymous reviewer strengthened the organization and interpretations.

REFERENCES

BATTEN, D.J. (1982) Palynofacies and salinity in the Purbeck and Wealden of southern England. In: *Aspects of Micropalaeontology* (Ed. Banner, F.T. & Lord, A.R.), pp. 278–308. Allen & Unwin, London.

BOULTER, M.C. & RIDDICK, A. (1986) Classification and analysis of palynodebris from the Palaeocene sediments of the Forties Field. *Sedimentology* **33**, 871–886.

CARATINI, C., BELLET, J. & TISSOT, C. (1983) Les palynofacies; representation graphique, interpret de leur étude pour les reconstitutions paleogeographiques. In: *Géochemie Organique des Sediments Marins d'Orgon a Misedor*, pp. 327–352. Centre National Recherche Scientifique, Paris.

CHEN MU & WU BAOLING (1979) On the discovery of lower Tertiary polychaetous tubes from the Jiyang district of Shandong Province, China. *Acta Oceanol. Sin.* **1**, 339–341 (in Chinese).

CHEN SHAOZHOU, GAO XINCHEN & QIU DONGZHOU (1982) A preliminary study on China's Eocene transitional facies. *Oil Gas Geol.* **3**, 343–350 (in Chinese with English abstr.).

CHEN, Y.Y., HARLAND, R., STOVER, L.E. & WILLIAMS, G.L. (1988) Fossil dinoflagellate taxa by Chinese authors, 1978–1984. *Geol. Surv. Can., Can. Tech. Rep. Hydrogr. Ocean Sci.* **103**, 40 pp.

CUSHING, E.J. (1967) Evidence for differential pollen preservation in Late Quaternary sediments in Minnesota. *Rev. Palaeobot. Palynol.* **4**, 87–101.

DEAN, W.E. & EGGLESTON, J.R. (1984) Freshwater oncolites created by industrial pollution. *Sediment. Geol.* **40**, 217–232.

FANNING, D.S. & FANNING, M.C.B. (1989) *Soil: Morphology, Genesis, and Classification*. Wiley, New York, 395 pp.

FU JIAMO, SHENG GUOYING & LIU DEHAN (1988) Organic geochemical characteristics of major types of terrestrial petroleum source rocks in China. In: *Lacustrine Petroleum Source Rocks* (Eds Fleet, A.J., Kelts, K. & Talbot, M.R.), Geol. Soc. London Spec. Publ. **40**, 279–289.

GU CHENGGAO (1988a) B3-3, an excursion guide to the Pingyi Basin, Shandong. In: *Tertiary Lacustrine Deposits, Shandong Province* (Eds Shuai Defu, Lin Qubin, Wang Weiming & Gu Chenggao), pp. 29–42. Guide Book Excursion B3, International Association of Sedimentologists, International Symposium on Sedimentology Related to Mineral Deposits, Beijing, China.

GU CHENGGAO (1988b) Characteristics and significance of Lower Tertiary caliches in the Pingyi basin, Shandong Province. *Acta Sediment. Sin.* **6**, 90 (in Chinese with English abstract).

GUAN SHICONG, WANG SHENG, ZHANG SHAOWEI, *et al.* (1981) A discussion on some questions in the study of Meso-Cenozoic continental hydrocarbon-bearing basins of China. *Oil Gas Geol.*, **2**, 314–320 (in Chinese with English abstract).

IRION, G. & MÜLLER, G. (1968) Mineralogy, petrology and chemical composition of some calcareous tufa from the Schwaebische Alb, Germany. In: *Carbonate Sedimentology in Europe* (Eds Muller, G. & Friedman, G.), pp. 157–171. Springer-Verlag, New York.

KLAPPA, C.E. (1980) Rhizoliths in terrestrial carbonates: classification, recognition, genesis, and significance. *Sedimentology* **27**, 613–629.

LEE, J.J., HUTNER, S. & BOVEE, E. (1985) *Illustrated Guide to the Protozoa*. Allen Press & Society of Protozoology, Lawrence, Kansas, 629 pp.

LEE, K.Y. (1989) Geology of petroleum and coal deposits in the North China basin, eastern China. *U.S. Geol. Surv. Bull.* **1871**, 36 pp.

LEE, K.Y. & MASTERS, C.D. (1988) Geologic framework, petroleum potential, and field locations of the sedimentary basins in China. *U.S. Geol. Surv. Misc. Inv. Map I-1952*.

LENTIN, J.K. & WILLIAMS, G.L. (1989) Fossil dinoflagellates: index to genera and species, 1989 edition. *Can. Geol. Surv., Contrib. Ser.*, 473 pp.

LENTIN, J.K., MAO SHAOZHI & YU JIANGZIAN (1988) Early Cretaceous non-marine dinoflagellates from east China. *Abstracts with Programs, American Association of Stratigraphic Palynologists Meeting*, Houston, Texas, p. 48.

MAO XIULAN (1983) Habitat of Gastropoda in 2nd Member of Shahejie Formation in northern part of Dongying depression. *Acta Petrol. Sin.* **4**, 29–35.

MASRAN, T.C. (1984) Sedimentary organic matter of Jurassic and Cretaceous samples from North Atlantic Deep-Sea Drilling sites. *Bull. Can. Petrol. Geol.* **32**, 52–73.

MURCHISON, D.G. (1987) Recent advances in organic petrology and organic geochemistry; an overview with some reference to 'oil from coal'. In: *Coal and Coal-Bearing Strata: Recent Advances* (Ed. Scott, A.C.), Geol. Soc. London Spec. Pub. **32**, 257–302.

PRICE, P.L. (1983) A Permian palynostratigraphy for Queensland. In: *Permian Geology of Queensland*, pp. 155–211. Geological Society of Australia, Queensland Division, Brisbane.

RISACHER, F. & EUGSTER, H.P. (1979) Holocene pisoliths and encrustations associated with spring-fed surface pools, Pastos Grandes, Bolivia. *Sedimentology* **26**, 253–270.

ROBBINS, E.I. (1987) Paleoecology. In: *Geological Investigations of the Vermillion Creek Coal Bed in the Eocene Niland Tongue of the Wasatch Formation* (Ed. Roehler, H.W.), U.S. Geol. Surv. Prof. Pap. **1134**, 75–104.

ROBBINS, E.I. & TEXTORIS, D.A. (1988a) Analysis of kerogen and biostratigraphy of core from the Dummit–Palmer No. 1 well, North Carolina. *U.S. Geol. Surv. Open File Rep.* **88–670**, 15 pp.

ROBBINS, E.I. & TEXTORIS, D.A. (1988b) Origin of Late Triassic coal in the Deep River basin of North Carolina (USA) (abstr.). *International Association of Sedimentologists, Abstracts of International Symposium on Sedimentology Related to Mineral Deposits*, Beijing, China, pp. 219–220.

ROBBINS, E.I. & TRAVERSE, A. (1980) Degraded palynomorphs from the Dan River (North Carolina)–Danville (Virginia) basin. In: *Geological Investigations of Piedmont and Triassic Rocks, Central North Carolina and Virginia* (Eds Price, V., Jr., Thayer, P.A. & Ranson, W.A.). Carolina Geological Society Field Trip Guidebook, Danville, Virginia, pp. X1–X11.

ROBBINS, E.I., WILKES, G.P. & TEXTORIS, D.A. (1988)

Coal deposits of the Newark rift system. In: *Triassic–Jurassic Rifting, Continental Breakup and the Origin of the Atlantic Ocean and Passive Margins* (Ed. Manspeizer, W.). Dev. Geotect. **22**, 649–682. Elsevier, Amsterdam.

RUDOLF, B. & HEROUX, Y. (1987) Chitinozoan, graptolite, and scolecodont reflectance as an alternate to vitrinite and pyrobitumen reflectance in Ordovician and Silurian strata, Anticosti Island, Quebec, Canada. *Bull. Am. Assoc. Petrol. Geol.* **71**, 951–957.

SHEN GUOYING, FAN SHANFA, LIN DEHAN, SU NENGZIAN & ZHOU HONGMING (1980) The geochemistry of n-alkanes with an even-odd predominance in the Tertiary Shahejie Formation of northern China. In: *Advances in Organic Chemistry 1979* (Eds Douglas, A.G. & Maxwell, J.R.), pp. 115–122. Pergamon, Oxford.

SHUAI DEFU (1988) B3-1 core workshop of Tertiary lacustrine deposits, Shengli Oilfield. In: *Tertiary Lacustrine Deposits, Shandong Province* (Eds Shuai Defu, Lin Qubin, Wang Weiming & Gu Chenggao), pp. 1–21. Guide Book Excursion B3, International Association of Sedimentologists, International Symposium on Sedimentology Related to Mineral Deposits, Beijing, China.

SHUAI DEFU, QIAN KAI, SONG YONGSHEN & GE RONG (1988) Stratigraphic–lithologic oil and gas pools in the Jiyang depression, China. In: *Petroleum Resources of China and Related Subjects* (Eds Wagner, H.C., Wagner, L.C., Wang, F.F.H. & Wong, F.L.), p. 329. Circum-Pacific Council for Energy and Mineral Resources, Earth Science Series **10**, Houston, Texas.

STAPLIN, F.L. (1977) Interpretation of thermal history from color of particulate organic matter — a review. *Palynology* **1**, 9–26.

SUNG ZHICHEN, HE CHENGQUAN, QIAN ZESHU, PAN ZHAORIN, ZHENG GUOQUANG & ZHENG YUEFANG (1978) *On the Paleogene Dinoflagellates and Acritarchs from the Coastal Region of Bohai*. Nanjing Institute of Geology and Paleontology, Academia Sinica, 190 pp. (in Chinese with English abstr.).

TISSOT, B.P. & WELTE, D.H. (1984) *Petroleum Formation and Occurrence*, 2nd edn. Springer-Verlag, New York, 699 pp.

TRAVERSE, A. (1988) *Paleopalynology*. Unwin Hyman, Boston, 380 pp.

WAN DEFA, SUN YONGCHUAN & ZHENG JUNMAO (1983) Depositional features and oil and gas distribution in some of the faulted-depression lake basins in north China. *Acta Petrol. Sin.* **4**, 13–21 (in Chinese with English abstr.).

WANG TIEGUAN, FAN PU & SWAIN, F.M. (1988) Geochemical characteristics of crude oils and source beds in different continental facies of four oil-bearing basins, China. In: *Lacustrine Petroleum Source Rocks* (Eds Fleet, A.J., Kelts, K. & Talbot, M.R.). Geol. Soc. London Spec. Publ., **40**, 309–325.

WILLIAMS, E.G. & KEITH, M.L. (1962) Relationship between sulfur in coals and the occurrence of marine roof beds. *Econ. Geol.* **58**, 720–729.

WRAY, J.L. (1977) *Calcareous Algae*. New York, Elsevier, 185 pp.

WU CHONGYUN (1983) Structural lake deltas and oil–gas distribution. *Acta Sediment. Sin.* **1**, 5–23 (in Chinese with English abstr).

WU CHONGYUN (1986) Sandbodies in lake basin. *Acta*

Sediment. Sin. **4**, 1–27 (in Chinese with English abstr.).

ZHANG MIMAN, ZHOU JIANJIAN & QIN DERONG (1985) Tertiary fish fauna from coastal region of Bohai Sea. *Mem. Inst. Vertebrate Palaeontol. Palaeoanthropol., Acad. Sin.* **17**, 55–60 (in Chinese with English abstr.).

ZHOU ZILI & DU YUNHUA (1986) Relationship between sedimentary facies of lacustrine carbonate rocks and oil and gas distribution. *Exp. Petrol. Geol.* **8**, 124–132 (in Chinese with English abstr.).

ZHOU ZILI & LU ZHANGMU (1987) The burial diagenesis and reservoir assessment of Tertiary clastic rocks in Shengli oil field, Shandong. *Earth Sci.* **12**, 311–319 (in Chinese with English abstr.).

ZHU HAORAN (1979) Microfossil algae from the Lower Tertiary Shehejia Formation of Bin Zian, Northern Shandong. *Acta Palaeontol. Sin.* **18**, 341–344.

Index

References to figures are shown in *italics* and
references to tables are shown in **bold**.